Methods in Enzymology

Volume 151
MOLECULAR GENETICS OF
MAMMALIAN CELLS

METHODS IN ENZYMOLOGY

EDITORS-IN-CHIEF

John N. Abelson   Melvin I. Simon

Methods in Enzymology

Volume 151

Molecular Genetics of Mammalian Cells

EDITED BY

Michael M. Gottesman

LABORATORY OF MOLECULAR BIOLOGY
NATIONAL CANCER INSTITUTE
NATIONAL INSTITUTES OF HEALTH
BETHESDA, MARYLAND

ACADEMIC PRESS, INC.

Harcourt Brace Jovanovich, Publishers

San Diego   New York   Berkeley   Boston
London   Sydney   Tokyo   Toronto

Academic Press, Inc.
San Diego, California 92101

United Kingdom Edition published by
ACADEMIC PRESS INC. (LONDON) LTD.
24-28 Oval Road, London NW1 7DX

Library of Congress Catalog Card Number: 54-9110

ISBN 0-12-182052-1   (alk. paper)

PRINTED IN THE UNITED STATES OF AMERICA
87 88 89 90   9 8 7 6 5 4 3 2 1

Table of Contents

Section I. Cell Lines Useful for Genetic Analysis

Section II. Special Techniques for Mutant Selection

Section V. Gene Regulation in Tissue Culture

Contributors to Volume 151

Article numbers are in parentheses following the names of contributors.
Affiliations listed are current.

SHIN-ICHI AKIYAMA (4), *Department of Cancer Chemotherapy, Institute of Cancer Research, Faculty of Medicine, Kagoshima University, 1208-1 Usuki-cho, Kagoshima 890, Japan*

KEVIN ALBRIGHT (19), *Experimental Pathology Group, Los Alamos National Laboratory, Los Alamos, New Mexico 87545*

MARTY BARTHOLDI (19), *Experimental Pathology Group, Los Alamos National Laboratory, Los Alamos, New Mexico 87545*

DAVID B. BROWN (26), *Department of Biology, Yale University, New Haven, Connecticut 06511*

PETER C. BROWN (7), *Department of Biological Sciences, Stanford University, Stanford, California 94305*

BARRY D. BRUCE (22), *Howard Hughes Medical Institute, University of California, San Francisco, California 94143*

SUSAN BUHL (5), *Department of Cell Biology, Albert Einstein College of Medicine, Bronx, New York 10461*

EVELYN CAMPBELL (19), *Experimental Pathology Group, Los Alamos National Laboratory, Los Alamos, New Mexico 87545*

CHARLES R. CANTOR (35), *Department of Human Genetics and Development, Columbia University, New York, New York 10032*

ADELAIDE M. CAROTHERS (34), *Institute of Cancer Research, Columbia University, New York, New York 10032*

C. THOMAS CASKEY (38), *Institute for Molecular Genetics, Department of Medicine, Biochemistry and Cell Biology, Howard Hughes Medical Institute, Baylor College of Medicine, Houston, Texas 77030*

LAWRENCE A. CHASIN (34), *Department of Biological Sciences, Columbia University, New York, New York 10027*

DOUGLAS CHRITTON (19), *Department of Surgery, Immunology Center, Loma Linda Medical Center, Loma Linda, California 92354*

PHILIP COFFINO (2), *Departments of Medicine and Microbiology and Immunology, University of California, San Francisco, San Francisco, California 94143*

FRANCIS S. COLLINS (35), *Departments of Internal Medicine and Human Genetics and the Howard Hughes Medical Institute, University of Michigan Medical School, Ann Arbor, Michigan 48109*

L. SCOTT CRAM (19), *Experimental Pathology Group, Los Alamos National Laboratory, Los Alamos, New Mexico 87545*

G. J. DARLINGTON (3), *Department of Pathology, Baylor College of Medicine, Houston, Texas 77030*

LARRY L. DEAVEN (19), *Experimental Pathology Group, Los Alamos National Laboratory, Los Alamos, New Mexico 87545*

JAN-ERIK EDSTRÖM (37), *Department of Genetics, University of Lund, S-22362 Lund, Sweden*

DAVID J. P. FITZGERALD (12), *Laboratory of Molecular Biology, National Cancer Institute, National Institutes of Health, Bethesda, Maryland 20892*

ROBERT FLEISCHMANN (29), *Laboratory of Molecular Biology, National Cancer Institute, National Institutes of Health, Bethesda, Maryland 20892*

C. MICHAEL FORDIS (27), *Laboratory of Molecular Biology, National Cancer Institute, National Institutes of Health, Bethesda, Maryland 20892*

ARLETTE FRANCHI (11), *Centre de Biochimie du CNRS, Faculté des Sciences, Université de Nice, Parc Valrose, 06034 Nice, France*

DEBORAH FRENCH (5), *Department of Cell Biology, Albert Einstein College of Medicine, Bronx, New York 10461*

GEORGE A. GAITANARIS (28), *Institute of Cancer Research, College of Physicians and Surgeons, Columbia University, New York, New York 10032*

SUSANNAH GAL (8), *Laboratory of Molecular Biology, National Cancer Institute, National Institutes of Health, Bethesda, Maryland 20892*

GERALD A. GILLESPIE (35), *Department of Human Genetics, Yale University School of Medicine, New Haven, Connecticut 06510*

STEPHEN P. GOFF (36), *Department of Biochemistry and Molecular Biophysics, Columbia University College of Physicians and Surgeons, New York, New York 10032*

MAX E. GOTTESMAN (28), *Institute of Cancer Research, Columbia University College of Physicians and Surgeons, New York, New York 10032*

MICHAEL M. GOTTESMAN (1, 9, 24), *Laboratory of Molecular Biology, National Cancer Institute, National Institutes of Health, Bethesda, Maryland 20892*

MARY E. HARPER (40), *Gen-Probe, San Diego, California 92121*

JOSEPH HIRSCHBERG (13), *Department of Genetics, Hebrew University of Jerusalem, Jerusalem 91904, Israel*

BRUCE H. HOWARD (27, 28, 29), *Laboratory of Molecular Biology, National Cancer Institute, National Institutes of Health, Bethesda, Maryland 20892*

HEDWIG JAKOB (6), *Unité de Génétique Cellulaire du Collège de France et de l'Institut Pasteur, 75724 Paris, Cedex 15, France*

ROLF KAISER (37), *Department of Radiation Biology, University of Bonn, D-5300 Bonn, Federal Republic of Germany*

MICHAEL E. KAMARCK (14), *Department of Exploratory Research, Molecular Therapeutics Inc., West Haven, Connecticut 06516*

THERESA KELLY (5), *Department of Cell Biology, Albert Einstein College of Medicine, Bronx, New York 10461*

YUN-FAI LAU (31), *Howard Hughes Medical Institute, and Departments of Physiology and Medicine, University of California, San Francisco, California 94143*

SIMON K. LAWRANCE (35), *Scripps Clinic and Research Foundation, La Jolla, California 92037*

ROGER V. LEBO (22), *Department of Obstetrics, Gynecology, and Reproductive Sciences, and Howard Hughes Medical Institute, University of California, San Francisco, California 94143*

PIN-FANG LIN (26), *Pharmaceutical Research and Development Division, Bristol-Myers Company, Wallingford, Connecticut 06492*

MARY LUEDEMANN (19), *Experimental Pathology Group, Los Alamos National Laboratory, Los Alamos, New Mexico 87545*

MENASHE MARCUS[1] (13), *Department of Genetics, Hebrew University of Jerusalem, Jerusalem 91904, Israel*

LISA M. MARSELLE (40), *Department of Anatomy, University of Massachusetts Medical School, Worcester, Massachusetts 01605*

MARY MCCORMICK (28, 29, 33), *Laboratory of Molecular Virology, National Cancer Institute, National Institutes of Health, Bethesda, Maryland 20892*

JOHN R. MCGILL (21), *Department of Obstetrics and Gynecology, The University of Texas Health Science Center, San Antonio, Texas 78284*

JULIE MEYNE (19), *Experimental Pathology Group, Los Alamos National Laboratory, Los Alamos, New Mexico 87545*

PAT MURPHY (26), *Department of Human Genetics, Yale University, New Haven, Connecticut 06510*

SUSAN L. NAYLOR (21), *Department of Cellular and Structural Biology, The Univer-*

[1] Deceased.

sity of Texas Health Science Center, San Antonio, Texas 78284

JEAN-FRANÇOIS NICOLAS (6), *Unité de Génétique Cellulaire du Collège de France et de l'Institut Pasteur, 75724 Paris, Cedex 15, France*

HIROTO OKAYAMA (32), *Laboratory of Cell Biology, National Institute of Mental Health, National Institutes of Health, Bethesda, Maryland 20892*

DAVID PATTERSON (10), *Eleanor Roosevelt Institute for Cancer Research, Denver, Colorado 80262*

JACQUES POUYSSÉGUR (11), *Centre de Biochimie du CNRS, Faculté des Sciences, Université de Nice, Parc Valrose, 06034 Nice, France*

DAN RÖHME (37), *Department of Genetics, University of Lund, S-22362 Lund, Sweden*

IGOR B. RONINSON (25), *Center for Genetics, University of Illinois College of Medicine, Chicago, Illinois 60612*

DAVID S. ROOS (7), *Department of Biological Sciences, Stanford University, Stanford, California 94305*

FRANK H. RUDDLE (26), *Department of Biology, Yale University, New Haven, Connecticut 06511*

PAUL J. SAXON (23), *Department of Microbiology and Molecular Genetics, University of California, Irvine, California 92717*

MATTHEW D. SCHARFF (5), *Department of Cell Biology, Albert Einstein College of Medicine, Bronx, New York 10461*

ROBERT T. SCHIMKE (7), *Department of Biological Sciences, Stanford University, Stanford, California 94305*

JERRY W. SHAY (17), *Department of Cell Biology, University of Texas Health Sciences Center at Dallas, Dallas, Texas 75235*

MICHAEL J. SICILIANO (15), *Department of Genetics, The University of Texas M. D. Anderson Hospital and Tumor Institute, Texas Medical Center, Houston, Texas 77030*

CASSANDRA L. SMITH (35), *Departments of Microbiology and Psychiatry, Columbia University, New York, New York 10032*

GILBERT H. SMITH (39), *Laboratory of Tumor Immunology and Biology, National Cancer Institute, National Institutes of Health, Bethesda, Maryland 20892*

ERIC J. STANBRIDGE (23), *Department of Microbiology and Molecular Genetics, University of California, Irvine, California 92717*

J. TIMOTHY STOUT (38), *Institute for Molecular Genetics, Department of Medicine, Biochemistry and Cell Biology, Howard Hughes Medical Institute, Baylor College of Medicine, Houston, Texas 77030*

FLOYD H. THOMPSON (20), *Arizona Cancer Center, University of Arizona, Tucson, Arizona 85724*

JEFFREY M. TRENT (20), *Arizona Cancer Center, University of Arizona, Tucson, Arizona 85724*

BRUCE R. TROEN (30), *Laboratory of Molecular Biology, National Cancer Institute, National Institutes of Health, Bethesda, Maryland 20892*

GAIL URLAUB (34), *Department of Biological Sciences, Columbia University, New York, New York 10027*

HOWARD B. URNOVITZ (16), *Medical Research Institute, San Francisco, California 94115*

ALEXANDER VARSHAVSKY (41), *Department of Biology, Massachusetts Institute of Technology, Cambridge, Massachusetts 02139*

CHARLES A. WALDREN (10), *Department of Radiology, University of Colorado Health Sciences Center, Denver, Colorado 80262*

SHERMAN M. WEISSMAN (35), *Departments of Human Genetics, Medicine, and Molecular Biophysics and Biochemistry, Yale University School of Medicine, New Haven, Connecticut 06510*

THEODOOR VAN DAALEN WETTERS (2), *Department of Microbiology and Immunol-*

ogy, University of California, San Francisco, California 94143

BILLIE F. WHITE (15), Department of Genetics, The University of Texas M. D. Anderson Hospital and Tumor Institute, Texas Medical Center, Houston, Texas 77030

WOODRING E. WRIGHT (18), Department of Cell Biology, The University of Texas Southwestern Medical School, Dallas, Texas 75235

MASARU YAMAIZUMI (26), Research Institute for Microbial Diseases, Osaka University, Osaka, Japan

BERNHARD U. ZABEL (21), Department of Pediatrics, University of Mainz, Mainz D-6500, Federal Republic of Germany

ULRICH ZIMMERMANN (16), Institute for Biotechnology, University of Würzburg, Röntgenring 11, 8700 Würzburg, Federal Republic of Germany

Menashe Marcus
(February 20, 1938–January 2, 1987)

This volume is dedicated to the memory of Menashe Marcus, a major contributor to the concept and substance of this book, who died on January 2, 1987 at the age of 48. Menashe was a scientific colleague, collaborator, and friend to many of the coauthors of this work. All who knew him were enriched by his kindness, generosity, wonderful sense of humor, and intellectual honesty. His professional life was spent at the Hebrew University in Jerusalem. He was dedicated to the advancement of biological research in Israel through his own work, his efforts to introduce precise scientific terminology into modern Hebrew, and through his many successful and devoted students. He maintained strong professional and personal ties with the scientific community in the United States, and did his postdoctoral work at the Massachusetts Institute of Technology, with sabbatical appointments at Columbia University College of Physicians and Surgeons, New York University School of Medicine, and the National Institutes of Health. His enthusiasm and vigorous support for the idea that the seeds sown in phage and bacterial genetics would bear fruit in the study of mammalian cells in culture has been borne out by the exciting developments of recent years. Guided by this precept, he pioneered techniques for the isolation and analysis of cell cycle mutants of mammalian cells. His scientific colleagues and friends join his wife Nima and his daughter Nufar in mourning his premature death. He leaves a legacy of scientific achievement which will be long remembered.

MICHAEL M. GOTTESMAN

Preface

The use of the tools of molecular biology to isolate, identify, and map a mutant gene, thereby defining an important process in cellular metabolism, is no longer the sole province of the microbiologist. The recent amalgamation of classical somatic cell genetics with recombinant DNA and gene transfer technology has resulted in new approaches especially useful for the study of mutant cells. This volume illustrates how special techniques in molecular biology can be applied to the study of mutant somatic cells in culture. Basic protocols for the manipulation of recombinant DNA can be found in other *Methods in Enzymology* volumes: Recombinant DNA, Parts A–F, Volumes 68, 100, 101, 153, 154, and 155.

The book is divided into five sections representing the chronological and conceptual development of molecular cell genetics. The first section describes the origins and use of several important tissue culture systems developed for the genetic analysis of both undifferentiated and differentiated cells. For additional discussion of cultured cell systems, the reader is referred to Cell Culture, Volume 58 of this series. The second section presents methodology useful for the isolation of mutant mammalian cells. The third section details new procedures for the mapping of mammalian genes defined either by somatic cell mutations or cloned DNA fragments. The fourth section describes novel techniques for the isolation of mutant genes, and the final section presents new approaches to the study of gene expression in cultured mammalian cells.

I would like to thank William Jakoby for suggesting this project to me, Nathan Kaplan for his enthusiastic endorsement, Ira Pastan for continued support and encouragement, and my wife, Susan, and children, Daniel and Rebecca, for their forbearance. Special thanks are due to Robert Fleischmann for critical comments on some of the manuscripts, to Joyce Sharrar for excellent secretarial help, to my other colleagues in the Laboratory of Molecular Biology in the National Cancer Institute who provided a sounding board for ideas, and to the many contributors to this volume for their timely and clearly presented contributions.

MICHAEL M. GOTTESMAN

METHODS IN ENZYMOLOGY

EDITED BY

Sidney P. Colowick and Nathan O. Kaplan

VANDERBILT UNIVERSITY
SCHOOL OF MEDICINE
NASHVILLE, TENNESSEE

DEPARTMENT OF CHEMISTRY
UNIVERSITY OF CALIFORNIA
AT SAN DIEGO
LA JOLLA, CALIFORNIA

METHODS IN ENZYMOLOGY

EDITORS-IN-CHIEF

Sidney P. Colowick and Nathan O. Kaplan

VOLUME XVIII. Vitamins and Coenzymes (Parts A, B, and C)
Edited by DONALD B. MCCORMICK AND LEMUEL D. WRIGHT

VOLUME XIX. Proteolytic Enzymes
Edited by GERTRUDE E. PERLMANN AND LASZLO LORAND

VOLUME XX. Nucleic Acids and Protein Synthesis (Part C)
Edited by KIVIE MOLDAVE AND LAWRENCE GROSSMAN

VOLUME XXI. Nucleic Acids (Part D)
Edited by LAWRENCE GROSSMAN AND KIVIE MOLDAVE

VOLUME XXII. Enzyme Purification and Related Techniques
Edited by WILLIAM B. JAKOBY

VOLUME XXIII. Photosynthesis (Part A)
Edited by ANTHONY SAN PIETRO

VOLUME XXIV. Photosynthesis and Nitrogen Fixation (Part B)
Edited by ANTHONY SAN PIETRO

VOLUME XXV. Enzyme Structure (Part B)
Edited by C. H. W. HIRS AND SERGE N. TIMASHEFF

VOLUME XXVI. Enzyme Structure (Part C)
Edited by C. H. W. HIRS AND SERGE N. TIMASHEFF

VOLUME XXVII. Enzyme Structure (Part D)
Edited by C. H. W. HIRS AND SERGE N. TIMASHEFF

VOLUME XXVIII. Complex Carbohydrates (Part B)
Edited by VICTOR GINSBURG

VOLUME XXIX. Nucleic Acids and Protein Synthesis (Part E)
Edited by LAWRENCE GROSSMAN AND KIVIE MOLDAVE

VOLUME XXX. Nucleic Acids and Protein Synthesis (Part F)
Edited by KIVIE MOLDAVE AND LAWRENCE GROSSMAN

VOLUME XXXI. Biomembranes (Part A)
Edited by SIDNEY FLEISCHER AND LESTER PACKER

VOLUME 85. Structural and Contractile Proteins (Part B: The Contractile Apparatus and the Cytoskeleton)
Edited by DIXIE W. FREDERIKSEN AND LEON W. CUNNINGHAM

VOLUME 86. Prostaglandins and Arachidonate Metabolites
Edited by WILLIAM E. M. LANDS AND WILLIAM L. SMITH

VOLUME 87. Enzyme Kinetics and Mechanism (Part C: Intermediates, Stereochemistry, and Rate Studies)
Edited by DANIEL L. PURICH

VOLUME 88. Biomembranes (Part I: Visual Pigments and Purple Membranes, II)
Edited by LESTER PACKER

VOLUME 89. Carbohydrate Metabolism (Part D)
Edited by WILLIS A. WOOD

VOLUME 90. Carbohydrate Metabolism (Part E)
Edited by WILLIS A. WOOD

VOLUME 91. Enzyme Structure (Part I)
Edited by C. H. W. HIRS AND SERGE N. TIMASHEFF

VOLUME 92. Immunochemical Techniques (Part E: Monoclonal Antibodies and General Immunoassay Methods)
Edited by JOHN J. LANGONE AND HELEN VAN VUNAKIS

VOLUME 93. Immunochemical Techniques (Part F: Conventional Antibodies, Fc Receptors, and Cytotoxicity)
Edited by JOHN J. LANGONE AND HELEN VAN VUNAKIS

VOLUME 94. Polyamines
Edited by HERBERT TABOR AND CELIA WHITE TABOR

VOLUME 95. Cumulative Subject Index Volumes 61–74, 76–80
Edited by EDWARD A. DENNIS AND MARTHA G. DENNIS

VOLUME 96. Biomembranes [Part J: Membrane Biogenesis: Assembly and Targeting (General Methods; Eukaryotes)]
Edited by SIDNEY FLEISCHER AND BECCA FLEISCHER

VOLUME 122. Vitamins and Coenzymes (Part G)
Edited by FRANK CHYTIL AND DONALD B. MCCORMICK

VOLUME 123. Vitamins and Coenzymes (Part H)
Edited by FRANK CHYTIL AND DONALD B. MCCORMICK

VOLUME 124. Hormone Action (Part J: Neuroendocrine Peptides)
Edited by P. MICHAEL CONN

VOLUME 125. Biomembranes (Part M: Transport in Bacteria, Mitochondria, and Chloroplasts: General Approaches and Transport Systems)
Edited by SIDNEY FLEISCHER AND BECCA FLEISCHER

VOLUME 126. Biomembranes (Part N: Transport in Bacteria, Mitochondria, and Chloroplasts: Protonmotive Force)
Edited by SIDNEY FLEISCHER AND BECCA FLEISCHER

VOLUME 127. Biomembranes (Part O: Protons and Water: Structure and Translocation)
Edited by LESTER PACKER

VOLUME 128. Plasma Lipoproteins (Part A: Preparation, Structure, and Molecular Biology)
Edited by JERE P. SEGREST AND JOHN J. ALBERS

VOLUME 129. Plasma Lipoproteins (Part B: Characterization, Cell Biology, and Metabolism)
Edited by JOHN J. ALBERS AND JERE P. SEGREST

VOLUME 130. Enzyme Structure (Part K)
Edited by C. H. W. HIRS AND SERGE N. TIMASHEFF

VOLUME 131. Enzyme Structure (Part L)
Edited by C. H. W. HIRS AND SERGE N. TIMASHEFF

VOLUME 132. Immunochemical Techniques (Part J: Phagocytosis and Cell-Mediated Cytotoxicity)
Edited by GIOVANNI DI SABATO AND JOHANNES EVERSE

VOLUME 133. Bioluminescence and Chemiluminescence (Part B)
Edited by MARLENE DELUCA AND WILLIAM D. MCELROY

Section I

Cell Lines Useful for Genetic Analysis

[1] Chinese Hamster Ovary Cells

By MICHAEL M. GOTTESMAN

Chinese hamster ovary (CHO) cells have been extensively used for genetic analysis in tissue culture since the pioneering work of Puck, who first isolated this cell line.[1] These cells have been used for the isolation of mutants affecting intermediary metabolism; DNA, RNA, and protein synthesis; membrane functions; and several more complex forms of cell behavior such as cell growth and endocytosis. A recent compilation of CHO mutants lists more than 80 classes of mutants isolated using this cell line.[2]

There are many reasons for the successful use of CHO cells in somatic cell genetics among which are (1) ease of growth with a doubling time of 12 hr and cloning efficiency in excess of 80%[3]; (2) simple karyotype with 21 large, easily recognized chromosomes[4]; (3) apparently high frequency of mutant phenotypes based on the "functional hemizygosity" of some of the CHO genome[5] as well as a high frequency of "segregation-like" events which unmask otherwise recessive mutations[6]; and (4) the ease with which CHO cells can be transfected with DNA.[7] Although these characteristics make CHO cells useful for the isolation of mutants affecting general cell functions, this line is not suitable for an analysis of most differentiated functions. There are two other disadvantages of these cells which should be borne in mind; namely, they are not derived from a fully inbred Chinese hamster line and hence mutant cell lines cannot be reintroduced back into the animal of origin, and they are not susceptible to infection by standard retroviruses which might be used as DNA vectors (see chapter by Goff [36]; this volume).

[1] T. T. Puck, S. J. Cieciura, and A. Robinson, *J. Exp. Med.* **108**, 945 (1985)

[2] M. M. Gottesman, *in* "Molecular Cell Genetics" (M. M. Gottesman, ed.), p. 887. Wiley, New York, 1985.

[3] M. M. Gottesman, *in* "Molecular Cell Genetics" (M. M. Gottesman, ed.), p. 139. Wiley, New York, 1985.

[4] M. J. Siciliano, R. L. Stallings, and G. M. Adair, *in* "Molecular Cell Genetics" (M. M. Gottesman, ed.), p. 95. Wiley, New York, 1985.

[5] L. Siminovitch, *Cell* **7**, 1 (1976).

[6] R. G. Worton and S. G. Grant, *in* "Molecular Cell Genetics" (M. M. Gottesman, ed.), p. 831. Wiley, New York, 1985.

[7] I. Abraham, J. S. Tyagi, and M. M. Gottesman, *Somatic Cell Genet.* **8**, 23 (1982).

History of CHO Lines

Hu and Watson introduced the Chinese hamster into the United States as a laboratory animal in 1948.[8] They were first bred seriously by Yerganian starting in 1951. In 1957, one of his partially inbred female hamsters was given to Puck, who established a fibroblastic cell line from the ovary of this animal.[1] The cell line was originally slightly aneuploid,[9] having either 23 or 21 chromosomes instead of the 11 pairs found in the Chinese hamster, and grew vigorously. One subline of the original isolate, called CHO-K1 (ATCC CCL 61) was maintained in Denver by Puck and Kao, whereas another subline was sent to Tobey at Los Alamos. This latter line was adapted to suspension growth by Thompson at the University of Toronto (CHO-S) in 1971 and has given rise to a number of Toronto subclones with similar properties including the line CHO Pro^{-5} used extensively by Siminovitch and numerous colleagues in Toronto, CHO GAT$^-$ of McBurney and Whitmore, subline 10001 of Gottesman at the NIH, and subline AA8 of Thompson. There are some differences in the karyotypes of the CHO-K1 and CHO-S cell lines, and CHO-S grows well in spinner and suspension culture, whereas CHO-K1 does not. Both sublines seem to give rise readily to mutant phenotypes. The methodologies described in this chapter were developed for work with the CHO-S sublines, but most of the methods, with the exception of suspension culture, can be used for CHO-K1 cell lines.

Growth of CHO Cells

CHO cells are proline auxotrophs, unlike most other cultured cell lines, and require medium containing this nutrient, such as Ham's F12 (for formulation, see Puck[9]) or α-modified Eagle's medium (α-MEM) without ribonucleoside or deoxyribonucleosides (for formulation, see Gottesman[3]), both of which are commercially available. These rich media must in addition contain other limiting nutrients for CHO cells, since they support more rapid growth of CHO lines than is possible in MEM alone supplemented with proline. We routinely use α-MEM supplemented with 10% fetal bovine serum (calf serum will work but will not support suspension growth) with penicillin (50 units/ml) and streptomycin (50 μg/ml). Fetal bovine sera must be prescreened and should support clonal growth of CHO

---

[8] G. Yerganian, *in* "Molecular Cell Genetics" (M. M. Gottesman, ed.), p. 3. Wiley, New York, 1985.

[9] T. T. Puck, *in* "Molecular Cell Genetics" (M. M. Gottesman, ed.), p. 37. Wiley, New York, 1985.

cells at a concentration of 0.5%. Our usual protocol for screening sera involves the following tests:

1. Determine the cloning efficiency of CHO cells in different serum concentrations. Plate 200 CHO cells in medium containing 10, 5, 2, 1, 0.5, and 0.2% serum. After 7–10 days clones should be visible at all serum concentrations with the possible exception of 0.2%. The clones can be more easily visualized by staining with 0.5% methylene blue in 50% ethanol.

2. Determine the doubling time of CHO cells in 10% serum. Plate 2×10^4 cells in eight 35-mm dishes or in eight individual wells of a 24-well multiwell dish (CoStar). After 16 hr, remove medium and add 1 ml of 0.25% trypsin, 0.2 M EDTA in PBS or Tris–dextrose buffer (TD buffer is NaCl, 8 g/liter; KCl, 0.38 g/liter; Na_2HPO_4, 0.1 g/liter; Tris–HCl, 3 g/liter; and dextrose, 1.0 g/liter adjusted to pH 7.4 with HCl). Incubate at 37° for 30 min and add the suspended cells to 9 ml isotonic cell counting medium for counting. Repeat the trypsinization and cell counts every 24 hr for 3 more days at which time the cell monolayers should be confluent. CHO cells should double every 12 hr. Failure to double at this rate suggests a problem with medium, serum, growth conditions (see below), or infection with a microorganism such as mycoplasma.

3. Test fetal bovine serum for ability to support growth in suspension (see below). Only one of three random fetal bovine serum samples will support optimal cell growth in suspension. Poor sera will result in clumping of cells.

4. Confirm that the appearance of the cells growing in the lot of serum being tested is the same as their appearance in other serum lots. The cells should be fibroblastic and nonvacuolated. Membrane ruffling and blebbing is quite common, especially after the cells are initially plated.

5. Confirm that cells growing in the lot of serum being tested have the same biochemical and genetic phenotypes as in previous lots of serum used in the laboratory. If extensive gene transfer studies are anticipated, serum lots should be tested for ability to support DNA mediated gene transfer at good frequency (see chapter by Fordis and Howard [27], this volume).

CHO cells grow optimally at 37°[3] and prefer a slightly alkaline pH (optimum pH is 7.4–7.8).[3] In bicarbonate-buffered medium such as α-MEM, CO_2 concentration should be approximately 5%. If higher CO_2 concentrations are used, as would be the case when CHO cells are cultivated in the same incubator with cells growing in MEM, the medium will be too acid and cells will not grow optimally.

CHO cells are transformed and will overgrow at high cell density and die. For this reason, it is essential to split cells every few days. For maintaining cultures, we split cells 1/50 to 1/100 every 3–4 days. If dense monolayer cultures of CHO cells are needed for biochemical analysis (i.e., DNA or RNA extraction, or preparation of extracts for enzymatic analysis), 5×10^5 cells should be plated 72 hr prior to harvesting in a 100-mm tissue culture dish containing 15 ml medium or 1×10^6 cells 48 hr prior to harvesting. For large quantities of cells. CHO cells can be grown in roller bottles (0.5 rpm) or on carrier beads in suspension (Cytodex beads, Pharmacia, 15 cells/bead, stirred at 20–30 rpm in a siliconized spinner flask). A maximum of approximately 10^8 cells can be grown in a 850 cm^2 roller bottle with 100 ml of medium or 10^8 cells can be grown on 10^6 beads in 100 ml medium.

For cloning and for certain mutant selections, CHO cells can be grown in semisolid medium such as agar (Difco, prescreened for toxicity) or agarose (Indubiose, Fisher). A cell suspension in 0.35% agar is poured over a bottom layer of 0.5% agar and colonies should appear within 7–10 days. A detailed protocol for this procedure has been published.[10] Suspension culture of CHO cells is also possible, either in Wheaton bottles in a gyrorotatory shaker bath at 160 rpm or in Spinner bottles.[3] We prefer the use of shaker baths since it does not require any special equipment or media. A monolayer of CHO cells is trypsinized as indicated above and 10^6 cells are inoculated into 20 ml of complete α-MEM in a 120-ml Wheaton bottle. The bottle is gassed with CO_2 to pH 7.4 and the screw-on top sealed with Parafilm. Rotation at 160 rpm is mandatory, since lower speeds allow the cells to settle and higher speeds result in cell lysis. Under these conditions, cells will double every 12–18 hr and will reach a maximum cell density of 10^6 cells/ml in 72 hr.

Mutagenesis of CHO Cells

For most selections, it is necessary to mutagenize cells to get a reasonable frequency of mutants. Because of its relative stability and ease of handling, we generally use ethylmethane sulfonate (EMS) as a mutagen. The following protocol should yield 10- to 100-fold increases in mutation rate:

1. Plate 5×10^5 CHO cells in each of three T-75 tissue culture flasks. Grow at 37° overnight. Each flask will be independently mutagenized and should give rise to independent mutants.

2. In a chemical fume hood, while wearing gloves, dilute 15 μl of EMS (Eastman Chemical Company) into 100 ml of complete α-MEM (final concentration 150 μg/ml) containing 10 μg/ml thymidine (to increase the

---

[10] M. M. Gottesman, this series, Vol. 99, p. 197.

mutation rate). In the absence of thymidine, it may be necessary to increase the concentration of EMS to 300 μg/ml. Remove the medium from the flasks and put 20 ml of EMS-containing medium into each flask. Incubate the cells overnight at 37°.

3. Remove the EMS-containing medium and dispose of this medium using standard techniques for dangerous chemical waste disposal. Trypsinize the cells, count them, and plate 200 cells from each flask, as well as cells from a flask with unmutagenized cells, in separate 100-mm tissue culture dishes to determine the percentage of cells that survived the mutagenesis procedure. Survival should be from 10 to 50% to reflect optimal mutagenesis.

4. There should be approximately 10^6 living cells in each flask. Plate these in a flask with complete α-MEM without EMS. Grow these cells for 3–10 days to allow expression of the mutant phenotypes. For each selection, it will be necessary to optimize the expression time, but for initial selections we usually grow the cells for 5 days in nonselective medium. Usually, cells will need to be split before the 5 days are over to allow optimal growth rates.

5. For a selection involving drug resistance, plate no more than 5×10^5 cells in a 100-mm dish with selective medium (see chapter on drug-resistant mutants [9], this volume). To monitor mutagenesis, we routinely use the frequency of ouabain-resistant mutants in the mutagenized population compared to the nonmutagenized cells. Ouabain-containing selective medium can be made by diluting 400 mM ouabain (Sigma) [(0.58 g ouabain + 1.5 ml dimethyl sulfoxide (DMSO)] 200-fold into complete α-MEM (final ouabain concentration, 2 mM). Ouabain-resistant colonies will appear in 7–10 days and can be stained with 0.5% methylene blue in 50% ethanol. The frequency of ouabain-resistant mutants should increase by a factor of 10- to 100-fold after EMS treatment.

Storage of CHO Cells

CHO cells are quite hardy and will survive most standard storage procedures. We use the following protocol for routine freezing:

1. Prepare a dense monolayer culture of cells (5×10^6/100 mm dish). Trypsinize and suspend at a density of 1×10^6 cells/ml in ice-cold medium. Add DMSO (Aldrich, Gold Label) to a final concentration of 7.5% (0.15 ml DMSO per 1.85 ml of cell suspension).

2. Freeze cells slowly to minimize damage from ice crystal formation. This is easily done by wrapping cells in an insulating material or placing them in Styrofoam and freezing them at -20° overnight. After they are

frozen they can be stored at $-70°$ (for up to 2 years) or at liquid nitrogen temperature, indefinitely. It is desirable to freeze multiple vials in multiple places since freezers have a tendency to defrost at the worst possible times. One way to protect against losing cells that have been defrosted is to freeze them in 10% glycerol. Although survival of cells frozen in glycerol is not as good as that of cells frozen in DMSO, cells frozen and defrosted in glycerol will survive at room temperature for several hours.

3. Defrost cells by rapid immersion in a $37°$ water bath and, as soon as the last trace of ice is gone, dilution into a 20-fold excess of complete medium. After the cells have attached (1–2 hr), the medium should be removed and replaced with fresh medium.

For short storage periods, CHO cells can be maintained in CO_2 tissue culture incubators at $30°$ for up to 2 weeks where their growth rate is so slow that they do not overgrow. This approach might be used under circumstances where a large number of clones is being tested for a particular phenotype and only a small percentage of them will be permanently stored. Master plates used in replica plating can also be stored in this manner (see chapter by Gal [8], this volume). It is also possible to store CHO cells for several weeks in a sealed, gassed flask at room temperature or in the refrigerator. At $4°$, cell survival under these conditions is about 50% every 24 hr. This property makes it possible to send flasks of CHO cells through the mail at all seasons with some assurance that living cells will be found within seven days after mailing them.

Special Techniques Involving CHO Cells

As mentioned above, a very large number of selective conditions have been devised to allow the isolation of Chinese hamster mutants. Some of these procedures are detailed in other chapters in this volume, such as the technique of replica plating (Gal [8]), selection of drug-resistant mutants (Gottesman [9]), suicide selections (Patterson and Waldren [10]; Pouysségur and Franchi [11]), formation of various types of hybrid cells (Shay [17]), and the isolation of temperature-sensitive mutants (Hirschberg and Marcus [13]).

[2] Cultured S49 Mouse T Lymphoma Cells

By THEODOOR VAN DAALEN WETTERS and PHILIP COFFINO

Introduction

S49 is a mouse T cell lymphoma that has been adapted to grow in culture. It has characteristics that are of general utility to investigators who make use of cultured cells. The cells grow quickly, with a generation time of approximately 16 hr.[1] They have a stable near-euploid karyotype.[2] They grow in stirred or stationary suspension culture. The cells do not adhere to the culture vessel and associate only loosely with each other. This obviates the need for trypsinization and greatly simplifies maintenance of cultures and the measurement of cell number and growth rate. The cells can readily be synchronized by centrifugal elutriation to study cell cycle timing and its control.[3,4] They can be grown in medium containing horse serum rather than the more expensive fetal calf serum. In addition, serum-free media have been developed for maintenance of these cells, a prerequisite for certain types of experiments. Isogenic mutant cell lines have been generated and characterized that are altered in diverse biochemical functions. The availability of these mutants can greatly enhance studies of the biological relevance of the mutated functions.

The characteristics described above are general ones that might appeal to investigators, otherwise indifferent to the particular nature of the line, who are seeking cells with desirable technical properties. S49 cells have in addition special properties that almost certainly reflect those of the normal T cell population in which the lymphoma arose. These include exquisite sensitivity to thymidine, to glucocorticoids, and to cyclic AMP.[5] The cells' vulnerability to these agents has been exploited in studies of nucleic acid metabolism and of hormonal responses.

Origin

The S49 lymphoma was induced at the Salk Institute in a BALB/c mouse by intraperitoneal injection of mineral oil.[2] Mice of this strain

---

[1] P. Coffino and J. W. Gray, *Cancer Res.* **38**, 4285 (1978).
[2] K. Horibata and A. W. Harris, *Exp. Cell Res.* **60**, 61 (1970).
[3] N. Kaiser, H. R. Bourne, P. A. Insel, and P. Coffino, *J. Cell. Physiol.* **101**, 369 (1979).
[4] V. E. Groppi and P. Coffino, *Cell* **21**, 195 (1980).
[5] P. Ralph, R. Hyman, R. Epstein, I. Nakoinz, and M. Cohn, *Biochem. Biophys. Res. Commun.* **55**, 1085 (1973).

respond to peritoneal irritation by producing lymphoid tumors, the majority of which are myelomas and the minority lymphomas. A female mouse was injected with phages at 3 and 4 months of age and with mineral oil (Bayol F) at 4, 6, and 8 months of age. The S49 tumor was discovered at 16 months of age and was maintained by serial subcutaneous injection into syngeneic hosts. The tumor was described as poorly tumorigenic on serial passage; after being adapted to culture it was still less tumorigenic, requiring subcutaneous injection of more than 10^7 cells to generate slow-growing tumors. The tumor was adapted to culture in July 1967 in its seventh transplantation generation by dissociating cells and maintaining them in Dulbecco–Vogt's modified Eagle's medium containing 10% heat-inactivated horse serum. Three weeks after initiation of culture the cells grew at the rapid rate that has since characterized them. A subclone designated S49.1 was generated by a cell cluster isolation technique. This subclone was subjected to several additional rounds of serial subcloning in soft agar at the University of California, San Francisco. A clone generated in this way and designated S49 24.3.2 is the wild-type progenitor of most of the mutant lines described below.[6]

Karyotyping soon after establishment indicated that S49 cells were euploid, i.e., contained no apparent marker chromosomes and had the normal mouse chromosome number of 40.[2] More recent analysis of S49 clone 24.3.2 using banding techniques not available earlier has demonstrated that the cells are pseudodiploid, with trisomy 1 and monosomy X. In addition, one chromosome 9 has an interstitial deletion of some of region 9E.[7,8]

Growth Conditions

Medium

S49 cells grow well in Dulbecco's modified Eagle's medium supplemented with 10% horse serum (inactivated by heating at 56° for 30 min), 584 mg/liter glutamine, 110 mg/liter sodium pyruvate, 3.7 g/liter $NaHCO_3$ and 3–4.5 g/liter glucose. (DMEM will refer hereafter to the above medium with all its components except serum.) Fetal calf serum can be substituted for heat-inactivated horse serum with an appreciable effect only on one's budget. The medium is buffered to pH 7.2–7.4 either by incuba-

[6] P. Coffino, H. R. Bourne, and G. M. Tomkins, *J. Cell. Physiol.* **85**, 603 (1975).
[7] U. Francke, *Cell* **22**, 657 (1980).
[8] L. McConlogue and P. Coffino, *J. Biol. Chem.* **258**, 12083 (1983).

tion in a 95% air/5% CO_2 atmosphere or by addition of HEPES buffer solution to 10 mM final concentration. Under these conditions, the culture doubling time is 16–18 hr and cells stop growing when their concentration approaches about $2-3 \times 10^6$ cells/ml.

Two defined media are available that will support S49 cell growth. One, described by Darfler et al.,[9,10] promotes a doubling time of about 23–29 hr. In this medium, density-dependent lag phase occurs at about $1.8-2 \times 10^6$ cells/ml. A complete formulation of this medium is available.[9,10] A second medium, HB101, commercially available from HANA biologicals (Berkeley, CA) promotes a doubling time of 18–20 hr and cells enter lag phase at about $2-2.5 \times 10^6$ cells/ml.

S49 cells resuspended in DMEM in the absence of serum will adhere tightly to the surfaces on which they rest. They can be dislodged in viable form only by further incubation in serum-containing medium for several hours. In some cases this can be used to advantage, for example, in DNA transfection experiments. Cell "stickiness" is evident in cultures being maintained in defined medium. This is not a serious problem because, in this case, the cells are easily dislodged by tapping the culture vessel.

S49 cells have been grown exponentially in culture volumes ranging from 100 μl to several hundred liters.[11] The only constraint on their passage during routine cell feeding or preparation of large volume cultures is their requirement for a rather narrow "window" of cell concentrations. The cells will cease exponential growth and, indeed, die if their concentration in nonconditioned medium is less than about $5-10 \times 10^4$ cells/ml or greater than 2×10^6 cells/ml. The lower constraint can be removed by growth in Darfler's defined medium or in DMEM that contains 50% by volume filtered, conditioned medium taken from exponentially growing S49 cells (optimally at a concentration of 10^6 cells/ml) and 50 μM 2-mercaptoethanol. S49 cells are more readily damaged by alkaline medium than most lines.

Cloning

S49 cells can form single-cell derived colonies when immobilized in medium made semisolid with agarose.[12] This property has proven invaluable in quantitating the effects of drugs on cell viability, in isolating the

[9] F. J. Darfler, H. Murakami, and P. A. Insel, *Proc. Natl. Acad. Sci. U.S.A.* **77,** 5993 (1980).
[10] F. J. Darfler and P. A. Insel, *J. Cell. Physiol.* **115,** 31 (1983).
[11] R. T. Acton, P. A. Barstad, R. M. Cox, R. K. Zerner, K. S. Wise, and J. D. Lynn, *in* "Cell Culture and Its Application" (R. T. Acton and J. D. Lynn, eds.), pp. 129–160. Academic Press, New York, 1977.
[12] P. Coffino, R. Baumal, R. Laskov, and M. D. Scharff, *J. Cell. Physiol.* **79,** 429 (1972).

hybrid products of cell–cell fusions, and in generating homogeneous populations of drug-resistant mutants.

S49 cell cloning can be carried out in nonconditioned medium over "feeder layers" of mouse embryo fibroblasts[12] or in 50% conditioned medium as described below. One hundred milliliters of cloning medium consists of

    10 ml heat-inactivated horse serum
    50 ml conditioned medium, obtained as described above
    5.5 ml 5.3% agarose (Seakem)
    1.0 ml 5.0 mM 2-mercaptoethanol
    33.5 ml DMEM

We use 5 ml cloning medium per 60-mm diameter plastic culture dish, therefore 100 ml cloning medium is sufficient for 20 dishes. Cells are plated by diluting them with 1.0 ml of cloning medium and layering them dropwise over a 4.0 ml prehardened base layer of cloning medium. Cloning efficiency using either the feeder layer or the conditioned medium method is usually 50–100%.

Colonies of cells become macroscopically visible after 6–7 days and are ready to transfer to liquid culture after 10–12 days. Clones can be retrieved from the plates with a micropipet under a dissecting microscope. The limitation on minimum cell concentration noted above necessitates passage through several culture volume stages. We transfer colonies to 200 μl volumes of medium in 96-well microtest plates and then, at confluence, to 2 ml and then 20–50 ml volumes before liquid nitrogen storage.

Notes

1. Solubilize 5.3% agarose in water by autoclaving. It can be kept liquid by immersion in a 44° water bath. The agarose stock can be kept at 4° and used repeatedly. Once hardened, however, agarose should *not* be remelted by autoclaving since this tends to change its concentration.

2. Agarose lots vary in their ability to support S49 clonal growth. We recommend testing several lots. Seakem has proven a reliable source.

3. S49 cells are very sensitive to excess agarose concentrations—an increase of 0.05%, i.e., from 0.29 to 0.34%, above the optimal can reduce cloning efficiency by 30–50%.

4. The complete cloning medium should be kept at 42–44° to prevent its hardening.

5. S49 cells are very sensitive to alkaline conditions. Take care during cloning to keep the pH from becoming alkaline either by reequilibrating the dishes in a CO_2 incubator or by addition of HEPES buffer to the cloning medium.

6. S49 cell colonies are translucent. Colonies can be seen and counted more easily if they are first stained with 2-(p-iodophenyl)-3-(p-nitrophenyl)-5-tetrazolium chloride hydrate (INT, Aldrich). This stain is metabolized within cells to an insoluble, dark brick-red compound whose accumulation is lethal but allows easy detection of clones.[13] INT is solubilized in water by autoclaving. Colonies are stained by addition of 1 ml per dish of a 1 mg/ml INT solution and overnight incubation at 37°.

Frozen Storage of S49 Cells

S49 cells can be stored frozen indefinitely and retain high viability. Exponentially growing cells are resuspended in filter-sterilized freezing medium (DMEM, 10% horse serum, 10 mM HEPES, pH 7.4, and 5% dimethyl sulfoxide) at a concentration of 10^7 cells/ml or greater. The suspensions are dispensed in 2 ml plastic NUNC vials (Almac Cryogenics), 1 ml per vial. The cells are slow frozen (a must!) simply and conveniently by transferring the vials to a small cardboard, low-temperature storage box lined with several paper towels. Three to four paper towels are packed over the vials to provide insulation, the lid replaced and the box placed in a −70° freezer overnight. At any convenient time thereafter the vials can be transferred to liquid nitrogen for storage.

Growing cultures are regenerated by quick-thawing the cells in a 37° water bath and adding the suspension to 30–50 ml of prewarmed medium. We generally allow a "recovery" time of 5–7 days after thawing before using the cells for experimentation.

Counterselection

We devised a selection method that enriches for cAMP-sensitive phenotypes among populations of cAMP-resistant mutant S49 cells.[14] This counterselection should be generally applicable to the isolation of variants that grow at a reduced rate under controllable conditions.

The counterselection scheme, as we used it, exploits two properties of growing S49 cells. (1) The ability of cAMP to arrest wild-type but not cAMP-resistant mutant cells in the G_1 phase of the cell cycle. (2) The extreme sensitivity of cells to white light after bromodeoxyuridine (BrdUrd) and Hoechst 33258 dye incorporation.[15]

[13] W. I. Schaeffer and K. Friend, *Cancer Lett.* **1**, 259 (1976).
[14] T. van Daalen Wetters and P. Coffino, *Mol. Cell. Biol.* **2**, 1229 (1982).
[15] G. Stetten, S. A. Latt, and R. L. Davidson, *Somatic Cell Mol. Genet.* **2**, 285 (1976).

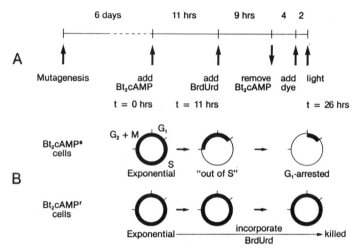

FIG. 1. Selection of Bt_2cAMP^s revertants.

Figure 1 depicts the counterselection strategy used to isolate revertants of S49 cell cAMP-dependent protein kinase (cA-PK) mutants. At time zero, 0.5 mM dibutyryl-cAMP (Bt_2cAMP) is added to exponentially growing mutant cells. Eleven hours later, when Bt_2cAMP-sensitive revertants have left S phase and are beginning to accumulate in the G_1 phase, 10 μM BrdUrd is added. Nine hours after that, the cells are resuspended in medium containing 10 μM BrdUrd but no Bt_2cAMP. S49 cells arrested in G_1 by Bt_2cAMP exhibit a 8-hr lag when that drug is withdrawn before entering S phase, therefore, revertants are temporarily protected from BrdUrd incorporation. At 24 hr, 1 μg/ml Hoechst 33258 dye is added and 2 hr later the cultures are placed over a white light source, such as a light box with fluorescent bulbs, for 10–15 min.[15] The cells are then either resuspended in medium containing 10 μM thymidine or plated on medium containing HAT (to eliminate coselected BrdUrd-resistant cells).

Reconstruction experiments demonstrate that cAMP-sensitive cells can be enriched 100-fold in populations of cAMP-resistant mutants. The procedure can be repeated to obtain a much larger cumulative enrichment. We used this protocol to isolate revertants of S49 mutants carrying lesions that affected the structure[14,16] and regulation[17] of cA-PK. Optimum use of the method to obtain mutants with alterations in their response to other effectors of cell growth will require measurement of the cell cycle kinetics

[16] T. van Daalen Wetters and P. Coffino, *Mol. Cell. Biol.* **3**, 250 (1983).

[17] T. van Daalen Wetters, M. P. Murtaugh, and P. Coffino, *Cell* **35**, 311 (1983).

of the response and appropriate modification of the timing of these manipulations.

Mutagenesis

Mutagenesis of S49 cells has generally been used by us to increase the frequency and predetermine the nature of mutations in selectable genes. In addition, we have described a mutagen screening system utilizing S49 cells that distinguishes general classes of mutagenic mechanisms, in particular, those that lead to base substitution and frameshift alterations.[18-20] In this system, the behavior of ICR-191 is consistent with a frameshift mechanism of action, whereas ethylmethane sulfonate (EMS) and N-methyl-N'-nitro-N-nitrosoguanidine (MNNG) act like base-substitution mutagens.

A quantitative comparison of the variations in survivorship and mutation frequencies with mutagen dose in S49 cells shows, for these mutagens and a variety of genetic makers, that a maximum number of mutational events is obtained when mutagen treatment results in cell survival frequencies of 10–30%. Such survival frequencies can be obtained by exposure of the cells to 0.75 μg/ml ICR-191 or 500 μg/ml EMS for 24 hr each or to 2 μg/ml MNNG for 4.5 hr. We measure survivorship by plating 100–200 cells per dish in nonselective medium immediately after mutagenesis.

Cultured mammalian cells require a period of time after mutagen treatment to express stable phenotypic alterations. For S49 cells this expression time is marker dependent, but has not been observed to exceed 6–7 days.[18] We therefore impose selective conditions on cells after this amount of time has elapsed following mutagenesis. If desired, the effectiveness of any individual mutagen treatment can be assessed by measuring the increase in frequency of cells resistant to 10 μg/ml 6-thioguanine. This drug selects for cells carrying lesions in the gene encoding hypoxanthine–guanine phosphoribosyltransferase (HGPRT). Mutation at the HGPRT locus is relatively unbiased with respect to base-substitution or frame-shift mutagens and is therefore generally useful as an estimator of mutagen effectiveness.

We have occasionally used other mutagens including X-irradiation and aflatoxin B_1. Conditions for their use have been described.[18,21]

[18] M. A. MacInnes, U. Friedrich, T. Van Daalen Wetters, and P. Coffino, *Mutat. Res.* **95**, 297 (1982).
[19] I. W. Caras, M. A. MacInnes, D. H. Persing, P. Coffino, and D. W. Martin, *Mol. Cell. Biol.* **2**, 1096 (1982).
[20] U. Friedrich and G. Nass, *Mutat. Res.* **110**, 147 (1983).
[21] U. Friedrich and P. Coffino, *Proc. Natl. Acad. Sci. U.S.A.* **74**, 679 (1977).

TABLE I
MUTANT OR VARIANT FORMS OF S49 CELLS

Function affected	Reference
Nucleotide Metabolism	
Adenosine kinase deficiency	22
Deoxycytidine kinase deficiency	23, 24
Thymidine kinase deficiency[a]	25
Uridine–cytidine kinase deficiency	26
Orotate phosphoribosyltransferase-OMP decarboxylase deficiency.	25, 27
Orotate phosphoriboxyltransferase-OMP decarboxylase, elevated levels.	25, 27
Hypoxanthine–guanine phosphoribosyltransferase substrate affinity alteration	28
Purine-nucleoside phosphorylase deficiency.	29
Ribonucleotide reductase alterations with abnormal responsiveness to dGTP	29–31
Ribonucleotide reductase alterations with abnormal responsiveness to dATP	30, 32, 33
Ribonucleotide reductase alterations with abnormal responsiveness to dTTP	34
Adenylosuccinate synthase deficiency	35, 36
CTP synthase refractory to inhibition by CTP	37
Deoxycytidylate deaminase deficiency	38, 39
AMP deaminase deficiency	40
IMP dehydrogenase alterations	41
Nucleoside transport deficiency	42
Nucleoside transport functions insensitive to complete inhibition by NBMPR, a potent inhibitor of nucleoside transport	43, 44
Glucocorticoid Response	
Altered ploidy of receptor gene	45, 46
Altered glucocorticoid binding by receptor	47–49
Altered nuclear transport of receptor	47–51
Independence of glucocorticoid sensitivity from other responses	5, 52
Cyclic AMP response	
Altered or deficient cAMP-dependent kinase[b]	6, 53–70
Revertants of kinase mutants	14, 16, 17
Mutants resistant to cAMP-induced cytolysis	67, 71
Mutants with altered or deficient adenylate cyclase[b]	72–82
Mutants with altered cAMP phosphodiesterase[b]	80, 83–85
Mutants deficient in β-adrenergic receptors	86
Mutants with altered cAMP transport	87
Miscellaneous	
Ornithine decarboxylase overproduction	8, 88, 89
Lectin resistance[a]	5, 90, 91
Surface antigen (thy) deficiency[a]	92, 93

[a] Available from ATCC.
[b] Available from UCSF.

S49 Cell Mutants

We present in Table I a list of mutant or variant forms of S49 cells that have been described in the literature.[22-93] We have tried to be comprehen-

[22] L. J. Gudas, A. Cohen, B. Ullman, and D. W. Martin, *Somatic Cell Genet.* **4**, 201 (1978).

[23] L. J. Gudas, B. Ullman, A. Cohen, and D. W. Martin, *Cell* **14**, 531 (1978).

[24] B. Ullman, L. J. Gudas, A. Cohen, and D. W. Martin, *Cell* **14**, 365 (1978).

[25] B. Ullman and J. Kirsch, *Mol. Pharmacol.* **15**, 357 (1979).

[26] B. Ullman, B. B. Levinson, D. H. Ullman, and D. W. Martin, *J. Biol. Chem.* **254**, 8736 (1979).

[27] B. B. Levinson, B. Ullman, and D. W. Martin, *J. Biol. Chem.* **254**, 4396 (1979).

[28] U. Friedrich and P. Coffino, *Biochim. Biophys. Acta* **483**, 70 (1977).

[29] B. Ullman, L. J. Gudas, S. M. Clift, and D. W. Martin, *Proc. Natl. Acad. Sci. U.S.A.* **76**, 1074 (1979).

[30] S. Eriksson, L. J. Gudas, S. M. Clift, I. W. Caras, B. Ullman, and D. W. Martin, *J. Biol. Chem.* **256**, 10193 (1981).

[31] B. Ullman, L. J. Gudas, I. W. Caras, S. Eriksson, G. L. Weinberg, M. A. Wormsted, and D. W. Martin, *J. Biol. Chem.* **256**, 10189 (1981).

[32] B. Ullman, S. M. Clift, L. J. Gudas, B. B. Levinson, M. A. Wormsted, and D. W. Martin, *J. Biol. Chem.* **255**, 8308 (1980).

[33] S. Eriksson, L. J. Gudas, B. Ullman, S. M. Clift, and D. W. Martin, *J. Biol. Chem.* **256**, 10184 (1981).

[34] M. A. Roguska and L. J. Gudas, *J. Biol. Chem.* **259**, 3782 (1984).

[35] B. Ullman, S. M. Clift, A. Cohen, L. J. Gudas, B. B. Levinson, M. A. Wormsted, and D. W. Martin, *J. Cell. Physiol.* **99**, 139 (1979).

[36] B. Ullman, M. A. Wormsted, M. B. Cohen, and D. W. Martin, *Proc. Natl. Acad. Sci. U.S.A.* **79**, 5127 (1982).

[37] B. Aronow, T. Watts, J. Lassetter, W. Washtien, and B. Ullman, *J. Biol. Chem.* **259**, 9035 (1984).

[38] G. Weinberg, B. Ullman, and D. W. Martin, *Proc. Natl. Acad. Sci. U.S.A.* **78**, 2447 (1981).

[39] G. L. Weinberg, B. Ullman, C. M. Wright, and D. W. Martin, *Somatic Cell Mol. Genet.* **11**, 413 (1985).

[40] M. Buchwald, B. Ullman, and D. W. Martin, *J. Biol. Chem.* **256**, 10346 (1981).

[41] B. Ullman, *J. Biol. Chem.* **258**, 523 (1983).

[42] A. Cohen, B. Ullman, and D. W. Martin, *J. Biol. Chem.* **254**, 112 (1979).

[43] B. Aronow, K. Allen, J. Patrick, and B. Ullman, *J. Biol. Chem.* **260**, 6226 (1985).

[44] B. Aronow and B. Ullman, *Proc. Soc. Exp. Biol. Med.* **179**, 463 (1985).

[45] S. Bourgeois and J. C. Gasson, *Biochem. Actions Horm.* **12**, 311 (1985).

[46] S. Bourgeois and R. F. Navy, *Cell* **11**, 423 (1977).

[47] K. R. Yamamoto, M. R. Stampfer, and G. M. Tomkins, *Proc. Natl. Acad. Sci. U.S.A.* **71**, 3901 (1974).

[48] R. Miesfeld, S. Okret, A. C. Wikstrom, O. Wrange, J. A. Gustafsson, and K. R. Yamamoto, *Nature (London)* **312**, 779 (1984).

[49] J. P. Northrop, B. Gametchu, R. W. Harrison, and G. M. Ringold, *J. Biol. Chem.* **260**, 6398 (1985).

[50] D. J. Gruol, D. K. Dalton, and S. Bourgeois, *J. Steroid Biochem.* **20**, 255 (1984).

[51] D. J. Gruol, E. S. Kempner, and S. Bourgeois, *J. Biol. Chem.* **259**, 4833 (1984).

[52] U. Gehring and P. Coffino, *Nature (London)* **268**, 167 (1977).

[53] V. Daniel, H. R. Bourne, and G. M. Tomkins, *Nature (London) New Biol.* **244**, 167 (1973).

[54] J. Hochman, H. R. Bourne, P. Coffino, P. A. Insel, L. Krasny, and K. L. Melmon, *Proc. Natl. Acad. Sci. U.S.A.* **74,** 1167 (1977).

[55] H. R. Bourne, P. Coffino, and G. M. Tomkins, *J. Cell. Physiol.* **85,** 611 (1975).

[56] J. Hochman, P. A. Insel, H. R. Bourne, P. Coffino, and G. M. Tomkins, *Proc. Natl. Acad. Sci. U.S.A.* **72,** 5051 (1975).

[57] P. A. Insel, H. R. Bourne, P. Coffino, and G. M. Tomkins, *Science* **190,** 896 (1975).

[58] L. C. McConlogue, L. J. Marton, and P. Coffino, *J. Cell Biol.* **96,** 762 (1983).

[59] R. A. Steinberg, *J. Biol. Chem.* **97,** 1072 (1983).

[60] R. A. Steinberg, *Mol. Cell. Biol.* **4,** 1086 (1984).

[61] R. A. Steinberg and D. A. Agard, *J. Biol. Chem.* **256,** 10731 (1981).

[62] R. A. Steinberg and D. A. Agard, *J. Biol. Chem.* **256,** 11356 (1981).

[63] R. A. Steinberg and P. Coffino, *Cell* **18,** 719 (1979).

[64] R. A. Steinberg and Z. Kiss, *Biochem. J.* **227,** 987 (1985).

[65] R. A. Steinberg, P. H. O'Farrell, U. Friedrich, and P. Coffino, *Cell* **10,** 381 (1977).

[66] R. A. Steinberg, T. van Daalen Wetters, and P. Coffino, *Cell* **15,** 1351 (1978).

[67] I. Lemaire and P. Coffino, *J. Cell. Physiol.* **92,** 437 (1977).

[68] V. Daniel, G. Litwack, and G. M. Tomkins, *Proc. Natl. Acad. Sci. U.S.A.* **70,** 76 (1973).

[69] Z. Kiss and R. A. Steinberg, *J. Cell. Physiol.* **125,** 200 (1985).

[70] C. S. Murphy and R. A. Steinberg, *Somatic Cell Mol. Genet.* **11,** 605 (1985).

[71] I. Lemaire and P. Coffino, *Cell* **11,** 149 (1977).

[72] H. R. Bourne, P. Coffino, and G. M. Tomkins, *Science* **187,** 750 (1975).

[73] H. R. Bourne, B. Beiderman, F. Steinberg, and V. M. Brothers, *Mol. Pharmacol.* **22,** 204 (1982).

[74] M. Shear, P. A. Insel, K. L. Melmon, and P. Coffino, *J. Biol. Chem.* **251,** 7572 (1976).

[75] J. Naya-Vigne, G. L. Johnson, H. R. Bourne, and P. Coffino, *Nature (London)* **272,** 720 (1978).

[76] G. L. Johnson, H. R. Kaslow, and H. R. Bourne, *Proc. Natl. Acad. Sci. U.S.A.* **75,** 3113 (1978).

[77] G. L. Johnson, H. R. Kaslow, and H. R. Bourne, *J. Biol. Chem.* **253,** 7120 (1978).

[78] L. S. Schleifer, J. C. Garrison, P. C. Sternweis, J. K. Northup, and A. G. Gilman, *J. Biol. Chem.* **255,** 2641 (1980).

[79] T. Haga, E. M. Ross, H. J. Anderson, and A. G. Gilman, *Proc. Natl. Acad. Sci. U.S.A.* **74,** 2016 (1977).

[80] M. R. Salomon and H. R. Bourne, *Mol. Pharmacol.* **19,** 109 (1981).

[81] P. A. Insel, M. E. Maguire, A. G. Gilman, H. R. Bourne, P. Coffino, and K. L. Melmon, *Mol. Pharmacol.* **12,** 1062 (1976).

[82] H. R. Bourne, D. Kaslow, H. R. Kaslow, M. R. Salomon, and V. Licko, *Mol. Pharmacol.* **20,** 435 (1981).

[83] H. R. Bourne, V. M. Brothers, H. R. Kaslow, V. Groppi, N. Walker, and F. Steinberg.

[84] V. M. Brothers, N. Walker, and H. R. Bourne, *J. Biol. Chem.* **257,** 9349 (1982).

[85] V. E. Groppi, F. Steinberg, H. R. Kaslow, N. Walker, and H. R. Bourne, *J. Biol. Chem.* **258,** 9717 (1983).

[86] G. L. Johnson, H. R. Bourne, M. K. Gleason, P. Coffino, P. A. Insel, and K. L. Melmon, *Mol. Pharmacol.* **15,** 16 (1979).

[87] R. A. Steinberg, M. G. Steinberg, and T. van Daalen Wetters, *J. Cell. Physiol.* **100,** 579 (1979).

[88] L. McConlogue and P. Coffino, *J. Biol. Chem.* **258,** 8384 (1983).

[89] L. McConlogue, M. Gupta, L. Wu, and P. Coffino, *Proc. Natl. Acad. Sci. U.S.A.* **81,** 540 (1984).

[90] P. Ralph, *J. Immunol.* **110,** 1470 (1973).

sive with respect to the variety of genetically marked cells that have been described. No effort has been made, however, to include all references to the utilization of these lines. In general, the criterion for inclusion is that the paper represent the original description of a line or a significant contribution to its genetic, cell biological, or biochemical characterization. Wild-type cells are available from the American Type Culture Collection, Rockville, Maryland or from the Cell Culture Facility of the University of California, San Francisco. In addition some mutant lines are available from these sources, as indicated in Table I.

Acknowledgments

This work was supported by grants from the NIH (CA29048), NSF (PCM78-07382), and the American Cancer Society (NP-477B).

[91] P. Ralph and I. Nakoinz, *J. Natl. Cancer Inst.* **51**, 883 (1973).
[92] R. Hyman, *J. Natl. Cancer Inst.* **50**, 415 (1973).
[93] I. S. Trowbridge, R. Hyman, and C. Mazauskas, *Cell* **14**, 21 (1978).

[3] Liver Cell Lines

By G. J. DARLINGTON

A variety of liver cell lines have been derived from tumorgenic and nontumorgenic hepatocytes and have been adapted to growth as permanent cell lines. Only a subset of those described in the literature will be presented here in great detail. Those which have been examined for only a small number of hepatospecific phenotypes or have not received wide distribution have not been covered within the scope of this chapter. While the culture and maintenance of normal diploid hepatocytes as a proliferating population have been difficult tasks, the growth of transformed hepatocyte-derived cell lines appears to be relatively uncomplicated. Techniques that are suitable for fibroblast culture are in general also suitable for hepatoma cell lines (although perhaps not optimal). The broad array of media employed for liver cell line culture suggests that no one formulation is dramatically superior to another. However, it is always prudent to measure a set of characteristic phenotypes when the cell line is established in one's laboratory, and to monitor these at various times throughout its culture *in vitro*.

Mouse Hepatoma Lines

A relatively small number of mouse hepatoma lines has been established *in vitro*. This chapter will describe two lines that have been well characterized with respect to their expression of differentiated function and are derived from the same transplantable tumor type, the BW7756 tumor carried in C57 leaden/J mice.

Hepa[1]

Hepa was originally adapted to *in vitro* growth from the BW7756, a tumor passaged subcutaneously in mice, by dissociating tumor chunks with trypsin and plating the cells *in vitro*. After varying periods in culture, the cells were reinoculated into the animal. The resulting tumor was processed in a fashion similar to the original tumor to reestablish the cells in culture. Alternate *in vivo – in vitro* passage resulted in the growth of a cell line which was capable of proliferation under standard conditions of cell culture. The cell line was originally isolated in Waymouth MAB 87/3 medium which had been devised for the serum-free cultivation of fetal mouse liver.[2] Our formulation contained 10% fetal bovine serum (FBS) and antibiotics. Subsequent cultivation of the Hepa cells has been done with a 3:1 mixture of minimum essential medium (MEM) and MAB 87/3 with no phenotypic changes detected. Other investigators have used Dulbecco's modified Eagle's medium (MEM) as the basal nutrient source in combination with FBS. Hepa cells can be cultured in a serum-free medium consisting of 3 parts MEM, 1 part MAB 87/3, plus $3 \times 10^{-8}\ M$ selenium. They continue to express albumin, α_1-antitrypsin, α-fetoprotein, and amylase, and they proliferate slowly in this protein-free medium. Assessment of the retention of differentiated function for periods longer than 4 weeks has not been made although growth in this defined medium can continue indefinitely. Under serum-free conditions, the addition of $10^{-7}\ M$ dexamethasone improves the attachment and growth of Hepa cells.

We routinely grow cells at 37° in a 5% humidified CO_2 environment and maintain the medium at a pH of 7.2. Long-term storage can be accomplished by freezing in growth medium containing 10% (v/v) fetal bovine serum and 10% (v/v) glycerol. The optimal time for harvesting cells to be frozen is during late log growth. These conditions are employed for all the cell lines maintained in our laboratory.

[1] G. J. Darlington, H. P. Bernhard, R. A. Miller, and F. H. Ruddle, *J. Nat. Cancer Inst.* **64**, 809 (1980).

[2] C. Waymouth, *in* "Tissue Culture" (C. V. Ramakrishnan, ed.), pp. 168–179. Dr. Junk Publ., The Hague, The Netherlands, 1965.

No special techniques are required for the maintenance of Hepa cells *in vitro*. Passage of the cells at a 1 : 8 dilution weekly is a convenient protocol. Hepa cells are removed from the growth surface by draining the supernatant medium, rinsing twice with 0.05% trypsin, and resuspending the cells in a serum-containing solution to stop the action of the enzyme. Trypsinization longer than 3–5 min results in membrane damage and low yield upon replating.

Cells attach satisfactorily to regular tissue culture plastic and Millipore membrane filters. Attachment to glass surfaces is not as efficient as for tissue culture plastic, and is improved by pretreating the glass with medium containing 10% serum for an hour prior to plating of the cells.

Hepa has a somewhat poor cloning efficiency (20–50%). The utilization of conditioned medium improves plating efficiency. For this purpose a 25% volume of medium conditioned for 48 hr by late log or early plateau Hepa cells combined with a 75% volume of fresh medium has been employed. Because the cells tend to clump, it is helpful to passage them daily for a few days before use in cloning experiments. This procedure facilitates obtaining a single cell suspension.

The expression of differentiated functions has been examined by a number of investigators and the phenotypes are enumerated in Table I.[1,3–6] It is apparent from this summary that Hepa retains several liver-specific properties, but lacks activity for some hepatic enzymes. As is typical of hepatomas, Hepa produces α-fetoprotein and other products characteristic of the fetal liver (e.g., aldolase A).

Mutagenesis of Hepa cells has been carried out by two laboratories in order to obtain variants for albumin production[7] and aryl hydrocarbon hydroxylase.[8] The latter author has examimed the effect of N-methyl-N'-nitro-N-nitrosoguanidine (MNNG), ethylmethanesulfonate (EMS), ultraviolet light, and ICR-191G on hypoxanthine phosphoribosyltransferase activity. The capability of selecting for benzo[a]pyrene resistance also enabled Hankinson to generate aryl hydrocarbon hydroxylase-deficient variants.

Identification of defects in secreted protein production by immuno-

[3] G. J. Darlington, C. C. Tsai, L. C. Samuelson, D. L. Gumucio, and M. H. Meisler, *J. Mol. Cell. Biol. 6,* 969 (1986).

[4] W. F. Benedict, J. E. Gielen, I.S. Owens, A. Niwa, and D. W. Nebert, *Biochem. Pharmacol.* **22,** 2766 (1973).

[5] H. Baumann and G. P. Jahreis, *J. Cell Biol.* **97,** 728 (1983).

[6] F. D. Ledley, unpublished observations.

[7] G. J. Darlington, J. Papaconstantinou, D. W. Sammons, P. C. Brown, E. Y. Wong, A. L. Esterman, and J. Kang, *Somatic Cell Genet.* **8,** 451 (1982).

[8] O. Hankinson, *Somatic Cell Genet.* **7,** 373 (1981).

TABLE I
Hepa PHENOTYPES

Hepa phenotypes	Designation	Reference
Albumin	+	1
Alcohol dehydrogenase (liver isoform)	−	1
Aldolase A	+	1
Aldolase B	−	1
α-Fetoprotein	+	1
Amylase AMY-1	+	3
Induction of AMY-1		
by dexamethasone	+	3
Amylase AMY-2	−	3
Induction of AMY-2		
by dexamethasone	+	3
Aryl hydrocarbon hydroxylase (AHH)	+	4
Inducibility of AHH		
by phenobarbitol	+	4
Ceruloplasmin	+	1
C3	−	1
Esterase 1	−	1
Esterase 2	+	1
Fibrinogen	−	1
Glycogen storage	−	1
Haptoglobin	+	5
Phenylalanine hydroxylase	−	6
Pseudocholinesterase	+	1
Transferrin	+	1
Tyrosine aminotransferase (TAT)	+	1
Inducibility of TAT by dexamethasone	−	1
Xanthine oxidase	−	1

overlay has been performed by Sammons et al.[9] to identify albumin negative sublines of Hepa 1a. The protocol is as follows (Fig. 1).

Cells are treated with concentrations of mutagen that results in 80–
90% killing. Surviving colonies are allowed to grow until they reach an
average diameter of 1 mm.

A solution of agarose (2% in serum-free medium, Seakem) is added to
an equal volume of medium containing 20% FBS and antiserum to the
secreted protein under study. Various concentrations of primary antibody
should be tried to optimize the results. For albumin, an abundant protein,
the most efficient titer of antibody was relatively high (1/16 to 1/64 dilu-

[9] D. W. Sammons, E. Sanchez, and G. J. Darlington, *In Vitro* **16**, 918 (1980).

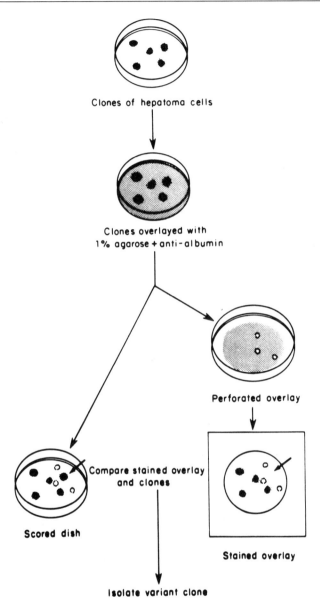

FIG. 1. General scheme for isolation of variants defective in synthesis or secretion of proteins, or both.

tion of antiserum). Proteins secreted in lower abundance may require higher dilutions in order to form a precipitation complex.

The agarose–antibody mixture is made at 42°, cooled to 39–40°, and pipetted onto colonies growing in a petri dish. The dishes are then incubated at 37° in a humidified CO_2 atmosphere for 18 to 24 hr.

Following this period the agarose patties are scored by using a small cork borer to mark the agarose and the dish in order to permit orientation after staining. Agarose patties are removed from the dish, and growth medium is added to the cells.

The agarose immuno-overlay is washed in buffer (310 mg boric acid, 47.5 mg sodium acetate, 8.8 g NaCl, and 200 ng sodium azide per liter H_2O, pH 8.6) with 3 successive changes at 2–3 hr intervals. Following washing, the immuno-overlay is either stained or incubated overnight with a second antibody to enhance the precipitin complex. Prior to staining the immunooverlay is dried onto a glass or mylar backing. We have utilized Coomassie blue, but silver staining should be applicable to this system. If a second antibody is required for visualization of the secreted protein, titration of the second antibody is also required to optimize the amplification of signal.

Hepa cells have been utilized in cell fusion studies generating both inter- and intraspecies hybrids.[10–14] For these studies, both Sendai virus and polyethylene glycol (50%) were employed as fusogens. The mouse hepatoma cells do not appear to have any unusual or adverse properties in response to either of these agents. An interesting property of Hepa–human diploid cell hybrids is the capability of these interspecific combinations to activate a nonhepatic human genome to produce human liver-specific products.[10,12] Human chromosomes segregate from hybrids formed between the murine line and human cells permitting the assignment of α_1-antitrypsin, a human liver-specific gene product, to chromosome 14.[11]

Preliminary analyses of the capacity of Hepa cells to serve as recipients in gene transfer studies have been carried out. Transient expression of chloramphenicol acetyltransferase (CAT) has been observed following a DEAE-dextran-mediated DNA transfer. The protocol used for DEAE-dextran-mediated DNA transfer into Hepa, Hep 3B2 a human hepatoma line, and EJ, a human bladder carcinoma line, is essentially that of Lopata *et*

[10] G. J. Darlington, H. P. Bernhard, and F. H. Ruddle, *Science* **185,** 859 (1974).

[11] G. J. Darlington, K. H. Astrin, S. P. Muirhead, R. J. Desnick, and M. Smith, *Proc. Natl. Acad. Sci. U.S.A.* **79,** 870 (1982).

[12] G. H. Darlington, J. K. Rankin, and G. Schlanger, *Somatic Cell Genet.* **8,** 403 (1982).

[13] J. F. Conscience, F. H. Ruddle, A. Skoultchi, and G. J. Darlington, *Somatic Cell Genet.* **3,** 157 (1977).

[14] O. Hankinson, *Somatic Cell Genet.* **9,** 497 (1983).

al.[15] This procedure has worked well for transient expression assays, as opposed to transfection of DNA for the isolation of stable transformants.

Transfection is performed on cells that are about 75% confluent; lower confluency is used for rapidly growing cells.

Reagents

100 mg/ml DEAE-dextran solution (Pharmacia, 50,000 MW)

2× HEPES buffered saline (HBS) (275 mM NaCl, 10 mM KCl, 1.4 mM Na$_2$HPO$_4$, 12 mM dextrose, 40 mM HEPES, pH 6.92)

10% DMSO in 1× HBS, made fresh

Mix 2 μl DEAE-dextran, 2 μg DNA, per ml serum-free medium in a sterile tube. Let stand 30 min at room temperature. Prior to addition of DNA, wash cells with serum-free medium, drain, and immediately add 2.4 ml of the DNA-DEAE-dextran mixture/60 mm petri dish. Incubate DNA and cells for 4–6 hr. After incubation, aspirate DNA solution and shock cells with addition of 2 ml of 10% DMSO solution/petri dish. Higher concentrations of DMSO may be tolerated by some cells. Remove DMSO after 2 min; timing is critical. Immediately wash cells with serum-free medium three times. Add fresh medium with serum. After 48–72 hr, harvest cells for assay. Stable transformants resulting from calcium phosphate-mediated transfer of the neomycin resistance gene via pSV2 Neo yields approximately 1 × 10^{-5} resistant colonies following selection in the cytotoxic drug, G418. This frequency is sufficiently high to utilize the differentiated Hepa cells for gene transfer studies.

Several different sublines of Hepa have been obtained and are diagrammed in Fig. 2. Those sublines which may be of particular interest to other investigators would include Hepa 1a, an HPRT-negative subline useful for cell hybridization studies,[9] HH, a ouabain-resistant, HPRT-negative interline hybrid with a 2 S chromosome number,[11,16] and the variants 19/2 and 1/C/1 which are deficient in the expression of albumin.[7]

BW[17]

A second mouse hepatoma line was isolated by Szpirer and Szpirer[17] also from the BW 7756 transplantable hepatoma. These investigators were successful in initiating a cultured cell line in a single plating of collagenase digested tumor tissue. The cultured cells were established and maintained in Dulbecco's modified Eagle's medium with 10% fetal bovine serum and antibiotics as additives. The BW cell line grows as a monolayer, as does

[15] M. A. Lopata, D. W. Cleveland, and B. Sollner-Webb, *Nucleic Acids Res.* 12, 5707 (1984).

[16] J. K. Rankin and G. J. Darlington, *Somatic Cell Genet.* 1, 1 (1979).

[17] C. Szpirer and J. Szpirer, *Differentiation* 4, 85 (1975).

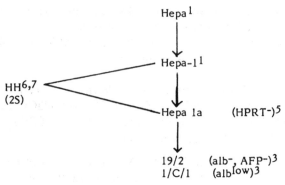

FIG. 2. Sublines of Hepa.

Hepa, and expresses a number of liver-specific phenotypes including albumin, α-fetoprotein, α_2-globulin, complement C_3, and transferrin.

BW cells or derivatives have been utilized in cell hybridization experiments.[18-20] The BW mouse hepatoma cells fused with adult rat hepatocytes generated hybrid cells which segregated rat chromosomes making the system potentially useful for genetic analysis of the rat genome.[20]

Several subclonal lines have been isolated from the BW cell population originally obtained from the tumor. Subclonal lines BW 1 and BW 2 retained the capability to secrete the 5 serum proteins produced by the uncloned parent BW. BW TG3 had an identical phenotype but was selected by growth in thioguanine and was shown to be deficient for HPRT.[17]

Rat Liver Lines

H4II EC3[21]

One of the best characterized and most utilized rat hepatoma cell lines is the H4II EC3 cell line established by Pitot *et al.* in 1964.[21] This cell line was obtained by explanting cellular material from the Reuber hepatoma H-35 in Eagle's minimal essential medium with 20% horse serum and 5%

[18] J. Szpirer and C. Szpirer, *Cell* **6**, 53 (1975).
[19] J. Szpirer and C. Szpirer, *J. Cell Sci.* **35**, 267 (1979).
[20] J. Szpirer, C. Szpirer, and J. C. Wanson, *Proc. Natl. Acad. Sci. U.S.A.* **77**, 6616 (1980).
[21] H. C. Pitot, C. Peraino, P. A. Morse, Jr., and V. R. Potter, *Natl. Cancer Inst. Monogr.* **13**, 229 (1964).

TABLE II
H4II EC3 PHENOTYPES

Phenotypes	Designation	Reference
Alanine aminotransferase	+	22
Alcohol dehydrogenase	+	23
Aldolase	+	24
Albumin	+	25
Fructose diphosphatase	+	26
Glucose-6-phosphate dehydrogenase	−	21
Glucokinase	+	21
Hexokinase	+	21
Histidase	+	21
Induction of TAT by dexamethasone	+	27
Intermediate filaments of the cytokeratin type	+	28
Ornithine transaminase	+	21
Phosphoenolpyruvate carboxykinase	+	26
Phenylalanine hydroxylase	+	29
Proline oxidase	−	21
Thymidine reductase	+	21
Tryptophan pyrrolase	−	21
Tyrosine aminotransferase (TAT)	+	21
Threonine dehydrase	+	21

beef embryo serum. These cells were reinoculated into a syngeneic host and a subsequent tumor resulted. Replating of this secondary tumor *in vitro* gave rise to epithelial colonies which were isolated from the fibroblast background and then seeded at low density. Successive clonal isolations gave rise to the cell line H4-II-E. Characterization of the cells for various enzymatic activities showed that the cultured line retained a number of liver-specific properties such as tyrosine aminotransferase and ornithine transaminase while other hepatic phenotypes were not found. Table II[21-29] lists some of the original phenotypes examined by Pitot and co-workers.

[22] R. S. Sparkes and M. C. Weiss, *Proc. Natl. Acad. Sci. U.S.A.* **70,** 377 (1973).
[23] R. Bertolotti and M. C. Weiss, *J. Cell. Physiol.* **79,** 211 (1972).
[24] R. Bertolotti and M. C. Weiss, *Biochimie* **54,** 195 (1972).
[25] J. A. Peterson and M. C. Weiss, *Proc. Natl. Acad. Sci. U.S.A.* **69,** 571, (1972).
[26] R. Bertolotti, *Somatic Cell Genet.* **3,** 365 (1977).
[27] J. A. Schneider and M. C. Weiss, *Proc. Natl. Acad. Sci. U.S.A.* **68,** 127 (1971).
[28] A. Venetianer, D. L. Schiller, T. Magin, and W. W. Franke, *Nature (London)* **305,** 730 (1983).
[29] D. F. Haggerty, P. L. Young, G. Popjak, and W. H. Carnes, *J. Biol. Chem.* **248,** 223 (1973).

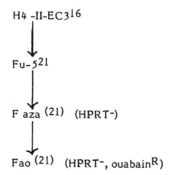

FIG. 3. Subclonal derivatives of H4II EC3.

Additional properties of the cells have been described by Weiss and co-workers and a number of these are also included in the table.

A large number of studies have been carried out on cell hybrids which employed derivative sublines of the H4II EC3 cells. Some combinations include the fusion of rat hepatoma cells with mouse L cells[27] and mouse hepatoma cells.[30] The analysis of differentiated gene expression utilizing these hybrid cell lines has led to a series of conclusions too diverse to enumerate here. However, a summary of much of this work has been published.[31]

Derivatives of the rat hepatoma line have also been utilized in gene transfer studies.[32] Transient expression assays following calcium phosphate precipitation mediated gene transfer showed that the cell line is capable of taking up and expressing sufficient quantities of DNA to be analyzed by measuring chloramphenicol acetyltransferase levels under the direction of various promoters. The authors comment that the rat hepatoma is not as efficient a recipient as some other cell lines, in particular the mouse hepatoma line described by Szpirer and Szpirer.[17]

The pedigree of cell lines derived from H4II EC3 is extensive and encompasses strains that carry a biochemical deficiency for hypoxanthine phosphoribosyltransferase as well as sublines which are termed by Weiss and colleagues as dedifferentiated. The major subclonal derivatives of the original H4II EC3 include FU5, Faza 9, an HPRT-deficient derivative of FU5, and finally FAO, a derivative of Faza 9 which is HPRT deficient and

[30] D. Cassio and M.C. Weiss, *Somatic Cell Genet.* **5,** 719 (1979).

[31] M. C. Weiss, R. Bertolotti, and J. A. Peterson, *Mol. Genet. Dev. Biol., Symp., 1971* p. 425 (1972).

[32] M. O. Ott, L. Sperling, P. Herbomel, M. Yaniv, and M. C. Weiss, *EMBO J.* **3,** 2305 (1984).

ouabain resistant[33] (Fig. 3). Dedifferentiated variants have been isolated from both biochemically marked sublines as well as from the original H4II EC3 cells. In addition, Killary et al.[34] have isolated a thymidine kinase-deficient strain of H4II EC3, a useful mutant for cell hybridization and gene transfer. FU5 and its derivatives have been maintained primarily in F12 medium with 5% fetal calf serum. Faza has been maintained in our laboratory in MEM/MAB 87/3 + 10% FBS for several years without apparent loss of hepatospecific properties.

HTC[35]

A second cell line which has been extensively studied by somatic cell geneticists is the HTC line originally isolated by Thompson et al.[35] The HTC was derived from an ascites tumor which had itself been derived from a solid hepatoma, number 7288C, induced and carried in male buffalo rats. The original isolate was grown in Swim's medium 77, supplemented with 20% bovine serum and 5% fetal bovine serum and antibiotics. A number of laboratories have utilized the HTC cells and a variety of media has been used for its culture including MEM and Dulbecco's modified MEM with 10% fetal bovine serum.

One liver-specific phenotype of interest in the HTC cells was tyrosine aminotransferase activity and the inducibility of this enzyme by dexamethasone and glucocorticoids. Rintoul et al.[36] have described the production of aldehyde dehydrogenase in HTC cells. HTC cells have been shown to produce α_1-acid glycoprotein which is inducible by dexamethasone.[37] These cells are reported to lack phenylalanine hydroxylase activity.[29] The production of plasminogen activator and its inhibition by dexamethasone has also been described.[38]

Studies involving cell hybridization have been done by a number of investigators, including Riddle and Harris,[39] Rintoul et al.,[36] and Thompson and Gelehrter.[40] Drug-resistant derivatives of HTC have been developed, in particular, the HPRT-deficient substrain HTC TG30 selected for resistance to 6-thioguanine by Riddle and Harris.[39]

[33] J. Deschatrette and M. C. Weiss, Biochimie 56, 1603 (1974).
[34] A. M. Killary, T. G. Lugo, and R. E. K. Fournier, Biochem. Genet. 22, 201, (1984).
[35] E. G. Thompson, G. M. Tomkins, and J. F. Curran, Proc. Natl. Acad. Sci. U.S.A. 56, 296 (1966).
[36] D. Rintoul, R. F. Lewis, Jr., and J. Morrow, Biochem. Genet. 9, 375 (1973).
[37] H. Baumann and W. A. Held, J. Biol. Chem. 256, 10145 (1981).
[38] S. A. Carlson and T. Gelehrter, J. Supramol. Struct. 6, 325 (1977).
[39] V. G. H. Riddle and H. Harris, J. Cell Sci. 22, 199 (1976).
[40] E. B. Thompson and T. D. Gelehrter, Proc. Natl. Acad. Sci. U.S.A. 68, 2589 (1971).

MH$_1$ C$_1$[41]

A third hepatoma line derived from the rat which has been relatively well characterized is that designated MH$_1$ C$_1$ established from the transplantable Morris hepatoma #7795 by Richardson *et al.*[41] The cells were directly plated from the tumor following dissociation with viokase. MH$_1$ C$_1$ grew as a monolayer in Hams F10 medium supplemented with 15% horse serum and 2.5% fetal calf serum. The original culture contained different cell types based on the morphological criteria and subsequent subcloning with mechanical selection for epithelioid cells resulted in the acquisition of the MH$_1$ C$_1$ subline which was further characterized for hepatic function. Staining with Oil Red O was positive indicating the presence of stored lipid. The MH$_1$ C$_1$ cells also synthesize and secrete albumin and have baseline levels of tyrosine aminotransferase that are inducible by hydrocortisone. Further analysis by Tashjian *et al.* showed that these cells produce the 9th component of complement and metabolize testosterone.[42] They also conjugate bilirubin[43] and possess activity for UDP glucuronyltransferase, although this is not stimulated by hydrocortisone. Phenylalanine hydroxylase activity is expressed.[29]

RL-PR-C[44]

A fourth cell line derived from rat has been described by Schaeffer.[44] The origin of this line was not from hepatomas as has previously been described, but rather a 3-day-old inbred Wistar/Lewis rat. The isolation procedure involved trypsin digestion of the liver with subsequent plating of the cells in the presence of hydrocortisone hemisuccinate in Hams F12 medium supplemented with 10% fetal bovine serum. After cloning, the RL-PR-C line was established. At the time of its description in 1980, the cells had undergone 326 population doublings and the karyotype of this strain showed a modal number of 42 which coincides with that of the normal animal. Table III lists the properties of the RL-PR-C line which indicate maintenance of the differentiated state *in vitro.*[44] A virtue of such a diploid, differentiated cell line is that gene dosage relationships should be similar to those of normal liver.

[41] U. I. Richardson, A. H. Tashjian, Jr., and L. Levine, *J. Cell Biol.* **40**, 236 (1969).
[42] A. H. Tashjian, Jr., F. C. Bancroft, U. I. Richardson, M. B. Goldlust, F. A. Rommels, and P. Ofner, *In Vitro* **6**, 32 (1970).
[43] H. E. Rugstad, S. H. Robinson, C. Yannoni, and A. H. Tashjian, Jr., *J. Cell Biol.* **47**, 703 (1970).
[44] W. I. Schaeffer, *Ann. N.Y. Acad. Sci.* **349**, 165 (1980).

TABLE III
RL-PR-C PHENOTYPES

RL-PR-C phenotypes	Designation	Reference
Albumin	+	44
α-Fetoprotein	+	44
Arylhydrocarbon hydroxylase	+	44
Catecholamine and Gpp (NH)p activation of adenylate cyclase	+	44
Cholera toxin-stimulated ADP	+	44
Ribosylation from NAD$^+$ and adenylate cyclase activity	+	44
Glucagon receptor-mediated activation of adenylate cyclase	+	44
Insulin receptors	+	44
Insulin activation of glycogen synthetase and stimulation of glycogenesis	+	44
Interaction with human antiinsulin receptor IgG	+	44
Ribonuclease II	+	44
Tyrosine aminotransferase	+	44

SV40 Transformed Hepatocyte Lines

Several cell lines have been established from primary rat hepatocytes by transformation with SV40. Among these are the WIRL-3C and WIRL-3B lines described by Diamond et al.[45] These cells grew out as clonal populations from an explant of hepatocytes taken from a 4-week-old male Wistar rat. Transformation of the WIRL-3B line by SV40 resulted in an indefinitely growing cell line. The WIRL-3 cell lines do not secrete albumin, but do produce α_2-globulin, and have glucose-6-phosphatase activity. They lack δ-aminolevulinic acid synthetase and phenobarbitol inducibility of aryl hydrocarbon hydroxylase.

Additional lines have been established by Chou from adult rat hepatocytes (RALA 255-10G)[46] and from rat fetal liver cells (RLA209-15)[47] by transformation of hepatocytes with a temperature-sensitive A mutant of Simian virus 40. The cell lines both produce albumin and transferrin and the fetal line produces α-fetoprotein. The synthesis of these three proteins

[45] L. Diamond, R. McFall, Y. Tashiro, and D. Sabatini, *Cancer Res.* **33**, 2627 (1973).
[46] J. Y. Chou, *Mol. Cell. Biol.* **3**, 1013 (1983).
[47] J. Y. Chou and S. E. Schlegel-Haueter, *J. Cell Biol.* **89**, 216 (1981).

by the cells can be modulated by adjusting the temperature at which the cells grow.

Isom *et al.* have also described the transformation of rat hepatocytes by Simian virus 40.[48] They have shown that the transformation process can be replicated a number of times to generate a variety of independently derived transformed rat hepatocyte lines. Some of the transformed derivatives were capable of producing albumin and of expressing tyrosine aminotransferase activity. The ability to transform primary hepatocytes with SV40 may allow the immortalization of hepatic cells from various developmental stages. If stage-specific gene expression were retained by such lines, it would be a potentially important source of investigative materials.

Human Hepatoma Lines

One of the first human hepatoma lines described was that designated SK Hep 1.[49] This line originated from an ascites effusion of an adenocarcinoma of the liver. The cells have been maintained as a monolayer culture in RPMI 1640 with $2mM$ glutamine, antibiotics, and 15% FBS[50] and in MAB 87/3 plus 10% FBS in our laboratory. Subsequent characterization of SK Hep 1 by Turner and Turner,[50] and by us shows that this monolayer line produces α_1-antitrypsin, but does not secrete albumin, α-fetoprotein, ceruloplasmin, or haptoglobin. We have also examined surface markers that suggest an endothelial origin for these cells rather than hepatic.

PLC/PRF/5

A second line designated PLC/PRF/5 was established by Alexander *et al.* from a primary hepatocellular carcinoma of an African male.[51] The cells were cultured in Dulbecco's modified Eagle's medium, supplemented with 10% fetal bovine serum. This line grows as a monolayer and has no unusual nutritional requirements. The PLC/PRF/5 line is capable of proliferation in the defined medium described above for Hepa cells.

PLC/PRF/5 has been characterized for expression of differentiated functions by Alexander *et al.*[51] and Knowles *et al.*[52] Table IV[52-57] lists the properties of the PLC/PRF/5 line described by these authors. α-Fetopro-

[48] H. C. Isom, M. J. Tevethia, and J. M. Taylor, *J. Cell Biol.* **85,** 651 (1980).

[49] J. Fogh and G. Trempe, *in* "Human Tumor Cells in Vitro"(J. Fogh, ed.), pp. 115–119. Plenum, New York, 1975.

[50] B. M. Turner and V. S. Turner, *Somatic Cell Genet.* **6,** 1 (1980).

[51] J. J. Alexander, E. M. Bey, E. W. Geddes, and G. Lecatsas, *S. Afr. Med. J.* **50,** 2124 (1976).

[52] B. B. Knowles, C. C. Howe, and D. P. Aden, *Science* **209,** 497 (1980).

[53] G. H. Darlington, D. R. Wilson, and L. B. Lachman, *J. Cell Biol.* **103,** 787 (1986).

[54] H. Saito, L. T. Goodnough, B. B. Knowles, and D. P. Aden, *Proc. Natl. Acad Sci. U.S.A.* **79,** 5684 (1982).

TABLE IV
PLC/PRF/5 Phenotypes

PLC/PRF/5 phenotypes	Designation	Reference
Albumin	+	52
α_1-Acid glycoprotein	+	52
α_1-Antitrypsin	+	52
α_1-Antichymotrypsin	−	52
α-Fetoprotein	+	53
α_2-HS-glycoprotein	−	52
α_2-Macroglobulin	+	52
α_2-Plasmin inhibitor	+	54
β-Lipoprotein	−	52
Complement C3	+	52
Complement C4	−	52
C′3 activator	−	52
Ceruloplasmin	+	52
EGF stimulation of ornithine decarboxylase	+	55
Fibrinogen	−	52
Gc-Globulin	−	52
Haptoglobin	−	52
Hepatitis B surface antigen	+	56
Insulin-like carrier protein	−	57
Plasminogen	+	52
Retinol-binding protein	−	52
Transferrin	+	52

tein has been detected by Laurell immunoelectrophoresis of concentrated supernatant medium in our laboratory. In addition to the expression of hepatospecific functions, this cell line has been intensively studied with respect to the integration of HBV genomes in the cellular DNA.[58–62]

[55] D. M. Moriarity, D. M. DiSorbo, G. Litwack, and C. R. Savage, Jr., *Proc. Natl. Acad. Sci. U.S.A.* **78,** 2752 (1981).

[56] G. M. MacNab, J. J. Alexander, G. Lecatsas, E. M. Bey, and J. M. Urbanowicz, *Br. J. Cancer* **34,** 509 (1976).

[57] A. C. Moses, A. J. Freinkel, B. B. Knowles, and D. P. Aden, *J. Clin. Endocrinol. Metab.* **56,** 1003 (1983).

[58] P. L. Marion, F. H. Salazar, J. J. Alexander, and W. S. Robinson, *J. Virol.* **33,** 795 (1980).

[59] Y. Shaul, M. Ziemer, P. D. Garcia, R. Crawford, H. Hsu, P. Valenzuela, and W. J. Rutter, *J. Virol.* **51,** 776 (1984).

[60] R. Koshy, S. Koch, A Freytag von Loringhoven, R. Kahmann, K. Murray, and P. H. Hofschneider, *Cell* **34,** 215 (1983).

[61] E. M. Twist, H. F. Clark, D. P. Aden, B. B. Knowles, S. A. Plotkin, *J. Virol.* **37,** 239 (1981).

[62] S. Koch, A. Freytag von Loringhoven, R. Kahmann, P. H. Hofschneider, and R. Koshy, *Nucleic Acids Res.* **12,** 6871 (1984).

Hep 3B[62]

A well-characterized human hepatoma derived from a liver tumor biopsy from an 8-year-old Black male was established by Aden *et al.*[63] Adaptation of the hepatoma cells to growth *in vitro* was done by culturing tumor minces on irradiated mouse cell feeder layers. After several months of culture the cells were able to grow in the absence of feeder cells and were maintained in MEM and 10% FBS. Hep 3B grows in monolayer and the cells undergo a rather long lag period if cultured at low density. Expansion is most efficient when the cells are split in ratios of 1:4.

Cell hybridization studies have been conducted with Hep 3B generating hybrids between the human hepatoma and a strain of mouse fibroblasts, the CAK cell line. The presence of human albumin has been documented in these hybrids. Polyethylene glycol was used as a fusogen at 48% rather than 50% as the higher concentration was toxic to Hep 3B.

Hep 3B and some subclonal derivatives have been characterized with respect to the production of hepatitis B surface antigen, albumin, α-fetoprotein, and a host of other liver-specific phenotypes as enumerated in Table V.[51–54,57,63–66]

Sublines of Hep 3B were isolated at the time the tumor was established. These have been shown to have different patterns of HBV integration and are therefore different clonal sublines albeit derived from the same tumor.[61]

3B2 has been used as a recipient in gene transfer studies by Ciliberto *et al.*[67] using calcium phosphate precipitated DNA. The frequency of stable transformants was not examined. Our experience suggests that DEAE-dextran results in better uptake of exogenous DNA than calcium phosphate mediated DNA transfer.

Hep G2[63]

A second hepatoma line established by Aden and Knowles is the Hep G2 strain initiated from a liver tumor biopsy of a 15-year-old Caucasian male from Argentina.[63] The method of cell line establishment was similar to that of Hep 3B in its utilization of feeder layers. Adaptation of Hep G2

[63] D. P. Aden, A. Fogel, S. Plotkin, I. Damjanov, and B. B. Knowles, *Nature (London)* **282,** 615 (1979).
[64] V. I. Zannis, J. L. Breslow, T. R. SanGiacomo, D. P. Aden, and B. B. Knowles, *Biochemistry* **20,** 7089 (1981).
[65] J. G. Haddad, D. P. Aden, and M. A. Kowalski, *J. Biol. Chem.* **258,** 6850 (1983).
[66] Y. M. Wen, K. Mitamura, B. Merchant, Z. Y. Tang, and R. H. Purcell, *Infect. Immun.* **39,** 1361 (1983).
[67] G. Ciliberto, L. Dente, and R. Cortese, *Cell* **41,** 531 (1985).

TABLE V

Hep 3B PHENOTYPES

Hep 3B phenotypes	Designation	Reference
Albumin	+	52,63
α-Fetoprotein	+	52,63
α_1-Acid glycoprotein	+	52
α_1-Antitrypsin	+	52
α_1-Antichymotrypsin	+	52
α_2-HS-glycoprotein	+	52
α_2-Macroglobulin	+	52
α_2-Plasmin inhibitor	+	54
Apoproteins apo A-I, apo A-II	+	64
Apo B, apo C-II, apo C-III	+	64
Apo E	+	64
β-Lipoprotein	+	51
C-Reactive protein stimulation by monocyte supernates		53
Complement C3	+	52
Complement C4	−	52
C'3 activator	+	52
Ceruloplasmin	+	52
Fibrinogen	+	65
Gc-Globulin	+	52
Haptoglobin	+	52
Hepatitis B surface antigen	+	63
Hepatitis B virus integration	+	66
Hepatitis B virus nuclear antigen	+	66
Insulin-like growth factor carrier protein	+	57
Plasminogen	+	52
Retinol-binding protein	+	52
Transferrin	+	52

to *in vitro* conditions free of feeder systems was also accomplished and the cells can be maintained in MEM plus 10% fetal bovine serum. Hep G2 will grow in the basal defined medium that consists of three parts MEM, one part Waymouth MAB87/3, and 3×10^{-8} M sodium selenite. The addition of insulin (1 μg/ml) improves the growth rate of the cells.

The analysis of the expression of differentiated functions for Hep G2 has been perhaps the most extensive of all of the hepatoma lines included in this summary. Table VI[52,54,57,63,64,66,68-77] compiles some of the charac-

[68] R. J. Andy and R. Kornfeld, *J. Biol. Chem.* **259**, 9832 (1984).
[69] D. S. Fair and B. R. Bahnak, *Blood* **64**, 194 (1984)

TABLE VI
Hep G2 Phenotypes

Hep G2 phenotypes	Designation	Reference
ADA binding protein	+	68
Albumin	+	52,63
α-Fetoprotein	+	52,63
α_1-Acid Glycoprotein	+	52
α_1-Antitrypsin	+	52
α_1-Antichymotrypsin	+	52
α_2-HS-glycoprotein	+	52
α_2-Macroglobulin	+	52
α_2-Plasmin inhibitor	+	54
Antithrombin III	+	69
Apolipoproteins, apo A-I, apo A-II	+	64
Apo B, apo C-II, apo C-III, apo E	+	64
Asialoglycoprotein receptor	+	70
β-Lipoprotein	+	52
Ceruloplasmin	+	52
Complement C3	+	52
Complement C3 precursor	+	71
Complement C4	+	52
C′3 activator	+	52
Complement C1r, C1s, C2	+	72
C3, C4, C5, factor B, C1 inhibitor, C3b inactivator, C6, C8	+	72
C1q, C7 C9	−	72
Estrogen stimulation of apolipoproteins apo C-II and apo A-1	+	73
Factor V	+	74
Factor X	+	69
Factor IX	−	69
Fibrinogen	+	52
Gelsolin	+	75
Gc-Globulin	−	52
Haptoglobin	+	52
Hepatitis B surface antigen	−	63
Hepatitis B virus integration	−	66
Hepatitis B virus nuclear antigen	−	66
Insulin-like growth factor carrier protein	+	57
Plasminogen	+	52
Prothrombin	+	69
Retinol-binding protein	+	52
Testosterone-estradiol-binding globulin	+	76
Thyroxine-binding globulin	+	76
Transcobalamin	+	77
Transferrin	+	52
Low-density lipoprotein receptors responsive to LDL	+	66

TABLE VII
HA 22T/VGH Phenotypes

HA 22T/VGH phenotypes	Monolayer	Aggregate	Reference
	Designation		
Albumin	−	+	78
α_1-Antichymotrypsin	−	+	78
α_1-Antitrypsin	−	+	78
α_2-HS-glycoprotein	−	+	78
α_2-Macroglobulin	−	+	78
α_1-Antichymotrypsin	−	+	78
β-Lipoprotein	−	+	78
Ceruloplasmin	+	+	78
Complement C3	+	+	78
Complement C4	+	+	78
C3 activator	+	+	78
Gc-Globulin	+	+	78
γ-Glutamyltransferase	+		78
Haptoglobin	−	+	78
Hepatitis B surface antigen	−		78
Plasminogen	−	+	78
Retinol-binding protein	−	+	78
Tyrosine aminotransferase	+		78
Alanine aminotransferase	+		78

teristics that have been examined in Hep G2. Of particular note is the absence of hepatitis B surface antigen or integrated viral DNA, and the retention of a large number of liver specific phenotypes.

Personal experience with the use of Hep G2 as a recipient for gene transfer suggests that stable transformants can be obtained after calcium phosphate-mediated DNA transfer at a frequency of 1×10^{-6}. Preliminary experiments in our laboratory utilizing a DEAE-dextran protocol for anal-

[70] A. L. Schwartz, S. E. Fridovich, B. B. Knowles, and H. F. Lodish, *J. Biol. Chem.* **256,** 8878 (1981).

[71] K. M. Morris, G. Goldberger, H. R. Colten, D. P. Aden, and B. B. Knowles, *Science* **215,** 399 (1982).

[72] K. M. Morris, D. P. Aden, B. B. Knowles, and H. R. Colten, *J. Clin. Invest.* **70,** 906 (1982).

[73] S. P. Tam, T. K. Archer, and R. G. Deeley, *J. Biol. Chem.* **260,** 1670 (1985).

[74] D. B. Wilson, H. H. Salem, J. S. Mruk, I. Maruyama, and P. W. Majerus, *J. Clin. Invest.* **73,** 654 (1984).

[75] H. L. Yin, D. J. Kwiatkowski, J. E. Mole, and F. S. Cole, *J. Biol. Chem.* **259,** 5271 (1984).

[76] M. S. Kahn, B. B. Knowles, D. P. Aden, and W. Rosner, *J. Clin. Endocrinol. Metab.* **53,** 448 (1981).

[77] C. A. Hall, P. D. Green-Colligan, and J. A. Begley, *Biochim. Biophys. Acta* **838,** 387 (1985).

ysis of transient expression of the albumin gene show that uptake is enhanced by trypsinizing the cells and adding the DEAE-dextran-DNA solution to the suspended G2 cells.

Ha 22T/VGF[78]

A final human hepatoma line was described in 1983 by Chang et al., the HA22T/VGF cell line.[78] This strain was established from a surgical specimen obtained from a Chinese male with hepatocellular carcinoma. The tumor was minced and plated in Dulbecco's modified Eagle's medium supplemented with 10% fetal bovine serum and antibiotics. Clones of epithelioid cells were selected for subculture.

The culture conditions for this cell line require comment. When grown as a monolayer, the cells express some hepatospecific gene products, but when cultivated as suspended aggregates in bacterial petri dishes where they do not attach, the production of liver-specific products increases both in kind and amount. This strain may therefore be a useful model for the analysis of cell–cell interactions with respect to the expression of differentiated function. Table VII[78] describes the hepatic phenotypes for HA22TVGH grown as a monolayer or as an aggregate culture.

[78] C. Chang, Y. Lin, T. W. O-Lee, C. K. Chou, T. S. Lee, T. J. Liu, F. K. P'Eng, T. Y. Chen, and C. P. Hu, *Mol. Cell. Biol.* **3**, 1133 (1983).

[4] HeLa Cell Lines

By SHIN-ICHI AKIYAMA

Establishment of HeLa Cell Lines

The HeLa cell was isolated from a biopsy of the cervical carcinoma of a 31-year-old black female, Henrietta Lacks (HeLa), by G. O Gey, W. D. Coffman, and M. T Kubicek in February, 1951. The original biopsy of the cervical carcinoma had been misinterpreted as epidermoid carcinoma. Jones et al. reexamined the specimen and suggest that HeLa is an aggressive adenocarcinoma.[1]

As the first established human cell line, the HeLa cell has contributed for over three decades to the development of virology, molecular and cell

[1] H. W. Jones, Jr., V. A. McKusick, P. S. Harper, and K.-D. Wuu, *Obstet. Gynecol.* **38**, 945 (1971).

biology, and also to the study of somatic cell genetics such as genetic control mechanisms and mutation at the cellular level.

Characteristics of the Cell Lines

Currently, there are 3 original HeLa lines in the American Type Culture Collection (ATCC) repository, namely HeLa (CCL 2), HeLa 229 (CCL 2.1), and HeLa S3 (CCL 2.2). HeLa (CCL 2) is considered most similar in characteristics to the cells described in the classic studies of Scherer et al.[2] Both HeLa 229 and HeLa S3 are derivatives of HeLa (CCL 2), but differ mainly in their virus susceptibility from the parental HeLa (CCL 2) cell line.[3] The derivation and some characteristics of these three HeLa cell lines are listed in Table I.

Genetic characteristics of HeLa cells have been extensively studied. HeLa cells express M, N, S, s, Tja, and HLA (A 28, A 3, BW 35) antigens and the following enzyme phenotypes: AK(1-1) for adenylate kinase, ADA(1-1) for adenosine deaminase, PGM(1-1) for phosphoglucomutase, 6PGD(A) for 6-phosphogluconate dehydrogenase, and G6PD(A) for glucose-6-phosphate dehydrogenase.[4]

Chromosome banding revealed marker chromosomes characteristic of HeLa cells in culture. Miller et al.[5] reported the four HeLa markers, 1 to 4 with marker 3 characteristically present in more than one copy per cell. Number 1 marker consists of the short arm and centromere of a No. 1 chromosome and an arm of the No. 3 chromosome. Number 2 marker is probably the short arm of a No. 3 chromosome and the long arm of a No. 5 chromosome. Number 3 marker is a small isochromosome in two or more copies, and the Number 4 marker is a "dull" short arm and long arm of a No. 9 chromosome or No. 18 chromosome with bright fluorescence. HeLa cells lack the Y chromosome.

The genetic characteristics of cell lines, such as the allozyme phenotype, and the marker chromosomes, are a useful way to estimate cellular identity or contamination. Many reports have documented instances of cross-contamination between cell cultures using the above chromosome markers. More than 40 established human cell lines have been reported to be

---

[2] W. F. Scherer, J. T. Syverton, and G. O. Gey, J. Exp. Med. 97, 695 (1953).

[3] American Type Culture Collection, "Certified Cell Lines" (R. Hay, M. Macy, A. Hamburger, A. Weinblatt, and T. R. Chen, eds.), 4th Ed. American Type Culture Collection, Rockville, Maryland, 1983.

[4] S. H. Hsu, B. Z. Schacter, N. L. Delaney, T. B. Miller, V. A. McKusick, R. H. Kennett, J. G. Bodmer, D. Young, and W. F. Bodmer, Science 191, 392 (1976).

[5] O. J. Miller, D. A. Miller, P. W. Allderdice, V. G. Dev, and M. S. Grewall, Cytogenetics 10, 338 (1971).

TABLE I
SOME CHARACTERISTICS OF HeLa CELL LINES[a]

Characteristic	HeLa (CCL 2)	HeLa 229	HeLa S3
Number of serial subcultures from tissue of origin	Unknown; 90–102 from culture received by W. F. Scherer, 1952	Unknown; 78–88 from culture sent to W. F. Scherer in 1952	Unknown
Morphology	Epithelial-like	Epithelial-like	Epithelial-like
Virus susceptibility	Susceptible to poliovirus type 1 and adenovirus type 3	Susceptible to adenovirus type 3 and vesicular stomatitis (Indiana strain) virus; This strain is 2 to 3 log less sensitive to poliovirus types 1, 2, and 3 than HeLa CCL 2	Susceptible to poliovirus type 1, adenovirus type 5, and vesicular stomatitis virus
Chromosome number	Mode 84 Range 58–179		Mode 68 Range 51–74
HeLa marker chromosomes	One copy of No. 1 One copy of No. 2 Four–five copies of No. 3 Two copies of No. 4		One copy of No. 1 One copy of No. 2 Two copies of No. 3 One copy of No. 4
Submitted to ATCC by	W. F. Scherer	J. T. Syverton	T. T. Puck

[a] From American Type Culture Collection.[3]

contaminated by the HeLa cell line and a list of cell lines with characteristics peculiar to HeLa cells has been reported by Nelson-Rees and Flandermeyer[6] (Table II).

Growth Conditions and Nutritional Requirements

HeLa cells are not so fastidious concerning medium, and many kinds of media with different components have been used for the genetic study of HeLa cells. Usually Eagle's minimal essential medium (MEM) or Dulbecco's modified Eagle's medium (D-MEM) with 10% newborn calf serum or 10% fetal calf serum, L-glutamine (2 mM), pH 7.3, is used for the maintenance of HeLa cells as monolayer cultures at 37° in an atmosphere

[6] W. A. Nelson-Rees and R. R. Flandermeyer, *Science* **171**, 96 (1976).

TABLE II
CELL LINES WITH CHARACTERISTICS PECULIAR TO HeLa CELLS[a]

Designation	Source
HeLa (adenocarcinoma cervix)	ATCC
HeLa (=CCL 2)	ATCC via A. Deitch
	A. Mukerjee
	V. Klement from Flow Labs., Inc.
	G. Gey
	Unlisted
	Four individuals, unlisted
	Grand Island Biological Co.
	N. Differante
HeLa 229 (=CCL 2.1)	ATCC
HeLa S3	G. Nette from E. Robbins
	L. Levintow
HeLa S3g	M. Griffin via G. Melnykovych
HeLa S3k	K. Kajievara via G. Melnykovych
KB (carcinoma, oral)	Unlisted
KB (=CCL 17)	ATCC
	S. Mak
	V. Klement from MBA
	H. Sussman
	E. Priori
	Commercial, unlisted
H.Ep.-2 (carcinoma, larynx)	Unlisted
H.Ep.-2 (=CCL 23)	ATCC
	Individual, unlisted
	P. Dent
	V. Klement from MBA
	M. Webber
	K. McCormick
H.Ep.-2 (clone)	K. V. Ilyin
AV3 (amnion)	Unlisted
AV3 (=CCL 21)	ATCC
AV3 (103)	I. Keydar from ATCC
AV3 (F-49-1)	P. Peebles from ATCC
L132 (=CCL5)(lung)	ATCC
L132 (G-38-7)	P. Peebles from ATCC
Chang liver (liver)	Unlisted
Chang liver (=CCL13)	ATCC
	R. Chang
	Individual, unlisted
HBT3 (carcinoma, breast)	P. Arnstein from R. Bassin
HBT-E (16c, clone of HBT-3)	R. Bassin
HBT-39b (carcinoma, breast) (clone 6)	P. Arnstein from E. Plata

(continued)

TABLE II *(Continued)*

Designation	Source
HEK (kidney)	Commercial, unlisted
	J. Rhim from C. Pfizer, Inc.
	C. Pfizer, Inc.
HEK/HRV (HEK, virus transformed)	S. Aaronson
MA160 (prostate)	The originators
MA160	P. Price, MBA
	M. Vincent, MBA
Prostate (=MA160)	Unlisted
SA4 (TxS-HuSa₁)(liposarcoma)	C. Pfizer, Inc.
SA4	D. Morton
RT4 (carcinoma, bladder)	J. Leighton via N. Abaza
Detroit 30A (carcinoma, ascitic fluid)	W. D. Peterson, Jr.
Detroit 98 (=CCL18-(sternal marrow)	ATCC
Detroit 98s (=CCL18.1)	ATCC
Detroit 98/AG (CCL18.2)	ATCC
Detroit 98AT-2 (=CCL18.3)	ATCC
Detroit 98/AHR (=CCL18.4)	ATCC
FL (=CCL62)(amnion)	ATCC
CaOV (carcinoma, ovary)	N. P. Mazurenko
J96 (leukemic blood)	T. A. Bektemirov
Jlll (monocytic leukemia) (=CCL24)	Commercial, unlisted
	Unlisted
	ATCC
T-9 (transformed normal diploid)	O. G. Andzaparidze
DAPT (astrocytoma, piloid)	A. O. Bykovsky
AO (amnion)	A. O. Bykovsky
KP-P₁ (carcinoma, prostate)	P. Lee via M. Glovsky
ElCo (carcinoma, breast)	R. Patillo
HCE (carcinoma, cervix)	D. Brown
CMP (adenocarcinoma, rectum)	Unlisted
CMPII C2	D. Rounds via J. Kim
JHT (placenta)	J. Cho via J. W. -Peng
OE (endometrium)	The originators
	P. Di Saia via L. Milewich
SH-2 (carcinoma, breast)	The originators
	G. Seman via R. Miller
SH-3 (carcinoma, breast)	The originators

(continued)

TABLE II *(Continued)*

Designation	Source
ESP₁ (Burkitt lymphoma, American)	P. Price, from E. Priori
	E. Priori
EB33 (carcinoma, prostate)	F. Schroeder
D18T (synovial cell)	D. T. Peterson
M10T (synovial cell)	D. T. Peterson
Detroit 6 (sternal marrow)	Unlisted
	Commercial, unlisted
Detroit 6 (=CCL3)	ATCC
Detroit 6 (clone 12) (=CCL3.1)	ATCC
Detroit 6 (=CCL3)	Child Research Center of Michigan or ATCC
Minnesota EE (esophageal epithelium)	Individual, unlisted
Minnesota EE (=CCL4)	ATCC
Intestine 407 (jejunum, ileum)(=CCL6)	ATCC
Intestine 407	Commercial, unlisted
Intestine 407 (=HEI=CCL6)	G. Spahn from ATCC
NCTC2544 (=CCL19)(skin) (epithelium)	ATCC
NCTC3075 (=CCL19.1)	ATCC
WISH (amnion)	Individual, unlisted
WISH (=CCL25)	ATCC
Girardi heart (heart)(=CCL27)	ATCC
TuWi (=CCL31)	ATCC
Wong-Kilbourne (conjunctiva) (=CCL20.2)	ATCC

[a] From Nelson-Rees and Flandermeyer.[6]

of 5% CO_2–95% air with >98% humidity. The antibiotics penicillin (50–100 units/ml) and streptomycin (50–100 μg/ml) are also added to the medium to eliminate or suppress microbial contaminants unless they affect any ultimate use intended for the cells. Plating efficiency of HeLa cells in the above culture medium with 10% FCS is approximately 80%.

Clone S3 is readily adaptable to growth in spinner culture with a medium deficient in Ca^{2+} or Ca^{2+} and Mg^{2+}.

HeLa cells also grow in a defined medium such as Ham's F12 medium supplemented with insulin, transferrin, hydrocortisone (aldosterone), fibroblast growth factor, and epidermal growth factor, at the same growth rate as that of cells in serum-supplemented medium.[7]

Storage of Cells

Exponentially growing cells are harvested with 0.25% trypsin solution and centrifuged at 150 g for 5 min and resuspended at 4° in culture medium containing 10% dimethyl sulfoxide (DMSO) to a concentration of more than 1×10^6 cells/ml. One milliliter of the cell suspension is dispensed into each sterile 2 ml polypropylene freezing tube (AS/Nunc, Roskilde, Denmark). The tubes are frozen at a controlled rate of 1°/min with a program freezer (Planer Produce Ltd., Middlesex, England). If a program freezer is not available the tubes can be wrapped in cotton and placed in a small paper box. After covering up the box, it is placed in a −80° freezer for more than 3 hr. The tubes are stored in liquid nitrogen after freezing.

To recover cells from frozen storage, the cells are thawed by removing the tubes rapidly from the liquid nitrogen refrigerator and immediately warming the bottom of the tubes to 37° in a water bath. Immediately after thawing, the tubes are removed from the water bath and wiped with 70% ethanol to sterilize the outside. To remove the freezing additive immediately, the tube contents are diluted with 10 ml of the culture medium, cells are sedimented, resuspended in the desired volume of culture medium, and placed in a culture dish.

Use of HeLa Cell Lines for Genetic Analysis

Cloning

In genetic studies, it is advantageous to use cloned populations of cells for their minimal genetic variability. Although there are several techniques for cloning, colony-forming method or dilution plating techniques are frequently employed for isolating single HeLa cells. It is recommended that cells be cloned two times to guarantee pure clones.

Cloning by colony formation is as follows,

1. Prepare single cell suspensions from monolayer cultures and dilute them to yield ultimately 1 ~ 10 colonies per plate considering the plating efficiency of the cells.

---

[7] S. E. Hutchings and G. H. Sato, *Proc. Natl. Acad. Sci. U.S.A.* **75,** 901 (1978).

2. Add the appropriately diluted suspensions to a 100-mm dish and incubate them at $37°$ in CO_2 incubator. The 100-mm dishes hold 10 ml of medium.

3. The cells are allowed to grow for 10 to 14 days into colonies without medium change.

4. Encircle the colonies on the plastic with a grease pencil and remove the medium from the plates and rinse them twice with PBS.

5. Place a sterilized steel cloning ring (8 mm in diameter, 10 mm high, and 1 mm thick walls) coated on one end with autoclaved silicone grease using sterilized forceps so that the ring encircles a colony of cells and the greased end forms a seal with the plastic.

6. Add 2–3 drops of trypsin solution to each of the rings and incubate the dishes for 5 ~ 10 min at $37°$ until the cells are rounded and ready to detach.

7. Pipet the trypsin solution gently up and down to detach the cells from dishes.

8. Transfer the suspended cells to small culture bottles with culture medium.

Mutagenesis

Many mutant sublines, such as bromodeoxyuridine-resistant mutants,[8] ouabain-resistant mutants,[9] an auxotrophic mutant,[10] and carbohydrate variants,[11] have been spontaneously isolated from HeLa cells. Considering that DNA is damaged at many other sites than the target gene by the mutagens, it is desirable to isolate mutants in the absence of mutagens. However, a mutagen will usually be indispensable to obtain the intended phenotype from a reasonable number of cell.

Mutagens used for HeLa cells are ethylmethane sulfonate (EMS),[12] *N*-methyl-*N'*-nitro-*N*-nitrosoguanidine (MNNG),[12] an acridine mustard mutagen (ICR-372),[13] and ethidium bromide (to increase the mutation frequency in mitochondrial DNA).[14] EMS has been commonly used for its simplicity to use and reliability for enhancing the mutation frequency.

The procedure for mutagenesis of HeLa cells and KB cells with EMS is as follows.

[8] S. Kit, D. R. Dubbs, and P. M. Frearson, *Int. J. Cancer* **1**, 19 (1966).
[9] B. Weissman and E. J. Stanbridge, *Cytogenet. Cell Genet.* **28**, 227 (1980).
[10] R. De Mars and J. L. Hooper, *J. Exp. Med.* **111**, 559 (1960).
[11] R. S. Chang, *J. Exp. Med.* **111**, 235 (1960).
[12] G. Milman, E. Lee, G. S. Ghangas, J. R. Melaughlin, and M. George, Jr., *Proc. Natl. Acad. Sci. U.S.A.* **73**, 4589 (1976).
[13] D. C. Wallace, C. L. Bunn, and J. M. Eisenstadt, *J. Cell Biol.* **67**, 174 (1975).
[14] C. M. Spolsky and J. M. Eisenstadt, *FEBS Lett.* **25**, 319 (1972).

TABLE III
EMS INDUCTION OF OUABAIN- OR COLCHICINE-RESISTANT MUTANTS

Cell	EMS (μg/ml)	Cell survival relative plating efficiency	Frequency[a]	
			Oua^{rb}	Ch^{rc}
HeLa CCL 2^d	0	1.00	$<2 \times 10^{-7}(0)$	—[e]
	55	0.81	$<1 \times 10^{-7}(0)$	—
	120	0.64	$1.5 \times 10^{-6}(18)$	—
	273	0.42	$4.4 \times 10^{-6}(22)$	—
	322	0.21	$2.2 \times 10^{-5}(43)$	—
KB	0	1.00	$5 \times 10^{-8}(2)$	$5 \times 10^{-8}(2)$
	100	0.75	$1.3 \times 10^{-6}(52)$	$9.0 \times 10^{-6}(360)$
	150	0.66	$3.3 \times 10^{-6}(132)$	$1.1 \times 10^{-5}(444)$
	200	0.59	$3.2 \times 10^{-6}(128)$	$7.7 \times 10^{-6}(308)$
	250	0.50	$3.1 \times 10^{-6}(124)$	$4.5 \times 10^{-6}(180)$
	300	0.38	$1.7 \times 10^{-6}(68)$	$2.5 \times 10^{-6}(100)$

[a] Frequency of ouabain- or colchicine-resistant colonies per cell assayed. Numbers in parentheses indicate actual number of mutant colonies scored.
[b] Oua^r, Ouabain-resistant colonies.
[c] Ch^r, Colchicine-resistant colonies.
[d] From Baker et al.[15]
[e] Not tested

1. Harvest exponentially growing cells and inoculate approximately 10^6 cells in each 100-mm dish.

2. Incubate the dishes overnight at 37°, and add 1% EMS (Eastman Kodak Co.) in PBS or serum-free medium to the culture to give a final concentration in the range of 100–250 μg/ml for HeLa (CCL 2) cells and KB cells, which are now thought to be HeLa, and incubate at 37° for 16 ~ 24 hr. Use gloves when handling EMS-containing solutions and make dilutions in a fume hood. HeLa S3 is more sensitive to EMS than HeLa (CCL 2), and the appropriate range of EMS concentration for mutagenesis of HeLa S3 is lower.[15] The correlation between cell survival rate and frequency of mutant induction of HeLa (CCL 2) and KB cells is presented in Table III. For efficient induction of mutants without immoderate damage to the cells, a surviving fraction of 0.3–0.6 is usually used.

[15] R. W. Baker, C. Van Voorhis, and L. A. Spencer, *Proc. Natl. Acad. Sci. U.S.A.* **76,** 5249 (1979).

TABLE IV

CHARACTERISTICS OF PARENTAL HUMAN CELLS AND HYBRIDS[a]

Cell line	Derivation	Oua[b] (D10 value)	HPRT	G6PD	Chromosome number	
					Mode	Range
HeLa (Stone)	Cervical carcinoma	S (3.5×10^{-8} M)	+	A	66	61–67
D98/AH-2	HeLa	S (1.8×10^{-8} M)	−	A	62	53–66
D98-OR	D98/AH-2	R (4.4×10^{-6} M)	−	A	62	57–66
HT1080	Fibrosarcoma	S	+	B	46	42–86
75-18-OR	Fibroblast	R	−	A	46	—
ESH7	D98-OR × HT1080	R	+	A/B	104	92–108
ESH20	D98/AH-2 × 75-18-OR	R	−	A	86	75–89
ESH38	D98-OR × HeLa (Stone)	R (10^{-6} M)	+	A	121	100–124

[a] From Weissman and Stanbridge.[9]

[b] Ouabain phenotype: S, sensitive; R, resistant. Numbers in parentheses indicate D10 Value for ouabain.

TABLE V

VARIANT CELL LINES DERIVED FROM HeLa AND KB CELLS

Character	Cell lines	Derivation	References
Nutritional variants			
Glutamine requiring	HeLa I-11a	HeLa I-11	10
Glutamine independent	HeLa I-11	HeLa S3-1	10
Carbohydrate	Ribose variants		11
	Xylose variants		11
	Lactate variants		11
Grow in protein- and lipid-free synthetic media	HeLa-P3	HeLa	a
Quantitatively different in their requirement of serum	HeLa S1	HeLa	b
	HeLa S3	HeLa	b
Drug resistant			
Actinomycin D	HeLa-R	HeLa	c
1-β-D-Arabinofuranosylcytosine	KB/araC	KB	d
8-Azaguanine	S3AG1	HeLa S3	13
Bromodeoxyuridine (Brd Urd)	HeLa BU-10	HeLa S3	8
	HeLa BU-15	HeLa S3	8
	HeLa BU-25	HeLa S3	8
	HeLa BU-50	HeLa S3	8
	HeLa BU-100	HeLa S3	8
Chloramphenicol	296-1	HeLa S3	14
Colchicine, multidrug resistant	KB-Ch^R	KB	18
Adriamycin, multidrug resistant	KB-A1	KB	e
Vinblastine, multidrug resistant	KB-V1	KB	e
Erythromycin	ERY2301	HeLa	f
Ethylmethane sulfonate (EMS)	HeLa A6	HeLa S3	15
Ouabain	D98-OR	D98/AH-2	9
6-Thioguanine	H23	HeLa	12
Toxin resistant	D98/AH-2	HeLa	17
Diphtheria toxin	KB-R2	KB	g
	KB-R2A	KB	g
Epidermal growth factor— Pseudomonas toxin	ET	KB	h
Altered virus susceptibility			
Poliovirus sensitive	HeLa I-3	HeLa S3	i
Poliovirus resistant	HeLa S3-1C	HeLa S3	i
Poliovirus resistant	"R"	HeLa S3	j
Others			
Alkaline phosphatase lacking	A clonal line of giant HeLa cells	HeLa	k
Ultraviolet sensitive	S-1M	HeLa S3	l
	S-2M	HeLa S3	l

[a] T. Takaoka and H. Katsuta, *Exp. Cell Res.* **67,** 295 (1971).

[b] T. T. Puck and H. W. Fisher, *J. Exp. Med.* **104,** 427 (1956).

3. Terminate the treatment by changing medium, and incubate at 37° for 5–10 days[15] after mutagen treatment to allow expression of induced mutations.

Hybridization

The application of cell fusion techniques has facilitated progress in the areas of somatic cell genetics and the genetic analysis of malignancy.

A commonly practiced method for promoting cell fusion is the use of polyethylene glycol (PEG) following the method of Davidson and Gerald.[16]

Mutant HeLa cell lines have been isolated for use in selecting viable hybrids. HeLa variants, D98/AH-2[17] and HeLa- BU[8] lack detectable hypoxanthine–guanine phosphoribosyltransferase (HPRT) and thymidine kinase (TK) activity, respectively. Weissmann and Stanbridge have isolated a double mutant (D98-OR) carrying a recessive mutant allele, HPRT⁻, and the codominant ouabain resistance allele.[9] When the opposite population for cell fusion has no genetic markers, D98-OR is convenient for studies involving cell hybrids.

The procedure below is used for making human × human hybrids in monolayer.

1. Plate cells (5×10^5 or 1×10^6) of each cell type together on a 60-mm plastic dish and incubate overnight at 37°.

2. Remove all the medium by aspiration, and add 3 ml of 50% PEG (Baker, PEG-1000) in α-modified Eagle's medium (α-MEM) (Flow).

3. Expose the cells to 50% PEG for 1 min at room temperature ($\sim 22°$), remove all the PEG, wash three times with 5 ml α-MEM containing 10% fetal bovine serum, and incubate for 24 hr in the same medium.

---

[16] R. L. Davidson and P. S. Gerald, *Methods Cell Biol.* **15**, 325 (1977).
[17] E. J. Stanbridge, *Nature (London)* **260**, 17 (1976).

---

[c] M. N. Goldstein, I. J. Slotnick, and L. J. Journey, *Ann. N.Y. Acad. Sci.* **89**, 474 (1960).

[d] T. Tsuruo, K. Naganuma, H. Iida, S. Sone, K. Ishii, E. Tsubara, S. Tsukagoshi, and Y. Sakurai, *Gann* **75**, 690 (1984).

[e] D.-W. Shen, C. Cardarelli, J.-L. Huang, N. Richert, S. Ishii, I. Pastan, and M. M. Gottesman, *J. Biol. Chem.* **261**, 7762 (1986).

[f] C.-J. Dorsen and E. J. Stanbridge, *Proc. Natl. Acad. Sci. U.S.A.* **76**, 4549 (1979).

[g] J. M. Moehring and T. J. Moehring, *Virology* **69**, 786 (1976).

[h] S. Akiyama, unpublished results (1983).

[i] J. E. Darnell, Jr., and T. K. Sawyer, *Virology* **11**, 665 (1960).

[j] M. Vogt and R. Dulbecco, *Virology* **5**, 425 (1958).

[k] P. Fortelius, E. Saksela, and E. Saxén, *Exp. Cell Res.* **21**, 616 (1960).

[l] K. Isomura, O. Nikaido, M. Horikawa, and T. Sugahara, *Radiat. Res.* **53**, 143 (1973).

4. Select the cells in α-MEM with HAT (1×10^{-4} M hypoxanthine, 4×10^{-7} M aminopterine, 1.6×10^{-5} M thymidine) and 0.1 μM ouabain.

5. Colonies derived from various fusions are surrounded by a steel ring and trypsinized from the dishes as described in the section on cloning. Hybridization frequencies for D98-OR $\times$ KB hybrids under these conditions are approximately 10^{-5}.[18] Some characteristics of parental human cells and hybrids are shown in Table IV.

Characteristics of Variants

A list of some interesting variant cell lines, derived from KB and HeLa cells, is presented in Table V.

Acknowledgments

The author would like to thank Dr. M. Kuwano (Oita Medical School, Oita, Japan) and Dr. M. M. Gottesman (NCI, NIH, Bethesda, MD) for their reading and helpful criticism of this manuscript.

[18] S. Akiyama, A. Fojo, J. A. Hanover, I. Pastan, and M. M. Gottesman, *Somatic Cell Mol. Genet.* **11**, 117 (1985).

[5] Somatic Cell Genetic Analysis of Myelomas and Hybridomas

By DEBORAH FRENCH, THERESA KELLY, SUSAN BUHL, and MATTHEW D. SCHARFF

Myeloma and Hybridoma Cells in Culture

As with other differentiated cell types, the generation of permanent cell lines that continue to produce antibodies proved to be impossible for many years. More recently cells at different stages in B lymphocyte differentiation have been converted to continuous tissue culture either by starting with malignant cells or by viral transformation. Such cells were used to study the production, assembly, and secretion of immunoglobulins[1] and to examine the somatic instability of immunoglobulin genes.[2]

[1] M. D. Scharff, *Harvey Lect.* **69**, 125 (1975).
[2] S. L. Morrison and M. D. Scharff, *CRC Crit. Rev. Immunol.* **3**, 1 (1981).

Mouse myeloma cell lines producing various classes and subclasses of immunoglobulin were particularly useful. Techniques were devised to identify clones of cells producing mutant immunoglobulin molecules which arose at a remarkably high frequency.[2] Nevertheless, such cultured immunoglobulin producing cells were of value to only a few investigators interested in that small area of somatic cell genetics. This changed dramatically in 1975 when Kohler and Milstein showed that the fusion of cultured mouse myeloma cells to spleen cells from immunized individuals resulted in the immortalization of normal antibody forming cells. The resulting hybrids (hybridomas) produced the antibody of the splenic B cell in culture and could be injected back into animals where very large amounts of antibody would accumulate in the serum and ascites fluid of tumor bearing animals. The hybridoma technology has revolutionized serology, made it possible to study the generation of antibody diversity, and holds enormous promise as a method for producing antibodies that can be used as *in vitro* and *in vivo* diagnostic reagents and in the therapy of human disease.

The study and genetic manipulation of hybridoma cells growing in culture have thus acquired great practical significance. As with parental myeloma cells, mouse, rat, and human hybridomas grow as suspended cells. The best hybridomas produce about 50 μg of antibody/10^6 cells/24 hr and usually have doubling times of 17–19 hr. They are commonly grown in petri or tissue culture dishes in CO_2 incubators but may also be propagated in large amounts in any of a variety of culture vessels.

Most monoclonal antibodies have been made by fusing one or another derivative of the P_3 cell line with mouse or rat spleen cells and largely reflect the growth characteristics of that cell line. A few mouse myelomas were originally made with derivatives of MPC 11 but these have similar characteristics. There are still considerable problems with making human monoclonal antibodies and an ideal fusion parent has not yet been found. In the sections that follow we will therefore limit our discussion to mouse and rat myelomas all of which have similar characteristics.

Growth Conditions for Normal Maintenance and Special Techniques

The nutritional requirements for the growth of myeloma and hybridoma cells can be determined by the cloning efficiency of the particular cell line. Most investigators use Dulbecco's modified Eagle's medium (DMEM) containing high glucose (4500 mg/liter) supplemented with glutamine, 1% nonessential amino acids (Gibco), 1% penicillin and streptomycin (pen-strep), and 10–20% heat-inactivated (56°, 30 min) fetal calf (FCS) or horse serum. In some laboratories, perhaps related to the water supply, higher cloning efficiencies can be achieved with commercially available liquid medium rather than medium reconstituted from commercial powder. For this reason we use commercially available liquid DMEM for cloning and other special techniques and prepare our own DMEM from commercially available powdered DMEM for routine mass culture.

The source of serum and its concentration in the medium often depend on the culture conditions, the types and potential uses of the monoclonal antibody, and whether the antibodies secreted into the medium will be purified prior to use. For example, some producers of FCS provide batches which are certified to have low endotoxin and hemoglobin levels, to have been filtered through 0.1-μm filters, to be free of mycoplasma, and to be characterized for many normal serum components. Most batches of FCS contain low, or even undetectable, levels of "natural" antibodies. This is important, since the presence of even small amounts of antibody reactive with the antigen being studied may interfere with screening for positive hybridomas or other immunoassays. Such antibodies are often present in calf or horse serum so most investigators choose to use FCS.

Many investigators use various types of feeders for cloning or fusions and this reduces the need of screening for special batches of serum. However, we have found it easier to screen for batches of FCS that support limiting dilution cloning of the parental myeloma that we use for our fusions or the myeloma cell line being used for other studies. Briefly, the limiting dilution assay and calculation of cloning efficiency are performed as follows:

1. Microtiter plates (flat-bottom) are pretreated with cloning medium (commercial liquid DMEM with glutamine, 1% nonessential amino acids, 1% pen-strep, and 20% FCS) (100 μl/well) which is removed by suction after 24 hr in the CO_2 incubator. It is not clear why this is useful but it does eliminate differences between batches of microtiter plates.

2. The cell line is diluted (0.30–0.10 cells/well) in complete medium plus 20% of the current "cloning" serum or the new lots to be tested. Five wells are plated with the next to last dilution and the cells/well counted immediately to confirm the accuracy of the cell counts and dilutions.

3. Cells are plated (0.30 cell/100 μl/well) in two 96-well, flat bottomed microtiter plates for each serum lot and incubated (37°, 8–10% CO_2) for 2–3 days after which the plates are fed with fresh medium (50 μl/well).

4. After 1 week, all wells are scored for positive growth and the absolute cloning efficiency and cloning efficiency relative to the current "cloning" serum are calculated. Only those batches of serum with a cloning efficiency greater than 50% and producing homogeneous sized clones are used.

6. If we are screening serum for fusions, batches giving high cloning efficiencies of the parental myeloma are compared with existing batches

for fusion efficiency. Some batches of FCS promote the outgrowth of attached cells which may inhibit the growth of hybridomas.

Several additional precautions are necessary in the preparation of medium used for cloning. Triton-containing filters cannot be used for sterilizing cloning medium since the detergent inhibits the growth of clones.[3] Medium exposed to normal fluorescent light has been reported to be toxic[4] and we have confirmed that medium shielded with aluminum foil will sometimes give higher cloning efficiencies. For hardy cell lines, or mass culture, many of these precautions are not necessary.

Mutagenesis

Different cell lines vary in their sensitivity to mutagenic agents, therefore killing curves should be carried out with each cell line. Dose ranges for the mutagens ICR-191, melphalan, nitrosoguanidine, and ethylmethane sulfonate (EMS) are given for the MPC 11 cell line (Table I).[5,6] An amount of mutagen is used that results in approximately 50% cell death with the reasoning that a low dose will result in a low frequency of mutagenesis whereas a higher dose will result in multiple mutations/cell and much cell death. Furthermore, very high doses of mutagen may activate repair mechanisms and decrease the frequency of mutants.

Depending on the needs of the experiment, higher or lower doses may be desirable. The degree of killing and the frequency of mutation are not necessarily directly related for all mutagens so the killing curve is at best a way of standardizing conditions. Once the desired dose of mutagen has been determined for each cell line, the cells (5×10^5 cells/ml in 5 ml of medium) are incubated with mutagen for $2-24$ hr at $37°$, $8-10\%$ CO_2. For nitrosoguanidine very short periods of mutagenesis in serum-free medium are used since it is unstable in serum. Other mutagens are incubated with the cells for $18-24$ hr. The cells are washed free of mutagen and resuspended in fresh medium for at least 24 hr to allow for segregation and expression of recessive mutations in daughter cells. It is worth noting that at higher doses of mutagen, cell division is delayed and longer segregation times are required. In fact, Penman and Thelly[7] recommend 7 days for segregation and expression.

[3] R. D. Cahn, *Science* **155**, 195 (1967).

[4] R. J. Wang, *In Vitro* **12**, 19 (1976).

[5] J.-L. Preud'homme, J. Buxbaum, and M. D. Scharff, *Nature (London)* **245**, 320 (1973).

[6] R. Baumal, B. K. Birshtein, P. Coffino, and M. D. Scharff, *Science* **182**, 164 (1973).

[7] S. W. Penman and W. G. Thelly, *Somatic Cell Genet.* **2**, 325 (1976).

TABLE I
EFFECT OF MUTAGENS ON THE INCIDENCE OF HEAVY-CHAIN NONSECRETING MUTANTS[a]

Cell line	Mutagen	Dose (μg/ml)	Percentage of cell survival	Incidence of variants[b]	Percentage	Reference
MPC-11	ICR-191	0	100	18/2104	0.86	6
		1	60	25/3635	1.54	
		2	25	110/3404	3.24	
		4	<1	15/229	6.55	
MPC-11	Melphalan	0	100	17/3777	0.45	5
		0.2	32	31/2336	1.33	
		0.4	28	100/5926	1.69	
		0.6	16	55/2961	1.86	
		0.8	9	31/1298	2.39	
MPC-11	Nitrosoguanidine	0	100	29/4993	0.58	6
		0.25	90	46/3828	1.21	
		0.5	68	33/4332	0.76	
		1.0	33	14/2404	0.58	
		2.0	8	10/1977	0.51	
MPC-11	Ethyl methanesulfonate	0	100	13/2168	0.60	6
		50	85	13/1594	0.81	
		100	45	10/1288	0.80	
		200	13	4/721	0.55	

[a] Frequencies were measured by the immunoplate assay (see Special Techniques: Immunoplate Assay for Detecting Mutants).
[b] Number of unstained per total.

Special Techniques

Immunoplate Assay for Detecting Mutants

Introduction. The immunoplate assay was one of the first methods used to detect mouse myeloma mutants.[8-10] Briefly, cells are cloned in soft agarose over feeder layers composed of rat embryo fibroblasts, 3T3 cells, or human fibroblasts. Whichever feeder is used, it must undergo contact inhibition. Feeder layers provide growth-promoting factors to the clones of hybridoma or myeloma cell lines and we have found that primary rat embryo fibroblasts give the best cloning efficiencies. Antibody reactive with the immunoglobulin that is secreted by the cloned cells is added to the plate and diffuses through the agarose forming an antigen–antibody pre-

[8] R. Coffino and M. D. Scharff, *Proc. Natl. Acad. Sci. U.S.A.* **68,** 219 (1971).
[9] P. Coffino, R. Laskov, and M. D. Scharff, *Science* **167,** 186 (1970).
[10] P. Coffino, R. Baumal, R. Laskov, and M. D. Scharff, *J. Cell. Physiol.* **79,** 429 (1972).

cipitate surrounding the clone. This precipitate appears as a collection of dark granules and specks under low or medium power with an inverted microscope (Fig. 1). Clones not surrounded by a precipitate are putative variants. We and others use this assay to look for mutants of hybridoma and myeloma cells, to clone hybridomas and identify subclones producing the desired antibody, to identify high producers, to stabilize unstable hybridomas such as the product of mouse and human fusions, and to clone switch variants enriched in a sib selection assay (see below).

Materials

Rat embryo fibroblasts: Eighteen-day rat embryos are removed and decapitated. After immersing in 70% ethanol, the hind legs are removed and placed in 1.0 ml of Earle's balanced salt solution (EBSS). The tissue is minced and placed in a flask containing 20 ml of trypsin solution (0.25% w/v in EBSS). The mixture is stirred at room temperature for 20 min and after the large fragments have settled (approximately 30 sec), only the supernatant from this first trypsinization is discarded. The trypsinization procedure is then repeated six times collecting and saving all of the supernatants each of which is added to DMEM containing enough serum to give a final concentration of 20%. The cell suspension is filtered through gauze (approximately 6 layers), centrifuged, and resuspended in DMEM con-

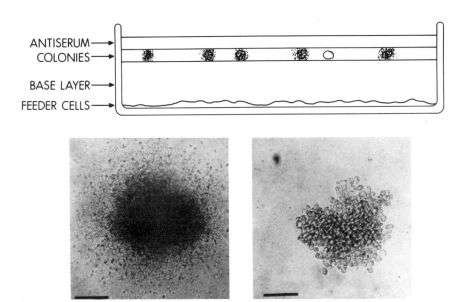

FIG. 1. The immunoplate assay with antibody overlay. The left-hand side shows a stained clone, while the right-hand side shows an unstained clone.

taining 15% FCS. The cells are added to three tissue culture flasks (Falcon #3024) and the medium is replaced after 2 hr. In approximately 4 days, the cells are removed by trypsinization and frozen (1×10^7 cells/ml) in DMEM supplemented with 10% DMSO and 20% FCS. Fibroblasts are grown in tissue culture flasks and passed every 3–4 days upon reaching confluency. Each batch of thawed cells is only used for approximately 6 weeks because the cells decrease in growth rate and production of the growth factors which are essential for the proliferation of clones in soft agar.

Cloning petri dishes: (Falcon #3002, 60-mm tissue culture dish).

Cloning medium: commercial liquid DMEM (4500 mg/liter glucose) with glutamine, 1% nonessential amino acids, 1% pen-strep, 20% heat-inactivated FCS.

Feeding medium: commercial powdered DMEM (4500 mg/liter glucose) with glutamine, 1% nonessential amino acids, 1% pen-strep, 10% heat-inactivated FCS.

Conditioned medium: Extra cloning plates are prepared containing a feeder layer and an agarose underlayer. Cloning medium (5.0 ml/plate) is added over the agarose underlayer and all plates are incubated for 48 hr at 37° in 8–10% CO_2. This "conditioned" cloning medium is removed, filtered, and stored at 4° and is stable for several weeks. It is our experience that medium conditioned directly by rat embryo fibroblasts inhibits growth whereas medium "filtered" through agarose supports growth and increases cloning efficiency. This "conditioned" medium is used for the recovery of clones picked from soft agar.

Sea plaque agarose (Marine Colloids Division, FMC Corp.): This type of agarose has the benefit of remaining in solution until the temperature is lowered to 19°. The agarose solution can therefore be maintained at 37° rather than 42° reducing the chance of killing the cells when they are suspended in the agarose. The optimum concentration of agarose must be determined for each lot. A 5–7% stock solution is prepared in distilled water (for example 3.0 g/50 ml H_2O). The agarose is dissolved by autoclaving (approximately 15 min) and a final solution containing 7, 8, 9, or 10 ml of the stock solution per 100 ml of cloning medium is tested to see which forms a gel that is loose enough to be almost fluid but firm enough to suspend the clones and not slide off the plate. If the clones do not "vibrate" in the agarose when gently jarred, the agarose is too concentrated. The agarose stock solution (5–7%) should not be autoclaved and reused more than three times since evaporation will change the agarose concentration. The use of agarose which is too "tight" is the most common cause of failure to obtain a high cloning efficiency and large clones.

Antiserum: Antiserum should be filtered before heat inactivation since antibody treated at 56° for 30 min may be trapped on the filter. Antiserum should be frozen in small aliquots and frozen and thawed as little as possible. To obtain a preliminary estimate of the amount of antiserum needed to form a visible precipitate in the overall technique, myeloma or hybridoma cells (1×10^7) are lysed in 50 μl of 0.5% NP-40 or Triton X-100 in cold isotonic buffer. The nuclei are removed by centrifugation and the cytoplasmic lysate is placed in the center well of an Ouchterlony plate. Two-fold dilutions of antiserum are added to the surrounding wells. If the antiserum at a one to four dilution still reacts well with the lysate, 100 μl of antiserum in 1.0 ml of agarose containing medium is used for each 60-mm plate. If the titer of the antiserum is higher, proportionately less is needed.

Antigen: Antigen only forms a visible precipitate in the overlay technique with secreted IgM or IgA which are highly polymeric. A polyvalent form of antigen must be used because a number of secreted immunoglobulin molecules are required to precipitate each antigen molecule. The size of antigen-induced precipitates is small and difficult to see but can be increased with the addition of subliminal amounts of anti-immunoglobulin to the mixture of cells and agarose when they are plated (Fig. 2).[11] The amount of subliminal antibody should not form a visible precipitate with the secreted immunoglobulin, therefore a control plate with subliminal antibody and cells alone must be included in the assay. We think the addition of subliminal antibody works because it slows the diffusion of the secreted immunoglobulin thereby increasing the local concentration of antibody reacting with the antigen, as well as increasing the amount of protein in the precipitate.

Myeloma or hybridoma cells: Prior to cloning, the cells should be in the logarithmic phase of growth. The optimum number of clones per 60-mm plate is between 1000 and 1500 and most cell lines that have been in culture for some time clone at efficiencies of 50–100%. Hybridomas that are only a few weeks old often have much lower cloning efficiencies.

Method of Cloning in Soft Agar
Feeder layer (rat embryo fibroblasts)

1. The fibroblast monolayer is washed with DMEM ($1\times$) to remove any nonadherent cells and serum.

2. Add 5 ml of trypsin solution (0.01% Bacto-trypsin, Difco 015361-EDTA, 2×10^{-3} M in EBSS) that has been warmed to 37°.

3. Rock the flask (approximately 1–2 min.) until the cells begin to round up and discard 4 ml of the trypsin solution.

4. Hit the flask on its side several times to remove the fibroblasts from

---

[11] W. D. Cook and M. D. Scharff, *Proc. Natl. Acad. Sci. U.S.A.* **74**, 5687 (1977).

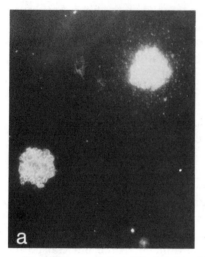

FIG. 2. Representative stained and unstained clones from the antigen or antibody overlay in the immunoplate assay. (a) An antigen (plus subliminal antibody) induced precipitate is visible around the clone in the upper right while the clone in the lower left is unstained. (b) An antibody-induced precipitate is visible around the clone on the left while the right clone is unstained.

the wall of the flask. The remaining 1 ml of trypsin should complete the removal of the cells from the plastic.

5. Resuspend the cells in 10 ml of feeding medium.

6. From a 10 ml pipet, 3–6 drops of resuspended fibroblasts are put into each 60-mm tissue culture dish containing 4 ml of feeding medium. The dish is rocked to distribute the cells.

7. The plates are incubated at 37°, 8–10% CO_2 overnight at which time the fibroblasts should be a quarter to half confluent. If the cells are fully confluent they will be less effective feeders.

8. The original flask of fibroblasts can be reused by adding back 1.0 ml of cells and 15 ml of feeding medium.

Cloning

1. Approximately 30 min before autoclaving the stock agarose solution (5–7%), warm the required amount of cloning medium to 37°.

2. Add the predetermined amount of stock agarose solution to the warm medium and keep at 37° until ready to use.

3. Suction the medium off of the fibroblast plates and add 5 ml of agarose-cloning medium to each plate.

4. Put the plates on a level plane at 4° for 10 min after which the plates

are warmed in a 37°, 8–10% CO_2 incubator for at least 10 min. At this stage the plates can be left overnight or the cells can be cloned directly.

5. Cells are harvested, counted, and diluted with one volume of cells to four volumes of agarose-cloning medium to a final volume of 1.0 ml per plate. This means that the cell layer contains a 20% lower (and looser) concentration of agarose than the underlayer. For example, if the cloning efficiency of a cell line is 100%, the cells are diluted to 5000 cells/ml. In a sterile tube, 200 μl of cells is mixed with 800 μl of agarose-cloning medium. When hybridomas are being cloned directly from their original wells following a fusion, 5 μl of resuspended cells is added to 200 μl of cloning media which in turn is added to 800 μl of agarose-cloning medium. If the subliminal antibody is being used, it is added to the cell mixture at this point. Cells are added to each plate in a dropwise fashion starting at the center of the plate and working toward the periphery in concentric circles. This procedure ensures an even distribution of cells over the agar underlayer.

6. Again, the plates are put on a level plane at 4° for 10 min after which they are incubated at 37°, 8–10% CO_2 in an incubator that is as humid as possible. To prevent drying of the plates, it is essential to maintain a very high humidity. This is easier to achieve in a water jacketed than a hot air incubator.

Antibody or antigen overlay

1. After 2–3 days, or when the clones reach the 8–16 cell stage, the plates are overlayed with heat-inactivated antibody or antigen in 1 ml of agarose-cloning medium. This procedure is not routinely done immediately after cloning since the antiserum is usually precious and should not be wasted on plates that do not have the desired number of clones or are contaminated. In addition, the growth of some cells is inhibited by antibody (even in the absence of complement) if it is present at the time of plating. Nevertheless, in many instances it is possible to add antisera or antigen with the cells.

2. Between 6 and 10 days after plating, the clones are examined using an inverted microscope for the lack of a visible antigen–antibody precipitate in the case of mutants or a visible precipitate in the case of hybridomas or switch variants (see Fig. 1).

3. As the clones get larger, the precipitate surrounding the clones may be solubilized by the large amounts of immunoglobulin secreted by the growing clone. Therefore clones should be marked when identified and picked approximately 8–10 days after plating. Clones can be marked by putting a red dot with a marking pen on the top of the plate over the clone. The red spot will then project down on the clone as it is visualized through

the inverted microscope. The top and bottom of the plate are aligned by marking lines on the side of each. More accurate marking can be done by placing autoclaved spangles (glitter obtained from a notion store) directly in the agar just above the clone using a small tweezer. Even if the agar rotates, the marker will move with the clone.

4. Clones are removed by using a bent-tipped finely drawn Pasteur pipet and gentle mouth-controlled suction. The opening of the pipet should be about the size of the clones.

5. Cells are deposited in a well of a 96-well flat bottomed culture dish containing 50 μl of cloning medium. If clones from a particular cell line are difficult to recover we use fibroblast conditioned medium (see materials) diluted 1:2 or 1:4 in cloning medium.

6. When the cells reach half confluence, fresh medium is added to each well. After several days of growth, supernatants from each clone can be tested and appropriate clones expanded into 24-well plates for further characterization.

A spectrum of mutants has been generated by the immunoplate assay,[8-17] however it has some serious drawbacks. First, it requires screening many wild-type clones for each mutant that is detected. Only 10,000 – 20,000 clones can be examined in each experiment, therefore only mutants that occur frequently will be identified. Second, the lack of an antigen – antibody precipitate may reflect any of a number of phenotypes. A particular clone may have lost the ability to synthesize or secrete either or both of the immunoglobulin polypeptide chains, may be secreting decreased amounts of wild-type immunoglobulin or may be producing a structurally altered immunoglobulin molecule. Every presumptive mutant recovered from the agarose must therefore be analyzed in detail. In addition, antigen overlay will not give a precipitate with IgG.

Nevertheless, as a positive assay for the detection of mutants, this method can be extremely useful. For example, a switch variant producing immunoglobulin of one of the IgG subclasses can be detected by using an antiserum in the overlay that is specific for that particular subclass. Only

[12] L. A. Wims and S. L. Morrison, *Mutat. Res.* **81**, 215 (1981).

[13] B. K. Birshtein, J.-L. Preud'homme, and M. D. Scharff, *Proc. Natl. Acad. Sci. U.S.A.* **71**, 3478 (1974).

[14] J.-L. Preud'homme, B. K. Birshtein, and M. D. Scharff, *Proc. Natl. Acad. Sci. U.S.A.* **72**, 1427 (1975).

[15] S. Koskimies and B. K. Birshtein, *Nature (London)* **264**, 480 (1976).

[16] T. Francus, B. Dharmgrongartama, R. Campbell, M. D. Scharff, and B. K. Birshtein, *J. Exp. Med.* **147**, 1535 (1978).

[17] S. L. Morrison, *Eur. J. Immunol.* **8**, 194 (1978).

subclones with a visible precipitate surrounding them would be identified, thus eliminating the necessity of picking and screening many clones. It is important to note that an overlay with a single monoclonal anti-IgG will not give a visible precipitate but that a pool of two or three may give a visible precipitate.

Presumed revertants of an IgM or IgA antigen-binding loss myeloma mutant can also be screened quickly using this assay since only clones surrounded by an antigen-induced precipitate would be selected. A double overlay can also be performed which makes the assay more discriminating. A clone not surrounded by a visible precipitate with, for example, an antiserum that reacts with a constant region domain can be marked and the plates overlayed again with a different antiserum reacting with other domains of the immunoglobulin molecules.[18] This procedure allows the identification of mutants and eliminates the possibility of a nonsecretor.

Sib Selection Technique for the Detection of Subclass Switching

Introduction. The technique of sib selection[19] can be applied to the detection of rare variants for which there is an antiserum or antigen which will bind the variant antibody but not the parental secreted immunoglobulin. The most common use is for the detection of hybridomas which have switched from expressing one class or subclass to another while retaining their original antigen binding.[20,21] An enzyme-linked immunoassay (ELISA) is used as a detection system and $2-3 \times 10^6$ cells can be screened. An advantage of the sib selection technique for the detection of subclass switching as compared to fluorescent-activated cell sorting[22,23] is that it does not require that the switch variant express surface immunoglobulin. The ELISA employed in our system can detect $0.5-2.0$ ng of IgA or the IgG subclasses.

The sensitivity of the ELISA for IgM to IgG subclass switch variants is determined by the number of cells plated in each well and the background of the assay due to nonspecific binding or cross-reactivity with the parental immunoglobulin. This level of sensitivity can be determined by reconstruction experiments in which various ratios of IgM and IgG producing

[18] D. E. Yelton and M. D. Scharff, *J. Exp. Med.* **156,** 1131 (1982).
[19] L. L. Cavalli-Sforza and J. Lederberg, *Genetics* **41,** 367 (1956).
[20] C. E. Muller and K. Rajewsky, *J. Immunol.* **131,** 877 (1983).
[21] G. Spira, A. Bargellesi, J.-L. Teillaud, and M. D. Scharff, *J. Immunol. Methods* **74,** 307 (1984).
[22] J. L. Dangl and L. A. Herzenberg, *J. Immunol. Methods* **52,** 1 (1982).
[23] B. Liesegang, A. Radbruch, and K. Rajewsky, *Proc. Natl. Acad. Sci. U.S.A.* **75,** 3901 (1978).

cells, for example, are plated starting with 1000 cells in each microtiter well.[21] The supernatant from each well is assayed for the IgG subclass as the cells reach confluence. The signal for the presence of each subclass, by days 5 and 7, is usually 2- to 6-fold the background level if only one cell of the 1000 plated produces IgG. Reconstruction experiments already performed suggest that more than 1000 cells can be inoculated into each well. However, 1000 cells/well allows a safe margin for variations in growth rate and IgG production in the rare IgG-producing cells. Initial screening for switch variants from myeloma or hybridoma cells requires analysis of at least 2×10^6 cells or 2000 wells. Some variants require the initial screening of $3-4 \times 10^6$ cells or 3000–4000 wells. A schematic of the sib selection assay is provided in Fig. 3.

Materials

Cloning medium: Commercial liquid DMEM (4500 mg/liter glucose) with glutamine, 1% nonessential amino acids, 1% pen-strep, 20% heat-inactivated FCS.

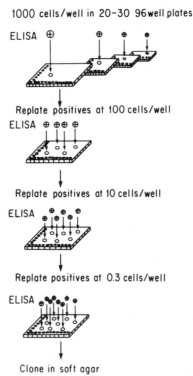

1000 cells/well in 20–30 96well plates

ELISA ⊕   ⊕   ⊕   •

Replate positives at 100 cells/well

ELISA ⊕ ⊕⊕ ⊕

Replate positives at 10 cells/well

ELISA ⊕ • • ⊕

Replate positives at 0.3 cells/well

ELISA ⊕ • • ⊕ •

Clone in soft agar

FIG. 3. A schematic of the sib selection assay.

Microtiter plates for sib selections: 96-well, flat-bottomed, tissue culture treated (sterile) plate.

Dispenser: 96-well (Handi Spense, Sandy Spring Instrument Co.).

D-Spenser 96-magazine, sterile.

Media reservoir, sterile.

Microplate washer.

Antibodies for ELISA: Rabbit, sheep, or goat anti-mouse IgM, IgG, and IgA. Alkaline phosphatase rabbit, sheep or goat anti-mouse IgM, IgA, whole IgG, IgG_1, IgG_{2b}, IgG_{2a}, and IgG_3.

All of the class and subclass specific antibodies need to be titered to determine the dilution which binds 75% of the immunoglobulin found in the media from secreting hybridomas. This dilution is used for adsorption to ELISA plates and guarantees an ample excess of coating to bind the small amounts of immunoglobulin in the medium. The probing or alkaline phosphatase-labeled antibody is used at the concentration recommended by the manufacturer. All antibodies are checked for specificity and lack of reactivity with each other and if they show inappropriate reactivities another lot of antibody is used. The amount of anti-IgG_3 in the commercial anti-whole IgG is usually low, therefore a mixture of the alkaline phosphatase anti-IgG subclass antibodies is used.

Phosphatase substrate tablets (p-nitrophenyl phosphate, disodium in 5 and 40 mg/tablet or 100 mg/capsule) (Sigma).

Tween 20(0.05%)/PBS (for washing).

1% BSA in PBS with 0.01 M sodium azide (for blocking ELISA plates and dilutions of antibody).

Bicarbonate substrate buffer: stock solutions, 0.01 M $MgCl_2$ (10X) and 0.05 M Na_2CO_3 (10X), are mixed (10 ml each) with distilled water (80 ml). The pH is adjusted to 9.8 with 6 N HCl.

ELISA reader.

Microtiter plates for ELISA: 96-well, flat-bottomed (nonsterile).

First Sib Selection

1. Myeloma or hybridoma cells (2×10^6) are resuspended in cloning medium (200 ml).

2. Cells are aliquoted (100 μl/well) with a 96-well multidispenser into 20 96-well plates.

3. The plates are incubated at 37°, 8–10% CO_2 until the cells are half confluent (approximately 5 days) at which time the cells are fed with cloning medium (50–75 μl/well).

4. When the cells are confluent the supernatants from each well are assayed in an ELISA for the detection of the immunoglobulin class or subclass of interest.

ELISA

1. Microtiter plates are coated with anti-μ, γ, or α antibody (approximately 1 μg/ml in PBS at 50 μl/well) using a 96-well multidispenser.

2. After a 1-hr incubation at room temperature the plates are washed with Tween 20/PBS using an automatic plate washer.

3. Free sites on the plates are blocked with 1% BSA in PBS (50 μl/well) for 1 hr at room temperature.

4. The plates can then be wrapped with Saran and stored at 4° for several weeks.

5. After the plates are washed with Tween 20/PBS, supernatants from each well of the sib selection plates are added (50 μl/well) to the precoated ELISA plates using a 96-well multidispenser that has been autoclaved and is kept in the hood. To avoid supernatant carry over between plates, a different 96-magazine can be used for each plate.

6. ELISA plates are incubated for 2 hr at room temperature or 37° and then washed.

7. Alkaline phosphatase-conjugated antibody (50 μl/well) is added and the plates are incubated for 2 hr at room temperature or 37° and then washed.

8. Substrate (50 μl/well) is added and the color allowed to develop until the positive controls reach a maximum OD of 1.0. Plates are read at an absorbance of 405 nm with an ELISA reader.

9. Positive wells greater than 4× the background are scored and all plates are fed with cloning medium (50 μl/well).

10. After 24 hr, supernatants from all positive wells are retested in the ELISA. If an alkaline phosphatase anti-whole IgG is used in the previous selection, the individual subclasses can be determined at this stage of the assay by using alkaline phosphatase-labeled antibodies to each subclass of Ig.

11. Two wells are chosen from each group of class and/or subclass switch variants for enrichment in a second sib selection.

Second Sib Selection

1. Cells are harvested from the 96-well plates, counted, and diluted to respread at 100 cells/well in a total of five plates/group.

2. The sib selection assay is repeated as above and the ELISA is developed with the anticlass or subclass antibody as determined in the first sib selection.

3. An enrichment of positive wells should occur at this stage which is a positive indication that the switches detected are real.

4. Again, two positive wells from each group are chosen for further enrichment.

Third Sib Selection

1. Cells are respread at 10 cells/well in a total of two plates/group.
2. The sib selection and ELISA assays are repeated and positive wells are chosen for a limiting dilution assay or cloning in soft agar.

Limiting Dilution Assay and Cloning in Soft Agar

1. In the third sib selection, almost every well should be positive for the desired switch. If the frequency of positive wells is less than 50–75%, cells from selected positive wells should be further enriched in a limiting dilution assay before cloning in soft agar. Otherwise, cells can be directly cloned from the third sib selection.

2. Cells are harvested, counted and diluted to approximately 1.0 cell/well in two 96-well plates/group.

3. Only one plate is assayed in an ELISA. Two plates are set up in case one plate becomes contaminated.

4. Two positive wells are chosen and the cells from those wells are cloned in soft agar (see above).

5. Instead of counting all of the cells for cloning, an aliquot (5 μl/cloning plate) of cells is directly taken from the well in the 96-well plate and added to 200 μl of cloning medium. Agarose-cloning medium (800 μl) is added and the cells are distributed over the agarose underlayer as previously described.

6. Cloning plates are usually overlayed with rabbit anti-mouse class or subclass specific antisera for the detection of only those subclones producing the immunoglobulin class or subclass of interest.

7. Clones surrounded by a visible precipitate are picked, expanded, and assayed. All subclones selected are cloned a second time before expansion and further characterization.

Other Techniques

A few other techniques[24] have been used to detect somatic cell variants of antibody-forming cells. The most successful has been cell sorting[22,23] which will be described elsewhere in this volume. Another technique which has been very effective was developed by Kohler and Shulman and their colleagues.[25] Using IgM producing hybridomas, antigen was attached to the surface of the parental cells. If complement was added, the cells secreting antigen-binding complement fixing wild-type antibody lysed themselves leaving only mutant cells to form viable clones.[25]

[24] D. J. Zack and M. D. Scharff, *in* "Single-Cell Mutation Monitoring Systems" (A. A. Ansari and F. J. DeSerres, eds.), p. 233. Plenum Press, New York, 1984.
[25] G. Kohler and M. J. Shulman, *Eur. J. Immunol.* **10**, 467 (1980).

Acknowledgments

The work described here was supported by grants from the NIH (AI-05231 and OIG-CA39838) and the NSF (PCM 8316150). D. F. is a fellow of the New York Heart Association and is currently supported by a grant from the NIAID (5T32-AI-07183).

[6] Mouse Teratocarcinoma Cells

By Hedwig Jakob and Jean-François Nicolas

Mouse teratocarcinomas are tumors which originate from the proliferation of a pluripotent stem cell, the embryonal carcinoma (EC) cell. In addition to these stem cells, the tumors contain a variety of differentiated tissues which resemble those of the early embryo.[1] Pluripotent stem cell lines have been established both from spontaneous and induced teratocarcinomas (EC cells[2-4]) and directly from preimplantation embryos (EK cells[5]). The karyotype of most EC and EK cells is diploid or pseudodiploid. EC and EK cells share the abilities to differentiate *in vitro,* to form teratocarcinomas *in vivo,* and to contribute to the formation of adult mice after blastocyst injection.[6,7] This last property most clearly indicates the similarity of EC and EK cells to early embryonic cells and justifies their use in studies of the molecular basis of mammalian embryogenesis.

EC and EK lines can be maintained as stem cells in culture for years but can be induced to differentiate under controlled conditions. Interestingly, the differentiated phenotypes that the cells acquire vary qualitatively from one EC cell line to another (Table I).[8-29] For example, some cell lines

[1] L. C. Stevens, *Dev. Biol.* **21,** 364 (1970).
[2] C. F. Graham, *in* "Concepts in Mammalian Embryogenesis" (M. I. Sherman and C. F. Graham, eds.), p. 315. MIT Press, Cambridge, Massachusetts, 1977.
[3] G. Martin, *in* "Development in Mammals" (M. J. Johnson, ed.), p. 225. North Holland Publ., Amsterdam, 1978.
[4] J. F. Nicolas, H. Jakob, and F. Jacob, *in* "Functionally Differentiated Cell Lines" (G. Sato, ed.), p. 185. Liss, New York, 1981.
[5] M. J. Evans and M. H. Kaufman, *Nature (London)* **292,** 154 (1981).
[6] V. E. Papaioannou and J. Rossant, *Cancer Surv.* **2,** 165 (1983).
[7] A. Bradley, M. Evans, M. H. Kaufman, and E. Robertson, *Nature (London)* **309,** 255 (1984).
[8] E. G. Bernstine, M. L. Hooper, S. Grandchamp, and B. Ephrussi, *Proc. Natl. Acad. Sci. U.S.A.* **70,** 3899 (1973).
[9] R. Gmur, D. Solter, and R. B. Knowles, *J. Exp. Med.* **151,** 1349 (1980).
[10] R. G. Oshima, J. McKerrow, and D. Cox, *J. Cell. Physiol.* **109,** 195 (1981).
[11] F. Kelly, O. Kellerman, F. Mechali, J. Gaillard, and C. Babinet, *Cancer Cells* **4,** 363 (1986).

give rise to endodermal differentiated progeny, whereas others differentiate to form cells that resemble neurons. EC cells also differ in their surface membrane properties,[30] their sensitivity to various treatments, and their adhesiveness to inner cell masses.[6] While the basis of this diversity is unknown, one attractive possibility is that it represents the stabilization of distinct states, that correspond to normal steps in the differentiation of multipotential embryonic cells. It is important to note, however, that the diversity of phenotypes emphasizes the need for caution in using EC cells or in interpreting results obtained with them: there is no EC cell that possesses all the properties one might hope for in a "universal" EC line, or that completely resembles an early embryonic cell.

EC cells have provided a rich source for the isolation of molecular probes for the study of early embryogenesis. They have been used extensively to study *in vitro* differentiation and to assess the effect of specific molecules on differentiation.[31,32] EC cells have also been used to analyze

[12] A. Linnenbach, K. Huebner, and C. M. Croce, *Proc. Natl. Acad. Sci. U.S.A* **77**, 4875, (1980).

[13] S. Y. Wang and L. J. Gudas, *J. Biol. Chem.* **259**, 5899 (1984).

[14] M. J. Rosenstraus, *Dev. Biol.* **99**, 318 (1983).

[15] J. F. Nicolas, P. Avner, J. Gaillard, J. L. Guénet, H. Jakob, and F. Jacob, *Cancer Res.* **36**, 4224 (1976).

[16] A. J. J. Reuser and B. Mintz, *Somatic Cell Genet.* **5**, 781 (1979).

[17] J. F. Nicolas and P. Berg, *Cold Spring Harbor Conf. Cell Prolif.* **10**, 469 (1983).

[18] H. Jakob, T. Boon, J. Gaillard, J. F. Nicolas, and F. Jacob, *Ann. Microbiol. (Inst. Pasteur)* **124B**, 269 (1973).

[19] J. L. R. Rubenstein, J. F. Nicolas, and F. Jacob, *Proc. Natl. Acad. Sci. U.S.A.,* **81**, 7137 (1984).

[20] P. A. McCue, K. I. Matthaei, M. Taketo, and M. I. Sherman, *Dev. Biol.* **96**, 416 (1983).

[21] S. E. Pfeiffer, H. Jakob, K. Mikoshiba, P. Dubois, J. L. Guénet, J. F. Nicolas, J. Gaillard, G. Chevance, and F. Jacob, *J. Cell Biol.* **88**, 57 (1981).

[22] G. R. Martin and M. J. Evans, *Proc. Natl. Acad. Sci. U.S.A.,* **72**, 1441 (1975).

[23] M. L. Hooper and C. Slack, *Dev. Biol.* **55**, 271 (1977).

[24] J. F. Nicolas, H. Jakob, and F. Jacob, *Proc. Natl. Acad. Sci. U.S.A.,* **75**, 3292 (1978).

[25] G. R. Martin, C. J. Epstein, B. Travis, G. Tucker, S. Yatziv, D. W. Martin, Jr., S. Clift, and S. Cohen, *Nature (London)* **271**, 329 (1978).

[26] M. W. McBurney and B. J. Strutt, *Cell* **21**, 357 (1980).

[27] M. W. McBurney and B. J. Rodgers, *Dev. Biol.* **89**, 503 (1982).

[28] E. M. V. Jones-Villeneuve, M. A. Rudnicki, J. F. Harris, and M. W. McBurney, *Mol. Cell. Biol.* **3**, 271 (1983).

[29] M. Darmon, M. H. Buc-Caron, D. Paulin, and F. Jacob, *EMBO J.* **1**, 901 (1982).

[30] G. Gachelin, *Biochim. Biophys. Acta* **516**, 27 (1978).

[31] F. Hyafil, C. Babinet, C. Huet, and F. Jacob, *Cold Spring Harbor Conf. Cell Prolif.* **10**, 197 (1983).

[32] R. Müller and E. F. Wagner, *Nature (London)* **311**, 438 (1984).

TABLE I
CHARACTERISTICS OF THE MOST USED EMBRYONAL CARCINOMA CELL LINES

Name	Origin mouse	Culture conditions[a]	Phenotype after retinoic acid treatment	Ability differentiate spontaneously	Contribution to chimaeras	Tumors[b]	Mutants[c]
F9	129/Sv	G	Endoderm; parietal and visceral	–	Unknown	N	F9[8], HPRT[-4]; TK[-9]; Oua[R 10]; Neo[R 11]; HSV1k[+d 12]; RA[-/13]; SSEA 1[-14]
PCC3	129/Sv	P	Multiple types	+	4/71	M	PCC3[15], HPRT[-15]; Oua[R 4]; Cap[R 4]; APRT[-16], Neo[R d17]; Xgpt[+d17]
PCC4	129/Sv	P	Multiple types	–	0/15	M	PCC4[18], HPRT[-4], Oua[R 4]; Cap[R 4]; Neo[R d17,19], Xgpt[+d17]; MLV[19], RA[-e20], HMBA[-20]
PCC7-S	RI[(C57 × 129)	P	Neuronal	Only neuronal	0/152	M	PCC7-S[21]; HPRT[-4]
Nulli-SSC1	129/Sv	G	Endoderm: parietal	–	Unknown	N	Nulli-SCC1[22]
PSA-1	129/Sv	F	Unknown	+	19/33	M	PSA-1[22]
PC-13	129Sv	G	Endoderm: parietal and visceral	–	1/66	M	PC13[8]; HPRT[-23]; Oua[R 23]
LT-1	LT	P	—	–	Unknown	N	LT-1[24]; HPRT[-24]; Neo[R d19]
LT 1.2A and 2D	LT	F	Unknown	–/	12/30	M	LT1-2A[25]
P10	C3H/He Ha	F	Endoderm: parietal	+	34/67	M	P10[26]
P19	C3H	P	Multiple types	–/	33/87	M	P19[27]; HPRT[-27]; Oua[R 27]; RA[-d28]; DMSO[-e28]
C17-S1-1003	C3H/He	P	Epithelioid	+	Unknown	M	C17-S1[29]

[a] Culture on P, plastic; F, feeder-layer; G, gelatin.

[b] M, Multipotential: respresentative tissues from all germ layers, N, nullipotential: only stem cells.

[c] HPRT⁻, Hypoxanthine phosphoribosyltransferase negative; TK⁻, thymide kinase negative; Oua[R], ouabain-resistant; Neo[R], neomycin-resistant; HSV1k+, Herpes simplex virus thymidine kinase positive; RA⁻, retinoic acid negative; SSEA 1⁻, stage specific embryonic antigen 1 negative; CAP[R], chloramphenicol resistant; APRT⁻, adenine phosphotransferase negative; Xgpt+, bacterial xanthine–guanine phosphotranferase positive; MLV, murine leukemia virus; HMBA⁻, hexamethylenebisacetamide negative; DMSO⁻, dimethyl sulfoxide negative. Reference numbers follow mutant names. In some cases, a subclone of the cell line was used for the injection. See specific references.

[d] Obtained by transfection or viral infection.

[e] Variants.

[f] Differentiation after aggregation.

the expression of gene regulatory elements in stem cells.[17,33-36] In addition, EC cells have been used in a number of studies that are not strictly developmental.[37-41] A general view of these subjects can be found in previously published work.[4,42-44]

In the present chapter we describe methods necessary to (1) maintain stem cells in an undifferentiated state in culture, (2) induce the cells to differentiate under controlled conditions, (3) isolate new stem cell lines directly from embryos, and (4) introduce genes by transfection or infection into EC cells.

Culture Conditions

Media

For optimal growth, EC cells are maintained in Dulbecco's modified Eagle's medium (DMEM) containing high concentrations of glucose (4.5 g/liter) and glutamine (0.584 g/liter) and serum (see below). Commercially available powdered medium is generally used to make up the medium. The powder is dissolved in the desired volume of quartz-distilled, Milli Q-filtered (Millipore), or apyrogenic quality water and supplemented with pyruvic acid (0.22 g/liter) and $NaHCO_3$ (3.7 g/liter for culture in a 10–12% CO_2 atmosphere). Penicillin (5×10^4 to 5×10^5 units/liter) and streptomycin (0.01 to 0.1 g/liter) can also be added. These antibiotics have no obvious deleterious effect on the cells. The medium is then filtered through a 0.22-μm membrane and finally serum is added just before use.

Serum. Medium is supplemented with 10–15% fetal calf serum. Newborn or calf serum is inadequate for maintaining EC cells in an undifferen-

[33] D. E. Swartzendruber and J. M. Lehman, *J. Cell. Physiol.* **65,** 179 (1975).

[34] N. M. Teich, R. A. Weiss, G. R. Martin, and D. R. Lowy, *Cell* **12,** 973, (1977).

[35] A. Rosenthal, S. Wright, H. Cedar, R. Flavell, and F. Grosveld, *Nature (London)* **310,** 415 (1984).

[36] R. W. Scott, T. F. Vogt, M. E. Croke, and S. M. Tilghman, *Nature (London)* **310,** 562 (1984).

[37] L. L. Johnson, L. J. Clipson, W. F. Dove, J. Feilbach, L. J. Maher, and A. Shedlovsky, *Immunogenetics* **18,** 137 (1983).

[38] T. Boon and A. Van Pel, *Proc. Natl. Acad. Sci. U.S.A.* **75,** 1519 (1978).

[39] D. C. Burke, C. F. Graham, and J. M. Lehman, *Cell* **13,** 243 (1978).

[40] J. N. Wood and A. G. Hovanessian, *Nature (London)* **282,** 74 (1979).

[41] D. Morello, F. Daniel, P. Baldacci, Y. Cayre, G. Gachelin, and P. Kourilsky, *Nature (London)* **296,** 260 (1982).

[42] R. L. Gardner (ed.), *Cancer Surv.* **2,** (1983).

[43] L. M. Silver, G. R. Martin, and S. Strickland (eds.), *Cold Spring Harbor Conf. Cell Prolif.* **10,** (1983).

[44] J. Forejt and P. Dráber (eds.), *Cell Differ.* **15,** (1984).

tiated state in culture: their use results in cell death and/or differentiation. Batches of fetal calf serum should be pretested on EC lines for five passages to monitor generation time (12 hr for F9 and PCC3 lines to 16 hr for PCC7-S 1009) and to be sure that no morphologic signs of differentiation appear. In addition, plating efficiency should be tested with the preselected serum (percentage of colonies formed after plating of 10^3 or 10^2 cells on a 6-cm-diameter plastic tissue culture dish). An efficiency of less than 10% is inadequate. PCC3 cells (Table I) are a good standard for testing the quality of the serum. Serum can be incubated at 56° for 30 min, to inactivate complement, before being added to the medium; this treatment does not change its qualities for EC cells. Filtered medium without serum can be stored at 4° for a few weeks. Fetal calf serum can be stored at $-20°$ for years.

Serum-Free Media. F9, C17-S-1003, and other EC cell lines have been grown for several passages in serum-free medium.[29,45,46] Generally DMEM and Ham's F12 or F10[47] are mixed in proportions varying from 1 : 1 to 3 : 1 and the mixture is supplemented with insulin (1 to 10 μg/ml), transferrin (5 to 50 μg/ml), and selenous acid (2.5×10^{-8} M). Sometimes human fibronectin is added at 5 μg/cm^2 of dish surface, to improve attachment of the cells to the plastic.[48] Some EC cells (PC 13) require exogenous lipid for multiplication; defined media can therefore be supplemented with low-density lipoprotein (LDL at 50 μg/ml) and high-density lipoprotein (HDL 50 μg/ml).[46]

Growth rates in defined medium are slower than in serum-containing medium, e.g., 28 hr instead of 12 hr for F9 cells. Some cells stop dividing and are induced to differentiate in defined media.[29]

Substrata

A number of EC cell lines (see Table I) grow directly on the plastic of tissue culture dishes or bottles. Others require a gelatin-coated surface for long-term survival although they can attach and grow directly on plastic for one or two passages. Finally, cell lines like PSA-1 or EK cells need to be grown on a layer of "feeder" cells.

Gelatin Coating of Petri Dishes. A 1% gelatin (gelatin for microbiology from Merck or swine skin, type II, gelatin from Sigma) stock solution is prepared in water, aliquoted and autoclaved at 110° for 30 min. To coat the plates, a 0.1% solution of gelatin is poured in the plates (4 ml/90 mm

[45] A. Rizzino and G. Sato,*Proc. Natl. Acad. Sci. U.S.A.* **75**, 1844 (1978).
[46] J. K. Heath, *Cancer. Surv.* **2**, 141 (1983).
[47] J. Paul (ed.), "Cell and Tissue Culture." Churchill-Livingstone, Edinburgh, Scotland, 1975.
[48] A. Rizzino and C. Crowley, *Proc. Natl. Acad. Sci. U.S.A.* **77**, 457, (1980).

plate). The plates are stored at 4° for at least 1 hr. The liquid is drained off and the dishes filled with medium.

Preparation of Feeder Layers. The mouse fibroblast line STO[49] is generally used to prepare feeder layers. The cells are grown to near confluency (5×10^6 cells/90-mm-diameter plate) and treated with a freshly prepared solution of mitomycin C (10 μg/ml) for at least 2 hr and no more than 4 hr. Alternatively, the cells can be irradiated directly in their medium with 6000 rads of γ-irradiation from a cobalt source. In either case, the cells are then washed with 2 rinses of PBS, detached with ATV solution (see below), and plated at 5×10^4/cm^2 in new dishes. The dishes can be used as soon as the feeder cells are attached (2 hr at 37°) or can be stored in medium at 37° for a maximum of 1 week.

Incubation

All cultures are incubated at 37° in a humidified air–CO_2 atmosphere. The CO_2 content is set at 12% with a $NaHCO_3$ concentration in the medium of 3.7 g/liter.

Harvesting of Cells

Pipetting. Cells which grow directly on plastic can be detached and replated by pipetting. The medium is drained off and 5 ml of fresh medium is added per 90 mm plate. A 2-ml pipet with a fine tip is used for repeated pipetting until a single cell suspension is obtained. Cells are then counted and replated at 1.5×10^4/cm^2.

Washing cells twice in PBS without Ca^{2+} and Mg^{2+} before trituration helps in obtaining single cell suspensions. Some lines, such as LT-1 or PCC7-S 1009, gradually become more difficult to dissociate by trituration, and need to be dissociated by a brief (1 min) trypsin treatment, one passage out of every five.

Trypsin Treatment. F9 cells as well as cells requiring feeder layers are detached by a mixture of EDTA and trypsin called ATV. ATV is prepared as follows: NaCl, 0.13 M; KCl, 5 mM; glucose, 1 g/liter; $NaHCO_3$, 7 mM; EDTA 0.5 mM; trypsin, 0.5 g/liter (grade 1 : 300 from Nutritional Biochemicals Corporation, Cleveland) in water. The solution is membrane filtered and can be stored at $-20°$ for weeks. To detach cells with ATV, remove the culture medium from the plates and gently pour 4 ml of ATV into each 90-mm-diameter dish. Incubate 1 to 7 min at room temperature (2 min for PSA-1; 7 min for F9 cells). Remove all but about 0.2 ml of the

---

[49] L. M. Ware and A. A. Axelrad, *Virology* **50**, 339 (1972).

ATV leaving the surface of the cells wet and wait another 2 min. Finally, add serum-containing culture medium and dissociate the cells by pipetting.

To detach EC and EK cells from feeder layers, ATV treatment must be short (2 min) in order to avoid dissociating the feeder cells.

It is better to avoid centrifugation of EC cells as much as possible because they adhere to each other very rapidly. When centrifugation is unavoidable use conical-shaped tubes (Sterilin 128 C) and spin at 1000 rpm for 5 min.

Density of Plating

High-density plating promotes proliferation, but EC cells lyse or begin to differentiate when they reach confluence. On the other hand, plating at very low densities is also usually deleterious. Therefore, cells must be replated frequently (every 48 hr) and at a controlled density (usually $5-10 \times 10^3$ cells per/cm^2 in 15 ml medium per 90-mm dish). The efficiency of cloning ranges from 10 to 80% depending on the cell line.

Storage of EC Cells

Cells are frozen in DMEM medium to which 15% (or more) serum and 10% DMSO are added. Cells (5×10^6) are suspended in 1 ml freezing medium in a small tube, kept at room temperature for 5 min, and then transferred to dry ice. Cells are then kept in the vapor phase of a liquid nitrogen freezer. EC cells have remained viable for more than 12 years under these conditions. If cells are to be kept in liquid nitrogen itself, stepwise freezing is required. The simplest way to accomplish this is to fill an ice bucket with cotton and place it at $-80°$. Then put the tubes coming from room temperature into the cotton. Leave the tubes overnight before transferring them to liquid nitrogen. Cell survival is less when cells are stored directly in liquid nitrogen as compared to storage in vapor phase.

Chemicals Used to Select Mutants

Hypoxanthine Phosphoribosyltransferase Mutants (HPRT⁻). Hypoxanthine phosphoribosyltransferase mutants are grown in medium supplemented with 8-azaguanine or 6-thioguanine. Stock solutions of azaguanine are prepared ($\times 20$ concentration at 37° in distilled water adjusted to pH 7.0 with 1 M NaOH). After membrane filtration, solutions are kept frozen at $-20°$. Mutant cells generally grow in 15 μg/ml. Stock solutions of thioguanine are prepared in pure DMSO at 24 mg/ml. The solution is kept at 4°. Mutant cells grow in 10 μg/ml.

Ouabain-Resistant Mutants (Oua^R). Due to low solubility ouabain cannot be prepared as an aqueous stock solution. It is dissolved directly in the medium before filtration ($3 \times 10^{-3} M$).

Chloramphenicol-Resistant Mutants (CAP^R). Stock solutions of chloramphenicol ($\times 200$) are prepared fresh each week in pure ethanol. They are then diluted in medium. Mutant cells are generally resistant to 10 μg/ml.

Thymidine Kinase Mutants (TK^-). Thymidine kinase mutants are grown in bromodeoxyuridine (Brd Urd). Aqueous stock solutions ($\times 100$) are prepared and stored at $-20°$. Cells are generally grown in 30 μg/ml. Solutions and the treated plates are sensitive to light and can be protected with aluminum foil.

Adenine Phosphoribosyltransferase Mutants (APRT^-). Adenine phosphoribosyltransferase negative mutants are grown in 10 to 50 μg/ml 8-azaadenine.

Neomycin-Resistant Cells (Neo^R). Neomycin-resistant cells are grown in 250 μg/ml of the pure drug G418 (Gibco). Note that the commercially available powder is only 40 to 60% pure drug and the dosage must be corrected to account for this. A solution at 20 mg/ml of powdered G418 is prepared in serum-free DMEM medium and membrane filtered. The stock solution is kept at 4° and diluted for use.

Bacterial Xanthine–Guanine Phosphotransferase Transfected Cells (Xgpt). HPRT$^-$ cells transfected with the bacterial *Xgpt* gene in an eukaryotic expression vector are grown in medium containing guanine and mycophenolic acid (Lilly Co). A 30$\times$ stock solution is prepared as follows: 200 mg of guanine is added to 100 ml of 0.1 M NaOH and heated to 60° to dissolve. The solution is cooled to room temperature and 75 mg of mycophenolic acid added. The solution is then aliquoted and kept at $-20°$.

Establishment of Embryo-Derived Stem Cells

All of the EC cell lines were isolated from spontaneous or embryo-derived transplantable tumors.[2-4] More recently methods have been described to obtain stem cell lines directly from early mouse embryos; these are called EK cells.[5,50-52] Although very similar to EC cells, EK cells contribute much more readily to somatic tissues of chimaeric embryos following injection into blastocysts and can give rise to germ line chimeras. For example, 20% of germ line chimeras were obtained with EK strains CP1 and CC1 derived from 129/Sv mice.[7] Another advantage of EK

[50] G. R. Martin, *Proc. Natl. Acad. Sci. U.S.A.* **78,** 7634 (1981).
[51] T. Magnuson, G. R. Martin, L. M. Silver, and C. J. Epstein, *Cold Spring Harbor Conf.* **10,** 671 (1983).
[52] H. R. Axelrod and E. Ladel, *Cold Spring Harbor Conf. Cell Prolif.* **10,** 665 (1983).

cells is that they can, in principle, be isolated from any given mouse strain, e.g., for investigation of developmental mutants.[51] EK cell lines from the following genotypes have already been established: 129, F_1 (CBA × C57BL), LT, tw^5, Rm, CFLP.

The procedure described to date is as follows.[53] Blastocysts are obtained from the uterine horns at 3.5 days of pregnancy. Individual blastocysts are explanted into individual wells of 24-well dishes (Costar, Falcon, or Nunc) containing a preformed feeder layer and 1 ml of medium. The culture medium is 80% DMEM (4.5 g/liter glucose + 3.7 g/liter $NaHCO_3$), pH adjusted to 7.2 in a CO_2 atmosphere, penicillin (5×10^4 unit/liter), streptomycin (0.01 g/liter), 10% fetal calf serum, 10% newborn bovine serum, 10^{-4} M β-mercaptoethanol, and 1% nonessential amino acids (100×, Gibco). A mixture of nucleosides at a final concentration of 30 μM each of uridine, adenosine, cytosine, guanosine, and 10 μM of thymidine is added before use. The blastocysts are allowed to hatch and to attach, a process which takes place within 48 hr. The blastocysts then spread out and the inner cell clump increases in size. After this clump has enlarged for 4 days, it is removed with the sealed end of a finely drawn out Pasteur pipet and its cells dissociated into several small aggregates with 0.25% trypsin, 10^{-4} M EDTA in PBS (without Ca^{2+} and Mg^{2+}) or Tris-buffered saline, pH 7.0. The aggregates are then transferred into wells (1 cm tissue culture) prepared with a layer of mitomycin C treated STO fibroblasts (see above). The dissociation step is repeated at 4 days intervals until nests of stem cells become apparent. They are then subcultured and cloned to establish lines. All passages are on feeder layers.

Cell lines have also been established from immunosurgically isolated inner cell masses (ICM)[52] prepared according to the method of Solter and Knowles.[54] The ICMs are cultured on feeder layers and handled like cultures from entire blastocysts.

Initial reports used medium conditioned by EC cells[50] or a delay of implantation *in vivo*,[5] but it is now clear that neither of these conditions is necessary.

In Vitro Differentiation of EC Cells

Spontaneous Differentiation

EC cells proliferate as stem cells when maintained in exponential growth, but produce differentiated cell types when the culture conditions

[53] E. J. Robertson, M. H. Kaufman, A. Bradley, and M. J. Evans, *Cold Spring Harbor Conf. Cell Prolif.* **10**, 647 (1983).
[54] D. Solter and B. B. Knowles, *Proc. Natl. Acad. Sci. U.S.A.* **72**, 5099 (1975).

are changed. The nature of these cells depends on the cell line used. The differentiated cells appearing in F9 populations are mainly endodermal; in PCC3 populations they are endodermal and fibroblastlike cells[15]; in PCC7-S they are usually neuronal. When induced to differentiate the same cell types appear in the culture but in greater number (Table I).

In Monolayers. When PCC3 cells are plated at $10^4/cm^2$ but then maintained in culture without passaging, many types of differentiated cells appear. Differentiation can be assessed by morphological criteria and by use of specific markers (see below).

Three days after plating the culture reaches confluence and a variable proportion of the cells die. Endodermal cells then appear, followed, after 12 days, by areas of neuronal tissue. Neural differentiation is in turn followed by the appearance of contractile cells. Cartilage appears around day 22 to 24 and this is followed by the appearance of adipocytes, zones of keratin, areas of contractile skeletal muscle, and pigmented retinal epithelial cells. This pattern of *in vitro* differentiation is highly reproducible.[15] Use of Petriperm (source: Heraeus) dishes is recommended.

In Aggregates. For most EC lines, aggregation promotes differentiation. In aggregates of F9, some of the external cells show endodermal characteristics.[55] Aggregation of cell lines such as PSA-1 gives rise to so-called embryoid bodies with an external layer of endoderm and an internal arrangement of mesodermal and ectodermal structures reminiscent of the normal mouse embryo.[56] Other cell lines like PCC7-S do not show morphological change while aggregated but do so 24 hr after redissociation of the aggregated cells and replating as a monolayer. Typical neurons are then formed.[21]

To form aggregates, cells harvested from a 2 day culture are centrifuged at 1000 rpm for 1 to 3 min. The cells are then kept for 10 min at 37°. Aggregates of different sizes are obtained by pipetting and plating onto Petri dishes that have not been treated for tissue culture (bacteriological Petri dishes). Direct plating of cells in nontreated dishes also results in formation of aggregates, which then grow in suspension.

Induced Differentiation

One of the most commonly used inducers is retinoic acid (RA)[57] either alone or in conjunction with dibutyryl-AMP (dBcAMP).[58-60] Other in-

[55] A. Grover, R. G. Oshima, and E. D. Adamson, *J. Cell Biol.* **96**, 1690 (1983).
[56] G. R. Martin, L. M. Wiley, and I. Damjanov, *Dev. Biol.* **61**, 230 (1977).
[57] A. B. Roberts and M. B. Sporn, *Retinoids* **2**, 210 (1984).
[58] S. Strickland and V. Mahdavi, *Cell* **15**, 393 (1978).
[59] A. M. Jetten, M. E. R. Jetten, and M. I. Sherman, *Exp. Cell Res.* **124**, 381 (1979).
[60] E. L. Kuff and J. W. Fewell, *Dev. Biol.* **77**, 103 (1980).

ducers that have been used include hexamethylenebisacetamide (HMBA),[61] dimethylsulfoxide (DMSO)[62] difluoromethylornithine (DMPO),[63] or 3-aminobenzamide. The cell types that are induced depend on the cell line, the drug concentration, and the culture conditions (monolayers or aggregates).

Retinoic Acid and Dibutyryl-cAMP. Retinoic acid (all-*trans,* Sigma) solutions are prepared either in ethanol or in DMSO. Stock solutions (1 mg/ml) in pure DMSO are aliquoted and stored at −80°. They are diluted to 50 μg/ml in DMSO and then further diluted in water or medium. dBcAMP is prepared as 0.1 M stock solution in water and kept at −20°.

F9 cells are induced to differentiate with RA as follows: if F9 is grown as a monolayer on gelatin-coated plates, until the cells are nearly confluent and then treated with 10^{-7} M RA and 1 mM dBcAMP, parietal endoderm-like cells appear after 2 to 6 days of treatment.[58] Growth without gelatin at high cell density and treatment with RA alone for 48 hr followed by treatment with RA plus 1 mM dBcAMP for 4 to 10 days elicits appearance of neuronal-like cells as well as endodermal cells.[60] Small (30–50 cells) aggregates treated with 5×10^{-8} M RA for 24 hr before subculture in fresh medium give rise to an outer layer of visceral endoderm as well as to parietal endoderm.[64] For P19 cells variation of the concentration of RA influences the types of differentiated cells produced: endoderm is formed at 10^{-9} M RA whereas nerve cells form at 5×10^{-7} M.[62]

Other Inducers. P19 aggregates treated with 0.5 or 1% DMSO give rise to cardiac or skeletal muscle. In contrast, in monolayer, there is no induction of differentiation by DMSO.[62] F9 cells are insensitive to induction by DMSO. PCC3/A/1[61] or Nulli-SCC1 respond as well to HMBA (5×10^{-3} M) as to RA treatment. The differentiated cells have not been well-characterized. The Nulli-SCC1 line was shown to present inducer-dependent phenotypic divergence.[65]

Mutants Defective for Induced Differentiation

Mutants of PCC4 Aza^{R} [20] and F9[13] cells have been obtained which fail to differentiate in response to RA or HMBA. Most of the PCC4 Aza^{R} cells noninducible by RA were shown to possess reduced levels of cellular binding protein for RA (cRABP). Some HMBA nonresponders were obtained from the same line. All of the RA^{-} and one of the HMBA^{-} mutants

[61] D. Paulin, J. Perreau, H. Jakob, F. Jacob, and M. Yaniv, *Proc. Natl. Acad. Sci. U.S.A.* **76,** 1891 (1979).

[62] M. W. McBurney, E. M. V. Jones-Villeneuve, M. K. S. Edwards, and P. J. Anderson, *Nature (London)* **299,** 165 (1982).

[63] M. Kelly, P. P. McCann, and J. Schindler, *Dev. Biol.* **111,** 510 (1985).

[64] B. L. M. Hogan, A. Taylor, and E. Adamson, *Nature (London)* **291,** 235 (1981).

[65] J. Schindler, R. Hollingsworth, and P. Coughlin, *Differentiation* **27,** 236 (1984).

TABLE II
DIFFERENTIATED PERMANENT CELL LINES DERIVED FROM EC CELLS

Name of line	Cell type	Origin	Derived from	Reference
PYS-2	Parietal endoderm	129	Teratocarcinoma OTT 6050)	66
PCD-1	Myocard	129	PCC3	67
PCD-2	Skeletal muscle	129	PCC3	67
3-TDM 1	Trophoblast	129	PCC3	15
3/A/1-D3	Fibroblast	129	PCC3	4
3/A/1-D1	Bone forming	129	PCC3	68
PSA-4 EB 22/20	Fibroblastic	129	PSA-4	69
PSA-4 EB 26/1	Fibroblastic	129	PSA-4	69
PSA-4 EB 28/5-9	Epithelial	129	PSA-4	69
C17-S1 T984	Skeletal muscle	C3H	C17-S1	70
C17-S1 1246	Fibroblast	C3H	C17-S1-1003	71
P19 END2	Parietal endoderm	C3H	P19	72
P19 EP1 7	Epithelioid	C3H	P19	72

were unresponsive to both inducers whereas three HMBA$^-$ mutants differentiated in response to RA.[20] Complementation analysis by cell fusion showed that the differentiation defective phenotypes are complementary and recessive.

Most of the F9 RA noninducible mutants remain inducible by HMBA.[13] They were shown to lack cRABP except one mutant, which manifested an impairment in collagen IV synthesis and secretion.

Differentiated Cell Lines Derived from EC Cultures

In general, the differentiated cells produced by EC cells grow for some limited number of generations and then stop proliferation. However, some permanent differentiated cell lines have been established. Table II[66-72] gives some of the best-characterized lines.

[66] J. M. Lehman, W. C. Speers, D. E. Swartzendruber, and G. B. Pierce, *J. Cell. Physiol.* **84,** 13 (1974).

[67] T. Boon, M. E. Buckingham, D. L. Dexter, H. Jakob, and F. Jacob, *Ann. Microbiol. (Inst. Pasteur)* **125B,** 13 (1974).

[68] J. F. Nicolas, J. Gaillard, H. Jakob, and F. Jacob, *Nature (London)* **286,** 716 (1980).

[69] H. M. Morgan, J. A. Henry, and M. L. Hooper, *Exp. Cell Res.* **148,** 461 (1983).

[70] H. Jakob, M. E. Buckingham, A. Cohen, L. Dupont, M. Fiszman, and F. Jacob, *Exp. Cell Res.* **114,** 403 (1978).

[71] M. Darmon, G. H. Sato, W. Stallcup, and A. J. Pittman, *Cold Spring Harbor Conf. Cell Prolif.* **9,** 997 (1982).

[72] C. L. Mummery, A. Feijen, P. T. van der Saag, C. E. Van den Brink, and S. W. de Laat, *Dev. Biol.* **109,** 402 (1985).

Markers to Characterize the *in Vitro* Differentiation of EC Cells

It is often necessary to measure accurately the percentage of differentiated cells present in a culture. This is because cultures of EC cells might be heterogeneous. The heterogeneity is due to their sensitivity to chemicals or physical factors which trigger differentiation. Thus, in inappropriate conditions (cell density, serum) morphologically distinct cells appear in culture. In Table III,[73-83] we have summarized some of the markers generally used to define a cell as a stem cell or as one of the differentiated cells most often encountered in the EC cultures. Many of these markers can be followed in individual cells by immunofluorescence using monoclonal antibodies. Quantification of the cell types in the mixed population can thus be obtained. Some of these markers are now available as DNA probes[84] allowing quantitation on the whole population at the DNA level.

As noted above, some of the EC phenotypes may correspond to distinct states of differentiation. Thus, PCC3, PCC4, LT1-2A, PSA-1 (see Table I) may correspond to slightly different states of differentiation of pluripotent cells. It is conceivable that cells corresponding to these various states coexist in culture. Markers to test this idea have not yet been found.

Mutagenesis

So far one of the main goals of the mutagenesis of EC cells has been to obtain markers for cell fusion experiments. Some attempts have also been made to isolate variants affected in their ability to differentiate (see above). In the future, the availability of new selective procedures and of methods to clone the mutated genes may encourage further genetic analysis.

[73] D. Morello, H. Condamine, C. Delarbre, C. Babinet, and G. Gachelin, *J. Exp. Med.* **152,** 1497 (1980).

[74] R. Kemler, D. Morello, and F. Jacob, *INSERM Symp.* **10,** 101 (1979).

[75] R. Kemler, *Fortschr. Zool.* **26,** 175 (1980).

[76] D. Solter and B. B. Knowles, *Proc. Natl. Acad. Sci. U.S.A.* **75,** 5565 (1978).

[77] R. Kemler, P. Brûlet, M. T. Schnebelen, J. Gaillard, and F. Jacob, *J. Embryol. Exp. Morphol.* **64,** 45 (1981).

[78] W. E. Howe and R. G. Oshima, *Mol. Cell. Biol.* **2,** 331 (1982).

[79] M. W. McBurney, E. M. V. Jones-Villeneuve, M. K. S. Edwards, and M. Rudnicki, *Cold Spring Harbor Conf. Cell Prolif.* **10,** 121 (1983).

[80] J. F. Nicolas, P. Dubois, H. Jakob, J. Gaillard, and F. Jacob, *Ann. Microbiol. (Inst. Pasteur)* **126A,** 3, (1975).

[81] C. M. Croce, A. Linnenbach, K. Huebner, J. R. Parnes, D. H. Margulin, E. Appela, and J. G. Seidman, *Proc. Natl. Acad. Sci. U.S.A.* **78,** 5754 (1981).

[82] D. Paulin, J. F. Nicolas, M. Yaniv, F. Jacob, K. Weber, and M. Osborn, *Dev. Biol.* **66,** 488 (1978).

[83] M. J. Sleigh, *Trends Genet.* **1,** 17 (1985).

[84] P. N. Goodfellow, *Cell Differ.* **15,** 257 (1984).

TABLE III
PRINCIPAL MARKERS OF EC CELLS AND THEIR EARLY DIFFERENTIATED DERIVATIVES

Marker	EC	Endoderm		Neural (glia and neurons)	Fibroblasts	Other somatic tissues	References
		Parietal	Visceral				
Immunological							
F9 Antisera IgM	+	+	−	−	−	−	73
ECMA-3Ma[a]	+	−			−	Yes[c]	74
ECMA-3 Ma	+	−	−	−	−	Yes	75
Anti-SSEA-1 Mo	+	+	+		−	Yes	76
TROMA-1	−	+	+	−	−	Yes	77
ENDO-A	−	+	+		−	Yes	78
TROMA-3	−	+	−	−	−	Yes	77
GFP				+ + +	−	Yes	79
Neurofilament	−	−	−	+ + +	+	Yes	79
H2-K	−	+	+	+	+	Yes	80
β2-Microglobulin	−	+	+		+	Yes	81
Actin organization[b]	Dis	Org	Org	Org	Org	Yes	82
Other							
Plasminogen activator	−	+	+ +			Yes	58
αFetoprotein	−	−	+			Yes	64
MLV Expression	−	−			+ +	+	34
Polyoma expression	−	−			+ +	+	83
SV40 expression	−	−			+	+	83

[a] Ma, Monoclonal antibody.
[b] Dis, Disorganized; org, organized.
[c] Yes signifies that at least some other differentiated tissues are positive.

Chemical Mutagenesis

Mutagenesis of PCC4 Aza^{R}1,[38] PCC3,[16] and F9[13] has been reported. In all cases exponentially growing cells were mutagenized by exposure to 3 to 5 μg/ml of N-methyl-N'-nitrosoguanidine for 1 to 18 hr. From 60 to 99% of the cells did not survive the treatment. HPRT^{-38} and APRT^{-16} variants as well as variants altered in ability to differentiate or tumorigenic properties have ben isolated. It is worth noting that Boon and Van Pel[38] reported that 22% of mutagenized clones are no longer able to produce tumors in syngeneic mice, unless the mice are irradiated before cell injection. This is due to a stable change in the properties of the EC cell surface.

Insertion Mutagenesis by Recombinant Retroviruses

King et al.[85] have reported the mutagenesis of F9 cells by insertion of M-MuLV in 131 Su III.[86] They exposed F9 cells to the virus by cocultivating them on mitomycin-treated 3T3 producer cells. A 1 day exposure resulted in 5 to 10 integration sites per F9 cell. They selected for HPRT$^-$ cells. F9 cells contain a single X chromosome and thus presumably a single copy of the X-linked HPRT gene. They found that infection increased the frequency of HPRT mutants produced by a factor of 10. The M-MuLV virus they used contains a bacterial transfer RNA suppressor gene Su III, that they used for retrieving the virus–cell DNA junction fragment from genomic DNA librairies. Other recombinant retroviruses[87,88] can also be used. A complete discussion of this approach is presented by Goff elsewhere in this volume. This method has so far not been used to mutagenize diploid genes.

Introduction of Genes into Embryonic Stem Cells

A variety of genes have been introduced into EC and EK cells either by transfection[12,17,32,36,89] or by infection with recombinant retrovisuses.[19,90] The goals of these experiments are to test the ability of regulatory elements of various genes to function in embryonic cells and, ultimately, to be able to express any foreign gene in these cells. Transient expression as well as stably transformed cells have been obtained using eukaryotic DNA vectors and retroviral vectors.

[85] W. King, M. D. Patel, L. I. Lobel, S. P. Goff, and M. Chi Nguyen-Huu, *Science* **228**, 554 (1985).

[86] L. I. Lobel, M. Patel, W. King, M. Chi Nguyen-Huu, and S. P. Goff, *Science* **228**, 329 (1985).

[87] C. L. Cepko, B. E. Roberts, and R. C. Mulligan, *Cell* **37**, 1053 (1984).

[88] W. Reik, H. Weiher, and R. Jaenisch, *Proc. Natl. Acad. Sci. U.S.A.* **82**, 1141 (1985).

[89] A. Pellicer, E. F. Wagner, A. El Kareh, M. J. Dewey, A. J. Reuser, S. Silverstein, R. Axel, and B. Mintz, *Proc. Natl. Acad. Sci. U.S.A.* **77**, 2098 (1980).

[90] J. Sorge, A. E. Cutting, V. D. Erdman, and J. W. Gautsch, *Proc. Natl. Acad. Sci. U.S.A.* **81**, 6627 (1984).

DNA Transfer into EC Cells

Transfection of EC cells has most often used the calcium phosphate precipitation method[91] (see chapter by Howard, this volume [27]). The cells are dissociated 24 hr before transfection and plated at $10^4/cm^2$ in 6-cm plates. Single cell suspensions are necessary for good results. Calcium phosphate is made by mixing 10 μg of DNA in 2× HEPES solution (see below) and 1 volume of 250 mM CaCl$_2$ solution with constant vortexing. The precipitate is left 25 min at 25°. The cells are washed twice with 1× HEPES solution, 0.5 ml of the calcium phosphate–DNA coprecipitate is gently poured on each 6-cm plate, and the plates left for 25 min at room temperature on a leveled surface. Four milliliters of DMEM with 10% fetal calf serum is then added to each plate, and the cultures are incubated at 37° for 15 hr. At this time, the calcium phosphate precipitate is easily visible on the cells. The medium of the plates is changed twice and the cells are incubated further. For selection procedures, the cells are dissociated 8 hr later and replated at 1 : 10. The selection is applied 1 or 2 days later. For assays of transient expression, the cells are harvested 48 hr after the treatment with the precipitate.

We found that EC cells' sensitivity to calcium phosphate varies with the cell line. PCC3 cells are relatively resistant but LT-1 and PCC4 are more sensitive and an incubation of more than 15 hr results in cell lysis. Glycerol shock or chloroquine does not increase the efficiency of this protocol.

Two times HEPES-Buffered saline medium is made of HEPES, 10 g/liter; NaCl, 16 g/liter; KCl, 0.74 g/liter; Na$_2$HPO$_4$·2H$_2$O, 0.25 g/liter; Dextrose 2 g/liter; pH 7.05. It is stored at −20° in plastic tubes.

Retrovirus Infection of EC Cells

EC cells are plated 18 hr before infection at $10^4/cm^2$ in regular medium. This medium is replaced by a medium containing no more than 5 μg/ml of polybrene (Sigma; filtered stock solution at 10 mg/ml in water is freshly prepared before use). Ten minutes later, an aliquot of the virus-containing supernatant is added and left for 4 to 18 hr. The cells are washed twice and incubated another day before being replated at 1 : 10. Selection is applied 1 or 2 days later. DEAE-dextran has been used instead of polybrene.[92-95]

[91] F. L. Graham and A. J. Van Der Eb, *Virology* **52,** 456 (1973).

[92] J. A. Levy, H. Jakob, D. Paulin, F. Kelly, J. C. Chermann, and F. Jacob, *Virology* **120,** 157 (1982).

[93] J. F. Nicolas, and J. L. R. Rubenstein, *in* "Vectors: A Survey of Molecular Cloning Vectors and Their Uses" (R. L. Rodriguez, ed.). Butterworths, 1987.

[94] J. F. Nicolas, J. L. R. Rubenstein, C. Bonnerot, and F. Jacob, *Cold Spring Harbor Symp. Quant. Biol.* **50,** 713 (1985).

[95] J. R. Sanes, J. L. R. Rubenstein, and J. F. Nicolas, *EMBO J.* **5,** 3133 (1986).

Section II

Special Techniques for Mutant Selection

[7] Amplification of Genes in Somatic Mammalian Cells

By ROBERT T. SCHIMKE, DAVID S. ROOS, and PETER C. BROWN

There is an ever-increasing number of reports of gene amplification as a means for overcoming growth constraints imparted by various drugs and physical agents. Gene amplification has also become a recurring theme for a variety of cellular oncogenes both in cell lines derived from tumors and clinical material (see Refs. 1–3 for reviews). Table I provides a partial list of examples of gene amplifications generated under laboratory conditions in response to various selection protocols.[4-24]

The study of gene amplification under experimental, i.e., laboratory, conditions is important for a number of reasons: (1) it provides an opportunity to examine mechanisms of gene amplification under controlled conditions where such processes might be understood as they occur in cancer biology and during evolution, (2) it provides for the possible understanding of mechanisms of resistance to cancer chemotherapeutic agents, including the isolation of specific DNA sequences (genes) that may be useful in predicting resistance to specific drugs, prior to or during chemotherapy; and (3) gene amplification provides an opportunity to obtain readily and to study the structure and regulation of a number of genes. This approach is particularly useful in the study of "housekeeping" genes which are normally expressed at very low levels in all cells.

This Chapter describes the experience from the author's laboratory with techniques we have employed to generate resistance phenomena as a

[1] R. T. Schimke, *Cell* **37,** 705 (1984).
[2] G. R. Stark and G. Wahl, *Annu. Rev. Biochem.* **53,** 447 (1984).
[3] J. L. Hamlin, J. D. Milbrandt, N. H. Heintz, and J. C. Azizkhan, *Int. Rev. Cytol.* **90,** 31 (1984).
[4] R. T. Schimke, F. W. Alt, R. P. Kellems, R. Kaufman, and J. R. Bertino, *Cold Spring Harbor Symp. Quant. Biol.* **42,** 649 (1978).
[5] G. M. Wahl, R. A. Padgett, and G. R. Stark, *J. Biol. Chem.* **254,** 8679 (1979).
[6] D. J. Chin, K. L. Luskey, R. G. W. Anderson, J. R. Faust, J. L. Goldstein, and M. S. Brown, *Proc. Natl. Acad. Sci. U.S.A.* **79,** 1185 (1982).
[7] D. G. Skalnik, D. B. Brown, P. C. Brown, R. L. Friedman, E. C. Hardeman, R. T. Schimke and D. Simoni, *J. Biol. Chem.* **260,** 1991 (1985).
[8] C.-Y. Yeung, D. W. Ingolia, C. Bobonis, B. S. Dunbar, M. E. Rise, M. J. Siciliano, and R. E. Kellems, *J. Biol. Chem.* **258,** 8338 (1984).
[9] M. Debatisse, B. R. Vincent, and G. Buttin, *EMBO J.* **3,** 3123 (1984).
[10] A. P. Young and G. M. Ringold, *J. Biol. Chem.* **258,** 11260 (1983).
[11] P. G. Sanders and R. H. Wilson, *EMBO J.* **3,** 65 (1984).
[12] L. Akerblom, A. Ehrenberg, A. Graslund, H. Lankinen, P. Reichard, and L. Thelander *Proc. Natl. Acad. Sci. U.S.A.* **78,** 2159 (1981).

TABLE I

SOME EXAMPLES OF SELECTION FOR GENE AMPLIFICATION EVENTS

Gene	Selection agent	Reference
Dihydrofolate reductase	Methotrexate	4
CAD (pyrimidine synthesis)	PALA	5
Hydroxymethylglutaryl-CoA reductase	Compactin	6, 7
Adenosine deaminase	Alanosine, adenosine, uridine	8
	Corformycin	9
Glutamine synthetase	Methionine sulfoximine	10, 11
Ribonucleotide reductase small subunit(M2)	Hydroxyurea	12
Histidyl-tRNA synthetase	Histidinol	13
Hypoxanthine–guanine phosphoribosyltranferase	Growth of leaky auxotroph	14
Multidrug resistance	*Vinca* alkaloids	15–17
	Actinomycin D	
Thymidylate synthase	Fluorodeoxyuridine	18
Ouabain-binding subunit of Na+K+-ATPase	Ouabain	19
Ornithine decarboxylase	Fluoromethylornithine 6-azauridine	20
UMP synthase		21
Leu 2 (surface protein)	Cell sorting for overproducer	22
Transferrin receptor	Cell sorting for overproducer	23
metallothionein II	Cadmium	24

[13] F. W. Tsui, I. L. Andrulis, H. Murialdo, and L. Siminovitch *Mol. Cell. Biol.* **5**, 2381 (1985).

[14] J. Brennard, A. C. Chinault, D. S. Konecki, D. W. Melton, and T. C. Caskey, *Proc. Natl. Acad. Sci. U.S.A.* **79**, 1950 (1982).

[15] J. R. Riordan and V. Ling, *Pharmacol. Ther.* **28**, 51 (1985); J. R. Riordan, K. Deuchars, N. Kartner, N. Alon, J. Trent, and V. Ling, *Nature (London)* **316**, 817 (1985).

[16] I. B. Roninson, H. T. Abelson, D. E. Housman, N. Howell, and A. Varshavsky, *Nature (London)* **309**, 626 (1984); A. T. Fojo, J. Whang-Peng, M. M. Gottesman, and I. Pastan, *Proc. Natl. Acad. Sci. U.S.A.* **82**, 7661 (1985); D.-W. Shen, A. Fojo, J. E. Chin, I. B. Roninson, N. Richert, I. Pastan, and M. M. Gottesman, *Science* **232**, 643 (1986).

[17] A. M. Van der Bliek, T. Van der Velde-Koerts, V. Ling, and P. Borst, *Mol. Cell. Biol.* **6**, 1671 (1986); K. W. Scotto, J. L. Biedler, and P. W. Melera, *Science* **232**, 751 (1986).

[18] P. K. Geyer and L. F. Johnston, *J. Biol. Chem.* **259**, 7206 (1984).

[19] J. R. Ash, R. M. Fineman, T. Kalka, M. Morgan, and B. Wise, *J. Cell Biol.* **99**, 971 (1984).

[20] L. McConlogue, M. Gupta, L. Wu, and P. Coffino, *Proc. Natl. Acad. Sci. U.S.A.* **81**, 540 (1984).

[21] J. J. Kanalas and D. P. Suttle, *J. Biol. Chem.* **259**, 1848 (1984).

[22] P. Kavathas and L. A. Herzenberg, *Nature (London)* **384**, 385 (1983).

[23] L. C. Kuhn, A. McCelland, and F. H. Ruddle, *Cell* **37**, 95 (1984).

[24] L. R. Beach and R. D. Palmiter, *Proc. Natl. Acad. Sci. U.S.A.* **78**, 2110 (1981).

result of gene amplification. Our experience has focused primarily on resistance to methotrexate and amplification of the dihydrofolate reductase gene in Chinese hamster ovary and mouse 3T6 cells. We anticipate that our experience may prove useful to other investigators working on very different systems. However, it is likely that the mechanisms by which drug treatments alter cell metabolism, or kill cells differ with the particular treatment undertaken. As will be discussed below, the effects of various treatments on cell physiology are critical in studying amplification processes. Also, we wish to emphasize the variability of cells in culture, particularly with respect to phenomena such as gene amplification, where the process under investigation necessarily involves genome instability. Even cell lines carried for long periods in the same laboratory without apparent change must be repeatedly subcloned and characterized. Thus this chapter will not attempt to provide specific protocols; rather, it will attempt to provide general guidelines.

On the Mechanism of Gene Amplification

An important consideration in developing protocols to obtain somatic mammalian cells with specifically amplified genes is the mechanism(s) of gene amplification. A current concept on the mechanism, by no means uniformly accepted nor entirely documented, is that gene amplification results from DNA overreplication–recombination within a single cell cycle.[2,25] Schimke et al.[26] have proposed that overreplication results from the accumulation of an increased capacity for initiation of DNA synthesis in cells in which DNA synthesis is partially inhibited, but where synthesis of RNA and (especially) protein synthesis can continue. According to this proposal, amplification events will be observed most clearly and frequently under conditions of selection where DNA synthesis is severely, but not completely inhibited. Note that in most cases cell death is likely to be extensive under these conditions. It also follows that drugs or treatments that preferentially inhibit DNA synthesis relative to RNA and protein synthesis are more likely to result in amplification events leading to resistance phenomena. Conversely, agents that inhibit protein synthesis without directly blocking DNA synthesis are unlikely to generate resistances as a result of gene amplification.

[25] A. Varshavsky, *Proc. Natl. Acad. Sci. U.S.A.* **75,** 5553 (1981).
[26] R. T. Schimke, A. B. Hill, and S. W. Sherwood, *Proc. Natl. Acad. Sci. U.S.A.* **83,** 2157 (1986).

On Multiple Mechanisms of Resistance

There are often several possible physiological and molecular bases for resistance to a single selecting drug (or other selection regimen) and multiple mechanisms may occur in the same cells. Changes from one mechanism to another may occur during progressive stepwise selection for higher resistance, or even during continual propagation of cells under a constant selection regimen. Methotrexate resistance can result from mutations in the dihydrofolate reductase gene,[27] alterations in methotrexate transport,[28] as well as amplification of the DHFR gene. Combinations of these resistances can occur in the same cells.[29,30] Beverley et al.[31] have shown that resistance to methotrexate in the parasitic protozoan *Leishmania tropica* can result from amplification of two completely different DNA sequences. We raise these points early in this discussion to indicate the complexities in analyzing various of the problems to be encountered in determining the mechanism(s) of resistance.

There are two general procedures for obtaining cells which overproduce a protein as a result of gene amplification. The method employed most frequently is to plate cells into selective media and subsequently clone cells for further analysis or selection. In certain instances where a specific fluorochrome is available (e.g., fluoresceinated methotrexate or a fluoresceinated antibody to cell surface proteins) populations of cells can be sorted by flow cytometric techniques to isolate the occasional, spontaneous amplification event.

Conditions for Selection of Cells as a Result of Amplification Events

Nature of Cells Employed

In our experience virtually all continuous cell lines can be made resistant to methotrexate (MTX) as a result of dihydrofolate reductase gene amplification, including Chinese hamster ovary (CHO), mouse 3T3, 3T6, L5178Y, and various rat hepatoma cell lines. With mouse 3T3, 3T6, L5178Y, and Chinese hamster ovary cells, progressive stepwise selection for methotrexate resistance results in *DHFR* gene amplification. However,

[27] D. A. Haber, S. M. Beverley, M. Kiely, and R. T. Schimke, *J. Biol. Chem.* **256,** 9501 (1981).

[28] F. M. Sirotnak, D. M. Moccio, L. E. Kelleher, and L. J. Goutas, *Cancer Res.* **41,** 4447 (1981).

[29] W. F. Flintoff, S. V. Davidson, and L. Siminovitch *Somatic Cell Genet.* **2,** 245 (1976).

[30] D. A. Haber and R. T. Schimke, *Cell* **26,** 355 (1981).

[31] S. M. Beverley, J. A. Coderre, D. V. Santi, and R. T. Schimke, *Cell* **38,** 431 (1984).

such a finding is not universal. For instance in a series of closely related rat hepatoma cell lines, one of the parental lines became resistant to low levels of MTX by a mechanism not related to *DHFR* gene amplification, and only subsequently, upon progressive selection, was the *DHFR* gene amplified. In all other of these cell lines, *DHFR* gene amplification was the dominant mechanism.[32] We make this point to indicate the variability from cell line to cell line with respect to the propensity for gene amplification.

Figure 1 shows the inherent variability of killing of subclones of mouse 3T6 fibroblast cells by methotrexate as determined by single-step clonogenic assays.[33] Thus it is important to begin selections with a *subcloned cell line*. Further, since cell lines can become progressively heterogeneous upon prolonged growth, it is necessary to carefully maintain frozen stocks of the subclones. Additionally, and if possible, it is extremely wise to employ the *same* serum stock throughout the experimental protocols. This may well require the maintenance of multiple stocks of serum, each dedicated to a set of experiments. We, as well as other, have found that many properties of cells change significantly when different serum lots are used, and this includes properties with respect to gene amplification events. Lastly, because of the changing properties of cells, either upon prolonged selection or as it occurs during repeated step-selections, it is important to freeze cells at different stages in the generation of the resistance for subsequent attempts to reconstruct the biochemical and molecular mechanism(s) of the selected phenotype.

Growth Phase for Selection

It follows from the central thesis (gene amplification requires DNA synthesis) that cells subjected to selection when they are in a state of active growth are most likely to undergo gene amplification. Thus we have found that amplification of the *DHFR* gene has the highest frequency when cells are placed under selective conditions in *early to mid log* growth.[33,34] There is an understandable tendency to grow cells to the highest density compatible with the studies to be undertaken, often bringing cells to late growth phase or beyond. We suggest that this tendency is counterproductive. It is important to note that adherent cells require variable periods of time after replating to progress into a rapid logarithmic growth phase, a variability that may be inherent to the cell line or due to the severity of the trypsiniza-

[32] C. Fougere-Deschatrette, R. T. Schimke, D. Weil, and M. C. Weiss, *J. Cell Biol.* **99,** 497 (1984).
[33] P. E. Brown, T. D. Tlsty, and R. T. Schimke, *Mol. Cell. Biol.* **3,** 1097 (1983).
[34] T. D. Tlsty, P. E. Brown, and R. T. Schimke, *Mol. Cell. Biol.* **4,** 1050 (1984).

tion regimen. Thus cell growth parameters must be well understood prior to applying selections.

Selection Regimens

What initial selection conditions to employ is a critical decision in deriving resistance by gene amplification. Both the type of resistance phenomena generated and the frequency of resistance are strikingly determined by the selection conditions employed. At a concentration of methotrexate 20 times the LD_{50} for the cells, the frequency of methotrexate-resistant colonies is 1×10^{-6} (Fig. 1), and such colonies have no amplified *DHFR* genes.[33] In contrast, cells selected at 5–10 times the LD_{50} have a higher frequency (1×10^{-5}) and 50% of the colonies have amplified *DHFR* genes. Overall, shallow, stepwise selections are more likely to result in rapid emergence of resistant variants than are large, single step selection regimens. This is documented in Rath *et al.*[35] In these experiments 3T6 cells were subjected to incremental step selections as a total population [as opposed to use of clonogenic methods (as in Fig. 1)]. When linear incremental steps were employed (40, 80, 120, 160, 200 μM methotrexate) starting with a concentration that was 2 times the LD_{50}, a population of cells resistant to 200 nM methotrexate emerged within 6–10 days. Given the fact that the normal cell doubling time of such cells is 18 hr, it is clear that resistance (and gene amplification) was generated in a large fraction of the initial population with relatively little cell death. When cells were subjected to selection in 200 nM methotrexate as a single step, the time required to generate a similar number of resistant cells was 45 days. The importance of the stepwise regimen is further indicated in the studies of Fougere-Deschatrette *et al.*[32] with various rat hepatoma cell lines. In one protocol, which involved a progressive 2-fold stepwise selection, all variants became resistant by a gene amplification mechanism. However, when the step selections were increased to 4-fold increments, only certain of the cell lines became resistant. Such results reemphasize the need for shallow increments in selection protocols. These observations are consistent with the hypothesis that a certain amount of DNA synthesis is required for gene amplification. Typically the ranges of drug concentrations most effective in selecting drug resistance due to gene amplification are those in which cell killing is in the range of 60–90%. We point out that such drug concentrations–selection pressures are within the range of biologically relevant phenomena, including killing of cells in cancer chemo-

[35] H. Rath, T. Tlsty, and R. T. Schimke, *Cancer Res.* **44**, 3303 (1984).

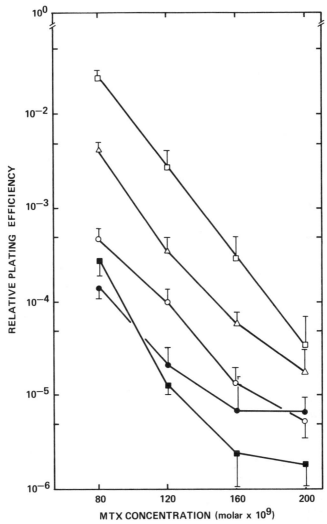

FIG. 1. Resistance to methotrexate as measured by colony formation. Clonally derived sublines of mouse 3T6 cells were trypsinized and plated into medium containing various concentrations of methotrexate. After approximately 3 weeks, colonies (50 cells) were fixed in methanol, stained with crystal violet, and counted. Symbols: parental 3T6 cells, clone 5, clone 6, clone 7, clone 10. Bars indicate standard deviations. This figure is reproduced from Ref. 33, which also includes a discussion of the plating densities ad culture conditions necessary to obtain accurate quantitative results.

therapeutic regimens, as opposed to the conditions employed in high dose, single step regimens.

An obvious question arises with any experiment in which a resistant cell line is being generated: How high a resistance level should be sought? In a number of instances, extremely high levels of gene amplification can be achieved, on the order of 600 or more copies of a gene.[7,36] Such high level amplification clearly simplifies the isolation of the relevant DNA sequences by recombinant DNA techniques (see below). However such high levels of amplification are far beyond the 5-fold amplification observed clinically.[37-39] In the case of certain drugs, e.g., methotrexate, the mechanism of action is highly specific and hence high levels of resistance and gene amplification may be obtained. However, with many selective agents the specificity of the drug may not be great, and hence attempts to generate high resistances may be thwarted by killing of cells by a variety of other mechanisms. Additionally, extensive overproduction of a protein may slow growth, or be lethal, and hence extensively amplified cells may not emerge from a cell population. It must also be kept in mind that the likelihood of generating resistance by multiple mechanisms is increased as the resistance level is increased. Thus the analysis of mechanism(s) of resistance may be far more complicated when selections are carried to high resistance levels.

Enhancement of the Frequency of Gene Amplification

The frequency of methotrexate resistance in CHO cells as estimated from single-step selections of the type shown in Fig. 1 is approximately 5×10^{-5} and resistance as a result of *DHFR* gene amplification in these cells involves approximately 5–10 copies of the gene.[33] However, employing flow cytometric techniques, Johnston *et al.*[40] estimated that the spontaneous frequency of a doubling of the gene copy number was of the order of 1×10^{-3}/cell generation. This difference between these estimates has been ascribed to the fact that many of the amplified genes are highly unstable (presumably small DNA segments) and would be lost during the growth required to generate detectable colonies.

[36] J. E. Milbrandt, N. H. Heintz, W. C. White, S. M. Rothman, and J. L. Hamlin, *Proc. Natl. Acad. Sci. U.S.A.* **78**, 6034 (1981).

[37] G. A. Curt, D. N. Carney, K. H. Cowna, J. Jolivet, B. D. Bailey, J. C. Drake, C. S. Kao-Shan, J. D. Minna, and B. A. Chabner, *N. Engl. J. Med.* **208**, 199 (1983).

[38] R. C. Horns, W. J. Dower, and R. T. Schimke, *J. Clin. Oncol.* **1**, 2 (1984).

[39] M. D. Cardman, J. H. Schornagel, R. S. Rivest, S. Srimatkandada, C. S. Portlock, T. Duffy, and J. R. Bertino, *J. Clin. Oncol.* **2**, 16 (1984).

[40] R. N. Johnston, S. M. Beverley, and R. T. Schimke, *Proc. Natl. Acad. Sci. U.S.A.* **80**, 3711 (1983).

The frequency of gene amplification as estimated from clonogenic procedures (Fig. 1) can be increased by a variety of treatments prior to placement of cells in selective conditions. Varshavsky[41] and Brown et al.[33] used hydroxyurea, an inhibitor of ribonucleotide reductase to increase the frequency approximately 10-fold. T1sty et al.[34] employed UV irradiation and treatment with N-acetoxyacetoaminofluorene to produce a similar 10-fold increase in the frequency of MTX resistance and gene amplification. In addition to these agents, we have used aphidicolin, an inhibitor of DNA polymerase α with the same result (see below). Most recently, we have found that hypoxia treatment can produce as much as a 1000-fold enhancement in the frequency of methotrexate resistance and DHFR gene amplification.[42] Barsoum and Varshavsky[43] discovered that the phorbol ester, TPA, markedly enhances the frequency of hydroxyurea-induced gene amplification when incorporated into the selection medium, and T1sty et al.[34] confirmed a similar effect of TPA following UV irradiation (a 100-fold increase in frequency over untreated cells and 10-fold over UV irradiation alone). Barsoum and Varshavsky have also reported[43] that the addition of various growth-promoting and mitogenic factors can enhance the frequency of gene amplification. These latter results may be understood as maintaining cells in a state of active growth and DNA replication, requisites for gene amplification events. In addition their studies may suggest an explanation for the perplexing variability in amplification properties of cells in different experiments, inasmuch as "growth factor" components of serum can be highly variable depending on serum lot and length of storage. Clearly this is an area deserving of further investigation.

The treatments that enhance gene amplification have several properties in common. (1) Each results in inhibition of DNA synthesis, and it is only within a short time window following the resumption of DNA synthesis that the enhanced frequency of gene amplification can be demonstrated. Thus if cells are plated into selective media immediately following treatment with carcinogens or UV light, the frequency of gene amplification is not enhanced. Similarily, if cells are plated into methotrexate at the time when hydroxyurea or aphidicolin is removed, no enhancement of amplification is observed. If cells are allowed to recover from the inhibition of DNA synthesis for long time periods (72 to 96 hr after treatment), the enhanced frequency can no longer be demonstrated. With each treatment, the optimal time for placing cells in selective conditions varies somewhat, and is dictated by the time necessary for recovery of DNA synthesis. With

[41] A. Varshavsky, Cell 25, 561 (1981).

[42] G. L. Rice, C. Hoy, and R. T. Schimke, Proc. Natl. Acad. Sci., 83, 5978 (1986).

[43] J. Barsoum and A. Varshavsky, Proc. Natl. Acad. Sci. U.S.A. 80, 5330 (1983).

hydroxyurea the optimal time for selection in 3T6 cells is 12 hr after drug removal,[33] whereas after the UV radiation the optimal time is approximately 24 hr.[34] (2) In those treatments in which the time of inhibition of DNA synthesis can be controlled, e.g., drugs and hypoxia, the duration of treatment is also a critical variable. Thus the longer the time when DNA synthesis is inhibited, the higher the frequency of drug resistance.[42] The length of time of treatment, however, must be balanced with the degree of toxicity of the treatment. In general we have found that pretreatments which result in survival of 10–30% of cells (when cloned in nonselective media) are optimal for demonstrating enhanced resistance phenomena.

Microtiter Analysis of Critical Concentrations for Enhancement of Amplification Events

In many instances the concentrations of selecting drug as well as the concentration–dose of agents employed to enhance drug resistances are critical, constituting a narrow "window" which is not inherently obvious. Therefore we have developed a simple and rapid screening method to study drug interactions as they relate to enhanced resistance phenomena.[44]

In the multiwell assay, cells are plated into microtiter plates at a sufficiently low density that the initial inoculum is virtually undetectable without cell proliferation (note that this density of cells is much higher than those used for colony assays). Drugs (or other treatments) are administered to the wells in appropriate increments. When the cells in the control well have grown to confluence, the entire plate is stained with crystal violet and quantitated either by eye or by a suitable scanning densitometer device. Typically 10^4 (24-well plate) or 10^3 (96-well plate) cells are plated per well and reach confluence at 10^5 and 10^4 cells/well, respectively. Depending on growth rates of the cells, the analysis can be made in 3–10 days. This general method allows for extremely rapid initial screening for drug concentrations and drug interactions compared with the long times required by colony formation methods. Furthermore, the study of drug effects on a population of cells rather than on individual colonies allows for a great reduction in culture area, and hence in the expense of a large study of drug interactions over a wide range of concentrations. For all drugs tested on mouse 3T6 fibroblasts, the LD_{50} obtained from multiwell assays is similar to that obtained from more laborious colony assays. 3T6 cells are routinely plated into the multiwell plates and allowed to attach for 18 hr. In the examples described below, we have studied the effects of various drugs on

---

[44] D. Roos and R. T. Schimke, *Proc. Natl. Acad. Sci. U.S.A.*, in press (1987).

resistance to methotrexate. In all cases the cells were subjected to serially increasing treatment with the drug in question in adjacent columns across the plate. After 18 hr in the test drug, methotrexate was added in increasing concentrations in the vertical dimension. The media was not changed for the remainder of the growth period.

Figure 2 shows representative effects of a variety of agents on growth of mouse 3T6 cells in increasing methotrexate concentrations. Panels A–D show the enhancement of resistance to methotrexate by hydroxyurea (A), aphidicolin (B), TPA (C), and PALA (D). Although there are certain differences in the effects of these agents, in general the most pronounced enhancement of methotrexate resistance occurs within a narrow range of concentrations where the agent itself borders on being cytotoxic. Likewise, the concentration range for the selecting drug, i.e., methotrexate, can be extremely narrow. This is particularly evident in the case of the relationship between treatment of cells with methotrexate and PALA, an inhibitor of the (fused) first three enzymes of pyrimidine synthesis, whose gene can itself be amplified in response to appropriate selection.[5] The relationship between methotrexate and PALA is complex: in addition to the ability of PALA to enhance methotrexate resistance, methotrexate can also enhance resistence to PALA (arrow in D). The concentrations of both methotrexate and PALA in which the enhancement is demonstrated are greater than would be predicted from the killing concentrations of the two drugs studied independently. Panel E shows that growing cells in increasing concentrations of BrdUrd has a striking effect on growth in methotrexate. This type is particularly interesting in view of the result of Biswas and Hanes[45] who showed that treatment of rat pituitary cells with BrdUrd generated colonies which had activated and amplified the prolactin gene. Inasmuch as amplification of the prolactin gene did not occur under any obvious selection pressure, one is tempted to speculate that the incorporation of BrdUrd into DNA perturbed DNA structure sufficiently, perhaps at sites of active gene transcription, to generate specific gene amplification (see Ref. 1 for discussion of possible relationships between transcription and replication).

We want to emphasize that it is not yet clear to what extent the enhancement of methotrexate resistance demonstrated in these multiwell assays results from gene amplification. However, the concentration windows determined from this technique are strikingly similar to those shown to enhance gene amplification in clonogenic assays of the effects of hydroxyurea, TPA, and ultraviolet irradiation (this latter data is now

[45] D. K. Biswas and J. D. Hanes, *Nucleic Acids Res.* **10,** 3995 (1982).

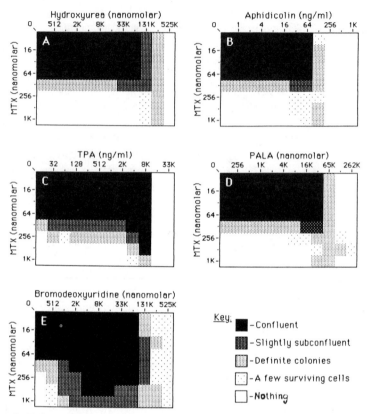

Fig. 2. Enhancement of methotrexate resistance by various drugs. Each well in five parallel 96-well microtiter plates was inoculated with 10³ mouse 3T6 cells. After 18 hr of growth in modified Eagle's medium containing 10% fetal calf serum, the wells were supplemented with various drugs at the indicated concentrations (serial 2-fold dilutions from right to left across the horizontal dimension). Eighteen hours later methotrexate was added in serial dilutions (from 1 μM) across the vertical dimension. The final result of this procedure is the formation of a matrix of drug concentrations, varying methotrexate in one dimension, and one of a variety of other cytotoxic agents in the other dimension. When the control well (upper left-hand corner of each plate) reached confluence (5 days in these experiments), all plates were rinsed in 100% methanol, stained with crystal violet, and air dried. Quantitation of cell proliferation was performed visually. Quantitation by scanning densitometry gives similar results. Details of this method can be found in Ref. 44.

shown in Fig. 2). Studies to determine the nature of resistance observed in multiwell assays are in progress. We suggest that the 24- or 96-multiwell assay is an appropriate, and sometimes essential, method for initial determinations of drug concentrations to be employed for any selection method.

Flow Cytometric Techniques for Isolating Overproducing Cell Variants

In our laboratory we have made extensive use of a fluorescein congugate of methotrexate for flow cytometric studies of cell population dynamics and population heterogeneity with respect to dihydrofolate reductase enzyme levels and gene amplification events.[46] Because there is a finite, spontaneous frequency of amplification of the dihydrofolate reductase gene[47] we were able to sort the 3–5% of most fluorescent CHO cells through 10 sequential cycles of sorting and growth, ultimately obtaining a cell population with 50 times the gene copy number of the original cells. Comparable strategies have been employed by Kavathas and Herzenberg[22] with a human Leu-2 surface antigen, and by Kuhn et al.[23] with a human transferrin receptor gene, employing specific antibodies tagged with fluoroscein to obtain cells that overproduce the protein product in a subset of cells by virtue of gene amplification (see chapter by Kamarck, this volume [14]). In both cases total human DNA was transfected into mouse L cells, and the rare cell in which the integrated DNA was expressed was initially isolated by flow cytometric techniques, following which overproducing cells were sorted for progressively increasing surface antigen content. This general strategy may be applicable in instances where a specific fluorochrome is available. It should be noted, however, that not all of the cell isolates that expressed the Leu-2 gene following transfection could be amplified. In addition attempts to transfect, express, and amplify other human surface markers in L cells have not been successful routinely.[48]

Methods for Documentation of Gene Amplification

This discussion below describes various general strategies which we and others have employed to demonstrate that a selected phenotype results from gene amplification.

Determination of Stable versus Unstable Phenotype

Gene amplification can result in either stable or unstable resistance phenotypes. So-called "stable" phenotypes have the amplified genes residing on chromosomes,[49] initially called "homogeneous staining regions" (HSRs).[47] In the unstable phenotype, the amplified genes reside on acen-

---

[46] R. J. Kaufman, J. R. Bertino, and R. T. Schimke, J. Biol. Chem. 253, 5852 (1978).

[47] J. L. Biedler and B. A. Spengler, Science 191, 185 (1986).

[48] L. Herzenberg, personal communication.

[49] J. N. Nunberg, R. J. Kaufman, R. T. Schimke, G. Urlaub, and L. A. Chasin, Proc. Natl. Acad. Sci. U.S.A. 75, 5553 (1978).

tromeric extrachromosomal "minute" chromosomes, often occurring as doublets.[50] The "stable" state of amplified, chromosomal genes is a relative term inasmuch as upon prolonged growth in nonselective conditions such genes can also be lost.[40,51] By virtue of the lack of centromeric regions, the minute chromosomes can be distributed unequally into daughter cells and undergo micronucleation, both resulting in rapid loss of the amplified genes under nonselective conditions.[52]

A "hallmark" of amplification is the unstable phenotype. In aneuploid cell lines, the amplified genes are characteristically extrachromosomal. In cell lines with a stable karyotype, the amplified genes are often present on one or more chromosomes. The reasons for these differences between cell types are not known. A direct approach to determining if amplification may be the basis of drug resistance is to ask if the resistance is stable or unstable, by growing the cells under nonselective conditions. The tendency for any investigator is to maintain the cells under selection conditions during which time the resistance phenomenon is documented, and prior to a study of the stability of the phenotype. We have found that in CHO cells (in which amplified genes reside on chromosomes in cell populations grown at a specific methotrexate concentration for 60 cell doublings) *if* the cell population is analyzed immediately upon attaining the resistance level, the vast majority of the cells have genes in a highly unstable state, i.e., extrachromosomal.[53] Thus the unstable state of a gene amplification phenotype may be missed if cells are maintained under selective conditions for long time periods prior to determination of stability–instability. For this reason we always endeavor to retain frozen stocks from several intermediate cultures.

Karyotypic Analysis

Karyotypic analysis of metaphase chromosome spreads, and the demonstration of minute chromosomes or expanded regions of chromosomes can be considered presumptive evidence for gene amplification (see chapter by Trent and Thompson [20], this volume). However, there are a number of cautionary notes. The size of minute chromosomes can be highly variable, and may (in rare instances) be so small that they may not be detected with the light microscope. Although in the original description

[50] R. J. Kaufman, P. C. Brown, and R. T. Schimke, *Proc. Natl. Acad. Sci. U.S.A.* **76**, 5669 (1979).

[51] J. L. Biedler, P. W. Melera, and B. A. Spengler, *Cancer Genet. Cytogenet.* **2**, 47 (1980).

[52] R. T. Schimke, P. C. Brown, R. J. Kaufman, M. McGrogan, an D. L. Slate, *Cold Spring Harbor Symp. Quant. Biol.* **45**, 785 (1981).

[53] R. J. Kaufman and R. T. Schimke, *Mol. Cell. Biol.* **1**, 1069 (1981).

of chromosomally amplified genes[47] the expanded region contained little or no banding properties and hence were called "homogeneously staining regions" (HSRs), in many cases of amplified genes, banding patterns are present (abnormal banding regions or ABRs), and may therefore be difficult to identify. In addition, extensive gene amplification may be present on a chromosome, but may not be detected.[32] This is due to the fact that the length of an amplified unit of DNA can vary greatly. Where studied, the linear structure of the amplified DNA is basically that of the unamplified DNA sequence but with variable sites of recombination.[31,54] When such amplified DNA sequences are extrachromosomal, i.e., minute chromosomes, we suggest that there is a lower limit on the size of the amplified unit below which these extrachromosomal sequences may be lost in the course of cell division (either during nuclear disassembly in mitosis, or possibly by leakage through nuclear pores). In contrast, there is probably no restriction on the size of chromosomally amplified genes beyond the minimum size of the gene and its associated controlling regions.

The ultimate karyotypic proof of gene amplification is the use of *in situ* hybridization techniques to localize the amplified DNA sequences (see chapter by Naylor *et al.* [21], this volume).

Demonstration of Protein Overproduction

Since functional drug resistance associated with gene amplification must result in overproduction of a normal protein, it is highly desirable to demonstrate protein overproduction directly. Proteins from (amplified) overproducing cell variants can often be purified readily (compared with the low protein levels available from sensitive cells), and be used to generate antibodies for metabolic studies and for possible use for cloning of genes in expression vector systems. Dihydrofolate reductase is a single polypeptide enzyme. In the case of hydroxyurea resistance, only the hydroxyurea-binding subunit of ribonucleotide reductase is overproduced and amplified.[12] Likewise ouabain resistance results from amplification of only the ouabain binding subunit of Na^+,K^+-ATPase.[19] Thus the assayable amount of an enzyme is not necessarily increased in all resistance–amplification phenomena.

Recombinant DNA Technology and Gene Amplification

The study of gene amplification often involves the cloning and characterization of the relevant gene imparting the selective advantage and adja-

[54] N. A. Federspiel, S. M. Beverley, J. W. Schilling, and R. T. Schimke, *J. Biol. Chem.* **259**, 9127 (1984).

cent flanking sequences. Two general and complementary approaches (cDNA cloning and genomic cloning) can be employed, and in both instances the quantitative differences in gene or mRNA content between parental and selected (amplified) cell variants can be employed in screening procedures.

cDNA Cloning

Screening for clones constituting amplified DNA sequences can be accomplished by a variety of methods. If the gene in question can complement mutational or drug-induced defects in E. coli[55] or mammalian cells,[56] cloning into suitable vector systems is the most rapid and easiest method. If a monospecific antibody for the overproduced protein is available, cloning into an E. coli expression system is a suitable approach. In most cases, however, differential screening will be necessary. The simplest approach is a differential screen of the cDNA library generated from the amplified cell line from RNA, using radiolabeled cDNA prepared from amplified and unamplified cell variants.

This differential screening technique is suitable for those mRNAs present in high abundance cells and has been used extensively for obtaining differentiation-specific cDNA sequences, where the protein in question constitues a significant amount of total protein. For many proteins, however, an initial hybridization selection of the overproduced cDNA (mRNA) is desirable. Such methods involve synthesizing cDNA from the overproducing cells, and hybridization of this cDNA with mRNA from the unamplified cells. The majority of cDNAs present in multiple copies will not form double-stranded (DNA:RNA) hybrids and can be isolated using hydroxylapatite. The desired cDNA can be further purified by hybridization with mRNA from the amplified cells (see Refs. 57 and 58 for examples).

Genomic Cloning

Differentially amplified DNA sequences can be screened in a fashion analogous to that employed for cDNA cloning. The method described by

[55] A. Y. C. Chang, J. N. Nunberg, R. J. Kaufman, H. A. Erlich, R. T. Schimke, and S. N. Cohen, Nature (London) 275, 617 (1978).
[56] D. J. Jolly, H. Okayama, P. Berg, A. C. Esly, D. Filpula, P. Bohlen, J. E. Shively, T. Hunkapillar, and T. Friedman, Proc. Natl. Acad. Sci. U.S.A. 80, 477 (1983).
[57] F. W. Alt, R. E. Kellems, and R. T. Schimke, J. Biol. Chem. 202, 1051 (1978).
[58] M. M. Davis, D. I. Cohen, E. A. Nielsen, M. Steinmetz, W. E. Paul, and L. Hood, Proc. Natl. Acad. Sci. U.S.A. 81, 2194 (1981).

Brinson et al.[59] employed differential screening of a genomic library with labeled DNA from nonamplified cell variants. The major obstacle, the presence of multiple repeated sequences in the genome, was circumvented by pretreatment of the colonies to be screened with a large excess of middle-repetitive DNA sequences isolated by C_0t analysis techniques. Roninson et al.[16] have developed a relatively rapid alternative approach for the isolation of amplified DNA sequences (see chapter by Roninson, this volume [25]). Briefly, their protocol involves radiolabeling DNA from amplified cells, hybridizing this labeled DNA to genomic DNA, restricting the DNA with a suitable restriction endonuclease, electrophoresing the DNA in agar, denaturing and partially renaturing fragments in the gel, and subsequently treating the gel with S1 nuclease. This general method depends on the accelerated rate of hybridization of amplified DNA sequences. The S1 treatment destroys those DNA fragments of different sizes whose hybridization in liquid is due to the presence of relatively small middle-repetitive elements. Hence the radiolabeled bands detected following S1 nuclease treatment constitute those DNA sequences that can hybridize faithfully across a long sequence and will include the amplified unit(s). Such bands can be eluted from the gel and cloned.

There are certain limitations to the use of these methods. Genomic cloning will generate the amplified DNA sequences, but such sequences do not necessarily constitute the entire domain of the amplified unit(s). Because cells employed for such studies have usually undergone multiple stepwise selection events, different amplification events may have occurred, involving slightly different units of amplification. In this case, the cloned subsets of amplified sequences may be present at different frequencies in the cell line studied. For further study, the DNA clones must be screened for expressed genes, using either radiolabeled cDNA obtained from mRNA, or with cDNA clones. However, the demonstration that a particular gene is amplified and expressed does not guarantee that the overexpressed gene imparts the resistance phenomenon. Van der Bliek et al.[17] have isolated 4 different overexpressed cDNA clones from CHO cells resistant to colchicine and have shown that they are linked in the amplified DNA sequence. Only one particular cDNA clone was represented in all of a panel of independently derived resistance variants, and it is most likely that this particular gene, when amplified, imparts the resistance phenotype. This is in keeping with the concept that the unit length that is amplified can vary significantly in different cell isolates, and *if* another gene is expressed within the amplified unit in a particular cell variant, it may simply be "piggy-backed" during the selection process.

[59] O. Brinson, F. Ardeshir, and G. R. Stark, *Mol. Cell. Biol.* **2**, 578 (1982).

The ultimate, and necessary, proof of an amplication event imparting a selectable phenotype can only come from the demonstration that when such a DNA sequence is introduced into a "sensitive" cell, the overproduction of its mRNA level and protein product produces the selectable phenotype.

Gene Quantitation Methods

Estimation of gene copy number is relatively easy in cell lines or tissue materials with large gene copy numbers (10-fold or greater). Estimates can be made by quantitation of restriction fragment hybridizations following gel electrophoresis directly (ref. 40, for example), or by hybridization of DNA fixed to a suitable support, such as nitrocellulose paper. For highly accurate gene quantitation where the gene copy number is 2- to 5-fold above the control DNA sample, quantitation is more difficult, although careful restriction fragment analysis can allow for quantitative data.[60] Two- to five-fold gene amplification is the range found in initial step selections (Fig. 1) and it is this range that has been reported to be clinically important in imparting methotrexate resistance to human tumors.[37-39] The problem is confounded when studying mixed cell populations derived from tumors. If there is a significant degree of heterogeneity in the cell population with respect to the degree of gene amplification, then a 5- to 10-fold amplification in a small subset of the cells may be masked. In the case of methotrexate resistance and dihydrofolate reductase gene amplification, the use of flow cytometric techniques to analyze specific protein–enzyme content in cell populations[46] may circumvent this problem. Similarily, in other instances of clinical drug resistance phenomenon where a specific fluorochrome can be developed, for instance with the multidrug resistance and overproduction of a surface, 170 kDa protein,[15] flow cytometric techniques may be useful to characterize a tumor cell population.

Perhaps the most accurate and sensitive method for gene copy estimates is quantitation by liquid hybridization techniques (C_0t analysis)[57] in which the data are plotted as the reciprocal of the percentage double-strandedness of the driver probe against the reciprocal of the C_0t.[61] Such a plot provides straight lines, the slope of which gives the difference in gene copy number. Unfortunately such methods require large amounts of DNA, are time consuming, and require a labeled single-stranded probe. Thus they are not always feasible for more routine gene copy number estimates.

[60] M. Ohlson, J. Feder, L. L. Cavalli-Sforza, and A. van Gabain, *Proc. Natl. Acad. Sci. U.S.A.* **82,** 4473 (1985).
[61] G. S. McKnight, P. Pennenquin, and R. T. Schimke, *J. Biol. Chem.* **250,** 8150 (1975).

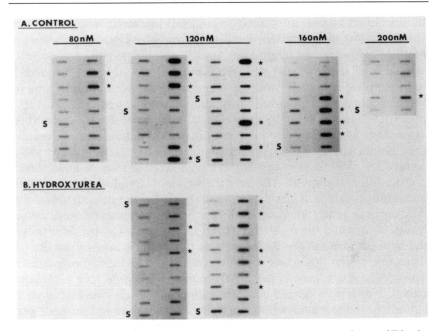

FIG. 3. Dihydrofolate reductase gene amplification in methotrexate resistance 3T6 colonies. Randomly picked colonies of 3T6 (subclone 5 from Fig. 1) resistant to 80 to 200 nM methotrexate were grown a minimum of 10–15 generations in the specified methotrexate concentration, and the extent of dihydrofolate reductase gene amplification was determined. On the left of each panel is shown hybridization to an α-fetoprotein probe, and the right side of each panel shows hybridization to a dihydrofolate reductase cDNA probe. The upper group of panels are control colonies not subjected to any pretreatment, and the lower panels are cells subjected to either 0.1 mM (left) or 1.0 mM hydroxyurea and selected at 160 nM methotrexate. S indicates control (sensitive) cells. Those colonies showing amplification of the dihydrofolate reductase gene (2–3 times sensitive ratios of dihydrofolate reductase/α-fetoprotein) are marked (*) to the right of each sample. Reproduced from Ref. 33, to which the reader is referred for details.

An alternative approach developed by Brown et al.[33] employs hybridization of radiolabeled probe to DNA fixed to nitrocellulose paper, as modified from the dot-blot method of Kafatos et al.[62] The major modification is the use of a plastic template which ensures a quantitative and uniform distribution of the DNA across the "slot," hence the name slot-blot. The uniform distribution of DNA, as well as the configuration of the slot, allows for accurate densitometric readings. A commercial modification of this apparatus is marketed by Schleicher and Schuell. The second modification of the technique is the preparation of DNA from a specific

[62] F. E. Kafatos, C. W. Jones, and A. Efstratiadis, Nucleic Acids Res. 7, 1541 (1979).

number of cells (generally 2×10^5) completely in a single tube. Cells can be stored at $-80°$ until sufficient samples are ready for DNA preparation and blotting hybridization. Duplicate "slots" are prepared from each DNA sample: one is hybridized with the labeled DNA sequence in question (dihydrofolate reductase) and the other is probed with another gene (or DNA sequence) not involved in the amplification event (such as α-feto-protein). The intensity of hybridization of each slot is determined and the ratio of the two values, relative to control cells, is a measure of differential gene amplification. The hybridization of the other sequence provides a control for differences in cell number or differential loss of DNA during preparation and transfer. This internal standard, coupled with the careful quantitative nature of the preparation allows for the detection of 2-fold differences in gene copy number. Figure 3 shows representative data from Brown et al.[33] and the reader is referred to this paper for specific details of the methods. One major difficulty with this method arises from the fact that unexpressed "pseudogenes" are often found in the same genome in addition to active gene copies. If there is a single gene (per haploid genome), but 7 pseudogenes containing sufficient nucleotide homology to hybridize on a slot analysis, then a 5-fold amplification of the "real" gene would only produce a 50% increase in hybridization intensity. Thus in cases where multiple pseudogenes are present, quantitative slot analyses must employ a sequence unique to the "real" gene which does not hybridize to the pseudogenes. In most case pseudogenes constitute mRNA-derived sequences, and hence probes derived from intervening sequences (but containing no repetitive elements) are suitable for use.

Ultimately what is needed are highly sensitive methods that allow for accurate quantitative estimates of specific nucleic acid sequences in individual cells (either RNA or DNA). Such methods are currently under development, and we anticipate major advances in this area over the next 10 years.

[8] Replica Plating and Detection of Antigen Variants

By SUSANNAH GAL

The identification and study of mutant cells defective in various aspects of cellular metabolism and growth have been invaluable contributions to our understanding of cell biology. Many cellular mutants have been isolated by mutagenizing a population of cells and then selecting those which survive a specific selection condition. However, there are a number of interesting mutants which cannot be isolated by direct selection. These include cells especially sensitive to a particular treatment (e.g., X ray or UV irradiation), cells expressing a conditional lethal phenotype such as temperature sensitivity, and cells expressing or missing specific antigens. Mammalian cell biology has in recent years borrowed the technique of replica plating from microbiology to screen large numbers of individual clones of cells for the phenotype of interest while still maintaining a master copy of the colonies. The technique involves plating 100–400 cells per dish and applying one of a variety of materials on top of the colonies so that some but not all of the cells from each of the colonies are transferred to the material from the dish. The replica can be screened for the desired phenotype while the master is maintained. When the colony of interest is identified by the particular assay, then one can return to the master dish and isolate cells from that colony. In this way, any test can be used to screen a population without requiring a selection-by-killing procedure for the desired phenotype.

Bacterial geneticists have used both felt and filter paper as materials for replica plating of colonies. Mammalian cell geneticists started replica plating with filter paper but have recently switched to nylon and polyester cloth disks for making more accurate replicas. We prefer polyester cloth disks for making replicas of living mammalian cell colonies for several reasons. Using polyester, one has the option of making multiple copies of a single master (up to 4 with CHO cells). The cloth can be purchased in several different pore sizes (1 to 17 μm), the largest of which allows cells to penetrate through several layers of the cloth and form multiple copies of the original colony. The polyester provides high fidelity copies for a variety of cell types and actually seems to work better than other materials for some cells. The polyester disks are recyclable by a simple acid-wash to remove the cells. The main disadvantages of using polyester cloth for replica plating are the sole source of material (Tetko, Inc, Elmsford, NY) and its expense. The cloth is purchased by the yard (40 in. wide) and costs about \$50/yard. One yard of material will yield about 100 circles for 100-mm dishes.

This chapter will describe how to use polyester cloth to make replicas of Chinese hamster ovary (CHO) cell colonies and how we process the replicas to screen for cells lacking a particular antigen. We have used this same technique for human HeLa S3 cells, mouse NIH 3T3, and Kirsten virus-transformed NIH 3T3 cells, and for monkey CV-1 cells. Each cell type has a different efficiency of replica plating but CHO cells give the best efficiency of replica formation. Finally, work in other laboratories using several other types of mammalian cells and a variety of methods for analyzing the replicas will be described.

Procedure for Making Replicas

Prepare a large number of glass beads (4 mm in diameter) by washing in 1 N HCl and rinsing several times in water in a 250-ml Erlenmeyer flask. Cover the flask with foil, autoclave, and dry the beads in a 200° oven. Between uses we also treat the beads with Clorox (1 : 1) overnight followed by the acid and water washes. The polyester cloth used in these studies is purchased from Tetko, Inc. (Elmsford, NY) and is designated PeCap HD 7-N, where N is the pore size in microns. The smaller pore sized material we have used (1 μm) makes excellent single replicas of CHO cells while the larger pore polyester (17 μm) is best for multiple copies from the same dish (up to 4 for CHO cells). Cut the desired number of polyester circles to the size of the dish and make a small notch on one side to aid in orienting the replicas. The polyester is rinsed in water, dried, and placed in a glass petri dish and autoclaved. The top two or three disks curl up along the edges during sterilization. To prevent this and to allow easy separation of the discs, one can interleave the polyester discs with Whatman filter paper (No. 50). To reuse the replicas, simply treat overnight with concentrated HCl followed by one rinse with concentrated HCl and extensive washing with phosphate-buffered saline (PBS) (Dulbecco), water, and ethanol. At least one worker has found some cytotoxic effect of new polyester on CHO cells and routinely treats new polyester with acid as well. The original description of this replica plating technique[1] suggested that the tissue culture dishes be coated with poly(L-lysine) (1 mg/100 ml in PBS) for 60 min prior to use to improve the adhesion of CHO cells to the master dish. In our hands this treatment was not necessary and required UV sterilization of the plates since poly(L-lysine) cannot be filtered or autoclaved. Growth of CHO cells and conditions for mutagenesis are described elsewhere in this volume (see chapter by Gottesman [1]). The procedure for making the replicas is described below and shown schematically in Fig. 1.

Day 1. Plate mutagenized cells at a density of 200–400 cells/100-mm tissue culture dish. The actual number of colonies obtained will depend on the plating efficiency of the cells. For 100-mm dishes, we use 10 ml of medium. Let the cells grow overnight to initiate colony formation.

Day 2. Have ready in the hood a pair of forceps in a small beaker of 95% ethanol, a Bunsen burner, the sterilized glass beads, and the sterile polyester discs. Aspirate the medium from the cells. Remove one polyester disk from the pile with flame-sterilized cool forceps and gently place onto the cells. If more replicas from the same dish are to be made, place another

---

[1] C. R. H. Raetz, M. M. Wermuth, T. M. McIntyre, J. D. Esko, and D. C. Wing, *Proc. Natl. Acad. Sci. U.S.A.* **79**, 3223 (1982).

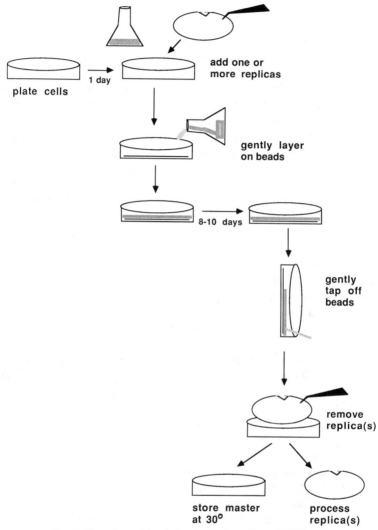

FIG. 1. Steps for making polyester cloth replicas of CHO cells.

polyester disk over the first trying to line up the notches of all the disks. The forceps should be dipped in alcohol and reflamed each time. It is best to align the notches by eye before putting the subsequent disks on the stack since it is very hard to move the disk once it is placed down on the dish. This process can be repeated for as many disks as are desired (up to a maximum of 5 which will yield 4 replicas for CHO cells). Once the last polyester disks have been placed on the cells, the glass beads are added to weight down the discs so they do not move once medium is added. Carefully uncover the Erlenmeyer flask containing the beads and flame sterilize the opening. Gently tap a small number of the glass beads onto the stack of polyester discs so that a single layer of the beads cover them. The glass beads are more easily tapped onto the dish if they have been dried prior to the procedure. Reflame the flask opening and cover with the foil top. Now add fresh medium to the cells. Mark on the bottom of the dish the position of the notch on the polyester disks so that the replica(s) and the master can be aligned. Return the plate to the incubator.

The cells are allowed to grow until large colonies (~ 1000 cells/colony) are present. One would like to have enough cells for adequate analysis of the desired phenotype but not so many that the colonies are no longer distinct. The time required for this cell growth will depend on the condition of the cells and the selection placed on them (if any) during this growing period. For CHO cells under no selective pressure, this takes 8 – 10 days. If a subsequent growth selection is desired, one may want fewer cells per colony and thus the growth period will be shorter.

Day 8 – 10. When the colonies have reached an appropriate size, the replicas are removed. Prepare dishes with an appropriate rinse solution (e.g., PBS) for the removed replicas, the forceps in ethanol, a Bunsen burner, and a large beaker for the used glass beads. First aspirate the medium and tap off the beads into the beaker. Remove the replicas one at a time with the flame sterilized cool forceps and place each in a separate dish. Label each dish as to plate designation and the replica's relative position in the stack. After all the replicas have been removed, cover the master cells with fresh medium using a very gentle flow so as not to disrupt the colonies. The master dish can be stored in an incubator at 30° to maintain colonies at a low growth rate while analysis of the replica(s) is taking place. CHO cells will remain viable and generally will not overgrow at this temperature for at least 1 week after the removal of the replicas. Some workers[2] have found longer storage of the master plates necessary and have frozen the dishes at − 70° in 7 ml medium plus 10% DMSO for at least 2 weeks. The master plate can be rapidly thawed by floating it on top of warm water.

Processing Replicas to Identify Antigen-Containing Cells

In our laboratory, we were interested in screening cells which had altered levels of a particular antigen, those with low or undetectable levels

---

[2] O. Kuge, M. Nishijima, and Y. Akamatsu, *Proc. Natl. Acad. Sci. U.S.A.* **82,** 1926 (1985).

to act as recipients for gene transfer and those with increased levels of the antigen from the transfected recipient population. The method we used to detect antigen in replicas combined a technique for fixing and permeabilizing cells as for immunocytochemistry[3] and a technique for detecting the antigen as for Western blots.[4] The exact steps are outlined below. The solutions needed for the first step are PBS, 3.7% formaldehyde in PBS, and 80% acetone in water. For the immunolocalization, several other solutions are needed including rinse solution (100 mM NaCl, 10 mM Tris, pH 7.4), soak buffer (rinse solution plus 3% BSA, 10% fetal bovine serum, and 0.01% NaN$_3$), and incubation buffer (rinse buffer plus 1% BSA and 0.01% NaN$_3$). The affinity purified rabbit antibody to the antigen is diluted 1 : 100 or 1 : 200 in incubation buffer. The first antibody is detected using either protein A or a goat anti-rabbit second antibody labeled with ^{125}I (Amersham, Corp.) both used at a concentration of 5 μCi in 30 ml incubation buffer.

The dishes with the polyester disks should be put onto a shaking platform during the processing to distribute the label more evenly and to prevent hot spots. The procedure is outlined below and illustrated in Fig. 2.

Fixing and Permeabilizing the Cells. Rinse the polyester disks several times in PBS. Then place them in 3.7% formaldehyde in PBS for 10 min at room temperature. Rinse the disks several times in PBS, then once in water. Add the 80% acetone and let the cells sit for 10 min at room temperature. Rinse the disks once with water then with PBS. The disks at this stage can sit for several hours with little loss of the antigens or the cells from the polyester. Some workers[2,3] use several rounds of freeze-thawing to lyse cells and follow their particular assay with an incubation in TCA to fix the cells[1,2,6].

Immunolocalizing Fixed Antigen. Rinse the disks from above in the rinse solution, then add the soak buffer and let the dishes sit for 1 hr at room temperature with shaking. Rinse the disks 4 times with incubation buffer each rinse lasting about 10 min. Add the first antibody solution and incubate for 1 hr. Rinse the disks 4 times with incubation buffer, each for 10 min, and transfer the disks to new dishes. The radioactive protein A or second antibody solution is added and the dishes are allowed to incubate for 1 hr and then the disks are rinsed several times with incubation buffer and finally with water and transferred to a fresh dish.

[3] A. R. Robbins, *Proc. Natl. Acad. Sci. U.S.A.* **76,** 1911 (1979).
[4] F. Cabral, I. Abraham, and M. M. Gottesman, *Mol. Cell. Biol.* **2,** 720 (1982).
[5] S. Gal, M. C. Willingham, and M. M. Gottesman, *J. Cell Biol.* **100,** 535 (1985).
[6] A. R. Robbins, R. Myerowitz, R. J. Youle, G. J. Murray, and D. M. Neville, Jr., *J. Biol. Chem.* **256,** 10618 (1981).

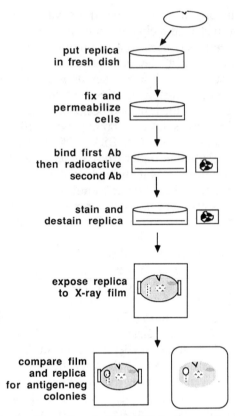

put replica
in fresh dish

fix and
permeabilize
cells

bind first Ab
then radioactive
second Ab

stain and
destain replica

expose replica
to X-ray film

compare film
and replica
for antigen-neg
colonies

FIG. 2. Steps for detecting antigen in CHO cell colonies on polyester cloth replicas.

Staining the Cells. The cells on the polyester disks are next stained for cellular protein with Coomassie blue G (Bio-Rad Laboratories) according to the technique of Raetz *et al.*[1] The disks are stained with the dye (0.05 g/100 ml) dissolved in methanol/water/acetic acid, 45:45:10 (v/v) for 1 hr. The disks are then destained with several changes of the same solvent.

At this point the polyester disks radiolabeled for antigen and stained for total protein are dried on filter paper and attached to a sheet of paper. The outline of the disk including the notch and an identification number can be indicated with radioactive ink. The sheet with the polyester disks should be covered with plastic wrap before putting it against X-ray film. We use preflashed X-Omat AR film (Kodak) and expose at $-70°$ with a Dupont Cronex screen. The length of time of exposure will depend on the amount of antigen present, the specific activity of the second antibody, as well as many other factors. The autoradiogram and the original polyester disks are then compared for colonies with altered antigen levels compared to Coomassie blue staining. The colonies of interest can then be picked from the master dish using any one of a number of isolation procedures. If necessary the colony can be rescreened using this technique or by another assay until the desired colony is purified.

Work in Other Laboratories Using Replica Plating

A number of other laboratories have used the replica plating techniques described in this chapter. These groups used a variety of different cell lines and assay systems to isolate mutants. In this last section, a few of the different cell lines and assay strategies that have been used successfully will be described to give the reader a feel for the general versatility of the technique.

Chinese hamster ovary (CHO) cells appear to form the most reliable replicas and are very amenable to genetic manipulations. Another advantage of the CHO cells is the ability to make multiple high fidelity copies (up to 4) with the polyester. Other cell types that have been used in our laboratory include mouse fibroblasts (NIH 3T3 and Kirsten virus-transformed NIH 3T3), human HeLa (S3), and monkey (CV-1) cells. None of these cell lines transfers with as high efficiency as the CHO cells in this system. These cells make acceptable single copies with the 1 μm pore size material, but do not form reliable copies beyond that. Other laboratories have found similar problems with non-CHO cells. Raetz et al.[1] described replica plating of a mouse myeloma line (SP210) and several hybridomas derived from the fusion of this line and mouse spleen cells. These workers supplement the growth medium with 2 mM $CaCl_2$ and 10% BSA and make a single replica with the 1 μm polyester cloth. Dr. A. Garcia-Perez (Laboratory of Kidney and Electrolyte Metabolism, NIH) working with kidney cells (pig, LLCPK$_1$ and dog, MDCK) (personal communication) and Dr. F. Amano (National Cancer Institute, NIH) with murine macrophages (personal communication) both indicate that these cells make single replicas with the 1 μm pore size polyester. Other workers have published work using replica plating of murine fibrosarcoma cells (HSDM$_1$C$_1$).[7]

Several laboratories have used replica plating to isolate a variety of

[7] E. J. Neufeld, T. E. Bross, and P. W. Majerus, J. Biol. Chem. 259, 1986 (1984).

TABLE I
MUTANT CHO CELLS ISOLATED BY REPLICA PLATING

Mutant phenotype	Screening procedure	References
Unable to synthesize phosphatidyl-choline	[^{14}C]Choline uptake	2
Receptor-mediated endocytosis defective	1. Low uptake of ^{35}S-labeled mannose 6-phosphate proteins	
	2. Toxin resistance	8, 9
Mannose 6-phosphate receptor deficient	Resistance of mannose 6-phosphate ricin	6
Lacking α-mannosidase	4-Methylumbelliferylmannoside hydrolysis	4
Protein secretion defective	^{35}S-Labeled secreted proteins bound to nitrocellulose membrane overlay	10, 11
γ-Ray sensitive	Growth after X-ray exposure	12
Lacking glucose-6-phosphate dehydrogenase	Histochemical stain for glucose-6-phosphate dehydrogenase activity	13

interesting CHO mutants. These mutant phenotypes and the methods of screening are summarized in Table I.[2,3,6,8-13]

Replica plating techniques could be used for the detection of cells receiving genes in DNA-mediated gene transfer experiments in a manner analogous to the use of the fluorescence-activated cell sorter (see chapter by Kamarck [14], this volume). Finally, there are a few examples in the literature of techniques for screening eukaryotic cells for specific DNAs.[14-16] The techniques described require the transfer of the cells to nitrocellulose or GeneScreen (New England Nuclear) membranes, treatment with alkali to degrade RNA, and then probing with the labeled DNA of interest. These techniques could presumably be applied to large colonies replicated on polyester sheets as described in this chapter. Replica plating is an extremely flexible technique whose utility in somatic cell genetics is just beginning to be realized.

[8] A. R. Robbins, S. S. Peng, and J. L. Marshall, *J. Cell Biol.* **96,** 1064 (1983).
[9] A. R. Robbins, and C. F. Roff, this series, Vol. 138, p. 458, 1987.
[10] A. Nakano, M. Nishijima, M. Maeda, and Y. Akamatsu, *Biochim. Biophys. Acta* **845,** 324 (1985).
[11] A. Nakano, and Y. Akamatsu, *Biochim Biophys. Acta* **845,** 507 (1985).
[12] T. D. Stamato, R. Weinstein, A. Giaccia, and L. Mackenzie, *Somatic Cell Genet.* **9,** 165 (1983).
[13] T. D. Stamato, and L. K. Hohmann, *Cytogenet. Cell Genet.* **15,** 372 (1975).
[14] A. Hayday, D. Gandini-Attardi, and M. Fried, *Gene* **15,** 53 (1981).
[15] B. Matz, *Anal. Biochem.* **144,** 447 (1985).
[16] J. Brandsma, and G. Miller, *Proc. Natl. Acad. Sci. U.S.A.* **77,** 6851 (1981).

[9] Drug-Resistant Mutants: Selection and Dominance Analysis

By MICHAEL M. GOTTESMAN

The isolation of somatic cell mutants resistant to specific drugs with known specificity has proved to be the simplest way to direct mutations to genes encoding proteins of special interest to the investigator. The great majority of somatic cell mutants isolated to date have been selected on the basis of drug resistance. For example, if a mutation affecting RNA polymerase II is desired, it is only necessary to select for resistance to the RNA polymerase II inhibitor α-amanitin, and multiple classes of RNA polymerase mutants are obtained, including those whose mutations render them temperature-sensitive for growth.[1] Similarly, selection for resistance to microtubule-depolymerizing agents such as Colcemid results in isolation of mutants with altered tubulin subunits which are also temperature-sensitive for growth.[2]

There are many advantages to the isolation of drug-resistant mutants. Since the drug targets the mutation to a specific protein or metabolic process, initial biochemical characterization of the mutant is much easier. Furthermore, drug-resistant mutants frequently have other properties which allow the investigator to begin the genetic analysis of a system which otherwise would be very complicated. The drug-resistant mutants in the examples given above are sometimes temperature-sensitive for growth which allows the isolation of temperature-resistant revertants.[3] These revertants may affect the original mutant protein, thereby proving that the mutant phenotype is due to this protein, or they may affect proteins which interact with the original mutant protein. Such "second site" mutations can be used to define an interacting network of proteins making up a complex metabolic pathway or structure (i.e., the transcription apparatus or the mitotic spindle). Another kind of conditional drug-resistant mutant may fail to grow in the absence of the drug (drug dependence).[4,5] In this instance, selection of drug-independent revertants can be used in the same way as selection of temperature-resistant revertants. Drug-resistant mu-

[1] C. J. Ingles, *Proc. Natl. Acad. Sci. U.S.A.* **75,** 405 (1978).

[2] F. Cabral, M. Sobel, and M. M. Gottesman, *Cell* **20,** 29 (1980).

[3] F. Cabral, I Abraham, and M. M. Gottesman, *Mol. Cell Biol.* **2,** 720 (1982).

[4] F. Cabral, *J. Cell Biol.* **97,** 22 (1983).

[5] C. Whitfield, I. Abraham, D. Ascherman, and M. M. Gottesman, *Mol. Cell. Biol.* **6,** 1422 (1986).

tants may be recessive or dominant in somatic cell hybrids. If they are dominant, drug resistance makes an excellent selectable marker in DNA-mediated gene transfer studies.

We have used Chinese hamster ovary (CHO) cells as a model system for the isolation of drug-resistant mutants. The basic principles of mutant isolation including growth of cells, mutagenesis, and cloning of cells have been described elsewhere in this volume and this series.[6,7] This chapter will present specific protocols for determining a suitable selecting concentration of drug to use in the isolation of single-step mutants. It has also been found that when cells are gradually adapted to increasing concentrations of a drug they frequently respond by amplification of the gene encoding the target protein for that drug. The classical example of this phenomenon is the amplification of the dihydrofolate reductase gene seen in the presence of methotrexate. Principles involved in the use of drugs to isolate mutants carrying amplified genes are given elsewhere in this volume.[8]

Determination of a Suitable Drug Concentration for Mutant Isolation

The isolation of single-step mutants resistant to a specific drug simplifies the characterization of the mutant, since only one biochemical target should be involved in generation of the phenotype. In order to isolate single-step mutants, it is necessary to choose the minimum concentration of a drug which will kill the great majority of cells and leave only the genetically resistant subpopulation. The reason for choosing the minimum drug concentration which will kill cells is that single-step drug resistance frequently involves subtle alterations in essential proteins which cannot tolerate major changes (such as the complete loss of a target protein). For example, we have found this to be the case for single-step mutants resistant to microtubule-depolymerizing agents, where the target protein is tubulin and only minor perturbations in the primary sequence of the protein can be tolerated if the cells are to survive.

The following procedure will usually yield drug-resistant mutants:

1. To determine a suitable selecting concentration of drug: plate 200 CHO cells in 2 ml of complete α-MEM in each of 12 wells of a multiwell tissue culture dish (CoStar) and 2×10^4 CHO cells in 2 ml of each of the remaining 12 wells of the dish. Add the selecting drug in increasing concentrations to each of the low-density and high-density wells. Stock con-

---

[6] M. M. Gottesman, this volume [1].
[7] L. Thompson, this series, Vol. 58, p. 308.
[8] R. T. Schimke, D. S. Roos, and P. C. Brown, this volume [7].

centrations of most drugs can be made in either DMSO (dimethyl sulfoxide, Aldrich, Gold Seal) or PBS (phosphate-buffered saline without calcium or magnesium), usually at 1–10 mg/ml. If the drug is dissolved in DMSO, be certain that the final concentration of DMSO in contact with the cells is not more than 0.5%. If the toxicity of the drug is entirely unknown, we usually use the following concentrations of drug: 0, 1 ng/ml, 3 ng/ml, 10 ng/ml, 30 ng/ml, 100 ng/ml, 300 ng/ml, 1 μg/ml, 3 μg/ml, 10 μg/ml, 30 μg/ml, and 100 μg/ml. If all or none of these doses are toxic, or if survival falls abruptly between any two drug concentrations, it may be necessary to repeat the killing curves with more appropriate drug concentrations. The reason for using two cell densities to do this testing is that frequently cells will be more resistant to drugs when they are plated at higher cell density. The higher cell density (2×10^4 cells/2 cm^2 well, or 1×10^4 cells/cm^2) is close to the density at which large-scale selections will be done and should mimic cell density under real selection conditions. In addition, having the higher cell densities will allow the killing curves to be calculated to survival levels below 1%.

2. After 7–10 days of incubation, remove medium from the plates and add 1 ml 0.5% methylene blue in water/ethanol (1:1). Stain for 10 min, aspirate the stain, and wash several times with deionized water. Colonies can be counted in each of the wells and the data can be graphically presented as shown in Fig. 1. In the example given here, the open circles represent killing of wild-type CHO cells by the adenosine analog, tubercidin. Above 10^{-9} M tubercidin there is a sharp drop-off in survival, with a new plateau reached at a survival of 10^{-3}. This plateau represents a population of mutant cells which appears at very high frequency when CHO cells are selected for resistance to tubercidin.[9] For most selections, this plateau will appear at a survival of 10^{-6} and will not be seen under conditions in which only 2×10^4 cells are plated.

To choose a concentration of drug suitable for selection, we find a drug concentration which will allow survival of 10% of the original population (LD$_{10}$). If this concentration is doubled, killing of the original population should be substantial. The LD$_{10} \times 2$ is usually a good concentration at which to begin selection. If the killing curve is not too steep, it may be necessary to use a concentration of drug which is 3-fold higher than the LD$_{10}$. The choice of this concentration can be confirmed by checking the high-density wells, where no cells should be growing at this drug concentration (except in the example shown in Fig. 1, where the mutant frequency is so high that several individual mutant clones would be seen at the selecting concentration of drug).

[9] M. S. Rabin and M. M. Gottesman, *Somatic Cell Genet.* **5,** 571 (1979).

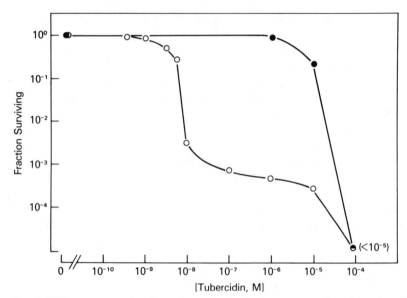

FIG. 1. Killing curves used to determine optimal selecting concentration of the adenosine analog, tubercidin. Cells were plated as described in the text and cloning efficiency in increasing concentrations of tubercidin was determined. Open circles represent wild-type (parental) CHO cells and closed circles represent a tubercidin-resistant mutant. Reproduced from Rabin and Gottesman.[9]

3. For selection of drug-resistant mutant cells, mutagenize cells as previously described in this volume.[6] It is best to mutagenize in initial selections, since this should increase the frequency of mutant cells in the population and enhance the possibility of detecting such a mutant. After a mutant has been isolated, it may be desirable to repeat the selection without mutagenesis to reduce the chance of coincident unrelated other mutations which could affect the phenotype of the cells.

4. For each independently mutagenized population of cells, plate 10 dishes each with 5×10^5 cells in 20 ml complete medium in 100 mm tissue culture petri dishes. Add medium containing drug at an appropriate concentration (selective medium) and incubate for 7–10 days at which point resistant clones should appear in some of all of the dishes. It is usually desirable to check the dishes after 2–3 days to be sure that the cells are being killed by the drug; if cells continue to grow, it is possible to increase the selecting concentration of the drug within the first few days and still obtain single-step mutants.

5. Pick resistant mutants. For CHO cells this is easily done by remov-

ing medium from the dish, circling each clone with a grease pencil or marker on the bottom of the dish, and adding a small drop of medium to the circled clone before scraping it off and aspirating it with a plastic 1 ml pipet. The aspirated cells are pipetted up and down into 2 ml of medium in a 24-well multiwell dish containing selective medium where they are allowed to grow for several days before recloning. Recloning is usually necessary and can be simply done by replating 200 cells in a 100-mm tissue culture dish with 20 ml of selective medium and repeating the picking and growing process.

Several other techniques for picking resistant clones have been devised, including the use of glass or stainless-steel cylinders placed on silicon stopcock grease and autoclaved and then used to surround the clone (for details, see chapter by Patterson and Waldren [10], this volume), or the use of a ring of silicon grease applied with the wide part of an Eppendorf pipet tip into which a trypsin solution can be added. These techniques are more time consuming, but have the advantage of allowing quantitative transfer of all the cells in the resistant clone to the multiwell dish for further growth.

6. After growth in small wells, resistant populations should be expanded to T-25 flasks for analysis and to be frozen for storage as described elsewhere in this volume.[6] If a great many clones are to be tested, it is feasible to carry these in 24-well multiwell dishes stored at 30° or grown at 37° and transferred weekly by scraping cells off the bottom of the dish and transferring a drop of medium with suspended cells to another multiwell dish.

7. Confirm the drug resistance of the doubly cloned cells by comparing killing curves of the resistant lines with the sensitive parent. Figure 1 (closed circles) shows the resistance characteristics of a clone which survived selection in tubercidin. This clone is almost four orders of magnitude more resistant to the selecting drug than the parent cell line. Resistance to tubercidin is due to high frequency loss of the enzyme adenosine kinase.[9] In the absence of this enzyme the tubercidin is not converted to tubercidin phosphate, which appears to be the toxic agent to tissue culture cells. As noted above, most selections for drug resistance do not give cells which are more than 2- to 3-fold resistant to the selecting agent in a single step. The general principle seems to be that cells are no more resistant than would be expected on the basis of the conditions of selection.

Problems Associated with the Isolation of Drug-Resistant Mutants

The most frequent problem encountered when isolating drug-resistant mutants is the high frequency of appearance of a phenotype commonly called "multidrug resistance." [10] This phenotype is characterized by single-step resistance to a great variety of hydrophobic selecting drugs such as colchicine, vinblastine, vincristine, actinomycin D, adriamycin, daunomycin, taxol, etc. Expression of resistance is associated with appearance of a 170,000 molecular weight glycoprotein (p-170) on the cell surface of all rodent and human cells so far analyzed.[10] Multidrug resistance also correlates with increased expression of the mRNA for a gene, termed *mdr1*, which is be the gene encoding p-170, which is amplified in multidrug-resistant cell lines showing high levels of resistance.[11] The genetic alterations in these cells results in decreased accumulation of the drugs to which the cells are resistant because of an active, energy-dependent efflux system. Study of these cells has indicated that the development of multidrug resistance can be phenotypically suppressed by addition to the medium of 0.01% Tween- 80[12] or the inclusion of a variety of calcium channel blocking agents such as verapamil ($5-20$ μg/ml, depending on cell sensitivity to this agent).[13] In the presence of these agents, multidrug-resistant mutants which arise are phenotypically suppressed, and only specific drug-resistance phenotypes are observed.

A second problem which occasionally arises, especially when the drug-resistant cells are only 2- to 3-fold more resistant to the selecting agent than the parent cells, is the need to be able to compare the resistant cells directly to the parent population from which it arose. For this reason, it is important not to carry cells for too long in tissue culture. Cells to be used for selection should be freshly defrosted from a cloned population of cells, and the resistant lines which derive from this selection should be compared with other freshly defrosted parent cells, not parental cells which have been in culture for several months where unknown selective pressures could change the phenotypes of these cells.

Finally, the experimenter should realize that mutagenized cells are likely to carry many mutations which could result in biochemical phenotypes unrelated to the drug resistance of the mutant cells. There are several ways around this problem including: (1) demonstration of a similar biochemical defect in several mutants derived from independent selections (different flasks of mutagenized cells); (2) demonstration of the loss of the biochemical defect in drug-sensitive revertants; (3) demonstration in so-

[10] V. Ling *in* "Molecular Cell Genetics" (M. M. Gottesman, ed.), p. 773. Wiley, New York, 1985.

[11] D.-W. Shen, J. E. Chin, N. Richert, I. Pastan, and M. M. Gottesman, *Science* **232**, 643 (1986).

[12] V. Ling and L. Thompson, *J. Cell. Physiol.* **83**, 103 (1974).

[13] A. Fojo, S.-I. Akiyama, M. M. Gottesman, and I. Pastan, *Cancer Res.* **45**, 3002 (1985).

matic cell hybrids that the drug resistance and the biochemical defect cosegregate; (4) demonstration of a conditional biochemical defect (such as temperature sensitivity of an enzyme) that correlates with a conditional drug resistance; and (5) cotransfer of the drug resistance and the biochemical defect to unmutagenized recipient cells. The ultimate proof that a specific gene imparts a drug resistance phenotype is the cloning of that gene and its transfer into a drug-sensitive cell. Strategies to achieve this goal are presented elsewhere in this volume. Taken together, these various criteria for proving that a biochemical defect is linked to a mutant gene can be considered to be the Koch's postulates of somatic cell genetics.

Genetic Analysis of Drug-Resistant Cells: Somatic Cell Hybrids

One useful way to characterize drug-resistant cells is to determine whether the mutation leading to drug-resistance is dominant or recessive. This is most easily done by forming somatic cell hybrids between parental cells and resistant mutants and testing the drug resistance of the hybrids. One difficulty with this approach is that hybrid cells tend to lose chromosomes at a rapid rate, and an individual hybrid has a substantial possibility of having lost either the wild-type allele or the mutant allele of interest. For this reason, many independent hybrid cells must be tested and the phenotype of the majority of the cells is assumed to reflect the dominance or recessiveness of the drug resistance mutation. One way to circumvent this problem is to select for hybrid cells using outside markers (selectable markers unrelated to the drug resistance of interest) in the presence or absence of varying amounts of the originally selecting drug. If the hybrids grow as well in the selecting drug as in the absence of the selecting drug, then the mutation is assumed to be dominant. The following protocol can be used to determine whether a drug-resistance mutation is dominant or recessive:

1. Two cell lines will be needed for hybridization. We use a wild-type CHO cell (drug-sensitive) which is a universal hybridizer because it is resistant to 2 mM ouabain (for selection of dominant ouabain-resistant mutants see Gottesman[6]) and sensitive to HAT medium (10^{-4} M hypoxanthine, 10^{-6} M aminopterin, and 10^{-4} M thymidine).[14] This cell line is hybridized to the drug-resistant CHO line of interest, which is ouabain-sensitive and HAT resistant (as are most cultured cells). Medium containing HAT and ouabain will select for hybrids between these two cell types.

---

[14] K. K. Jha and H. L. Ozer, *Somatic Cell Genet.* **2**, 215 (1976).

2. In separate wells of a 24-well tissue culture dish (CoStar) in 2 ml of nonselective medium plate the following: (a) 6×10^5 drug-sensitive cells (ouabain-resistant, HAT-sensitive), (b) 6×10^5 drug-resistant cells (ouabain-sensitive, HAT-resistant, (c) 3×10^5 drug-sensitive cells (ouabain-sensitive, HAT-resistant) and 3×10^5 drug-sensitive cells (ouabain-resistant, HAT-sensitive), and (d) 3×10^5 drug-sensitive (ouabain-resistant, HAT-sensitive) and 3×10^5 drug-resistant (ouabain-sensitive, HAT-resistant) cells. In (c) and (d), premix the cells before plating. These cells will form a dense monolayer when incubated for 16 hr.

3. After 16 hr, aspirate the medium, and add 1 ml of 50% (w/w) polyethylene glycol 1000 (Baker). Prepare this solution by autoclaving 50 g of polyethylene glycol 1000 and adding 50 ml of medium without serum. The polyethylene glycol 1000 should come from a freshly opened bottle, since it is highly hygroscopic and cannot be accurately weighed once exposed to atmospheric water. In practice, it is best to open a bottle and aliquot 50 g portions into several smaller bottles for future use. The solution will be fairly acid (phenol red appears yellow), but this does not interfere with its ability to promote hybridization. Some investigators recommend neutralization and filtering.

4. After 2 min, quickly aspirate the PEG and wash the cell monolayers three times with 2 ml of medium without serum.

5. Incubate for 24 hr, at which time many heterokaryons should be visible. Trypsinize cells and resuspend in nonselective medium. Plate dilutions in 100-mm tissue culture dishes with 15 ml ouabain–HAT medium, and in ouabain–HAT medium containing increasing concentrations of the original selecting drug up to the selecting concentration. In a typical experiment, hybrid cells would be suspended in 2 ml of medium and 0.5, 0.1, and 0.01 ml of the cell suspension plated under appropriate selective conditions.

6. Incubate the hybrids for 2–3 weeks at which time hybrid clones should appear in one or more of the dilutions in ouabain–HAT medium. No clones should be seen from the wells in which cells were fused to themselves [(a) and (b)]. Cells from well (c) should form ouabain–HAT-resistant hybrids, but these hybrids should not grow in ouabain–HAT medium containing the original selecting drug of interest. Cells from well (d) (the experimental fusion) should make hybrids which grow in ouabain–HAT medium and will grow in ouabain–HAT medium plus the original selecting drug only if the mutation causing drug resistance is dominant. The reason for using different drug concentrations is that few mutations in mammalian cells are completely dominant in somatic cell hybrids. Hybrids carrying incompletely dominant mutations will grow at intermediate drug concentrations.

For CHO cells, using this technique, hybrids arise at a frequency of approximately 0.01–0.1% of cells plated in selective medium. Since viability of these cells may be as low as 10% after polyethylene glycol treatment and growth under dense conditions, viable hybrids appear at an overall frequency of approximately 1% of input viable cells.

If the mutation of interest is dominant, it can be transferred by DNA-mediated gene transfer into an appropriate recipient cell (for details of this procedure, see the chapter by Fordis and Howard [27], this volume). Successful completion of such a transfer will confirm the dominant nature of the mutation and can be used as a technique for cloning the gene of interest.

[10] Suicide Selection of Mammalian Cell Mutants

By DAVID PATTERSON and CHARLES A. WALDREN

Introduction

The principle of suicide selection of mutant cells was used for over 20 years in microbial genetics prior to its application to mammalian somatic cells.[1,2] In this procedure, a population of cells, most of which display wild-type growth characteristics, and a few rare ones with additional mutant nutritional requirements, is placed in a selective medium in which the former but not the latter can grow. An agent that will kill growing cells is then added. We have used bromodeoxyuridine–visible light, first developed by Kao and Puck,[3,4] as the killing agent to isolate different mutants. This procedure depends on the differential incorporation of the thymidine analog, bromodeoxyuridine (BrdUrd), into the DNA of growing but not into nongrowing cells in selective conditions sufficient for growth of wild-type but not mutant cells. After incorporation of BrdUrd, the cell culture is exposed to "visible light," a component of which in the range of 313 nm results in photolysis of BrdUrd-containing DNA and death of wild-type (growing) but not mutant (nongrowing) cells. Growth in nonselective conditions will then allow clonal growth of mutant cells. Such selection has allowed for the isolation of a wide variety of mutants with new nutritional

[1] B. D. Davis, *J. Am. Chem. Soc.* **70**, 4267 (1948).
[2] J. Lederberg and N. Zinder, *J. Am. Chem. Soc.* **70**, 4267 (1948).
[3] T. T. Puck and F.-T. Kao, *Proc. Natl. Acad. Sci. U.S.A.* **58**, 1227 (1967).
[4] F.-T. Kao and T. T. Puck, *Proc. Natl. Acad. Sci. U.S.A.* **60**, 1275 (1968).

requirements[3-5] or of mutants with altered temperature sensitivity.[6] Other investigators have modified this procedure, for example by using a different light source which emits primarily in the region of 313 nm at which BrdUrd absorbs efficiently,[7] by adding agents to enhance the effectiveness of BrdUrd,[8,9] or by substituting other suicide agents for the BrdUrd.[10] We and others have altered the timing of the various steps in the procedure to carry out particular selections.[7-9,11] Here we describe in detail only the standard "BrdUrd–light" protocol and discuss the parameters which may require adjustment for particular cell types and applications. An earlier protocol is presented in Ref. 12.

Material and Methods

The typical mutagenesis protocol for isolation of auxotrophic mutants from CHO (see Fig. 1 for a schematic representation of this protocol) is as follows.

Day 1: Mutagenesis of the Cell Population

1. Inoculate each of two 100-mm plates with 3×10^6 cells/plate in 10 ml of medium F12 (complete medium) supplemented with 10% fetal calf serum (FCS).[13]
2. Incubate 2–6 hr to allow attachment and resumption of growth.
3. Treat with mutagen. (Dose as determined from survival curve.)
4. About 16 hr later, remove mutagen, wash 2× and readd complete medium.

Day 3: Recovery from Mutagenesis and Expression of Mutation

1. Check plates and replace the medium with complete medium. Plates should be watched carefully. When confluent, they should be subcultured.
2. When plates become confluent, trypsinize off the cells and dispense 1/4 of the cell suspension into each of two 100-mm plates containing

[5] D. Patterson, "De Novo Purine and Pyrimidine Biosynthesis." Wiley, New York, 1985.
[6] D. Patterson, C. Waldren, and C. Walker, Somatic Cell Genet. 2, 113 (1976).
[7] E. H. Y. Chu, N. C. Sun, and C. C. Chang, Proc. Natl. Acad. Sci. U.S.A. 69, 3459 (1972).
[8] G. P. Stetten, R. L. Davidson, and S. A. Latt, Exp. Cell Res. 108, 447 (1977).
[9] I. E. Scheffler and G. Buttin, Cell. Physiol. 81, 199 (1973).
[10] T. Kusano, M. Kato, and I. Yamane, Cell Struct. Funct. 1, 393 (1976).
[11] D. Patterson, D. B. Vannais, and W. Laas, J. Cell Physiol. 116, 257 (1983).
[12] F.-T. Kao and T. T. Puck, Methods Cell Biol. 8, 23 (1974).
[13] R. G. Ham, Proc. Natl. Acad. Sci. U.S.A. 53, 288 (1965).

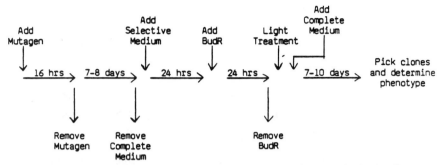

FIG. 1. Standard time course for BrdUrd–light selection of mutants in CHO cells.

10 ml of complete medium. Discard 1/4 of the cell suspension; leave 1/4 in the original plate. Add 10 ml of complete medium to the original plate. Resume incubating until they reach confluence.

Day 8: Enrichment for Mutants—BrdUrd Light Treatment

1. Harvest 3–100 mm plates, keeping the cells from each separate. Add the cells from each plate to a separate sterile tube containing an equal volume of deficient medium F12D + 10% dialyzed serum [F12D + FCM(10)].[4] (FCM designates the macromolecular fraction of calf serum. Generally we prepare this by dialysis.)
2. Count the cells.
3. Add 3.0 ml of F12D + FCM10 to each of 30–60 mm plates. Number each plate.
4. Add 3×10^4 cells to each of the 30 plates; it is best to use cells from one 100 mm plate for no more than 15–60 mm plates.
5. Add remaining cells to a 60-mm plate with F12 + FCS(10) as a reserve "farm" plate.

Day 9

1. Twenty-four hours after plating the cells in F12D + FCM(10), add bromodeoxyuridine (BrdUrd) to a final concentration of 10^{-5} M to each plate *except one:* The plate without BrdUrd is a control. *Do not put the light on in the hood while working with BrdUrd. Work in light as dim as possible.*

Day 10

1. Twenty-four hours after addition of the BrdUrd, remove the medium by aspiration. Add 3.0 ml saline G/plate. Then expose all the

plates except one to visible light for 60 min. Be sure to expose the plate which did not get BrdUrd. Monitor the temperature and use a fan if necessary to keep the plates cooled to less than 38°.

Day 15

1. Check plates for contamination; discard any contaminated.
2. Check remaining plates for pH, number of dead cells, general microscopic appearance. If necessary, change medium.

Day 17: Preliminary Nutritional Screening and Cloning of Presumptive Mutants

1. Examine the plates for colonies. Pick colonies, using the cloning cylinder trypsinization method, which are large (several hundred cells) into 35-mm plates containing 2 ml complete medium and separate 35-mm plates containing 2 ml deficient medium.

Day 22

1. Examine each plate for growth.
2. Discard 35-mm plates in which extensive and equal growth has occurred in both complete and selective medium.
3. Carefully reanalyze clones which displayed differential growth in complete and selective medium by dispensing a small number (~ 100) of cells into new sets of plates containing complete and selective medium.
4. Dispense ~ 10 cells for subcloning into 60-mm dishes containing complete medium.

Day 29

1. Examine test plates. Discard all clones which grew in F12D + FCM(10).
2. The clones which did not grow in F12D + FCM(10) but did grow in complete medium can be presumed to be mutants and must be tested further and frozen away.

Choice of Cells and Growth Conditions

It must be noted that although the BrdUrd–light method has provided a powerful means for isolating many novel mutants, it does not select absolutely against wild-type cells. Rather it allows enrichment of the proportion of mutants in the population, often by as much as 10,000-fold, so that their detection and isolation becomes feasible. The Chinese hamster ovary cell, CHO-K1, has a number of characteristics that make it particularly useful for the BrdUrd–light procedure. These properties require careful consideration when choosing other cell types to be used to isolate mutants.

First, the cells should have a high cloning efficiency (>25%) in both complete and selective medium. This is important since, at a number of stages in the procedure, a certain fraction of cells inevitably will be lost. If, for example, either heat- or cold-sensitive mutants are sought, the cells must have good cloning efficiencies at both temperatures. It is also useful to employ cells which have a rapid generation time (less than 24 hr), since this allows the experiment to be completed in a reasonable time of a month or less.

Selection of auxotrophic mutants permits the use of a wide range of complete (nonselective) and selective media combinations. In the protocol given here, the complete and selective media were chosen to optimize the number of mutants isolated without regard to nutritional requirements. F12D, which when supplemented with FCM is a minimal growth medium for CHO-K1, is medium F12[13] but without vitamin B_{12}, alanine, aspartic acid, glutamic acid, glycine, hypoxanthine, *myo*-inositol, lipoic acid, and thymidine. It is available from Irvine Labs (Irvine, CA). It is also possible to use other more selective combinations of media. For example, glycine, which is not a component of F12D selective medium, may be added so that any glycine-requiring auxotrophs will be able to incorporate BrdUrd and be eliminated, even before the final steps in nutritional screening. It may be possible to omit other agents, such as certain vitamins, from the F12 medium to broaden the spectrum of mutants isolated. It is also sometimes useful, on the other hand, to supplement the nonselective medium with additional nutrients, as, for example, in the isolation of uridine-requiring mutants where uridine was added to the complete medium.[11] In any case, it is important to employ only the macromolecular fraction of fetal calf serum in the selective medium since normal (undialyzed) serum contains a large number of undefined low-molecular-weight components as well as variable amounts of most of the nutrients present in the complete F12 medium, but which are omitted from the selective medium. In addition, thymidine must be left out of the selective medium since it competes with the BrdUrd for incorporation into DNA thereby seriously reducing the lethal effects of the exposure to light.

Treatment with Mutagenic Agents

In our hands it is necessary to treat CHO cells with a mutagen in order to obtain mutants in significant number. Virtually no spontaneous mutants have been isolated using the BrdUrd–light procedure on populations of unmutagenized cells. Our empirical observation is that the maximum number of mutants is isolated after doses of mutagen at which 10–70% of treated cells survive. In order to ascertain this dose, it is necessary to construct a single cell survival curve with the mutagen to be used.[14,15] A dose that kills approximately 30% is convenient and productive. Conditions for survival curves for the mutagen should duplicate those used in mutagenesis. Mutagenesis is carried out in complete medium including 10% fetal calf serum. Cells ($3-4 \times 10^6$ can be mutagenized in a single 100-mm petri dish, of which $\sim 10^6$ or so will survive. Other cell inocula have been tried with less success. In our standard procedure, cells are treated with chemical mutagen for a 16-hr time period; then it is removed. The cells should be plated several hours prior to the addition of the mutagenic agent (4–6 hr for CHO) to allow time for attachment and resumption of growth. This is a particularly convenient experimental protocol since the cells to be mutagenized can be plated in the morning and mutagen added in the late afternoon; mutagen can then be removed early the next morning.

The mutagenic treatment that we usually try first is ethylmethane sulfonate (EMS) at $1-3$ mM which produces mutants in CHO-K1 at a reasonably high frequency ($10^{-5}-10^{-4}$) without inducing a long lag in growth or hypersensitivity to trypsin treatment and subculturing. X ray, although relatively inefficient, is useful for obtaining mutants with very low reversion rates.

The Recovery Period

After removal of mutagen and addition of fresh, complete medium, a recovery period is essential before application of the light selection treatment. The duration of the recovery period which leads to optimal mutant yield has been studied in some depth for certain loci and mutagens (see Ref. 16, for example), but this analysis is not always possible, especially if one is attempting to isolate a spectrum of mutants, each of which may require different recovery. A limitation of this procedure is that for a mutant to be isolatable it must not make DNA in selective medium but it must then be able to survive and resume growth when transferred to nonselective conditions. This situation may well be unattainable for certain mutants in some cells. We generally proceed by observing the muta-

[14] F.-T Kao and T. T. Puck, *J. Cell Physiol.* **74**, 245 (1969).
[15] F.-T. Kao and T. T. Puck, *J. Cell Physiol.* **78**, 139 (1971).
[16] A. W. Hsie, J. P. O'Neill, D. B. Cough, J. R. SanSebastian, P. A. Brimer, R. Machanoff, J. C. Fuscoe, J. C. Riddle, A. P. Li, N. L. Forbes, and M. H. Hsie, *Radiat. Res.* **76**, 471 (1978).

genized cells under the microscope for a number of days. When they appear healthy and can be subcultured without undue loss of viability, the cells are split approximately 1 – 3 into additional 100-mm dishes, again in complete medium, and watched carefully. Remove from mutagen treatment takes various amounts of time depending on the particular agent, its dose, and the cell type. Therefore, it is important to assess recovery by empirical procedures. When the cells appear to have regained their normal growth characteristics, we assume, for the purposes of mutant isolation, that recovery is completed. We presume that the recovery period allows fixation of mutations and expression of recessive mutations as, for example, by dilution through cell division of required enzymes. Obviously certain mutants are lost such as those with a reduced rate of growth. It is critical that the cells are growing at a normal rate before the BrdUrd – light selection is applied or a number of problems may arise. First, a large number of cells may be growing slowly, which even though not mutant, will escape the BrdUrd treatment and confound the selection of mutants. Second, cells may be hypersensitive to the BrdUrd light treatment, even if they are mutant, and be lost. Our experience is that the duration of the recovery period must be determined empirically for each mutagenesis protocol in order to optimize the number of mutants isolated.

Bromodeoxyuridine Treatment

BrdUrd is light-sensitive and must be stored in light-excluding containers and added to the cells in dim or "yellow" light conditions to minimize exposure to 313 nm light. The object of the BrdUrd selection protocol is to maximize the destruction of wild-type cells while preserving as many of the mutant cells as possible. We have found that a dose of BrdUrd that reduces survival of light-exposed wild type cells to 10^{-4} adequately meets this requirement. The first step of the procedure is to distribute the mutagenized populations into the selective medium. It is useful to keep the populations of cells derived from the individual mutagen-treated plates separate so that mutants arising from the individual plates are known to be independent.

It is convenient to distribute $2 - 3 \times 10^4$ cells in selective medium into each of about 100 60-mm petri dishes. Thus, at the end of the selections, each dish will contain approximately 2 – 3 wild-type colonies and, if the induced mutant frequency is on the order of 10^{-5}, 25 mutants or so will be found in the 100 selection plates. It is not productive to deposit more than 3×10^4 cells per 60-mm dish as this seems to result in increased survival of wild type cells. Generally, we incubate the cells in selective medium for 24 hr before applying BrdUrd. This allows time for depletion in mutants of pools of required nutrients and permits the resumption of optimal growth by wild-type cells. It is often useful and sometimes necessary to alter the incubation time in selection medium in order to obtain some kinds of mutants. For example, certain kinds of mutants, such as those requiring pyrimidine, die very rapidly during starvation so that it is essential to minimize the starvation period. Generally, a 24-hr starvation period works well with CHO cells. After the 24-hr pretreatment period, BrdUrd is added to each plate to a final concentration of 10^{-5} M. In our hands, in several different cell types, increasing the BrdUrd concentration does not necessarily increase its lethal effects on wild-type cells. The dose of BrdUrd used is extremely important and must be determined carefully. Various investigators have attempted to increase the differential killing of wild-type oats vs mutant cells. Since BrdUrd does not select absolutely against wild-type cells, the addition of synergistically reacting agents such as Hoechst 33258 dye or inhibition of thymidylate synthase by addition of fluorodeoxyuridine may be helpful in some cases. We have not investigated these protocols in detail. After 24 hr of incubation with BrdUrd, the cells are exposed to visible light by any one of a number of protocols. We have designed two special illumination boxes, one using "cool white" bulbs and the other with "black light" bulbs. We have also carried out successful mutant searches using the "cool white" fluorescent lamps that are supplied in Lab Con. Co. (R) hoods. (The spectrum of these sources is given in Fig. 2.) Our illumina-

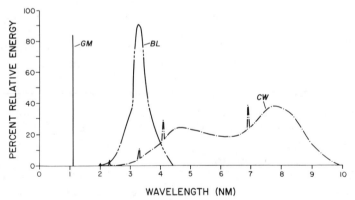

FIG. 2. Spectrum produced by various light sources. The component at 313 nm is responsible for the photolysis of BrdUrd-containing DNA so that the exposure required for cool white (CW) killing is 20–30 times that of black light (BL). The exposure period in our standard cool white light box is 1 hr where the measured light intensity on the plates, measured with an incident (not reflection) light meter, is ASA/ISO 100, 1/60-F/11 (~320 Lx). The standard germicidal (GM) emission if shown for comparison. We employ General Electric bulbs F15T8CW or F15T8B1. GM is a standard germicidal lamp.

tion boxes contain eight bulbs, four above and four below a Lucite shelf on which the petri dishes are placed about 5 cm from the lamps.

Several investigators have found it helpful in some circumstances to use black light or fluorescent sun lamps in place of standard cool white fluorescent lamps.[7] In any case it is important to carry out preliminary experiments to establish appropriate parameters of illumination. The proper light treatment should reduce the fraction of surviving BrdUrd-treated cells by a factor of 10^4-10^5 compared to cells that have not incorporated BrdUrd. It is also necessary to monitor the temperature during exposure to light and to use a fan to cool the cells if the temperature rises above 38°.

After light exposure, the selective medium containing BrdUrd is removed, the cells are wshed and complete medium is added to the plates. The cells are then incubated until colonies arise. The cells require extensive manipulation during this period so steps must be taken to minimize contamination; antibiotics such as penicillin and streptomycin or gentamicin (R) are commonly added and occasionally Fungizone is as well. These agents may be removed as desired after clones are picked and mutants identified.

Recovery from Mutagenesis and Selection of Clones

If the procedure has gone well to this point, incubation of the plates for 7–10 days will result in the growth of from 1 to 15 clones on each 60-mm dish. If many more clones than this arise, something has gone wrong, the chance of isolating mutants is severely reduced and it is usually easier to begin again. In any case, the clones that do survive will have quite heterogeneous growth rates and therefore sizes. Sometimes it may be worthwhile to allow plates with no or few clones visible after 7–10 days to incubate for considerably longer periods (up to a month) at which time colonies may become visible. Moreover, it is important with CHO cells not to manipulate the plates more than necessary during the 7–10 day recovery period because this will cause production of satellite colonies which will confuse the subsequent isolation of individual clones.

Picking and Analyzing Clones

A variety of techniques are available for picking clones of CHO and other mammalian cells. The one most often used in this laboratory employs stainless steel or aluminum cloning cylinders. These cylinders are easily cut from tubing stock of an appropriate and convenient size. An inner diameter opening of 4–5 mm and a well thickness of 2 mm works well. We examine by eye the 60-mm selection plates and circle on the undersurface of the plate with a marking pen each clone chosen to be picked. Clones can be readily seen when they are viewed looking up through the bottom of the dish which is held obliquely to a light source. The medium is then withdrawn from the dish and a sterile cloning cylinder, handled with sterile forceps, is dipped in sterile silicone grease and placed so as to encircle the colony. Take care to seat the cylinder firmly so that a tight seal is formed between the cylinder and petri dish. A few drops (2–3) of trypsin are then added to the interior of the cloning cylinder. The cylinders can be cleaned for use by shaking them in xylene followed by immersion in acetone to get rid of the xylene, then placed in small beakers and autoclaved. Silicone grease is conveniently prepared by spreading a smooth layer of it a few (~5) millimeters thick in a 60- or 100-mm *glass* petri dish. It is sterilized by autoclaving. Use your finger or a small spatula to spread the layer. After allowing a few minutes for trypsin to act, the trypsin containing the dispersed cells is removed with a Pasteur pipet. It is often useful to scrape the cells off the dish with the tip of the pipet rather than waiting for the trypsin to dislodge them. A drop of cell suspension is placed into a 35-mm petri dish containing complete medium, and a second drop added to another 35-mm petri dish containing selective medium. It is possible for an experienced worker to pick 30–50 clones in an hour. Several petri dishes, each containing two or three clones may be processed simultaneously. A different Pasteur pipet should, however, be used to disperse each clone, although one pipet may be used for addition of the initial trypsin solution to several cloning cylinders. Generally we pick between 75 and 150 clones. The plates containing the clones are then returned to the incubator and allowed to grow. Usually the plates can be examined within 5 days. Only those plates in which growth is clearly positive in both selective and nonselective media are safely discarded at this point. The growth of a few cells in selective medium does not necessarily mean that a clone is not mutant since some slight contamination with wildtype cells often occurs during the relatively crude procedure used to pick the clones. In addition, it has been our experience that cells harvested by traditional means with trypsin either from mass cultures or by clonal selection often do not show appropriate nutritional responses when plated directly into the selective medium. It is usually necessary either to dilute the cells substantially or to wash them by centrifugation, apparently to remove residual nutrients carried along with the cells in the harvesting procedure. Therefore, any clone showing apparent differences between growth in complete and selective medium is carefully reanalyzed by plating measured aliquots of approximately 100 cells in 35-mm dishes of complete and selective medium. Additionally, approximately 10 cells are placed into complete medium in a 60-mm plate for recloning. After 5–7 days of growth, the plates are examined for differences in clonal growth in complete and selective medium. At this point, clones showing equal growth in both conditions are discarded. All clones that grow in complete but not in selective media are then recloned from the 60-mm dishes and each newly selected subclone is again tested to verify differential growth. It is very important at this stage to grow up to a fairly large culture (10^6 cells or so) of each presumptive mutant clone for storage in liquid nitrogen. Further analysis may be undertaken as desired to define, for example, precisely the nutritional requirements in each clone or the biochemical lesion in the mutant.

Acknowledgments

This work was supported by NIH Grants AG00029, HD17449, GM34041, HD02080, CA36447, and MOD 15-10. This is ERICR contribution #633.

[11] Use of Tritium-Labeled Precursors to Select Mutants

By JACQUES POUYSSÉGUR and ARLETTE FRANCHI

Successful isolation of mammalian cell mutants altered in a given function relies almost exclusively on the degree of specificity and on the potency of the screening test used. A killing selection is certainly a method of choice for rapid mutant isolation. However it is very often difficult to meet the "comfortable" situation of having, for selection, a toxic compound analogous to an intermediate in the pathway of interest. In this chapter, we illustrate with four examples that the use of tritiated-labeled precursors overcomes this difficulty and provides a powerful and general method of selection for mutants defective in membrane transport and in metabolic pathways. This method, called radiosuicide and first used with bacterial genetics,[1,2] is highly selective since the substrate itself of the pathway to "mutate" is the killing agent. The principle is based on the killing effects generated by the radiolytic decay of the incorporated labeled precursor. Therefore any mutation able to reduce the uptake and/or the metabolism of the radioactive toxic precursor will be selected upon storage of the

[1] F. Harold, R. Harold, and A. Abrams, *J. Biol. Chem.* **240**, 3145 (1965).
[2] J. Cronan, T. Ray, and P. Vagelos, *Proc. Natl. Acad. Sci. U.S.A.* **65**, 737 (1970).

METHODS IN ENZYMOLOGY, VOL. 151

labeled cells. The wide variety of mutant cell lines derived from the application of this method demonstrates its usefulness.[3-8]

Glucose Transport and Glycolytic Defective Mutants: Selective Agent: 2-Deoxy[³H]glucose

Materials

2-Deoxy[1-³H]glucose (15–25 Ci/mmol)

The Chinese hamster lung fibroblast line O23 used in this study is a ouabain-resistant subclone of the established cell line CCL 39 (ATCC). These cells are maintained in DME medium supplemented with 10% fetal calf serum and have a doubling time of about 12 hr.

Cells were mutagenized with ethylmethane sulfonate (EtMes), 0.25 μl/ml, for 16 hr in regular medium.

Principle and Procedure of Selection

The glucose analog, 2-deoxyglucose, enters fibroblasts *via* the D-glucose facilitated diffusion transport system and accumulates inside the cell, mainly as 2-deoxyglucose 6-phosphate. A mutation that would decrease either hexokinase or glucose transport activity would give the cell resistance against the toxicity of ³H-labeled 2-deoxyglucose by decreasing the intracellular radiotoxic pool. The procedure of selection used and outlined in Fig. 1 is as follows. The mutagenized cell population grown for 5 to 6 generations is trypsinized and incubated at a cell density of 8×10^6 cells/ml for 1 hr at 37° in glucose- and serum-free DME medium containing 2-deoxy[1-³H]glucose (100 μCi/ml, 17 Ci/mmol). The cells are centrifuged to eliminate the free labeled sugar, and resuspended in cold medium containing 10% dimethyl sulfoxide (10^6 cells/ml) before storage in liquid N_2. Under these conditions the intracellular radioactive pool has reached a value corresponding to about 10^4 decays per cell per day. This amount of incorporated radioactivity reduces viability rapidly since only 0.1% of cells

[3] L. Thompson, R. Mankowitz, R. Baker, J. Till, L. Siminovitch, and G. Whitmore, *Proc. Natl. Acad. Sci. U.S.A.* **66**, 377 (1970).

[4] L. Medrano and H. Green, *Cell* **1**, 23 (1974).

[5] M. Finkelstein, C. Slayman, and E. Adelberg, *Proc. Natl. Acad. Sci. U.S.A.* **74**, 4549 (1977).

[6] A. Dantzig, C. Slayman, and E. Adelberg, *Somatic Cell Genet.* **8**, 509 (1982).

[7] J. Pouysségur, A. Franchi, J. C. Salomon, and P. Silvestre, *Proc. Natl. Acad. Sci. U.S.A.* **77**, 2698 (1980).

[8] A. Franchi, P. Silvestre, and J. Pouysségur, *Int. J. Cancer* **27**, 819 (1981).

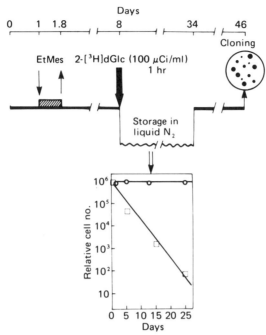

FIG. 1. Schematic representation of the tritium suicide method for the selection of glucose uptake deficient mutants. Periodically, samples were thawed and the cells were plated to analyze survivors. (□) Cells incubated with the radioactive sugar; (○) cells incubated with 10 μM unlabeled 2-deoxyglucose. Reprinted from Pouysségur et al.[7]

survive after 2 weeks of storage (Fig. 1). It is interesting to note that this loss of cell survival is generated by the radioactive damaging process, since a parallel cell population incubated with the same concentration of unlabeled deoxyglucose retained full viability (Fig. 1). After 25 days, when survival has decreased by a factor of 10^4, clones are isolated and tested for their ability to transport 2-deoxyglucose. One-third of the clones analyzed were found to have less than 50% of the wild-type hexose uptake activity.

Biochemical Characteristics of the Mutants

One of the most markedly affected clones, DS7, was analyzed in more detail. Its ability to take up 2-deoxyglucose was reduced to 10% of the parental value after an uptake of 10 min (Fig. 2A). Whereas 80–90% of the deoxy sugar taken up in 2 min was phosphorylated in the parental strain (dashed line), 90% of the radioactive pool present in DS7 was in the free form. Initial rates of uptake of the nonmetabolizable glucose analog, 3-O-

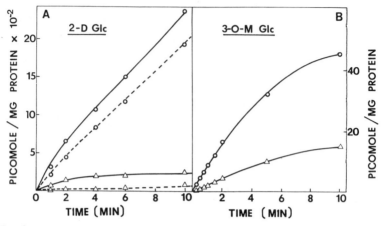

FIG. 2. Uptake and phosphorylation of 2-deoxyglucose (A) and transport of 3-*O*-methylglucose (B) in 023 cells (O) and DS7 (△) cell lines. Dashed lines represent the intracellular phosphorylated 2-deoxyglucose. Reprinted from Pouysségur *et al.*[7]

methylglucose, showed that glucose transport activity was decreased by 4- to 5-fold in DS7 cells (Fig. 2B). In contrast the transport of other metabolites such as phosphate or amino acid was unaffected, demonstrating the specificity of the 2-deoxyglucose tritium suicide for selection of glucose uptake deficient mutants. Further investigation of the mutant DS7 revealed that one glycolytic enzyme activity was missing: the phosphoglucose isomerase (pgi) (glucose-6-phosphate isomerase). As a result of this glycolytic block, DS7 accumulates a large intracellular pool of glucose 6-phosphate which in turn exerts a feedback inhibition on hexokinase and induces a severe "down-regulation" of the hexose transporter.[7] When the mutant is incubated in glucose-free medium, the glucose 6-phosphate pool returns to normal and glucose transport activity progressively increases to reach the same activity as in wild-type cells. Therefore, DS7 presents an altered glucose transport activity secondary to the pgi defect.

It is clear from this observation that to specifically select mutants altered in glucose transport, an improvement of the suicide technique would be to preincubate the cells in glucose-free medium for 5 to 10 hr prior to incorporation of the tritiated deoxy sugar. Although glucose is an essential carbon and energy source we predict that glucose transport mutants with only 5–10% of residual transport activity should grow at an almost normal rate. Indeed, DS7 which utilizes 15-fold less glucose than the parental strain for an identical protein increment, has a doubling time of 17 hr instead of 12 for the wild type. Also, thanks to the glucose passive diffusion process across the membrane, mutants with no detectable glucose-facilitated transport activity should be able to grow at a reasonable rate in a glucose-rich medium (50 mM).

Mutants Defective in Glucose Oxidative Pathway: Selective Agent: [^{3}H]Glucose

Materials

[6-^{3}H]Glucose

Chinese hamster lung fibroblast line O23, subclone of CCL 39 (ATCC)

Fischer Rat fibroblast cell line FR 3T3 isolated from secondary embryo fibroblasts[9]

Principle and Procedure of Selection

Glucose is metabolized into intermediates by the pentose phosphate route, the oxidative pathway, and glycolysis. The glycolytic pathway is the main route of glucose utilization in fibroblasts and the end product of this pathway, lactic acid, is secreted into the medium. Therefore we expected that cells, incubated with tritiated glucose and selected for resistance to radiolysis, would belong at least to two classes of mutants: class I, in which glucose transport is reduced and class II, in which glycolysis is strongly stimulated. Indeed, an increase in glycolytic rate is expected to protect the cells by converting [^{3}H]glucose into the excreted [^{3}H]lactic acid by-product.

Exponential cultures of mutagenized O23 cells (see above) were trypsinized and incubated at a cell density of 2×10^6 cells/ml for 3 hr at 37° in DME medium containing 1 mM of D-[6-^{3}H]glucose (100 μCi/ml). At the end of the incubation period, the suspension of cells was centrifuged to remove the nonincorporated radioactive material and stored in liquid N_2. Under these conditions a radioactivity of 1350 decays per cell per day was incorporated. After 2 months, cell viability was reduced by a factor of 10^4. Among the resistant clones analyzed, 20% have reduced by 2-fold their incorporation of labeled glucose into acid precipitable material.

Biochemical Characteristics of the Mutants

GSK3, a representative clone of the [^{3}H]glucose suicide selection, was unable to oxidize glucose, pyruvate, glutamate, and succinate. The rate of

[9] R. Seif and F. Cuzin, J. Virol. **24**, 721 (1977).

$^{14}CO_2$ produced from the oxidation of [6-^{14}C]glucose was reduced more than 20-fold whereas the glycolytic flux was stimulated 2.6-fold.[8,10] Consistent with this pleiotypic blockade in oxidation of Krebs cycle intermediates is the finding that GSK3 cells have a very low oxygen uptake rate. In many respects these mutant cells resemble the respiration defective mutants isolated by Scheffler's group.[11] Their energetic metabolism is strictly dependent on glycolysis. Indeed, GSK3 cells do not survive when glucose is replaced by poor glycolytic substrates (fructose, galactose) and lyse within 1 hr when starved of glucose. This last point is very important to consider if one wants to design a protocol of selection which favors one class of mutants more than the other. For example, to select specifically respiration-defective mutants, the concentration of tritiated glucose and therefore the time of incubation has to be precisely monitored to avoid glucose starvation. However, glucose starvation for 1–2 hr provides a very simple test to either counterselect respiration-deficient mutants or to rapidly get access to their frequency by evaluating the percentage of colonies which do lyse. Taking into account the properties of GSK3 we applied the [^3H]-glucose suicide selection to the normal fibroblast cell line FR3T3 (1.6 × 10^6 cells/ml, incubated with 0.75 mM [^3H]Glc at 200 μCi/ml for 2.5 hr). Under these conditions, representing 10^4 decays per cell per day, the two respiration-deficient mutants, GSR16 and GSR24, were isolated after 5 and 10 days of storage, respectively, in liquid N_2. Like GSK3 these cells rapidly lyse in glucose-free medium and display an increase in glucose transport activity and glycolytic rate of 3- and 6-fold, respectively.[8,10]

Mutants Defective in the Transport of Neutral Amino Acids: Selective Agents: α-[*methyl*-^3H]Aminoisobutyric Acid and [^3H]proline

A very brief account of the procedure of selection and of the properties of the mutants isolated by Adelberg's group will be reported here. For more details the reader should refer to the original papers.[5,6] The mouse lymphocytic cell line GF14 mutagenized with EtMes was incubated at 4 × 10^6 cells/ml for 50 min at 37° in Earle's balanced salt solution containing 8.0 μM [*methyl*-^3H]AIB at 2.5 Ci/mmol. By the end of this incubation period the intracellular pool of the nonmetabolizable amino acid analog produced 1330 decays per cell per day. After 4 months of storage in liquid N_2, survival had decreased by a factor of 10^4, and two clones out of 200 survivors, GF17 and GF18, were found to have a significant reduction in

[10] J. Pouysségur, A. Franchi, P. Silvestre, *Nature (London)* **287**, 445 (1980).
[11] C. Ditta, K. Soderberg, F. Landy, and I. Scheffler, *Somatic Cell Genet.* **2**, 331 (1976).

AIB uptake. Biochemical analysis revealed that the K_m for AIB transport is unchanged and that the V_{max} is reduced to 16% in GF17 and to 22% in GF18, compared with the parental cell line GF14. Further investigations demonstrated that the mutants are specifically defective in the transport system for which the major substrates are small neutral amino acids (AIB, L-alanine, L-serine, and L-proline).

Another related technique combining [³H]proline suicide and replica plating was applied to a proline-requiring subclone of CHO cells.[6] First, the cells were starved of amino acids for 1.25 hr and subsequently incubated at 3×10^6 cells/ml in DME medium containing [³H]proline (25 μCi/ml, 0.7 μM) for 1.75 hr at 37°. In that particular experiment cell-associated radioactivity produced 4190 decays per cell per day. The loss of survival upon storage at −70° is represented in Fig. 3A. On day 24, the tritium suicide was arrested, a time reducing cell survival by more than 10^4. It is interesting to note (Fig. 3B) that if the cells are stored at −70° with the

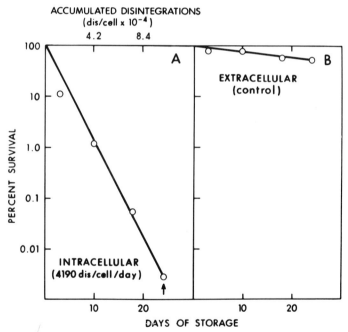

FIG. 3. Decrease in survival due to tritium suicide during storage (A) cells were incubated with [³H]proline (25 μCi/ml, 0.7 μM) for 1.75 hr. Survival is given as a function of both storage time and the total number of disintegrations in the cell. (B) cells were incubated in the absence of labeled substrate and frozen in prechilled medium containing [³H]proline (14 μCi/ml). Reprinted with permission from Dantzig et al.[6]

same amount of tritiated proline but kept in the extracellular space, cell killing is minimal. This experiment clearly illustrates that in order to exert their radiolytic effect the tritiated precursors must be incorporated into the cell.

Survivors of that selection were screened for proline uptake by replica plating. One mutant, CHY-1, which took up proline with a 2-fold reduced rate, was found specifically altered in the transport of neutral amino acids (A system). The K_m and V_{max} for [methyl-^{3}H]AIB transport were respectively increased 3.5- and 1.5-fold in CHY-1 cells, suggesting the existence of a mutation in the structural gene coding for the A transport protein.

Concluding Remarks

With these four examples of tritium suicide selection, we have shown that it is possible to devise a selection procedure allowing mutant isolation in a reasonable time period with either metabolizable or nonmetabolizable precursors. Among the negative selection methods used, that of the tritium suicide has the following advantages: (1) any substrate, required or not for growth, can be made lethal to the cell by virtue of its tritium label; (2) the selection can be made almost absolutely specific for transport mutants, by the use of non- or poorly metabolizable substrates such as AIB or 2-deoxy-glucose. However, one of the major concerns about the design of the experimental protocol is the length of storage time at low temperature. First, we have seen that to reduce to 10^{-4} the viability of the parental cells, an accumulation of about 10^5 disintegrations/cell is required (Fig. 3). Therefore, to reduce the time of storage to less than 2 months a minimum of incorporation giving 1 dpm/cell is needed. Second, too long a period can result in a total loss of the mutant cells in the population. From the exponential decay functions for the parent and the mutant

$$P_1 = P_0 \, e^{-k_p t} \quad \text{and} \quad M_1 = M_0 \, e^{-k_m t}$$

where P_0 and M_0 are the numbers of parent and mutant cells initially present in the cell population, P_1 and M_1 are the numbers after the elapsed time t, k_p and k_m are the rate constants for the parent and mutant, one can easily calculate that a mutant which incorporates either 50 or 25% of the parental radioactivity (ratio $k_p/k_m = 2$ or 4), will be respectively 100-fold or 10000-fold more resistant than the parental population to the tritium suicide test.[6] Therefore if one predicts that the mutants to isolate might have only moderate reduction in the rate of uptake of the tritiated precursor it will be essential to interrupt the time of storage and therefore the suicide when viability has dropped to about 10^{-2} and to repeat the same suicide cycle 2 or 3 times.

Acknowledgments

We are grateful to Drs. A. Le Cam and K. Seuwen for critical reading of the manuscript. This work has been supported by grants from the Centre National de la Recherche Scientifique (LP 7300) and the Institut National de la Santé et de la Recherche Medicale.

[12] Construction of Immunotoxins Using *Pseudomonas* Exotoxin A

By DAVID J. P. FITZGERALD

Microbial toxins are potent biological agents that are often capable of killing mammalian cells by enzymatically inactivating normal cellular function. If it were possible to harness the activities of toxins and redirect them to selectively kill or damage discrete populations of cells, powerful reagents might become available for use both as tools for cell biology and cell genetics and as novel drugs for the treatment of cancer or other diseases. Toward this end *Pseudomonas* exotoxin A (PE) was chemically coupled to a variety of monoclonal antibodies to make hybrid molecules termed immunotoxins (for a review of immunotoxins, see Refs. 1 and 2).

PE is a 68,000 Da bacterial protein toxin produced by *Pseudomonas aeruginosa*. This toxin inhibits protein synthesis of mammalian cells by virtue of its ability to ADP-ribosylate elongation factor-2 (EF-2). To gain access to EF-2, PE must translocate from outside the cell across a cellular membrane to the cytoplasm. The first step in this process is the binding of PE to a cell surface component that we can term a "PE-receptor." Following binding, PE enters cells by receptor-mediated endocytosis and has been localized by electron microscopy first in coated pits, then in receptosomes (endosomes), the Golgi, and lysosomes.[3-5] However, it is not clear yet how or where the toxin traverses one of these membrane-limited compartments to gain access to the cell cytoplasm. Hybrid molecules made using PE should be constructed to deliver PE to a similar intracellular compartment.

PE-immunotoxins have been constructed using antibodies that bind to

[1] I. Pastan, M. C. Willingham, and David J. P. FitzGerald, *Cell* **47**, 641 (1986).
[2] J. A. Cumber, J. A. Forrester, B. M. J. Foxwell, W. C. J. Ross, and P. E. Thorpe, this series, Vol. 112, p. 207.
[3] D. J. P. FitzGerald, R. E. Morris, and C. B. Saelinger, *Cell* **21**, 867 (1980).
[4] R. E. Morris, M. D. Manhart, and C. B. Saelinger, *Infect. Immun.* **40**, 806 (1983).
[5] M. D. Manhart, R. E. Morris, P. F. Bonventre, S. Leppla, and C. B. Saelinger, *Infec. Immun.* **45**, 596 (1984).

cell surface antigens of mammalian cells.[6-9] Ideally the immunotoxin should bind cells only via the antibody. However, many toxins, including PE, have a cell-binding capacity of their own. This must be inactivated or removed or else the immunotoxin will have two binding domains and will therefore have little or no selectivity. In the case of PE, the binding domain is inactivated following the reaction with 2-iminothiolane (see Ref. 8 and below).

Materials and Methods for Immunotoxin Construction

The following describes a typical conjugation procedure that can be used to couple PE to any monoclonal antibody of the IgG class. No conditions to make immunotoxins with IgM or IgA have yet been worked out.

Purified PE was purchased from Swiss Serum and Vaccine Institute, Berne, Switzerland. Purified monoclonal antibodies were prepared by conventional procedures. Cysteine and NAD were obtained from Sigma and 2-iminothiolane-HCl and 5,5'dithiobis(2-nitrobenzoic acid) (DTNB) were obtained from Pierce Chem. Co.,

1. Typically 5 mg of PE is dissolved in 0.15 M KPO$_4$, 1 mM EGTA, pH 8.0 (buffer A). Usually 400 μl of buffer A is added to 5 mg of lyophilized PE and then desalted on a PD10 (9 ml, bed volume) column (Pharmacia) to remove the lactose which is colyophilized with PE. The running buffer is also buffer A. Three fractions are collected: fraction one is 2.8 ml, fraction two is 2.0 ml, and fraction three is 1.0 ml. Most of the PE is contained in fraction two. The optical density of each fraction is measured at 280 nm (an OD of 1.2 is equivalent to 1 mg/ml of PE). If necessary, additional buffer A is added to fraction two until the OD is 1.7.

2. To PE in fraction two the following are added: 5 μl of NAD (66 mg/ml in buffer A) per ml of PE, and then 5 μl of iminothiolane (IT) (130 mg/ml in buffer A) per ml of PE. The reaction mixture is incubated at 37° for 1 hr.

3. The PE reaction mixture is loaded on an HPLC gel filtration column (Bio-Rad, TSK-250, 600 × 21.5 mm). An FPLC (Pharmacia) system is

[6] D. J. P. FitzGerald, I. S. Trowbridge, I. Pastan, and M. C. Willingham, *Proc. Natl. Acad. Sci. U.S.A.* **80,** 4134 (1983).
[7] D. J. P. FitzGerald, T. A. Waldmann, M. C. Willingham, and I. Pastan, *J. Clin. Invest.* **74,** 966 (1984).
[8] R. Pirker, D. J. P. FitzGerald, T. C. Hamilton, R. F. Ozols, M. C. Willingham, and I. Pastan, *Cancer Res.* **45,** 751 (1985).
[9] R. Pirker, D. J. P. FitzGerald, T. C. Ozols, W. Laird, A. E. Frankel, M. C. Willingham, and I. Pastan, *J. Clin. Invest.* **76,** 1261 (1985).

used to pump the buffer at 4 ml/min. The column is run in 0.15 M KPO$_4$, 1 mM EGTA, pH 7.0 (buffer B).

A gel filtration step is preferable to simple desalting, since the addition of IT causes a small amount of PE to aggregate. The aggregates are seen at the void volume and constitute 5% of less of the total amount of PE. PE elutes from this column (at 144 ml) as a single peak, typically collected in 12 ml.

4. The protein concentration is determined by measurement at OD$_{280}$. Then the number of new SH groups per molecule of PE is determined. To the PE fraction 5,5'-dithiobis(2-nitrobenzoic acid) (DTNB) is added to a final concentration of 1 mM. A stock solution of DTNB at 100 mM in 1.0 M KPO$_4$ (dibasic salt ~ pH 9.3) is made fresh on the day of use. The OD at 412 is determined against a blank, containing buffer and DTNB at 1 mM. The OD at 412 divided by 0.0136 gives the number of moles of SH/ml. Usually the number is exactly *3*. However, we have noted values from 2.6 to 3.2.

5. To remove excess DTNB, PE is desalted over a PD10 column. Before this can be accomplished the volume must be reduced from 12 ml to less than 1 ml. This is done using Centricon 30 microconcentrators (Amicon). Typically PE is concentrated to approximately 0.5 ml. PE is desalted on a column equilibrated with buffer A and three fractions collected: of 2.8, 2.5, and 1.0 ml. The majority of the protein is contained in the second 2.5 ml fraction. PE is now ready to be coupled by disulfide exchange to a monoclonal antibody having a free sulfhydryl.

6. A "new" sulfhydryl group is introduced in the monoclonal antibody by reaction with IT. The antibody is desalted into buffer A by passing it over a PD10 column equilibrated with buffer A. Using microconcentrators the antibody is then concentrated to approximately 0.5 ml and sufficient IT added from a 1.3 mg/ml stock solution to give a 2-fold molar excess of reagent over antibody. The reaction is allowed to proceed for 1 hr at 37° after which the antibody is desalted (PD10 column, buffer A). Ideally, there should be one new SH per antibody. This is evaluated by removing an aliquot of antibody and adding DTNB to a final concentration of 1 mM (see step 4).

7. The activated PE and antibody are mixed at either 37° or RT and the reaction followed by the appearance of thionitrobenzoate (TNB) (OD$_{412}$) The TNB is displaced from PE as disulfide bonds are formed.

8. When the OD$_{412}$ ceases to increase, the pH is lowered from 8.0 to 7.0 by the addition of 1.0 M monobasic KPO$_4$—add $\frac{1}{10}$ the volume of the reaction volume. Cysteine (1.2 mg/ml) in buffer B is then added until the OD$_{412}$ ceases to increase. This displaces any remaining TNB from PE (see Fig. 1 for schematic of how PE-immunotoxins are prepared).

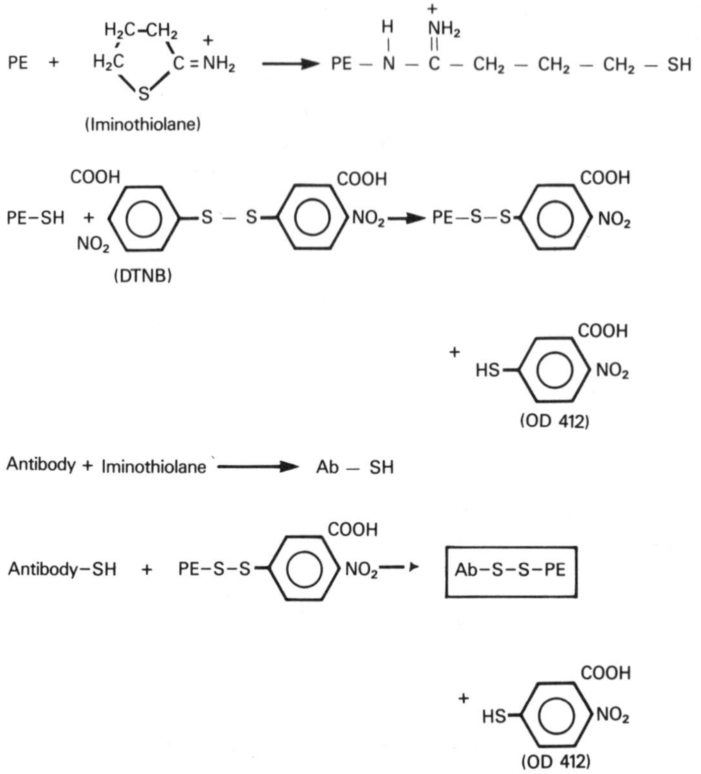

FIG. 1. Schematic outline for construction of PE-immunotoxin.

9. The reaction mixture is then chromatographed on the TSK-250 column. The following discrete products are routinely recovered: a 2:1 antibody to PE conjugate which elutes with a peak at 86 ml, a 1:1 antibody to PE conjugate which elutes at 106 ml, the unreacted antibody which runs at 118 ml, and the unreacted PE which comes off at 144 ml. Material eluting at the void volume (76 ml) is poorly characterized but may include 3:1 antibody:PE conjugates and any aggregates. TNB elutes at ~190 ml.

Buffer Preparation

We prepare immunotoxins for use in clinical and preclinical testing. Therefore, they must be free of pyrogens and microbial contamination. To ensure high quality in this regard we use sterile pyrogen-free water (Abbott Laboratories) to make all buffers used in the chemical coupling and subsequent chromatography. HPLC-grade water may not be suitable to make conjugates for clinical use but would be satisfactory when used for conjugates needed for cell culture experiments. Buffers A and B are routinely degassed and filtered (0.22 μm) immediately prior to use. Also, trace metals can wreak havoc with oxidation–reduction reactions, and for this reason we include EGTA in all our buffers.[10]

Evaluation of Immunotoxins

The final product (we consider the 1 : 1 Ab-PE conjugate as the desired product) runs as a single band of 210,000 Da on a nonreducing SDS–PAGE. Depending on the degree of separation there may be some free antibody present or some 2 : 1 antibody-PE conjugate (usually less than 10% of either). There should never be any free PE present in the final preparation.

Cytotoxic activity of the immunotoxin is usually assayed by determining the percentage inhibition of protein synthesis in immunotoxin-treated cells compared with control-treated cells. Immunotoxin preparations or control solutions are added to cells for approximately 24 hr. At the end of this period [^{3}H]leucine is added at a final concentration of 2 μCi/ml for a further 1 hr. The cells are then solubilized in 0.1 M NaOH, cell protein is precipitated by the addition of excess 12% trichloroacetic acid (TCA), the precipitates washed by centrifugation with 6% TCA, and the cpm/number of cells or cpm/μg cell protein determined. The percentage of protein synthesis inhibition is then calculated. If toxicity is mediated entirely via antibody binding, the addition of excess antibody should reduce the level of toxicity. While the addition of excess homologous antibody should reduce toxicity (usually by 100- 1000-fold), the addition of an irrelevant antibody (of the same IgG subclass) should not influence toxicity in any way. This latter evaluation guards against possible Fc-mediated uptake of immunotoxins.

Whereas inhibition of protein synthesis is a valid test of immunotoxin activity it may not always predict eventual cell death. Cytotoxic activity that results in cell death is often measured by inhibition of colony formation or a similar assay.[11] This assay can be used to determine concentrations of immunotoxin suitable for selection of mutant cell lines.

To inhibit protein synthesis, PE ADP-ribosylates EF-2. The preservation of this activity is essential if an active immunotoxin is to be made.

[10] P. W. Riddles, R. L. Blakeley, and B. Zerner, this series, Vol. 91, p. 49.
[11] S. Akiyama, M. M. Gottesman, J. A. Hanover, D. J. P. FitzGerald, M. C. Willingham, and I. Pastan, *J. Cell. Physiol.* **120,** 271 (1984).

When an immunotoxin appears to have little or no cytotoxic activity toward target cells, it is important to confirm the retention of full ADP-ribosylation activity, so as to be certain the conjugation reactions have not inactivated the enzymatic activity of the toxin.

The following assay is used for measuring ADP ribosylation activity (for additional information see Ref. 12):

To a 5 ml test tube the following are added:

a. 360 μl Tris, 0.05 M, pH 8.2
b. 20 μl DTT 1.0 M
c. 20 μl rabbit reticulocyte preparation (see Ref. 13)
d. 50 μl of immunotoxin (1–100 μg/ml) or 50 μl of PE at known concentration 1–25 μg/ml
e. 50 μl [^{14}C]NAD 1 μCi/ml (Amersham, CFA.497)

The reaction proceeds for 30 min at 37° and is then stopped by the addition of 0.5 ml of cold 12% TCA. The protein precipitate is collected either on filters or by centrifugation. The ADP-ribosylation activity of the immunotoxin is compared to PE standards.

Suggestions and Alternative Methodologies

Steps 1–9 describe the chemical coupling of PE to a typical monoclonal antibody. High purity of the antibody sample is important. Antibodies prepared by DEAE chromatography run the risk of being contaminated with transferrin. Inadvertently coupling PE to a small amount of TF could have potentially undesirable consequences and might be difficult to detect in the final preparation. When antibodies are prepared by ion-exchange chromatography, it may be necessary to also use a gel filtration step prior to beginning the coupling with PE.

Step 3 recommends a gel filtration step. Desalting may be satisfactory in some situations. If the latter is chosen, glycine could be used to neutralize any excess IT prior to the desalting step.

Steps 3 and *9* make use of a 600 × 21.5 mm HPLC gel filtration column. This column costs approximately $3000.00 and may not be within everyone's budget. We have also used the 300 × 7.5 mm column which is one-fifth the price of the larger column. The resolution is not as good and smaller protein loading must be used but satisfactory results are still possible. Additional information describing the use of DTNB and IT is found in Ref. 10 and 14.

[12] B. H. Iglewski and J. C. Sadoff, this series, Vol. 60, p. 780.
[13] R. Arlinghaus, J. Shaeffer, J. Bishop, and R. Schweet, *Arch. Biochem. Biophys.* 125, 604 (1968).
[14] J. M. Lambert, R. Jue, and R. R. Traut, *Biochemistry* 17, 5406 (1978).

The HPLC columns were attached to an FPLC pumping system. However, any high or medium pressure pumping system would be appropriate.

An EF-2 preparation from wheat germ can be substituted in place of the rabbit reticulocyte preparation.[12]

Summary

The above methodology has been used to prepare a variety of PE-immunotoxins. These reagents are potent cytotoxic agents for cells in culture that bind the appropriate antibody[6-9] and nontoxic when cells do not bind the antibody.[7,9] PE-immunotoxins can be used to select for mutant cultured cells which lack the target of the antibody (see chapter by Gottesman [9] on drug-resistant mutants). Recently, we have also demonstrated *in vivo* activity when a PE-immunotoxin was used to inhibit the growth of human ovarian cancer cells in a nude mouse model of ovarian cancer.[15,16] Also PE-immunotoxins are currently being evaluated in Phase 1 clinical trials for the treatment of adult T-cell leukemia.

Acknowledgments

This work was carried out in the laboratory of Mark Willingham and Ira Pastan. Their support is gratefully acknowledged. The technical expertise of M. Gallo and A. Rutherford is also acknowledged.

[15] D. J. P. FitzGerald, M. C. Willingham, and I. Pastan, *Proc. Natl. Acad. Sci. U.S.A.* **83,** 6627 (1986).
[16] M. C. Willingham, D. J. FitzGerald, and I. Pastan, *Proc. Natl. Acad. Sci. U.S.A.* **84,** 2474 (1987).

[13] Isolation of Temperature-Sensitive Mutants

By Joseph Hirschberg and Menashe Marcus[1]

Isolation and characterization of mutants which are defective in biochemical, cellular, or developmental processes provide a powerful tool in the analysis of the regulation of these processes at the molecular and cellular levels. A special group are the conditional mutants — usually temperature sensitive. Not only do they increase the number of genes in which mutations can be introduced but they also open the way to isolate mutants defective in indispensable functions.

Temperature-sensitive *(ts),* usually heat-sensitive, mutants have been

[1] Deceased.

isolated in various organisms and also in cultured cell lines.[1a] Such mutants have been proven to be very useful in physiological studies of mammalian cells grown in culture.[2]

The following protocols for isolating ts mutants in mammalian cells, have been adapted to the Chinese hamster cell line E-36, which is a derivative of V-79.[3] These cells are grown as monolayer cultures at 37° in an atmosphere of 5% CO_2 in Dulbecco's modified Eagle's medium (DMEM), supplemented with 10% fetal calf serum, L-glutamine (2 mM), penicillin (100 unit/ml), and streptomycin sulfate (100 μg/ml).

Under these conditions the average cell cycle time is 12.5 hr and the duplication time of a logarithmic culture is 13.1 hr.[4] Applying this protocol to other cell lines should take into consideration the cell cycle kinetics of the studied cell line, as it may differ considerably from that of E-36.

Choosing the Permissive and Restrictive Temperatures

It is very important that cell growth and culture proliferation of wild-type *(wt)* cells be normal at both the restrictive and permissive temperatures. This should be experimentally determined for each type of cell line to be used. When possible, it is desirable to choose cell lines that exhibit a fast rate of growth even at low temperature (33–34°) and yet can grow well at high temperature (39.5–40.5°). For the line E-36, 34 and 40° have been selected as permissive and restrictive temperatures, respectively. Duplication time of the culture is 18 hr at 34° and 14 hr at 40°.[4] Nishimoto and Basilico[5] who used BHK-13 cell line, encountered many "leaky" mutants in their previous attempts to isolate ts mutants. In order to overcome this problem, they performed the selection steps at 37.5° hoping that ts mutants that express their mutations at this temperature would be "tight" mutants at 40°.

Mutagenesis

Cultures should be grown at the restrictive temperature for at least 10 generations prior to mutagenesis, in order to eliminate variant cells whose growth is markedly impaired or slow under these conditions.

Mutagenesis has been successfully performed with different agents. It is

[1a] L. Siminovitch, *in* "Genes, Chromosomes and Neoplasia" (F. E. Arrigi, P. N. Rao, and E. Stubblefield, eds.), p. 157. Raven Press, New York, 1981.

[2] C. Basilico, *Adv. Cancer Res.* **24,** 223 (1977).:

[3] F. D. Gillin, D. J. Roufa, A. L. Beaudet, and C. T. Caskey, *Genetics* **72,** 239 (1972).

[4] J. Hirschberg, Ph.D. thesis. The Hebrew University of Jerusalem, Jerusalem, Israel, 1982.

[5] T. Nishimoto and C. Basilico, *Somatic Cell Genet.* **4,** 323 (1978).

advisable, however, to use more than one mutagen if one wishes to obtain a wide spectrum of different *ts* mutants, since the efficiency of mutagens in inducing mutations varies dramatically with different genes.[6] It is difficult but important to optimize the mutagenic treatment, especially when a nonselective, replica plating technique is used, as every such experiment is not only expensive but also time consuming. Recently a new and rapid technique, which allows estimation of the efficiency of a mutagenic treatment within 2–3 days, has been described.[7]

To a logarithmic cell culture add ethylmethane sulfonate (EMS, Sigma) to a final concentration of 0.4 mg/ml. Incubate for 16 hr at 34° (the permissive temperature). Wash the cells 3 times with serum-free medium. Add fresh medium and incubate for 2 days at 34°. The surviving fraction of cells after this mutagenic treatment is about 10–20%. Some cell lines exhibit high mortality after treatment with EMS. These cell lines should be treated with lower concentrations of the mutagen. Resuspend the cells with trypsin solution (0.025% trypsin in Puck's saline A) or in trypsin-EDTA (ethylenediaminetetraacetic acid) solution (0.005% trypsin; 0.002% EDTA in modified Puck's saline A) and replate in a new flask at a 2-fold higher concentration. Incubate for 24 hr at 34°. At this point the dead cells, that do not attach to the culture dish, are removed with the medium.

Negative Selection of *ts* Mutants

Since temperature sensitivity for growth is not selectable by itself, *ts* mutants can be obtained by a "mutant enrichment" step that is based on selective killing of *wt* cells.[8–11]

Shift exponentially growing cells after mutagenesis from 34 to 40°. Incubate for 18–24 hr to allow arrest of growth and cell division of *ts* cells. Add to the growth medium 5-fluorodeoxyuridine (FdUrd) (25 μg/ml) together with uridine 40 μg/ml. Other DNA "poisons," such as 5-bromodeoxyuridine (BrdUrd)[12] or cytosine arabinoside,[11] are equally efficient. Further incubate at the restrictive temperature for 18 to 24 hr in order to kill the nonarrested, DNA synthesizing, *wt* cells. Exposure to light is re-

[6] J. E. Cleaver, *Genetics* **87**, 129 (1977).
[7] A. Ronen, J. D. Gingerich, A. M. V. Duncan, and J. A. Hendle, *Proc. Natl. Acad. Sci. U.S.A.* **81**, 6124 (1984).
[8] C. Basilico and H. K. Meiss, *Methods Cell Biol.* **8**, 1 (1974).
[9] C. Basilico, *J. Cell. Physiol.* **95**, 367 (1978).
[10] T. Nishimoto and C. Basilico, *Somatic Cell Genet.* **4**, 323 (1978).
[11] L. H. Thompson, R. Mankovitz, R. M. Baker, J. A. Wright, J. E. Till, L. Siminovitch, and G. F. Whitmore, *J. Cell. Physiol.* **78**, 431 (1971).
[12] P. M. Naha, *Nature (London)* **223**, 1380 (1969).

quired at this point when BrdUrd is used as the killing agent.[12] Wash the plate twice with serum-free medium, add fresh medium (5–10 μg/ml thymidine should be added to the medium if FdUrd or BrdUrd is used as the killing agents). Shift back to 34°. Incubate until colonies are formed.

Since the rate of killing reaches 99%, it is essential to change medium after a few days, to get rid of debris of disintegrating cells. Isolate colonies and check for temperature sensitivity. If the frequency of *ts* mutants among the emerging colonies is too low, two (or more) cycles of mutant "enrichment" can be carried out with 36–48 hr intervals between them.

Isolation of *ts* Mutants by Replica Plating

The greatest disadvantage of the negative selection method is that it requires a relatively long exposure of cells to the restrictive conditions. In fact, in order to achieve reasonable killing frequencies of *wt* cells, the culture must be incubated at the restrictive temperature for three to four generation times. As a result, all those *ts* mutants that lose viability rapidly when they become arrested at the high temperature are lost. In order to recover all potential *ts* mutants, one has to avoid any selection step which involves exposure of cells to the restrictive temperature. Screening of conditional mutants by replica plating is a well known procedure in micro-organisms.[13] However, applying similar techniques for mammalian cells is much harder because of technical problems[14,15] (see chapter by Gal [8], this volume, for a discussion of replica plating).

We have designed and constructed a multisyringe injector, which fits commercially available microtest plates with 96 flat-bottomed wells (Nunc). This device, shown in Fig. 1, enables plating and harvesting of cell cultures in all 96 wells of a microtest plate simultaneously.[16] Sterilization of this device is achieved by autoclaving or by rinsing the syringes once with 70% ethanol followed by three rinses with sterile distilled water. All manipulations of cells in the microtest plates described here were carried out with the multisyringe device. A similar device is now commercially available (Bellco).

Harvest cells after mutagenic treatments, as described, and dilute in medium to a concentration of 20 cells/ml. Inoculate 0.15 ml of this suspension into each of the 96 wells of the microtest plates, and incubate at 34° until clones are formed. The mean number of three "mutagenized"

[13] J. Lederberg and E. M. Lederberg, *J. Bacteriol.* **63**, 399 (1952).
[14] T. D. Stamato and L. K. Hohman, *Cytogenet. Cell Genet.* **15**, 372 (1975).
[15] J. D. Esko and C. R. H. Raetz, *Proc. Natl. Acad. Sci. U.S.A.* **75**, 1190 (1978).
[16] J. Hirschberg and M. Marcus, *J. Cell. Physiol.* **113**, 159 (1982).

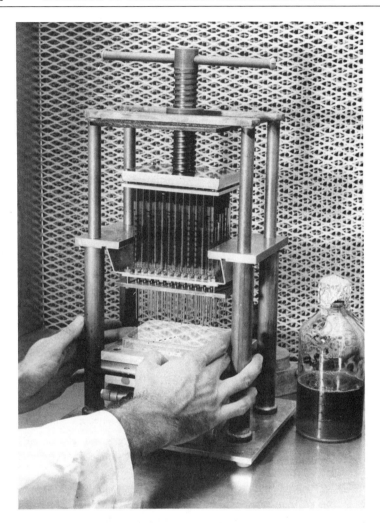

FIG. 1. A multisyringe injector adapted to microtest plates with 96 wells. The syringes are made of glass and have a volume of 1 ml. Each syringe has a stainless-steel cannula. The syringes are operated by turning the handle of the injector. The volume of fluid removed or added by each syringe is accurately controlled.

cells per well has been found to give rise to the highest number of wells with a single colony. This number may be different for different cell lines.

After 10 days of incubation screen the microtest plates under a microscope and mark all wells where only one colony of cells (a clone) has appeared.

Remove the medium from the wells. Add 0.05 ml of 0.025% trypsin solution in Puck's saline A to each well and resuspend the cells. Inoculate all the clones to two identical "sister" microtest plates in 0.2 ml of fresh medium. Incubate one microtest plate of each pair at 34° and the other at 40°. Screen marked wells in the 40° grown cultures for the ability of cells to grow at the high temperature and compare with the sister clones grown at 34°. Our experience indicates that about 1% of the surviving clones are temperature sensitive mutants for growth.[16] These *ts* clones can be further screened for a specific physiological or biochemical phenotype.

[14] Fluorescence-Activated Cell Sorting of Hybrid and Transfected Cells

By MICHAEL E. KAMARCK

The introduction of chromosomes and DNA into somatic cells by whole cell hybridization and transfection has supported studies on gene mapping, cloning, and function. The selection of host cells containing donor genetic material most often has depended on the complementation of defective genes, antibiotic resistance, or altered phenotype. Genes coding for cell surface exposed molecules also have been introduced into somatic cells, and their expression assessed with specific antibodies and indirect immunofluorescence. Using the fluorescence-activated cell sorter (FACS) individual cells have been isolated based on the expression of these surface antigen genes.

FACS selection of hybrid and transfected cells has facilitated a genetic analysis of the cell surface. The expression of surface antigens on hybrid cells allows the mapping of surface antigen genes and the development of hybrid cell lines homogeneous for the presence of a specific human chromosome and its surface antigens.[1,2] Using transfection protocols, it is possible to introduce cloned DNA sequences into recipients and to assess their contribution to the surface antigen array on the host cell. This approach has been useful in identifying functional members of large multi-

[1] M. E. Kamarck, J. A. Barbosa, and F. H. Ruddle, *Somatic Cell Genet.* **8**, 385 (1982).
[2] P. G. M. Peters, M. E. Kamarck, M. Hemler, J. Strominger, and F. H. Ruddle, *Somatic Cell Genet.* **8**, 825 (1982).

gene families, such as the major histocompatibility complex.[3,4] Finally, the entire genome of a donor can be introduced into host cells by transfection, resulting in a "library" of transfectants containing the introduced genome in randomized fragments. Sorter selection of the recipient cells expressing a surface antigen allows the antigen-coding genes to be isolated in host cells,[5,6] and ultimately cloned.[7-9] This chapter details methods which we have used for the selection of hybrid and transfected cells by flow cytometry and sorting.

Cell Lines and Biochemical Selection

Donor human chromosomes and DNA are introduced into derivatives of the mouse L cell[10] with specific biochemical defects in purine or pyrimidine biosynthesis. For the formation of human × mouse cell hybrids we have used the cell line LA9,[11] which is defective in the expression of the enzyme hypoxanthine phosphoribosyltransferase (HPRT, EC 2.4.2.8). Cloned genes or genomic DNA is introduced into the cell line Ltk−,[12] which is defective in the expression of the thymidine kinase gene (TK; ATP:thymidine 5'-phosphotransferase, EC 2.7.1.21).

L cell lines are maintained in Dulbecco's modified Eagle's medium (DMEM, Gibco, Grand Island, NY) supplemented with 10% fetal calf serum (FCS) (K.C. Biological, Lenexa, Kansas) without antibiotics. Both L cell lines grow in monolayer culture, are removed from flasks using phosphate-buffered saline (PBS) lacking Ca^{2+} and Mg^{2+} with 0.03% EDTA, and are passed at dilutions of 1 : 10 to 1 : 20. These cell lines can be established as suspension cultures in the same culture medium when passed at cell concentrations of 2×10^5 cells/ml into glass spinner flasks. Stock lines are regularly tested for the presence of mycoplasm using a commercial test kit (Gen-Probe, San Diego, CA). Following cell hybridization or transfection,

[3] G. A. Evans, D. H. Margulies, R. D. Camerini-Otero, K. Ozato, and J. G. Seidman, *Proc. Natl. Acad. Sci. U.S.A.* **79,** 1994 (1982).

[4] J. A. Barbosa, M. E. Kamarck, P. A. Biro, S. M. Weissman, and F. H. Ruddle, *Proc. Natl. Acad. Sci. U.S.A.* **79,** 6327 (1982).

[5] P. Kavathas and L. A. Herzenberg, *Proc. Natl. Acad. Sci. U.S.A.* **80,** 524 (1983).

[6] L. C. Kuhn, J. A. Barbosa, M. E. Kamarck, and F. H. Ruddle, *Mol. Biol. Med.* **1,** 335 (1983).

[7] P. Kavathas, V. P. Sukhatme, L. A. Herzenberg, and J. R. Parnes, *Proc. Natl. Acad. Sci. U.S.A.* **81,** 7688 (1984).

[8] L. C. Kuhn, A. McClelland, and F. H. Ruddle, *Cell* **37,** 95 (1984).

[9] D. R. Littman, Y. Thomas, P. J. Maddon, L. Chess, and R. Axel, *Cell* **40,** 237 (1985).

[10] W. R. Earle, *J. Natl. Cancer Inst.* **4,** 165 (1943).

[11] J. W. Littlefield, *Exp. Cell Res.* **41,** 190 (1966).

[12] S. Kit, D. Dubbs, L. Pierarski, and T. Hsu, *Exp. Cell Res.* **31,** 297 (1963).

the antibiotics penicillin (100 U/ml) and streptomycin (100 μg/ml) are added to culture medium until unique lines have been expanded, and frozen stocks established. The elimination of antibiotics in standard culture work serves to monitor tissue culture technique. Regular use of antibiotics also obscures the presence of mycoplasm infection. For permanent storage, cells are suspended in 90% standard tissue culture medium, with 10% dimethyl sulfoxide (DMSO) (ATCC, Rockville, MD). Cells are aliquoted into freezing vials at concentrations of 5×10^6 cells/ml and placed in the vapor phase of liquid nitrogen.

The L cell lines should be cloned before experimental use. Extensive aneuploidy and heterogeneity of L cell populations have been observed. A unique parental genotype should therefore be established by cloning before hybrid fusion. For transfection experiments, subclones should be established and transfection efficiencies assessed with a selectable plasmid prior to experimental use. Transfection frequencies of one in 300 cells per 50 pg of selectable marker can be achieved with Ltk⁻ clones.[6]

Both HeLa cells and primary cell fibroblasts have been used as a source of chromosomes and genomic DNA for these experiments. HeLa cells should be cloned and maintained identically to the mouse parental cell lines. Primary cell fibroblasts will grow readily in the standard culture media when passed at low dilution, but are difficult to clone and adapt to spinner flask.

The selectable biochemical marker for the formation of whole cell hybrids is the human enzyme HPRT whose gene is present on the distal tip of the human X-chromosome at Xq26–28. Following fusion, hybrids are selected with medium containing HAT (1×10^{-4} M hypoxanthine, 4×10^{-7} M aminopterin, and 1.6×10^{-5} M thymidine) and 0.05 mM ouabain. All surviving hybrids contain the human X-chromosome, or a fragmented X-chromosome which includes Xq26–28. Hybrids can also be selected on the basis of the thymidine kinase deficiency of Ltk⁻ cells. Human × Ltk⁻ hybrids which survive HAT and ouabain treatment will express the human thymidine kinase gene located at 17q21-22.

Genomic and cloned DNA is introduced into cells by cotransfer with a plasmid containing a selectable marker.[13] The selectable plasmid for the cotransfection of Ltk⁻ cells, pTKX-1, contains the herpes simplex virus–thymidine kinase gene (HSV-tk) 3.5 kb *Bam*HI fragment cloned in pBR322.[14] Cells stably transfected with this plasmid will survive treatment with HAT medium. L cell transfectants can also be obtained using the

[13] M. Wigler, S. Silverstein, L.-S. Lee, A. Pellicer, T. Cheng, and R. Axel, *Cell* **16**, 777 (1979).
[14] L. W. Enguist, G. F. Van de Woude, M. Wagner, J. R. Smiley, and W. C. Summers, *Gene* **7**, 335 (1979).

plasmid pSV2-*neo,* which carries the bacterial neomycin resistance gene.[15] L cells transfected with pSV2-*neo* are selected in medium containing 1 – 1.2 mg/ml of the antibiotic G418 (GIBCO, Grand Island, NY).

Hybrid Fusion

Mouse LA9 (HPRT⁻) and human primary fibroblasts (1 × 10⁷ of each) are removed from flasks and washed in DMEM without serum. The cells are mixed together in a 15 ml centrifuge tube and pelleted at 1500 rpm for 10 min. To the cell pellet add 1 ml of warmed 45% polyethylene glycol 1450 (J. T. Baker, Jackson, TN) in DMEM. This addition takes place over 1 min with gentle disruption of the pellet by the pipet tip. The tube is warmed at 37° for 1 min and the pellet is again diluted with 1 ml of DMEM. After 1 min at 37°, slowly add 10 ml of warm DMEM. Centrifuge the cells and resuspend in DMEM with 10% FCS at a cell concentration of 2 × 10⁶ cells/ml. Add 0.1 ml of cells to each well of a 96-well plate. The next day add 0.1 ml of HAT medium plus 0.05 mM ouabain to each well. Feed these wells at 4 day intervals. Aliquots from hybrids should be expanded for freezing as quickly as possible and cloned. Human chromosome segregation is extremely rapid in the first month after hybrid formation. Characterization of human chromosomes present in the hybrids is performed by methods detailed in this volume.[16,17]

DNA Mediated Gene Transfection

DNA Preparation

The isolation of plasmid and genomic DNA for use in transfection protocols has been extensively described in other volumes of this series[18] and in this volume.[19]

Cotransfection with Selectable Plasmid and Cloned DNA

Twenty-four hours prior to gene transfer the cells should be plated at a density of 1.2 × 10⁶ cells per 75 cm² tissue culture flask. For the expression of a cloned gene two to four flasks are used in each transformation. This

---

[15] F. Colbere-Garapin, F. Horodniceanu, P. Kourilsky, and A. Garapin, *J. Mol. Biol.* **150,** 1 (1981).

[16] M. Siciliano and B. F. White, this volume [15].

[17] J. M. Trent and F. H. Thompson, this volume [20].

[18] A. McClelland, M. Kamarck, and F. H. Ruddle, this series, in press (1986).

[19] C. M. Fordis and B. H. Howard, this volume [27].

ensures the generation of a variety of independent transfectants, even if cotransfection frequencies happen to be low. For each flask 100 ng of plasmid pTKX-1 DNA is mixed with 2.5 μg of cloned DNA and 20 μg of mouse L cell carrier DNA. The donor DNA solution is diluted with 250 mM calcium chloride, 25 mM HEPES, pH 7.12 to twice the final DNA concentration desired. The DNA sample should comprise between 1 and 5% of this volume. The DNA is carefully mixed and is added dropwise to an equal volume of 280 mM NaCl, 25 mM HEPES, 1.5 mM Na$_2$HPO$_4$, pH 7.12. Constant mixing is achieved by bubbling a gentle stream of air through the solution using a 1 ml sterile plastic pipet attached to a filtered air source. A DNA–calcium phosphate precipitate forms immediately, but the tube should be left undisturbed for 30 min at 20° to allow additional precipitate to form. The precipitate is dispersed by brief shaking and is added directly to the culture medium in the recipient cell flask and incubated at 37°. Good results are obtained with the addition of 1.5 ml of precipitate to each T75 flask.

The medium is replaced with fresh nonselective medium after 20 hr. At 60 hr after transfection, selective medium is added and is changed every 2–3 days. If immediate assessment of cloned gene expression is desired, it may be made between 60 and 84 hr posttransfer. At this time gene expression has been observed in between several percent and 50% of the transfected cells.[20] Cell cultures remaining under selection will display visible colonies at 8 to 10 days after transfer, and after 2 weeks should be detached and replated in new flasks. This will considerably reduce the problems caused by cellular debris during the cell sorter selection. Frozen cell stocks should be established as soon as possible following cell expansion.

Cotransfection with Selectable Plasmid and Genomic DNA

Twenty-four hours prior to gene transfer the cells should be plated at a density of 2.6×10^6 cells per 150 cm^2 tissue culture flask. For each flask 200 ng of plasmid pTK-X1 DNA is mixed with 80 μg of human genomic DNA and diluted into 1.5 ml of 250 mM calcium chloride, 25 M HEPES at pH 7.12. The DNA is carefully mixed and is added dropwise to an equal volume of 280 mM NaCl, 25 mM HEPES, 1.5 mM Na$_2$HPO$_4$, pH 7.12. Mixing of these solutions and precipitate formation proceed exactly as described for cotransfection with plasmid and cloned DNA. Following precipitate formation, it is hand shaken and 3 ml added directly to the T150 flask at 37°. Selection, transfer, and maintenance of transfected cells are performed exactly as described for the cotransfer of cloned genes. It is

---

[20] J. A. Barbosa, Ph.D. thesis. Yale University, New Haven, Connecticut, 1983.

crucial to immediately establish frozen stock of the transfectants to minimize the impact of faster growing colonies on the population.

The genomic DNA used for transfection consists of contiguous fragments greater than 100 kilobase pairs (kbp) in size. The ratio of 200 ng of plasmid to 80 μg of genomic DNA, a molar ratio of approximately 1 to 40, ensures that the majority of tk$^+$ colonies contain unselected genomic DNA. The amount of genomic DNA transferred into an individual cell may range from 500 to 50,000 kilobase pairs.[6] If each transfected cell clone contains a random fragment of 1000 kbp, then the human haploid genome (3×10^6 kbp) would be covered by 3000 HSV-tk positive recipient cells. A total of 20 to 25 flasks are used in each transfection experiment to ensure the generation of greater than 20,000 independent clones, which cover the genome with some redundancy.

Independent pools of HAT resistant mass populations should be immediately established if the recovery of distinct positive transfectants is desirable. We commonly pool 3000 to 5000 tk$^+$ colonies for frozen stock and cell sorter analysis.

Fluorescence Labeling

Hybrid cells and transfectants generated by the introduction of cloned gene segments contain antigen-positive subpopulations of significant size (greater than 10% of the population). These cells can be split and passaged at relatively high dilution without danger of eliminating an important cell population. In contrast, genomic cell transfectants must be passaged at relatively low dilution.

Hybrid and transfectant cells are expanded to approximately 10^7 cells in T-flasks prior to sorter analysis. The cells are detached from the flask by incubation in PBS without Ca^{2+} and Mg^{2+} containing 0.03% EDTA at 37°. The cells are washed twice in ice cold DMEM containing 2% FCS. Cells ($1-3 \times 10^6$) are incubated with saturating amounts of monoclonal antibody or antiserum in 50 μl DMEM with 2% serum for 1 hr at 4° with regular resuspension. Control staining of 1×10^6 cells is performed with irrelevant hybridoma antibody or normal serum as appropriate. The cells are washed twice and incubated for 1 hr at 4° with a fluorescein conjugated second antibody directed against the first antibody (Cooper Biomedical, Malvern, PA). Excellent specificity for mouse monoclonal antibodies can be obtained with fluoresceinated anti-heavy-chain subclass reagents (Southern Biotechnology Associates, Birmingham, AL). If propidium iodide (PI) is added for the detection of dead cells, it is included at 0.5 μg/ml during this incubation. Finally, the cells are centrifuged through 3 ml of horse serum and resuspended at 2×10^6 cells/ml in 12×75 mm test tubes for the FACS. The cells are stored at 4° prior to analysis and will maintain high viability for up to 6 to 8 hours if necessary.

Microscopic examination before FACS evaluation serves to monitor cell viability and staining intensity, and is critical in the interpretation of FAC results. Cells are examined by microscopy using epifluorescence at magnifications of approximately 500×. Excellent sensitivity and resolution of FITC staining are achieved using oil immersion with a Planapochromat 63/1.3 objective (Carl Zeiss, Inc. Thornwood, NY).

FACS Parameters for Analysis and Sorting

Fluorescence analysis and sterile cell sorting is performed on a FACS IV (Becton-Dickinson FACS Systems, Mountain View, CA). Fluorescein and propidium iodide dyes are excited by an argon-ion laser producing 400 mW at 488 nm on light mode. A Zeiss 580 nm dichroic mirror (46 63 05) is used to separate fluorescence signals to two photomultiplier tubes (PMT). Signals to the fluorescein PMT (QL30, EMI Gencon Inc., Plainview, NY) also pass through a 530/30 dichroic filter (Becton-Dickinson), and signals to the PI PMT (QL20, EMI) through a 625/35 dichroic filter (Becton-Dickinson). Logarithmic conversion of PMT voltage outputs for histogram display are performed with integral FACS electronics. Stream alignment is optimized by the use of standardized fluorescence microspheres of 5.2 μm diameter (Polysciences, Inc., Warrington, PA). In addition to maximizing the fluorescence signal, optimal scatter alignment is critical for distinguishing live and dead cells (see below).

Cell analysis and sorting are performed at rates of 2000 to 2500 cells/sec with an 80 μm nozzle at transducer frequencies of approximately 20 kHz. Use of this larger nozzle will reduce problems caused by debris plugging the nozzle orifice. Following an initial analysis of 40,000 to 100,000 cells, sorting windows are established. Subpopulations of greater than 1%, as found in hybrids and cloned DNA transfectants, are sorted from the center of a peak to minimize contamination from overlapping cell populations. Special considerations in the sorting of subpopulations consisting of less than 0.5% of the population are detailed below. Three consecutive droplets are sorted for each selected cell, one on either side of the droplet predicted to contain the cell of interest. Electronics which abort stream charging if a second cell is contained within the three droplet grouping are routinely used, unless the desired cells make up less than 0.5% of the population.

Each nozzle used for sorting will have a characteristic break-off point, droplet shaping at break off, and droplet delay requirement. These parameters should be closely monitored during an experiment and from day to day as a measure of the integrity of the nozzle orifice. The optimal drop delay should be determined for each nozzle. This is best done by assessing recovery of standardized beads which have been sorted using single drop charging.

Sterile Cell Sorting

Viable cell sorting requires the sterile maintenance of all FACS tubing carrying air or fluid into the open system. Gas pressure is maintained by a standing N_2 tank with a downline 0.22 μm air filter. Sheath fluid is delivered from a 10L pressure vessel filled with sterile PBS lacking Mg^{2+} and Ca^{2+} and delivered through a 0.22 μm filter. Before and after every experiment the system is flushed with a 1:250 dilution of 7X detergent (Flow Laboratories, McLean, VA) delivered from a separate pressure vessel. Before samples are run through the machine, sample delivery tubing is extensively washed with 70% ETOH.

Cells are sorted initially into DMEM, with 10% FCS, penicillin, streptomycin, and garamycin (GIBCO, Grand Island, NY) into 35 × 10-mm petri dishes. Cells rapidly sink in these petri plates and attach to the bottom surface. Microscopic observations of sorted cells should reveal a "landing pad" of several millimeters into which all the cells have settled. This indicates accurate phase alignment during sorting, but will produce a cell density problem if a large number of cells are deposited into each petri dish. For this reason it is useful to disperse the cells following sorting. Multiple petri plates are sorted for each sample and sorted cells can be expanded into T-flasks within a week. Cells are maintained in medium containing antibiotics until aliquots are placed in liquid N_2 storage.

It is critical to extensively wash tubing between samples, as a limited number of cells are lodged in the tubing during an analysis and may reappear in a subsequent sample. This cross-sample contamination is an important consideration when sorting exceptionally rare cells, particularly if positive controls are run at the outset of an experiment.

Live/Dead Cell Discrimination

Cell viabilities of greater than 98% can be standardly achieved with the protocols outlined above. Dead cells in the population, however, will nonspecifically take up the fluoresceinated second antibody and produce a fluorescently positive peak of cells. These dead cells are readily distinguished from specifically stained cells by microscopy, but are indistinguishable on machine analysis. This does not pose a practical problem in sorting hybrids or transfectants generated from cloned material because of

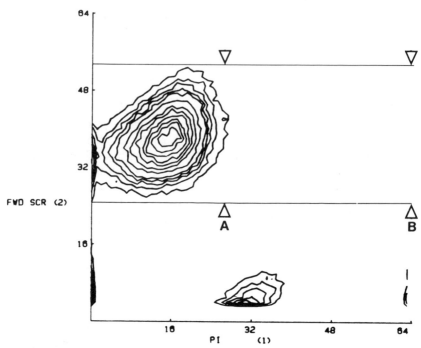

FIG. 1. FACS contour plot distributions of forward scatter and propidium iodide (PI) signals produced by a transfected cell population. Two-dimensional contour plots were generated by the Consort 40 computer using Becton-Dickinson software. FACS optical configuration is described in the text. Dead cells appear in the lower center of the plot with significantly lower scatter signals and higher PI staining. They constitute approximately 5% of the cell population. Living cells are bounded by the horizontal lines and can be further bracketed by vertical PI gates set at arrow A or B (see text).

the large size of the subpopulations. In the isolation of genomic transfectants, where positive cell frequency may range from 10^{-3} to 10^{-4}, dead cells can easily obscure specific staining in the first selection passes. These cells can be excluded from the fluorescence histogram by examining (or "gating" on) only those cells which display a distinctive scatter profile, and exclude 0.5 μg/ml propidium iodide which is taken up by dead cells.[21]

Figure 1 presents a two-dimensional contour analysis of forward scatter and propidium iodide staining of a transfected cell population. Dead cells possess distinguishable scatter and PI fluorescence from the majority of live cells in the population and appear as the peak in the lower right. "Gating"

[21] J. L. Dangl, D. R. Parks, V. T. Oi, and L. A. Herzenberg, *Cytometry* **2,** 395 (1982).

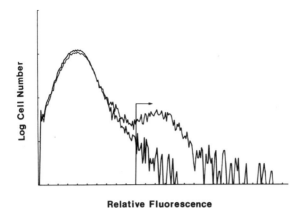

Relative Fluorescence

Fig. 2. Indirect immunofluorescence and FACS analysis of unsorted L-cell transfectants. Cells were stained with a mixture of anti-HLA-A, B, C (W6/32) and anti-4F2 antibodies and fluoresceinated second antibody. Detection of fluorescence was made either on live cell gated or ungated populations. FACS parameters are described in the text and cell numbers are presented on a log scale for easier visualization. The sorting window defined by the arrow includes 0.5% of the gated cells and approximately 5% of the ungated cells. The gated population is the population of cells which does not show the secondary peak of fluorescence.

can be accomplished by analyzing only those cells within the region between the horizontal lines and left of the arrows labeled "A." But, it is clear from Fig. 1 that live/dead discrimination of L cells can be based on scatter alone (area between the horizontal lines bounded by the arrows at "B"). By eliminating propidium iodide staining from the analysis a fluorescence photomultiplier tube is available for quantitation of additional antigens. A comparison of gated and ungated transfected cells analyzed by indirect immunofluorescence is shown in Fig. 2. It demonstrates that the effect of removing dead cells from the analysis is to decrease the background in the region of the curve where signals from specifically stained cells are expected.

Recovery of Rare Cells in Sorting

Even with efficient live cell gating, antigen-positive cells which occur at frequencies of 0.01–0.1% in the population do not appear as a distinct peak on the first analysis (Fig. 2). It is therefore necessary to sort the brightest 0.5% of the cells to assure recovery of antigen-positive cells, and to obtain sufficient cells so that cell expansion and reanalysis can be performed rapidly. Cells to the right of the arrow in the gated population of Fig. 2 represent the 0.5% of the population sorted in this experiment.

Figure 3a presents the analysis of this population of transfectants following the initial enrichment sort. In this example, the original genomic population was stained and sorted for the expression of both human HLA and 4F2 antigens. It can be seen that two groups of antigen-positive cells are visible after the first sort, and that each is present in a different fraction of the population (Fig 3a). These cells were sorted a second time, again using a mixture of both antibodies. Following this second sort almost all the cells are positive for transfected antigens (Fig. 3b), a larger fraction expressing 4F2 rather than HLA (compare Fig. 3c and d). The data presented in Fig. 3 demonstrate that antigen expressing transfectants can be recovered with an initial frequency of 0.05% or less. In addition it illustrates the simultaneous selection of different "primary" antigen transfectants by using mixtures of antibodies. This considerably improves the overall efficiency of the procedure since the tissue culture expansion of sorted cells is particularly time consuming.

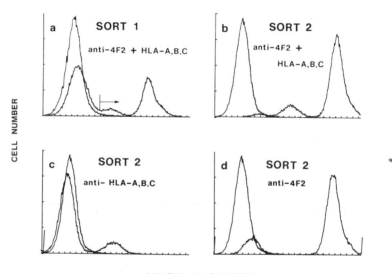

FIG. 3. Indirect immunofluorescence and FACS analysis of L cell transfectants sorted for HLA-A, B, C and 4F2 expression. SORT 1: (a) Cells were expanded into T-75 cm² flasks and reanalyzed with gating using a mixture of anti-4F2 and anti-HLA-A, B, C antibodies. The control curve (large first peak in each graph) was labeled without the first antibody. This population was resorted using the window designated by the arrow. SORT 2: (b) Indirect immunofluorescence with anti-HLA and anti-4F2 compared to control. (c) Indirect immunofluorescence of cell population shown in (b) with anti-HLA alone compared to control. (d) Indirect immunofluorescence of cell population shown in (b) with anti-4F2 alone compared to control.

Once sufficient homogeneity has been achieved, transfectants stained with a single antibody can be subcloned by limiting dilution into 96-well microtiter dishes. Alternatively, transfectants can be subcloned using the FACS single cell deposition system (Becton-Dickinson, Mountain View, CA). A single positive cell contained in three drops is sorted through a narrow stainless-steel collar before being deposited in a well of a 96-well microtiter tray containing standard tissue culture medium. After each cell is sorted, the microtiter plate is moved, and the sorting logic retriggered for the next positive cell. Problems can occur from the build up of stray droplets within the collar, which catch subsequent droplets and prevent them from being deposited correctly into the wells. It is easy to remove such drops by passing a Q-tip soaked in ethanol through the collar before initiating sorting. To confirm the accurate set-up of the single cell cloner we standardly sort 50 cells through the collar onto a designated spot on the 96-well tray holder. This amount of material is visible as a minute drop when correct sorting parameters have been established. Cells of designated antigen phenotype can be cloned with high efficiency using this method.

Sorting L Cells for the Expression of the HLA Surface Antigens

The major histocompatibility antigens (MHC), designated HLA-A, B, C in man, provide the major barrier to tissue transplantation and are responsible for the MHC restriction of T cell-mediated cytotoxicity. The HLA-A, B, C antigens are highly polymorphic, integral membrane glycoproteins of 44,000 Da, expressed on the surface of most tissues in noncovalent association with the 12,000-Da, water-soluble polypeptide β_2-microglobulin (β_2m). The proper expression of HLA heavy chains in the membrane is dependent on this association with β_2m and can be achieved by association with heterologous mouse β_2m in somatic cell hybrids.[1] We have studied the expression of the HLA antigens in somatic cell hybrids, transfectants with cloned HLA genes, and genomic transfectants from genomic DNA. This work illustrates the application of the methods detailed in this paper.

Hybrid Cells

A human × mouse hybrid cell line was examined for the expression of HLA-A, B, C antigens (coded by chromosome 6), and human β_2-microglobulin (coded by chromosome 15). The FACS analysis presented in Fig. 4a demonstrates that hybrid cells expressing HLA can be specifically identified by indirect immunofluorescence. Expressing and nonexpressing cells were sterilely sorted based on HLA expression, and were allowed to grow

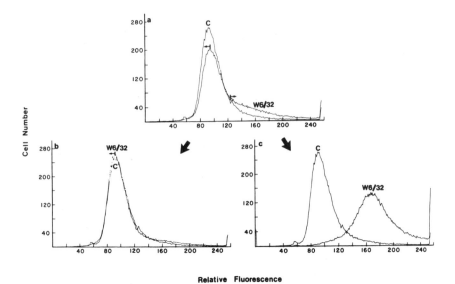

Relative Fluorescence

FIG. 4. Analysis and cell sorting of human × mouse hybrid cell line FRY 4 for the expression of HLA-A, B, C antigens. Cells were stained by indirect immunofluorescence with an anti-HLA-A, B, C antibody (W6/32) and analyzed by the FACS. The parameters for cell sorting are indicated by the arrows in the first panel. (a) FRY 4, (b) FRY 4.HLA⁻, (c) FRY 4.HLA⁺.

up in tissue culture before reanalysis. Figure 4b and c demonstrates the production of somatic cell hybrid lines that were homogeneously positive or negative for the HLA surface antigen.[1] Karyotype analysis revealed that human chromosome 6 was not present in HLA-A, B, C-negative cells, but could be detected in greater than 85% of those hybrids which expressed the antigen. The fact that no antigen-negative cells contained the antigen-coding chromosome demonstrates that HLA-A, B, C are constitutively expressed on somatic cell hybrids.

The expression of human cell surface antigens on human × mouse hybrids allows the rapid characterization and manipulation of hybrid genotype at the cell surface using chromosome-specific antibodies and flow sorting. A large number of surface antigen antibodies are available for this purpose.[22] This hybrid cell selection protocol makes it possible to monitor continual human chromosome loss, and to produce genetically homogeneous human × mouse hybrid lines. This procedure has also been used to

[22] C. Partridge, U. Francke, K. Kidd, and F. H. Ruddle, Human Gene Library, Yale University. Supported by National Institutes of Health, grant GENE-GM29111.

obtain chromosome map positions for surface antigen genes. Using the FACS, an antigenically heterogeneous hybrid cell population can be fractionated into two subpopulations, each of which is either homogeneously antigen positive or negative. The genomes of these sublines should differ from each other by the single human chromosome that codes the surface antigen gene. This strategy has been used to obtain a number of surface antigen gene map positions.[2,23]

Transfection with Cloned Genes

Southern blot analysis of human genomic DNA from the JY cell, hybridized with an HLA-B7 cDNA probe, revealed 15 to 20 unique gene segments which could be mapped to the histocompatibility locus.[4] To identify which members of this multigene family were functioning genes, Ltk⁻ cells were cotransfected with genomic clones and pTK-X1 as described, and HAT-resistant mass populations were characterized by indirect immunofluorescence and FACS analysis. Of the 23 clones tested, only two (JYB3.2 and JY158) produced HLA positive transfectants. Results obtained with genomic clone JY158 are presented in Fig. 5. As determined by surface expression, greater than 95% of the thymidine kinase-positive cells have also been transfected with the HLA surface antigen gene. Similar positive results were obtained with clone JYB3.2. This contrasts with the total absence of expression observed with all other genomic clones, including a known pseudogene. Thus, using this approach the presence of functioning gene clones can be readily assessed. These HLA-positive transfectants were also tested with allospecific antibodies to identify the product of JYB3.2 as HLA-A2 and JY158 as HLA-B7.[4]

Transfection with Genomic DNA

DNA-mediated gene transfer of genomic DNA and selection in mammalian cells provides an alternative to conventional cloning of messenger RNA.[24] Figure 3c demonstrates the recovery of "primary" transfectants expressing the HLA surface antigens. When examined by Southern blot hybridization, these primary transfectants can be shown to contain the human HLA genes.[6] In addition to the surface antigen gene, "primary" transfectants contain several thousand kbp of irrelevant donor DNA which can be quantitated based on species-specific repetitive sequences. The large

[23] J. Schroder, B. Nikinmaa, P. Kavathas, and L. A. Herzenberg, *Proc. Natl. Acad. Sci. U.S.A.* **80,** 3421 (1983).

[24] F. H. Ruddle, M. E. Kamarck, A. McClelland, and L. C. Kuhn, *in* "Genetic Engineering" (J. K. Setlow and A. Hollaender, eds.), Vol. 6, p. 319. Plenum Press, New York, 1984.

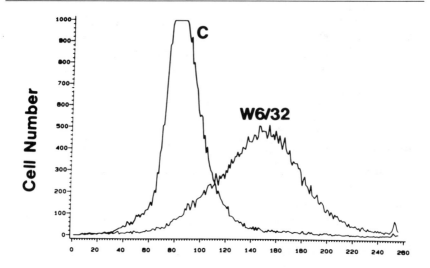

Relative Fluorescence

Fig. 5. Indirect immunofluorescence and FACS analysis of HLA expression of a HAT resistant population derived from a cotransfection of Ltk⁻ cells with human genomic gene JY158 and HSV-tk. The panel presents specific staining of the transfected cell population with an anti-HLA-A, B, C antibody (W6/32) compared to control.

amount of DNA incorporated into the majority of primary transfectants necessitates the use of this material as donor DNA for a second round of gene transfer and cell sorting. This is performed identically to the first round. The secondary transfectants may contain from between 30 and 100 kbp of donor DNA, and serve as starting material for the cloning of the surface antigen gene in phage or cosmid vectors. These procedures have been detailed in several chapters in this volume and in another volume of this series.[18]

A number of surface antigen genes have been cloned by this strategy and include the human transferrin receptor,[8] nerve growth factor receptor,[25] T cell antigen T8,[7] and the human myeloid antigen gp150.[26] Other workers have cloned the T4 and T8 T cell antigens using this transfection approach, but performing initial selections with an antibody based rosetting assay rather than the FACS.[9,27] It is noteworthy that of the genes

[25] M. V. Chao, M. A. Brothwell, A. H. Ross, H. Koprowski, A. A. Lanahan, C. R. Buck, and A. Sehgal, *Science* **232**, 518 (1986).
[26] A. T. Look, S. C. Peiper, M. B. Rebeutisch, R. A. Ashman, M. T. Rousell, C. W. Rettenmier, and C. J. Sherr, *J. Clin. Invest.* **75**, 569 (1985).
[27] P. J. Maddon, D. R. Littman, M. Godfrey, D. E. Maddon, L. Chess, and R. Axel, *Cell* **42**, 93 (1985).

cloned by this method, only the transferrin receptor is normally expressed on fibroblasts. Using the transfection/selection procedures outlined here it should be possible to clone a variety of receptor genes by using fluoresceinated ligands as well as species-specific antibodies.

Summary

Mouse cells containing human surface antigen genes introduced by cell hybridization or DNA transfection can be labeled by indirect immunofluorescence and isolated using the FACS. As illustrated in this chapter, this methodology facilitates genetic studies of the human cell surface ranging from the initial chromosome mapping of a surface antigen gene to its isolation and cloning.

Acknowledgments

The author is indebted to John T. Hart and Donna Babbitt for expert technical assistance and comments on this manuscript.

Section III

Genetic Mapping and Analysis

[15] Isozyme Identification of Chromosomes in Interspecific Somatic Cell Hybrids

By MICHAEL J. SICILIANO and BILLIE F. WHITE

Nucleotide sequence differences between homologous enzyme loci of different species result in electrophoretically informative amino acid substitutions in the encoded enzymes. Therefore, the products of enzyme loci from different species will generally have different electrophoretic mobilities following electrophoresis in a suitably buffered gel matrix. These electrophoretically different forms of the same enzyme (isozymes) may be visualized directly in the gel matrix by histochemical staining following electrophoresis (zymogram technique).[1]

In interspecific somatic cell hybrids, with chromosomes from both species encoding the same enzyme, one observes two bands representing the isozymic form of each species. Where the enzyme under study is composed of more than one polypeptide subunit, additional numbers of intermediately migrating heteropolymeric bands are seen since subunits of the same enzyme produced in the same cell combine randomly into multimeric molecules.[2,3] For dimeric enzymes (composed of two subunits), in addition to the homodimer representative of the isozyme of species A (AA) and the homodimer of species B (BB), one additional intermediately migrating heterodimeric band is produced composed of one subunit for each species (AB). For trimeric enzymes a total of four isozymes are seen (AAA, AAB, ABB, and BBB) and for tetrameric enzymes five-banded patterns are the rule (AAAA, AAAB, AABB, ABBB, BBBB). Since, under the conditions of sample preparation and electrophoresis, the intermediately migrating hybrid molecules can only be produced in interspecific somatic cell hybrids, their presence represents evidence that the cells being studied are true hybrids (Fig. 1).

In interspecific somatic cell hybrids the chromosomes of one of the two species are slowly, and more or less randomly, lost.[4] It is therefore possible to assign isozyme loci to that species' chromosomes, or syntenic groups, by correlating the presence of the segregating species' isozymes with the retention of specific chromosomes and/or by correlating the absence of the isozymes with the loss of the chromosomes. Such concordant segregation

[1] R. L. Hunter and C. L. Markert, *Science* **125**, 1294 (1957).
[2] C. L. Markert, *Science* **140**, 1329 (1963).
[3] C. R. Shaw, *Brookhaven Symp. Biol.* **17**, 117 (1964).
[4] M. C. Weiss and H. Green, *Proc. Natl. Acad. Sci. U.S.A.* **58**, 1104 (1967).

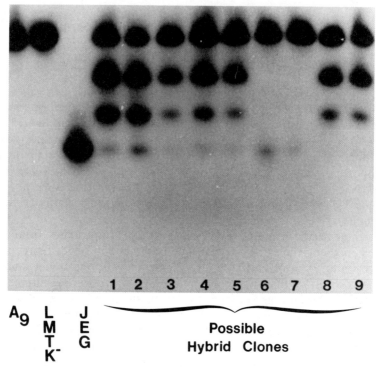

FIG. 1. Zymogram showing the forms of nucleoside phosphorylase (NP) produced in two mouse cell lines (A9 and LMTK⁻), the human choriocarcinoma cell line (JEG), and nine putative hybrid clones produced by fusion between LMTK⁻ and JEG. The anodal end of the gel is toward the top. Since NP is a trimeric enzyme producing a four-banded pattern in heterozygous cells, true hybrids should express the typical mouse band, human band, and two intermediately migrating heteropolymeric hybrid bands. The pattern is skewed in intensity toward the mouse form, since human chromosomes are segregating in this fusion scheme. Clones 1, 2, 3, 4, 5, 8, and 9 are clearly true hybrids because in order for the heteropolymeric hybrid bands to be produced, both mouse and human NP subunits need to be produced in the same cell. Clones 6 and 7, although producing both mouse and human homopolymers, are likely merely mixed cultures and not true hybrids, since the intermediately migrating hybrid bands are not seen.

analysis (see review by Ruddle[5]) established isozyme markers on virtually every chromosome of all species whose cells could be interspecifically hybridized and whose chromosomes would segregate in an interspecific cell fusion scheme. Isozyme markers present on the chromosomes of different species are complied in such volumes as *Genetic Maps.*[6]

[5] F. H. Ruddle, *Adv. Hum. Genet.* **3,** 173 (1972).
[6] S. J. O'Brien, "Genetic Maps 1984." Cold Spring Harbor Lab., Cold Spring Harbor, New York, 1984.

Therefore isozymes can be used as markers to identify which chromosomes of the segregating set are present in interspecific somatic cell hybrids. This has certain advantages over the other two major methods of making such determinations—direct chromosome analysis, and hybridization with chromosomally assigned gene probes after DNA from hybrid cells is cut with restriction enzymes, electrophoresed, and blotted onto nitrocellulose filters.[7] Isozyme procedures are much less labor intensive so that large numbers of cell hybrids (approximately 30) can be very rapidly tested (one week by a single person) for all the chromosomes. Furthermore, chromosomal segments generated by rearrangements which often take place in hybrid cells and which are not readily identifiable by direct karyological analysis can be detected in hybrid cells by the judicious selection of isozyme markers.

Here we shall present the specific procedures that have proven successful in our laboratory for using isozyme markers to detect chromosomes in somatic cells. For a more general discussion of isozyme principles and techniques, the reader is referred to Harris and Hopkinson[8] and/or Siciliano and Shaw.[9]

Separation of Isozymes

Unless otherwise indicated all procedures are carried out between 0 and 4°. With one exception, all isozymes are resolved by using the technique of starch gel electrophoresis. The key to success in using this technique is the ability to slice a 1-cm-thick slab of starch gel, through which samples (30 at a time) have been electrophoresed, into as many as 10 1-mm slab slices. Each slice can be stained histochemically for a different enzyme. Therefore, the entire battery of enzymes can be assayed from only 0.1 ml of packed hybrid cells.

Sample Preparation

Approximately 0.1 ml of washed (in isotonic saline), packed, cultured cells is diluted in a homogenizing medium (1 : 4, cells : medium) and rigorously ground either manually in an all glass, or with a motorized Teflon and glass, homogenizer. Homogenizing medium contains 0.01 M Tris–HCl, pH 7.5, 0.001 M 2-mercaptoethanol, and 0.001 M EDTA. Homogenates are then cleared by centrifugation (two 30-min spins at 12,000 g are

[7] E. M. Southern, *J. Mol. Biol.* **98**, 503 (1975).

[8] H. Harris and D. A. Hopkinson, "Handbook of Enzyme Electrophoresis in Human Genetics." North-Holland, Amsterdam, 1976.

[9] M. J. Siciliano and C. R. Shaw, *in* "Chromatographic and Electrophoretic Techniques" (I. Smith, ed.), p. 185. Heinemann, London, 1976.

usually sufficient). Cleared supernatants may then be applied directly to gels.

Electrophoresis

Starch gel electrophoresis (first developed by Smithies to gently separate different forms of hemoglobin[10]) is used almost exclusively because of the adequate separation and resolution of isozymic forms and the ability to slice the gels into many slabs. One tries to use the buffer system which will give the best resolution of a particular isozyme system while at the same time trying to limit the number of different buffer systems so that many isozymes can be viewed as the result of a single gel run. The one exception to our exclusive use of starch gels is to resolve the isozymes of β-glucuronidase (GUSB) which we can resolve well only on Cellogels. The following buffer systems are used to resolve the enzymes that will be described.

Buffer Systems. In each case the electrode buffer is used directly in the buffer tanks at the anodal and cathodal ends of the gel. The gel buffer is mixed with the starch. Adjust pHs with either 40% NaOH or concentrated HCl. The following different buffer systems are used.

    a. Electrode: 0.13 M Tris, 0.043 M citrate, pH 7.0.
       Gel: Dilute 40 ml of above up to 600 ml with water.
    b. Same as above, but 40 mg NAD is added to the molten gel prior to degassing.
    c. Same as above except 40 mg NADP instead of NAD is added.
    d. Electrode: 0.1 M Tris, 0.1 M maleic acid, 0.01 M EDTA, 0.01 M $MgCl_2$, pH 7.4.
       Gel: Dilute above 1 : 14 with water.
    e. Electrode: 0.5 M Tris, 0.016 M EDTA, 0.65 M borate, pH 8.0.
       Gel: Dilute 60 ml of above up to 600 ml with water.
    f. Same as above, but 40 mg NADP is added to the molten gel prior to degassing.
    g. Same as above except 40 mg ATP instead of NADP is added.
    h. Electrode: 0.1 M Tris, 0.05 M borate, pH 8.6 and add 30 mg NADH to each buffer box.
       Gel: Dilute 120 ml of above up to 600 ml with water and add 30 mg NADH.
    i. Electrode: 0.1 M NaH_2PO_4, pH 6.5.
       Gel: Dilute 60 ml of above up to 600 ml with water.
    j. Cellogel buffer: 0.04 M Tris brought to pH 6.0 with citric acid.

Gel Preparation. Starch gels are prepared using 85 g of Connaught starch and mixing it with 600 ml of gel buffer in a 2000-ml-thick walled

[10] O. Smithies, *Biochem. J.* **61**, 629 (1955).

flask. (The ideal amount of starch may vary by as much as 10% with lot number. The point is to produce a tough "rubbery" gel which can be sliced and handled — not too mushy and not too grainy.) This mixture is heated over a gas flame with constant swirling. First the mixture thickens and then begins to thin again as it comes to a boil. At this point remove it from the flame since overcooking produces fragile gels. If a cofactor (NAD, NADH, NADP, or ATP) is to supplement the gel, it is added here. The molten mixture is then degassed with a vacuum pump (to a pressure of about 25 mm Hg), 2 drops of 2-mercaptoethanol are added, and gently mixed in. The mixture is then slowly and carefully poured into the form avoiding the production of any bubbles (Fig. 2a). A variety of horizontal and vertical starch-gel forms are available (Buchler Instruments, Fort Lee, NJ). The top of the form, containing the sample slot maker (a second slot maker can be spaced 9 cm apart from the first to run a double set of samples on the same electrophoretic run), is then pressed on, squeezing out the excess molten gel material (Fig. 2b). After cooling, 30 min at room temperature followed by 30 min at 4°, the ends and top (with the slot makers) of the form can be removed leaving the gel set in the remainder of the form.

Cellogels come prepared (Kalex Scientific Co., Manhasset, NY) and need only be equilibrated for 15 min in 200 ml of buffer ("j" above).

Electrophoretic Run. Slots in the cooled starch gel may be better separated to prevent cross contamination of samples by laying a thin line of melted petroleum jelly between them. Cleared supernatant samples are best filled into the slots with a Pasteur pipet that has been drawn out into a fine tip (Fig. 2c). After all the slots in the gel have been filled they can be sealed off by covering them with melted petroleum jelly. The entire surface of the gel is then covered with a clear plastic wrapping material (Saran Wrap, Dow Chemical Company, Midland, Michigan, 48640). The gel is then placed in the verticle apparatus located in a cold room or sliding glass door, "beer cooler" type refrigerator. Filter paper wicks from the tanks containing the electrode buffer are applied to both ends (anodal and cathodal) of the gel. In verticle systems, which we use exclusively, one end of the gel actually gets immersed in the anodal buffer tank (Fig. 3). Electrodes are wired to a constant voltage power supply (e.g., Heathkit IPW-17m Heath Company, Benton Harbor, Michigan, 49022) and a current of 200 V is applied. This can be run overnight (16–18 hr). Shorter runs (5 hr at 400 V) also give adequate resolution.

After electrophoresis, the gel is removed from the form, trimmed, and laid flat on a well-washed and rinsed, smooth-surfaced table. Using piano wire (0.008 in diameter) strung tightly to a Büchler gel slicer, slices as thin as 1 mm may be removed from the undersurface of the gel (Fig. 3). As many as 10 slices may therefore be gotten from a gel 1 cm thick. Each slice

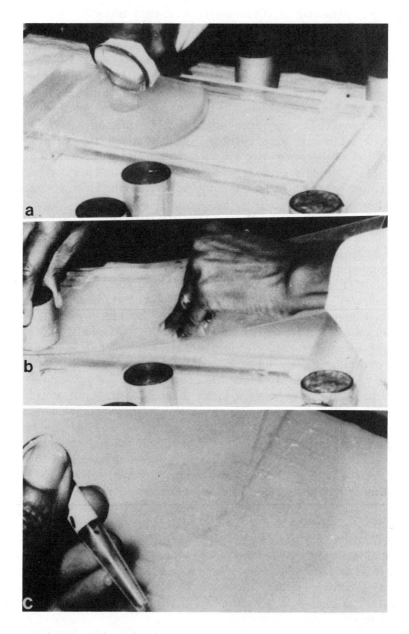

FIG. 2. (a) Molten, degassed starch–buffer mixture is shown being poured into a gel form. Cylindrical objects around the edge are lead weights which will be placed on the form top after the gel is poured. (b) Gel form top containing the sample slot maker (held and hidden by the worker's left hand) is shown being pressed onto the freshly poured molten starch buffer mixture. The top is applied from one end and pressed on with the aid of lead weights toward the other end to squeeze out all excess molten gel material. (c) After the top had been removed from the cooled gel, the sample slots, separated by strips of petroleum jelly, are shown being filled with a drawn out Pasteur pipet. From Siciliano and Shaw.[9]

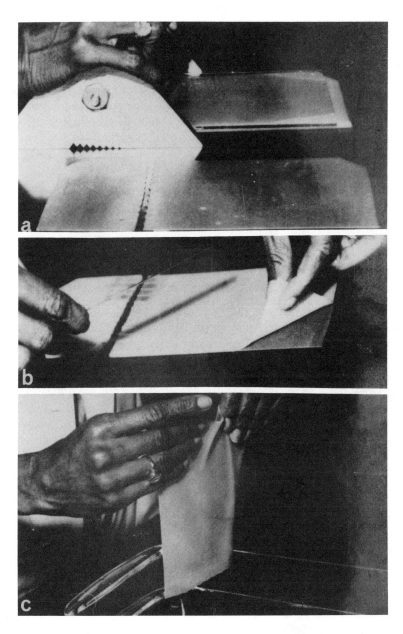

FIG. 3. (a) Gel laid flat on a smooth table surface and covered by a glass plate to add weight to help keep it pressed to the table top is shown being sliced by a gel slicer to cut slices of 1 mm thickness off the undersurface. In the foreground is such a slice. (b) After the slicer has been moved through the gel, the upper piece is shown being removed to be laid down on the table to be sliced again. (c) A 1-mm-thick gel slice is shown being lifted into a Pyrex dish for histochemical staining. From Siciliano and Shaw.[9]

may then be put into a shallow staining dish and stained for a different enzyme.

When Cellogel membrane is used it is removed from its buffer presoak, excess buffer is blotted off (Whatman, 3MM chromatographic paper), samples are applied with an applicator, and it is run on a Kalex electrophoresis chamber at 200 V, 6 mA for 1 hr. After electrophoresis the gel is removed and stained as directed below for GUSB.

Visualization of Isozymes

Isozymes are histochemically visualized by applying the appropriate reagents to the cut surface of a gel slab slice. The biochemical rationale and original references for the visualization procedures are generally contained in Harris and Hopkinson.[8] For marker methods identified after that volume was published, we will site the original references for rationale.

General Methods: Colorimetric Reactions, Agar Overlay, Autoradiography

The histochemical procedures usually result in either colored, blanched, fluorescent, or defluorescent bands of enzyme activity in and/or on the gel. In most cases it is sufficient merely to apply the reagents to the cut surface of the gel and incubate (37° until the colorimetric reaction has taken place (15 min to 2 hr) and then view the gel either under white or ultraviolet light. In several cases it is necessary to limit the diffusion of the reagents while the reaction is taking place. This is accomplished by fixing the reagents into an agar overlay. To do this 1.5 g of agar is added to 100 ml of water and the mixture is boiled for 1 hr or microwaved for 2–3 min. The agar is then placed in a 50–60° waterbath and allowed to cool down to that temperature before use. When ready, the appropriate amount of agar solution is poured into the remainder of the histochemical stain mix, gently swirled in with the rest of the ingredients, and evenly poured onto the cut surface of the gel in a staining dish (much care is taken not to produce bubbles). The overlay is allowed to set for 1 or 2 min before placing the gel in the incubator.

For a few reactions autoradiography is used. In these reactions, a ^{14}C-labeled substrate is phosphorylated. The reaction products are picked up on DEAE-81 paper, unreacted substrate washed off, and autoradiography is conducted to visualize the site of the reaction. The specific steps of this procedure are as follows:

   a. Discard cathodal piece of the gel slice (the enzymes that this procedure visualizes all move anodally) and trim off the anodal end beyond which the enzyme migrates.
   b. Cut the DEAE-81 paper to the trimmed gel slice size and apply to the cut surface of the gel.

c. Place gel and paper on an oversized piece of plastic wrap.
d. Evenly distribute the total contents (5 ml) of the reaction mixture onto the paper.
e. Fold up the plastic wrap so that the gel and soaked paper are snuggly wrapped.
f. Incubate the package for 2 hr at 37°.
g. Remove paper and allow to dry overnight.
h. Wash paper in 200 ml of water for 15 min with sporadic agitation, and repeat for 5 more washes.
i. Allow paper to dry and then spray with 14% PPO in ether until saturated.
j. When dry, apply paper to X-ray film in cassette and let incubate the appropriate number of days at −60°.
k. Develop film.

The Enzymes

The recipes for the histochemical stains used to visualize the isozymes which serve as markers for human chromosomes are given below. Since most have also been chromosomally assigned in other species, these markers are the ones that are generally useful in interspecific somatic cell hybrid experiments. For each isozyme system we shall identify the letter(s) of the preferred buffer system(s) (listed above) for optimal resolution of the species forms in interspecific hybrids and list the ingredients of the histochemical stain to be added to the cut surface of the gel following electrophoresis. We shall also identify the genetic loci that code for the different forms, indicate the chromosomes they mark in humans, and give some indication of their relative migration in the buffer systems used to resolve them when compared to species whose cells are often fused with human in such experiments—mouse (as produced in L cells) and Chinese hamster *(Cricetulus griseus)*. Abbreviations for isozyme genetic loci will follow the recommendations of the Committee for the International System for Human Gene Nomenclature.[11] It should be noted that there are other enzyme markers known than are listed here, and other methods exist for the enzymes listed than are described here. However, we have included only those markers and methods which have given the most consistent, reliable, and reproducible results in our laboratory.

1. Acid phosphatase (ACP, EC 3.1.3.2). Best resolved on buffered systems a, b, or c. Stain ingredients:
10 mg, 4-methylumbelliferyl phosphate
5 ml, 0.05 M citrate/0.05 M phosphate

---

[11] T. B. Shows and 24 others, *Cytogenet. Cell Genet.* **25**, 96 (1979).

Incubate for 30 min and view under long wave UV light. Sprinkle surface with NH_4OH to intensify fluorescent bands of ACP activity. The product of human *ACP1* (chromosome 2) will be just anodal to the origin, well resolved from the more anodal mouse form, and even more well resolved from the Chinese hamster form which is at least the same distance again more anodal than the mouse form. Bands on the cathodal portion of the gel are likely the products of mouse or hamster *ACP2* which are not informative.

2. Adenine phosphoribosyltransferase (APRT, EC 2.4.2.7). Best resolved on buffer systems a, b, or c. Stain ingredients:

   5 mg, phosphoribosylpyrophosphate
   0.4 ml, 100 mg $MgCl_2$/ml water
   0.5 ml, 0.5 M Tris–HCl, pH 7.1
   5 μCi, [^{14}C]adenine
   4 ml, water

Distribute the above solution over DEAE-81 paper applied to the gel surface and continue as described in autoradiography procedures. Develop film after 2 days in the freezer. The film will be exposed in the zones of APRT activity. The product of human *APRT* (chromosome 16) is slightly anodal to the origin and well resolved from the very anodal mouse bands. In true mouse × human somatic cell hybrids, a clear heterodimer band of activity will be observed. The enzyme, assayed by these methods, is uninformative in human × Chinese hamster fusion schemes because hamster and human forms comigrate.

3. Adenosine deaminase (ADA, EC 3.5.4.4). Best resolved on buffer systems e, f, or g. Stain ingredients:[12]

   40 mg, adenosine
   50 mg, sodium arsenate
   2 ml, 10 mg NBT/ml water
   4 ml, 0.2 M Tris–HCl, pH 8.0
   1 ml, 1 mg PMS/ml water
   7 ml, water
   1.6 units, xanthine oxidase
   5 units, nucleoside phosphorylase
   15 ml, molten 1.5% agar

Zones of ADA activity appear as dark blue bands. The product of the most common form (type *1*) of human *ADA* (chromosome 20) migrates ano-

---

[12] Abbreviations for reagents commonly used in tetrazolium histochemical staining reactions are: PMS, phenazine methosulfate; MTT, 3-(4,5-dimethylthiazol-2-yl)-2,5-diphenyltetrazolium bromide; NBT, nitro blue tetrazolium.

dally and is well resolved from the more anodal most common Chinese hamster form which comigrates with the mouse form. The product of the human *ADA*2* allele is also well resolved from the rodent forms since it is less anodal in migration than human ADA*1. The product of a less common Chinese hamster allele is intermediate in migration between the more common form (and the mouse form) and human ADA*1. This is noteworthy since Chinese hamster ovary cells (CHO), which are very common fusion partners in somatic cell hybridization schemes, are homozygous for this less common form.[13]

4. Adenosine kinase (ADK, EC 2.7.1.20). Best resolved on buffer systems a, b, or c. Combine

0.25 ml, 0.2 *M* Tris–HCl, pH 8.0
35 mg, ATP
0.1 ml, 100 mg MgCl$_2$/ml water
4.5 ml, water
5 μCi, [^{14}C]adenosine

and, with a Pasteur pipet, evenly distribute the reaction solution over DEAE-81 paper which has been applied to the cut surface of the gel. Continue with autoradiographic procedures as described above. Expose film for 6 days. Mouse ADK is most anodal with Chinese hamster cathodal to it. The product of human *ADK* (chromosome 10 marker) is cathodal to the Chinese hamster product.

5. Adenylate kinase (AK, EC 2.7.4.3). Use buffer system g. Combine and pour over cut surface of gel:

20 mg, ADP
15 mg, NADP
45 mg, β-D(+)-glucose
2 ml, 5 mg MTT/ml water
1 ml, 1 mg PMS/ml water
0.5 ml, 100 mg MgCl$_2$/ml water
5 ml, 0.2 *M* Tris–HCl, pH 8.0
170 units, hexokinase
80 units, glucose-6-phosphate dehydrogenase
15 ml, molten 1.5% agar

Look for blue bands on a white background. The product of human *AK1* (chromosome 9) is the slowest anodally migrating isozyme seen on the gel. The product of human *AK2* (chromosome 1) is slightly more anodal to human AK1. Both forms are well resolved from the usually more anodally

---

[13] R. L. Stallings and M. J. Siciliano, *Somatic Cell Genet.* 7, 295 (1981).

migrating products of Chinese hamster (as seen in CHO cells) and mouse *AK2* loci. The exception is a slow form polymorphism of Chinese hamster AK2 which is present in certain cell lines of that species and which overlaps human AK2.[13] Also well resolved are the much faster rodent forms of AK1 (the mouse form migrates faster than the Chinese hamster form).

6. Aminoacylase 1 (ACY1, EC 3.5.1.14). Best resolved on buffer systems e, f, or g. Stain ingredients:
    50 mg, N-acetyl-L-methionine
    10 mg, peroxidase
    0.5 ml, 0.1 M MnCl$_2$
    0.5 ml, 25 mg o-dianisidine/ml water
    0.3 ml, 15 mg snake venom/ml water
    15.0 ml, 0.2 M Tris–HCl, ph 8.0
    20 ml, molten 1.5% agar
Incubate until brown bands of activity appear (1 hr). The product of human *ACY1* (chromosome 3) is most anodal with mouse bands intermediate in mobility between human and hamster. Presence of the heterodimer indicates the gene of the segregating set of chromosomes.

7. Creatine kinase (CK, EC 2.7.3.2). Best on buffer system g. Stain ingredients:
    15 mg, phosphocreatine
    30 mg, ADP
    35 mg, β-D(+)-glucose
    5 mg, NADP
    2 ml, 0.5 M Tris–HCl, pH 7.0
    8 ml, water
    2 ml, 5 mg MTT/ml water
    1 ml, 1 mg PMS/ml water
    0.5 ml, 100 mg MgCl$_2$/ml water
    70 units, hexokinase
    40 units, glucose-6-phosphate dehydrogenase
    15 ml, 1.5% molten sugar
Incubate until blue bands appear. The product of human *CKBB* (chromosome 14 marker) is informative only when it is not produced by the cells of the fusion partner whose chromosomes are not segregating since electrophoretic mobilites are generally similar in human, mouse, and Chinese hamster. CKBB bands are the most anodal bands that will appear. AK bands will also appear with this staining system so that another slice off the same gel needs to be stained for AK as a control.

8. Diaphorase[14] (DIA, EC 1.8.1.4, dihydrolipamide dehydrogenase). Use buffer system h. Stain ingredients:

20 mg, NADH
2 mg, FAD
50 ml, 0.025 M Tris–HCl, pH 8.5
3 ml, 5 mg MTT/ml water
2 ml, diaphorase solution (10 mg of 2,6-dichlorophenol-indophenol in 10 ml 0.025 M Tris–HCl, pH 8.5)

Dark blue bands on the cathodal portion of the gel indicate the product of human *DIA4* which marks chromosome 16. Homogenates from mouse and hamster cell lines do not produce an equivalent activity. Occasionally bands appear on the anodal portion of the gel which may be the products of *DIA1* (chromosome 22). Since we have not found these bands to be consistently expressed in cultured cells, we do not recommend DIA1 as a reproducable marker for chromosome 22.

9. Enolase (ENO, EC 4.2.1.11). Use buffer system e. Stain ingredients:

25 mg, D(+)-2-phosphoglyceric acid
50 mg, ADP
30 mg, NADH
1 ml, 0.2 M Tris–HCl, pH 8.0
4 ml, water
0.5 ml, 100 mg $MgCl_2$/ml water
20 units, pyruvate kinase
40 units, lactate dehydrogenase

View yellow fluorescent stain over the gel under UV light. The product of human *ENO1* (chromosome 1) appears as a dark, nonfluorescing band. The mouse homodimer migrates just anodal to the origin while the Chinese hamster band is well resolved and considerably more anodal. The human form is slightly cathodal to the origin. Production of an intermediately migrating heterodimer indicates the presence of the gene of the segregating chromosome set.

10. Esterase D (ESD, EC 3.1.1.1, carboxylesterase). Use buffer system a. Stain ingredients:

5 mg, 4-methylumbelliferyl acetate
2 ml, acetone
8 ml, 0.5 M $NaHPO_4$ buffer (19 g $NaH_2PO_4 \cdot H_2O$ and 17.3 g $Na_2HPO_4 \cdot 7H_2O$ up to 500 ml with water) adjusted to pH 7.0

[14] Y. H. Edwards, J. Potter, and D. A. Hopkinson, *Biochem. J.* **187**, 429 (1980).

Cut 3MM Whatman chromatographic paper to size of gel, place over gel, and soak with above mixture. View under UV light. When bright fluorescent bands appear remove the paper. Intensify bands by sprinkling surface with NH$_4$OH. Both human and Chinese hamster homodimers migrate twice the distance anodally than do mouse homodimers. The product of human *ESD* (chromosome 13) can be seen as the intermediately migrating heterodimer. This system is not informative in human × Chinese hamster hybrid cells since their forms comigrate.

11. Galactokinase (GALK, EC 2.7.1.6). Buffer systems a, b, or c can be used. Combine

    10 mg, ATP
    0.08 ml, 100 mg MgCl$_2$/ml water
    2 ml, 0.5 *M* Tris–HCl, pH 7.1
    3 ml, water
    15 μCi, D-[1-^{14}C]galactose

Evenly distribute above, with a Pasteur pipet, over DEAE-81 paper which has been placed over the cut surface of the gel slice. Continue as described in the autoradiographic procedures above. Expose film for 14 days. The product of human *GALK* (chromosome 17) is just anodal to the origin whereas the mouse form migrates approximately 6 cm anodally. The Chinese hamster form is also well resolved approximately 1 cm cathodal to the mouse form.

12. Glucose-6-phosphate dehydrogenase (G6PD, EC 1.1.1.49). Use buffer system f. Stain ingredients:

    100 mg, glucose 6-phosphate
    15 mg, NADP
    2 ml, 10 mg NBT/ml water
    1 ml, 1 mg PMS/ml water
    0.5 ml, 100 mg MgCl$_2$/ml water
    10 ml, 0.2 *M* Tris–HCl, pH 8.0
    40 ml, water

Look for blue bands. The product of the most common form (*b* allele) of human *G6PD* (X chromosome) migrates slower than the mouse or Chinese hamster forms which comigrate on this system. The product of the *a* allele (present in HeLa cells), while more anodal than the *b* polypeptide, is still sufficiently resolved from rodent forms. The presence of a heterodimer band marks the chromosome of the segregating set.

13. Glucose-6-phosphate isomerase (GPI, EC 5.3.1.9). Use buffer systems a, b, or c (actually almost any buffer system will do for this wonderful enzyme). Stain ingredients:

80 mg, fructose 6-phosphate
10 mg, NADP
2 ml, 5 mg MTT/ml water
1 ml, 1 mg PMS/ml water
0.5 ml, 100 mg MgCl$_2$/ml water
5 ml, 0.2 M Tris–HCl, pH 8.0
80 units, glucose-6-phosphate dehydrogenase
15 ml, 1.5% molten agar

The product of human *GPI* (chromosome 19) migrates 2.5 cm toward the cathode and is well resolved from the less cathodal mouse form. The Chinese hamster form is very well resolved from either of these since it is slightly anodal. The presence of a heterodimer indicates the gene of the segregating set in hybrid cells.

14. α-Glucosidase (GAA, EC 3.2.1.20). Use buffer system e. Stain ingredients:

10 mg, 4-methylumbelliferyl-α-D-glucopyranoside
5 ml, 0.05 M citrate/0.05 M phosphate, pH 4.0

Incubate 30 to 45 min. View fluorescent bands under UV light (fluorescence can be increased by raising gel pH with NH$_4$OH). The product of human *GAA* (chromosome 17) migrates resolvably less anodally than does the Chinese hamster product. The mouse form is difficult to separate from the human form.

15. β-Glucuronidase (GUSB, EC 3.2.1.31). Done on Cellogel as described above. Stain ingredients:

6 mg, 4-methylumbelliferyl-β-D-glucuronide trihydrate
6 ml, 0.5M sodium acetate, pH 5.0

Cover Cellogel with 3MM filter paper and soak with the above solution. Incubate for 30 to 45 min, remove filter paper, and view under UV light for bright fluorescing bands (intensify by spraying with 0.5 M sodium carbonate). The product of human *GUSB* (chromosome 7) is less anodal in migration than either of the rodent forms (which are not readily separable on this system). The presence of heterotetramers indicates the presence of the gene of the segregating set of chromosomes.

16. Glutamate-oxaloacetate transaminase (GOT, EC 2.6.1.1, aspartate aminotransferase). Use gel buffer d. Stain ingredients:

50 mg, L-cysteinesulfinic acid
20 mg, NAD
25 ml, 0.5 M Tris–HCl, pH 7.5
0.4 ml, 0.1 M α-ketoglutaric acid, pH 7.5
1 ml, 5 mg MTT/ml water

1 ml, 1 mg PMS/ml water
50 units, α-glutamate dehydrogenase
25 ml, 1.5% molten agar

Incubate and look for blue bands on white background. The product of human *GOT1* (chromosome 10) is further anodal than the rodent forms (which are not readily separable from each other on this system). The presence of the heterodimer indicates the gene of the separating set. Cathodal bands are the products of *GOT2* which are not informative in this system.

17. Glutathione peroxidase (GPX, EC 1.11.1.9). Use buffer system a. Stain ingredients:

15 mg, reduced glutathione
5 mg, NADPH
1 mg, EDTA
5 ml, 0.1 M $K_2HPO_4 \cdot KH_2PO_4$, pH 7.0
0.06 ml, 0.7% *tert*-butyl hydroperoxide solution
50 units, glutathione reductase

View yellow fluorescence overall under UV light. The product of human *GPX1* (chromosome 3) will produce a dark band most anodally on the gel, with the hamster form about half the distance anodally and the mouse form intermediate in migration between human and hamster forms. A smear between any two forms reveals the presence of the gene of the segregating set of this tetrameric enzyme.

18. Glutathione reductase (GSR, EC 1.6.4.2). Use buffer system e. Stain ingredients:

30 mg, oxidized glutathione
10 mg, NADPH
1 ml, 0.2 M Tris–HCl, pH 8.0
4 ml, water

View yellow fluorescence overall under UV light. The product of human *GSR* (chromosome 8) leaves a dark band of defluorescence on the anodal portion of the gel. The Chinese hamster form migrates twice the distance anodally and the mouse form is intermediate in migration between the human and hamster forms. The presence of the heterodimer indicates the presence of the gene from the segregating set of chromosomes in somatic cell hybrids.

19. Glyceraldehyde-3-phosphate dehydrogenase (GAPD, EC 1.2.1.12). Use buffer system b. To stain combine

270 mg, fructose 1,6-diphosphate
2 ml, 0.2 M Tris–HCl, pH 8.0

3 ml, water
100 units, aldolase
and incubate at 37° for 1 hr, and then add to remaining ingredients:
25 mg, NAD
75 mg, sodium arsenate
2 ml, 10 mg NBT/ml water
1 ml, 1 mg PMS/ml water
10 ml, 0.2 M Tris–HCl, pH 8.0
35 ml, water

Incubate until blue bands appear on a white background. The product of human *GAPD* (chromosome 12) migrates cathodally while the rodent forms show only slight migration toward the anode. Mouse and hamster forms cannot be separated on this system. Heterotetrameric bands intermediate in migration indicate the presence of the gene from the segregating set of chromosomes.

20. Glyoxalase I (GLO1, EC 4.4.1.5). Use buffer system d. Combine
  10 mg, reduced glutathione
  6 ml, 0.2 M pH 7 phosphate buffer (7.8 g $NaH_2PO_4 \cdot H_2O$ and 6.965 g Na_2HPO_4 brought to 500 ml with water)
  0.3 ml, methyl glyoxal (40% aqueous solution)

and apply to filter paper overlaying the gel slice. Incubate at 37° for exactly 35 min. Remove the filter paper and overlay the gel with 1.5% molten agar containing 1 ml of 0.1 N potassium iodide. The regions of glyoxylase activity will become dark blue-black as the pale blue background fades to white. The product of one of the two common human *GLO1* alleles, *2* (chromosome 6), is resolvably faster migrating toward the anode than the product of the *1* allele. The product of the human *2* allele comigrates with the Chinese hamster form of GLO1 and the product of the human *1* allele comigrates with the mouse form. In appropriately constructed interspecific somatic cell hybrids, the presence of the heterodimer indicates the gene from the segregating set of chromosomes.

21. Hexosaminidase (HEX, EC 3.2.1.52, *β-N*-acetylhexosamidase). Use buffer system i. Stain ingredients:
  10 mg, 4-methylumbelliferyl-2-acetamido-2-deoxy-β-D-glucopyrano-side
  5 ml, 0.05 M citrate/0.05 M phosphate, pH 4

View fluorescent bands under UV light which may be intensified by sprinkling with NH_4OH. The product of human *HEXA* (chromosome 15) produces a broad anodal band of intense fluorescence which is quite distinguishable from the faint, thin rodent bands. This system therefore clearly indicates the presence of human chromosome 15 in hybrids segregating human chromosomes but is not informative for rodent forms. The product of human *HEXB* (chromosome 5) migrates resolvably further cathodally than the rodent forms (which cannot be distinguished from each other on this system).

22. Isocitrate dehydrogenase (IDH, EC 1.1.1.42). Use buffer system c. Stain ingredients:
  15 mg, NADP
  10 ml, 0.2 M Tris–HCl, pH 8.0
  8 ml, 0.1 M sodium isocitrate, pH 7.0
  2 ml, 10 mg NBT/ml water
  1 ml, 1 mg PMS/ml water
  0.5 ml, 100 mg $MgCl_2$/ml water
  32 ml, water

Look for blue bands on a white background. The product of human *IDH1* (chromosome 2) migrates further anodally than the rodent forms (mouse and Chinese hamster products cannot be resolved on this system). The presence of the heterodimer indicates the gene product from the segregating set of chromosomes. The cathodal portion of the gel contains the products of *IDH2*. Chinese hamster IDH2 has a slower migration which can barely be resolved from mouse in interpsecific hybrids between these two species. Human IDH2 is intermediate in migration from hamster and mouse forms and cannot be well resolved from either.

23. Inosine triphosphatase (ITPA, EC 3.6.1.19). Good on buffer systems e, f, or g. Combine and soak into the cut surface of the gel:
  20 mg, inosine 5′-triphosphate
  5 ml, 0.2 M Tris–HCl, pH 8.0
  0.1 ml, 2-mercaptoethanol
  1 ml, 100 mg $MgCl_2$ml water

Incubate at 37° for 2 hr, rinse gel , and apply molybdenum reagent (600 mg of ascorbic acid in 9 ml of 2.5 g ammonium molybdate/8 ml H_2SO_4/92 ml water) in the following way: apply a few milliliters of reagent and let it stay on the gel for 10 min (until blue bands begin to appear). Rinse off the reagent and repeat the process until all 9 ml of reagent is used or dark blue bands are readily apparent against a white background. The product of human *ITPA* (chromosome 20) migrates much faster anodally than the Chinese hamster form yet is still readily resolvable than the even faster mouse form. The presence of the gene from the segregating set of chromosomes is indicated by the heterodimeric molecule.

24. Lactate dehydrogenase (LDH, EC 1.1.1.27). Use buffer systems a, b, or c. Stain ingredients:

25 mg, NAD
8 ml, 1 M sodium lactate, pH 7.0
10 ml, 0.2 M Tris–HCl, pH 8.0
2 ml, 10 mg NBT/ml water
1 ml, 1 mg PMS/ml water
35 ml, water

Look for blue bands on a white background. The product of human *LDHA* (chromosome 11) is cathodal and well resolved from the anodal mouse form and the even more anodal Chinese hamster form (the mouse and Chinese hamster forms are well resolved from each other). The presence of heterotetramers indicates the gene product from the segregating set of chromosomes in interspecific hybrids. The product of human *LDHB* (chromosome 12) migrates much more toward the anode than any of the LDHA bands of any of the species but cannot be resolved from the LDHB forms of the mouse or hamster. When human cells that express LDHB are hybridized to rodent cells which do not produce significant amounts of the enzyme (most of the long-term cell lines preferentially produce LDHA to LDHB), the presence of human chromosome 12 can be detected by the observation of heterotetramers between the location of the human LDHB homopolymer and the human and/or rodent LDHA homopolymers.

25. Malate dehydrogenase (MDH, EC 1.1.1.37). Use buffer system a or b. Stain ingredients:

25 mg, NAD
10 ml, 0.2 M Tris–HCl, pH 8.0
5 ml, 1 M sodium malate, pH 7.0
2 ml, 10 mg NBT/ml water
1 ml, 1 mg PMS/ml water
35 ml, water

Look for blue bands on white background. The product of human *MDH1* (chromosome 2) is less anodal than the fast migrating rodent forms (which are not resolvable from each other). The intermediately migrating heterodimer indicates the presence of the gene from the segregating set of chromosomes in interspecific hybrids. The cathodal portion of the gel will have the product of *MDH2* gene which is not sufficiently different in migration between the three species under discussion to be informative in such mapping studies.

26. Malic enzyme [ME, EC 1.1.1.40, malate dehydrogenase (oxaloacetate-decarboxylating) (NADP$^+$)]. Use buffer system f. Stain ingredients:

15 mg, NADP
10 ml, 0.2 M Tris–HCl, pH 8.0
8 ml, 1 M sodium malate, pH 7.0
2 ml, 10 mg NBT/ml water
1 ml, 1 mg PMS/ml water
0.5 ml, 100 mg $MgCl_2$/ml water
35 ml, water

Produces blue bands on a white background. The product of human *ME1* (chromosome 6) is the fastest anodally and well resolved from the less anodal Chinese hamster form which is very well resolved from the mouse form located only slightly anodal to the origin. The presence of heterotetramers indicates the presence of the gene from the segregating set of chromosomes.

27. Mannose-6-phosphate isomerase (MPI, EC 5.3.1.8). Use buffer system a, b, or c. Stain ingredients:

50 mg, mannose 6-phosphate
15 mg, NADP
4 ml, 0.2 M Tris–HCl, pH 8.0
2 ml, 5 mg MTT/ml water
1 ml, 1 mg PMS/ml water
0.5 ml, 100 mg $MgCl_2$/ml water
100 units, phosphogluconate isomerase
80 units, glucose-6-phosphate dehydrogenase
18 ml, 1.5% molten agar

Look for blue bands on a white background. The product of human *MPI* (chromosome 15) is resolvably slower anodally than Chinese hamster but comigrates with the mouse form on this system.

28. α-Mannosidase; A (MANA, EC 3.2.1.24). Use buffer system e. Stain ingredients:

10 mg, 4-methylumbelliferyl-α-D-mannopyranoside
5 ml, 0.05 M citrate/0.05 M phosphate, pH 4

View under UV light for appearance of fluorescent bands which may be intensified, after they appear, by sprinkling the surface of the gel with NH_4OH. The product of human *MANA* (chromosome 15) comigrates with the Chinese hamster form and is resolvably slower migrating toward the anode than the mouse form.

29. Nucleoside phosphorylase (NP, EC 2.4.2.1, purine-nucleoside phosphorylase). Use buffer system e, f, or g. Stain ingredients:

100 mg, inosine
100 mg, sodium arsenate
10 ml, 0.2 M Tris–HCl, pH 8.0

2 ml, 10 mg NBT/ml water
1 ml, 1 mg PMS/ml water
1.6 units, xanthine oxidase
20 ml, 1.5 molten agar
Look for blue bands on a white background. The product of human *NP* (chromosome 14) is slower anodally than the form from Chinese hamster. The hamster form is resolvably slower than the mouse form. Heterotrimers indicate the gene from the segregating set of chromosomes in somatic cell hybrids.

30. Peptidases [(PEP, EC 3.4.11.* or 3.4.13.*); the asterisk means that any of those specific peptidase activities whose digits would go in that place in the EC number will be visualized by this procedure]. Use buffer system e, f, or g. Combine and pour over the cut surface of the gel:
Peptide substrate
    40 mg, L-valyl-L-leucine (VL) or
    20 mg, L-leucyl-L-alanine (LA), or L-leucyl-L-proline (LP), or L-leu-
        cylglycylglycine (LGG)
and
    10 mg, peroxidase
    0.5 ml, 0.1 M MnCl$_2$
    0.5 ml, 25 mg o-dianisidine/ml water
    0.3 ml, 15 mg snake venom/ml water
    15 ml, 0.2 M Tris–HCl, pH 8.0
    20 ml, 1.5% molten agar
Look for brown bands on a white background. With the LA substrate the products of human *PEPC* (chromosome 1), *PEPA* (chromosome 18), and *PEPS* (chromosome 4) may be visualized. VL will also visualize the *PEPA* and *PEPS* products, often with greater intensity. Of these PEPC is the fastest in anodal migration. The human form is double banded (the more cathodal of the two closely spaced bands is more intense) and well resolved from the more anodal mouse form. The single thin Chinese hamster form migrates between the two human bands; however, in hybrids in which human chromosomes are segregating and chromosome 1 is retained, the distinctive pattern for human PEPC can be discerned. PEPA shows up as less anodal than PEPC. The human form is less anodal than the comigrating hamster and mouse forms and can be detected by the presence of the heterodimer. PEPS is the slowest anodally migrating system. The human form is well resolved between the faster migrating Chinese hamster and slower migrating mouse forms. In interspecific somatic cell hybrids retaining the gene from the segregating set of chromosomes, a smear representing the poorly resolved heterotetrameric bands between the two species types will be observed. The LGG substrate visualizes the PEPB isozymes. The

product of human *PEPB* (chromosome 12) comigrates anodally with the Chinese hamster form but is well resolved from the less anodal mouse form. The LP substrate visualizes the *PEPD* isozymes. The product of human *PEPD* (chromosome 19) is well resolved, further anodal to the Chinese hamster form. However, it is not well resolved, on this buffer system, from the slightly more anodal (to the human) mouse form. The presence of the heterodimer indicates the gene product of the segregating set of chromosomes.

31. Phosphoglucomutase (PGM, EC 5.4.2.2). Use buffer system a, b, or c. Stain ingredients:

    300 mg, glucose 1-phosphate (grade III disodium salt from Sigma Chemical Co., St. Louis, MO 63178)
    15 mg, NADP
    10 ml, 0.2 M Tris–HCl, pH 8.0
    2 ml, 5 mg MTT/ml water
    1 ml, 1 mg PMS/ml water
    0.5 ml, 100 mg $MgCl_2$/ml water
    80 units, glucose-6-phosphate dehydrogenase
    20 ml, 1.5% molten agar.

Also use a slice from buffer system d. To it add

    100 mg, glucose 1-phosphate (#6123-02 from Koch-Light Limited, Haverhill, Suffolk, England)
    5 mg, NADP
    3 ml, 0.2 M Tris–HCl, pH 8.0
    2 ml, 5 mg MTT/ml water
    1 ml, 1 mg PMS/ml water
    0.5 ml, 100 mg $MgCl_2$/ml water
    40 units, glucose-6-phosphate dehydrogenase
    22 ml, 1.5% molten agar
    1 ml, 0.16 mg/ml glucose 1,6-diphosphate

Look for blue bands on a white background. Buffer system a, b, or c best resolves the anodally migrating product of human *PGM2* (chromosome 4) from the less anodal products of both human *PGM1* alleles— *1* (slower) and *2* (faster)—which mark human chromosome 1. This system also resolves the human forms from the more anodal Chinese hamster PGM1 and even more anodal hamster PGM2. Mouse PGM1 comigrates with human PGM2 so that these gene products are best resolved on buffer system d. This system also resolves mouse PGM2 from the slightly less anodal hamster PGM2. Buffer system d, and the staining mixture described for it, also best resolves and visualizes the most anodal PGM isozymes—the products of *PGM3*. The products of both common alleles of human *PGM3* (*1* whose product is faster, and *2*) are slower and well resolved from the hamster form while the mouse form is not usually expressed in cultured cells.

32. 6-Phosphogluconate dehydrogenase (PGD, EC 1.1.1.44). Use buffer system f. Stain ingredients:
   50 mg, 6-phosphogluconic acid
   15 mg, NADP
   10 ml, 0.2 M Tris–HCl, pH 8.0
   2 ml, 10 mg NBT/ml water
   1 ml, 1 mg PMS/ml water
   0.5 ml, 100 mg $MgCl_2$/ml water
   40 ml, water

Look for blue bands on white background. The product of human *PGD* (chromosome 1) is slower anodally than the well resolved Chinese hamster form. The mouse form comigrates with the human form. The heterodimer indicates the presence of the gene of the segregating set of chromosomes in interspecific somatic cell hybrids.

33. Phosphoglycolate phosphatase[15] (PGP, EC 3.1.3.18). Use buffer system a, b, c. Stain ingredients:
   30 mg, phosphoglycolic acid
   10 mg, $MgSO_4$
   100 mg, inosine
   10 ml, 0.2 M Tris–HCl, pH 8.0
   2 ml, 10 mg NBT/ml water
   1 ml, 1 mg PMS/ml water
   1.6 units, xanthine oxidase
   5.0 units, nucleoside phosphorylase
   20 ml, 1.5% molten agar.

Look for blue bands on a white background. The product of human *PGP* (chromosome 16) is well resolved at only one-half the anodal mobility of mouse and hamster forms (which comigrate). The presence of heterotetramers indicates the gene from the segregating set of chromosomes.

34. Pyruvate kinase (PK, EC 2.7.1.40). Use buffer system a. Stain ingredients:
   15 mg, phospho(enol)pyruvate
   30 mg, ADP
   15 mg, fructose 1,6-bisphosphate
   40 mg, $MgSO_4$
   40 mg, KCl

[15] R. F. Barker and D. A. Hopkinson, *Ann. Hum. Genet.* **42,** 143 (1978).

8 ml, 0.2 *M* Tris–HCl, pH 8.0
10 mg, NADH
160 units, lactate dehydrogenase

View under UV light and look for dark bands against a yellow fluorescent background. The cathodally migrating product of human *PKM2* (chromosome 15) is well resolved from the anodal rodent forms and may be detected in hybrids by the presence of intermediately migrating heterotetramers.

35. Superoxide dismutase (SOD, EC 1.15.1.1). Use buffer system e, f, or g. Stain ingredients:
    100 mg, sodium arsenate
    10 ml, 0.2 *M* Tris–HCl, pH 8.0
    3 ml, 10 mg NBT/ml water
    1 ml, 1 mg PMS/ml water
    0.8 ml, 100 mg $MgCl_2$/ml water
    40 ml, water

Leave the gel in bright light after immersing in stain. Look for blanched, white regions against a background of blue stain. (SOD also can be visualized on gels where a tetrazolium reaction is used to detect enzymes such as NP and ADA.) The product of human *SOD1* (chromosome 21) is well resolved more anodal to mouse and Chinese hamster forms (which comigrate). The presence of the heterodimer indicates the gene from the segregating set of chromosomes.

36. Triose-phosphate isomerase (TPI, EC 5.3.1.1). Use buffer system a, b, or c. Stain ingredients:
    25 mg, NAD
    100 mg, sodium arsenate
    5 ml, 0.2 *M* Tris–HCl, pH 8.0
    1 ml, 10 mg dihydroxyacetone phosphate/ml stock solution (as per Sigma Chemical Co., catalogue #D 7878, St. Louis, MO)
    2 ml, 5 mg MTT/ml water
    1 ml, 1 mg PMS/ml water
    820 units, glyceraldehyde-3-phosphate dehydrogenase
    15 ml, 1.5% molten agar

Look for blue bands on a white background. The anodally migrating product of human TPI (chromosome 12) is resolvably more anodal than the mouse form and less anodal than the Chinese hamster form. The presence of the heterodimer indicates the expression of the gene from the segregating set of chromosomes.

TABLE I
CHROMOSOMAL REGIONS OF THE HUMAN GENOME MARKED BY ISOZYME LOCI[a]

Chromosome regions	Marker loci	Chromosome regions	Marker loci
1 pter-p36.13	ENO1	12 p13	GAPD, TPI1
1 p36.2-p36.13	PGD[b]	12 p12.2-p12.1	LDHB[d]
1 p34	AK2	12 q21	PEPB[e]
1 p22.1	PGM1	13 q14.1	ESD[e]
1 q25 or 1 q42	PEPC	14 q13.1	NP
2 p25, or p23	ACP1	14 q32	CKBB[d]
2 p23	MDH1	15 q11-qter	MANA[e]
2 q32-qter	IDH1	15 q22-q25.1	HEXA
3 p21	ACY1	15 q22-qter	MPI, PKM2
3 p13-q12	GPX1	16 pter-q11	PGP
4 p14-q12	PGM2	16 q12-q22	DIA4
4 p11-q12	PEPS	16 q22	APRT[e]
5 q13	HEXB	17 q21-q22	GALK
6 p21.31-p21.1	GLO1[c]	17 q23	GAA
6 q12	ME1, PGM3	18 q23	PEPA
7 cen-q22	GUSB	19 p13.2-cen	PEPD
8 p21	GSR	19 cen-q13	GPI
9 p34	AK1	20 p	ITPA
10 cen-q24	ADK	20 q13.2-qter	ADA
10 q25.3	GOT1	21 q22.1	SOD1
11 p1203-p1208	LDHA	22	None
		X q28	G6PD

[a] From the 1985 catalogue of mapped genes in P. J. McAlpine, T. B. Shows, R. L. Miller, and A. J. Pakstis, *Cytogenet. Cell Genet.* **40**, 8 (1985).
[b] Not informative in hybrids made with mouse cells.
[c] Not informative in hybrids made with Chinese hamster cell unless the human cell parents are homozygous for the *1* allele, and not informative in hybrids made with mouse cells unless the human cell parents are homozygous for the *2* allele.
[d] Informative only when the isozyme of the nonhuman fusion partner is not expressed.
[e] Not informative in hybrids made with Chinese hamster cells.

Regions of the Human Genome Marked by Isozyme Loci

As is shown in Table I, the isozyme markers described here are capable of detecting the presence of almost every human chromosome in interspecific somatic cell hybrids in which human chromosomes are segregating. An obvious exception is our failure to report a marker for chromosome 22. As mentioned earlier, we are only reporting methods for markers with which we have had some considerable reproducible success. For that reason, in the discussion of diaphorase isozymes, we did not recommend *DIA1* as a marker for chromosome 22. Some have reported the ability to resolve the product of human *aconitase 2 (ACO2)*, which has been assigned to chromosome 22,[16] in human–rodent hybrids. We have not found that this gene product is expressed in cultured cells consistently enough to depend on it as a reproducible marker for that chromosome. In that situation, and in cases where because of the nature of the fusion scheme, a marker for a chromosome is not informative (e.g., in human × Chinese hamster hybrids, the human and hamster products of the only human marker for chromosome 13, *ESD*, comigrate), DNA probes may be used to visualize restriction fragments which will distinguish human from rodent sequences and which are specific to certain chromosomes. Methods for such procedures are presented elsewhere in this series (Vols. 68, 100, and 101). However, since those methods are considerably more labor intensive, utilize expensive isotopes and reagents, and take at least twice as long to give an answer than the isozyme procedures, the search for additional isozyme markers to cover all segments of the human genome remains a useful enterprise.

Acknowledgments

We appreciate the excellent technical assistance of Pam Watson and Rosie Creswell. This work has been supported in part by NIH Grants CA 04484 and CA 34936, as well as gifts from the Kleberg Foundation, the Exxon Corporation, and Mr. Kenneth D. Muller.

[16] P. Meera Khan, L. M. M. Wijnen, and P. L. Pearson, *Cytogenet. Cell Genet.* **22,** 212 (1978); C. A. Slaughter, S. Povey, B. Carritt, E. Solomon, and M. Bobrow, *Cytogenet. Cell Genet.* **22,** 223 (1978); R. S. Sparkes, T. Mohandas, M. C. Sparkes, and J. D. Shulkin, *Cytogenet. Cell Genet.* **22,** 226 (1978).

[16] Principles of Electrofusion and Electropermeabilization

By ULRICH ZIMMERMANN and HOWARD B. URNOVITZ

Biological membranes exhibit limited permeability to both electrolytes and nonelectrolytes. Throughout evolution, selective transport systems have been developed to achieve and control uptake and release of specific solutes: charged or uncharged, low and high molecular weight.

Increasing membrane permeability is of significant interest for biotechnological applications with noted emphasis on modification of cellular metabolic pools and the genetic information of cells. The procedure must be performed in a reversible manner and within acceptable time constraints. During the process, intracellular substances (genes, proteins, and subcellular organelles, etc.) must remain within the confines of the membrane boundaries.

Cellular modification can be achieved by fusing two separate cells of differing origins. The resultant hybrid contains valuable properties of both parents within one cell. The effects of cytoplasmic mixing can be studied in short-term cultures. Clonal expansion of selected hybrids permits cultivation of genes and gene products. Fusion is based on the dual characteristics of the cell membrane: liquid properties in the membrane plane and solid properties normal to the membrane surface. Fusion induces membrane perturbations perpendicular to the membrane plane. These perturbations increase membrane fluidity leading to an intermingling of the adjacent surfaces. In single cell suspensions, membrane-impermeable compounds are driven through membranes via perturbations in the direction of the vector normal to the cell membrane.

There are similar requirements for cell fusion and direct mass transfer through membranes. Physical techniques for reversible membrane perturbations would be predictably superior to chemical procedures due mostly to the vectorial properties of physical forces.

The past 15 years have shown that electric field pulses can be used both for entrapment of membrane-impermeable substances and cell fusion.[1-17] The required membrane perturbations are induced by external electrical

[1] U. Zimmermann, F. Riemann, and G. Pilwat, *U.S. Patent* 4,154,668 (1979), filed in 1974.

[2] U. Zimmermann, J. Vienken, and G. Pilwat, *Bioelectrochem. Bioenerg.* **7**, 553 (1980).

[3] U. Zimmermann, P. Scheurich, G. Pilwat, and R. Benz, *Angew. Chem., Int. Ed. Engl.* **20**, 325 (1981).

[4] U. Zimmermann and J. Vienken, *J. Membr. Biol.* **67**, 165 (1982).

[5] U. Zimmermann, *Biochim. Biophys. Acta* **694**, 227 (1982).

[6] U. Zimmermann, *Trends Biotechnol.* **1**, 149 (1983).

[7] U. Zimmermann, *in* "Targeted Drugs" (E. Goldberg, ed.), p. 153. Wiley, New York, 1983.

[8] U. Zimmermann and J. Vienken, *in* "Cell Fusion: Transfer and Transformation" (R. F. Beers and E. G. Bassett, eds.), p. 171. Raven Press, New York, 1984.

[9] U. Zimmermann, J. Vienken, G. Pilwat, and W. M. Arnold, *Ciba Found. Symp.* **103**, 60 (1984).

[10] U. Zimmermann, K.-H. Büchner, and W. M. Arnold, *in* "Charge and Field Effects in Biosystems" (M. J. Allen and P. N. R. Usherwood, eds.), p. 293. Abacus, Tunbridge-Wells, England, 1984.

[11] U. Zimmermann, J. Vienken, and G. Pilwat, *in* "Investigative Microtechniques in Medicine and Biology" (J. Chayen and L. Bitensky, eds.), Vol. 1, p. 89. Dekker, New York, 1984.

[12] W. M. Arnold and U. Zimmermann, *in* "Biological Membranes" (D. Chapman, ed.), Vol. 5, p. 389. Academic Press, London, 1984.

fields. High-intensity, short-duration pulses cause reversible increases in membrane permeability and fluidity. Reversible, locally restricted electrical breakdown of the cell membrane leads then to both electropermeabilization and cell electrofusion. The field pulse techniques can be applied universally to fusion of all cells (and artificial liposomes). The techniques also relate both to the intracellular incorporation of a diverse array of compounds and the release of intracellular substances into the external medium.

The complexity of each individual cell system dictates the requirement for understanding the basic physical properties governing the procedure. Each cell system responds to electric fields in its own unique and reproducible manner. The experimental conditions must be tailored for each unique cell type and system. This chapter describes guidelines for choosing the correct boundary conditions for electrofusion and electro-mass transfer through biological membranes. Although it is not possible to provide detailed protocols which can be used for any cell type, these guidelines should allow the investigator to determine which boundary conditions are optimal for the system under study.

Adjustment of the Breakdown Parameters

Reversible electrical breakdown of the cell membrane is the fundamental step in electrofusion and electropermeabilization (also termed electroinjection, or, less appropriate, electroporation).[15] Local breakdown in the cell membrane occurs if the breakdown voltage is exceeded.[1,2] The breakdown voltage of most cell membranes (for exceptions, see Ref. 5) is about 1 V at room temperature and about 2 V at 4°.[3,5] If two membranes are arranged in series, the breakdown voltage of the whole barrier is correspondingly 2 V. The breakdown voltage depends on the duration time of the field pulse.[3] As a rule, for very short pulses (less than 1 to 10 μsec) the breakdown voltage assumes, at room temperature, a value of 1 V. For longer pulse durations, a value of about 0.5 V is assumed. For very

[13] U. Zimmermann and J. Vienken, in "Hybridoma Technology in Agricultural and Veterinary Research" (N. J. Stern and H. R. Gamble, eds.), p. 173. Rowman & Allanheld, Totowa, New Jersey, 1984.

[14] U. Zimmermann, J. Vienken, J. Halfmann, and C. C. Emeis, Adv. Biotechnol. Processes 4, 79 (1985).

[15] U. Zimmermann, Rev. Physiol. Biochem. Pharmacol. 105, 175 (1986).

[16] U. Zimmermann, in "Hybridoma Formation" (A. H. Bartal and J. Hirschaut, eds.). Humana Press, Clifton, New Jersey, in press, 1986.

[17] U. Zimmermann, in "Membrane Fusion" (J. Wilschut and B. Hoekstra, eds.). in press.

long pulse durations (generally above 50 to 100 μsec), the breakdown translates into irreversible destruction of the cell membranes leading to cell death.

Temperature and pulse length values for breakdown voltage are known for many cells and can be determined by the appropriate equipment.[5,15] Experience shows that the assumption of 1 V breakdown voltage is a reasonable approximation for calculation of the field parameters in electrofusion (performed at room temperature). Optimum incorporation of membrane impermeable substances is achieved at 4°, therefore, a value of 2 V has to be assumed for electropermeabilization studies. The required field strength, E_c, of the external pulse for reaching the membrane breakdown voltage, V_c, can be obtained by the following two equations[5]

$$V_c = 1.5aE_c \cos \theta \tag{1}$$

$$\frac{1}{\tau} = \frac{1}{R_m C_m} + \frac{1}{aC_m(\rho_i + 0.5\rho_e)} \tag{2}$$

where a is the radius, θ, angle between a given membrane site in a field direction, R_m, specific membrane resistance, C_m, specific membrane capacitance, ρ_i, specific intracellular resistivity, ρ_e, specific extracellular resistivity, and τ, relaxation time.

Equation (1) describes the stationary state, i.e., the membrane potential which is generated in response to a field pulse of field strength, E. The equation is derived for spherical cells; for ellipsoid cells a similar, but more complicated relationship is obtained. Equation (1) shows that the field strength required for the generation of the breakdown voltage depends on the radius, a, of the cell (Fig. 1). Smaller cells need a higher field strength (assuming the same value of the breakdown voltage). The difference in the critical field strength may be quite substantial if the size distribution of the cell population is highly variable.

The potential difference across the membrane exhibits an angular dependence [Eq. (1) and Fig. 1]. For a given cell size and field strength, the generated membrane potential difference has its maximum value at membrane sites oriented in the field direction and its minimum value (always zero) at sites located perpendicular to the field vector. The generated membrane potential is proportional to the cosine of the angle. Breakdown of membrane sites (oriented by a certain angle to the field vector) requires higher (supracritical) field strengths than those located in the field direction.

Let us assume a sufficiently strong field pulse is applied where the breakdown voltage of the smallest cell is reached for membrane sites in the field direction ($\cos \theta = 1$). Under these conditions, the largest cells would

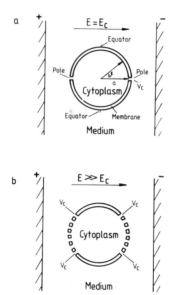

FIG. 1. Schematic diagram of a cell exposed to high electric field strengths [Eq. (1)]. (a) Critical field strength (E_c) produces breakdown only at the poles (membrane sites oriented in field direction). (b) Supracritical field strength produces membrane breakdown over larger areas (as compared to a). Although the schematic demonstrates "pores" of uniform sizes and distribution, the membranes are in fact hyperpermeable in the direction of the electric field. For membrane sites oriented perpendicular to the field direction, the potential, V, is always zero.

have larger membrane areas subjected to breakdown (taking into consideration the cell radius and angle dependence of the field strength and the generated membrane potential). If the membrane area of the larger cells reaches a certain proportion of the total surface area, irreversible destruction of the cells occurs. The release of intracellular substances into the medium significantly alters the success of the electrofusion or electroinjection process. At the other extreme, field pulses of too low intensity cannot sustain fusion between adherent cells of substantially different sizes because the smaller cells fail to break down. In electropermeabilization studies, the incorporation of substances (particularly macromolecules) is greatly reduced under the above conditions.

These considerations have to be modified through Eq. (2). The build-up of a membrane potential difference, in the presence of an external electric field, is a relaxation process, i.e., the process needs some time before the steady-state value is reached.[12,15]

A simple exponential relaxation process is completely described by the relaxation time, τ, given by Eq. (2). This process takes into account the time dependent build-up of a membrane potential in response to an external field pulse. The relaxation time is the time to reach 63% of the steady-state value. Experience shows a 5-fold increase in the relaxation time is required to approximate the final value.

Inspection of Eq. (2) shows that the relaxation time consists of two terms: the RC term which reflects electrical membrane properties and the term, $aC_m(\rho_i + 0.5\rho_e)$, which reflects the electrical properties of the solutions facing the membrane. The capacitance is generally of the order of 1 $\mu F/cm^2$. The membrane resistance of most cells is larger than 100 Ω cm^2. Thus, the RC term is on the order of 100 μsec to 1 msec.

The magnitude of the second term depends mainly on the ionic strength of the medium in which field pulse application is performed.

Electropermeabilization experiments are carried out in more conductive solutions (containing significant amounts of electrolytes).[1,2,18–20] Solutions should contain isotonic concentrations of electrolyte, or at minimum, a 30 mM univalent electrolyte concentration with the corresponding amount of nonelectrolytes. The specific resistivities of these solution are in the range of 10^2 to 10^3 Ω cm.

With an average internal resistivity value of 10^2 Ω cm, the solution term assumes values of 50 to 300 nsec for cell radii between 3 μm (mammalian cells) and 20 μm (plant protoplasts). Comparing the magnitude of the two terms, the RC term can be neglected. Only the solution term controls the relaxation time of the membrane potential build up. Values of 50 to 200 nsec show that the steady-state value of the membrane potential is approximately reached if pulse lengths of 1 to 5 μsec are applied.

Electrofusion is normally performed in weakly conductive solutions (see below). The specific external resistivity is in the range of 10^4 to 10^5 Ω/cm. Under these conditions (using the same values for the internal resistivity and for the radii), the solution term assumes values on the order of 1 to 20 μsec. The RC term can therefore no longer be neglected and must be taken into account. The total relaxation time is therefore in the range of tens of microseconds.

Compared to electropermeabilization, electrofusion pulse lengths have to be longer in order to reach the steady state value of the membrane potential difference [see Eq. (1)]. According to the calculations, electrofusion pulse lengths of 20 to 50 μsec (depending on the cell radius) can be

[18] U. Zimmermann, F. Riemann, and G. Pilwat, *Biochim. Biophys. Acta* **436**, 460 (1986).
[19] J. Vienken, E. Jeltsch, and U. Zimmermann, *Cytobiologie* **17**, 182 (1978).
[20] H. Stopper, U. Zimmermann, and E. Wecker, *Z. Naturforsch.* **402c**, 133 (1985).

applied without irreversible destruction of the cells as compared to electro-transfer of genes and proteins. Experiments with eggs or very large cells require pulse lengths of hundreds of microseconds or even milliseconds, pulse lengths[15] which undoubtedly lead to death of "normal sized" cells. Once the pulse has been applied, the actual generated voltage in large eggs is relatively smaller after a certain time interval.

The radius dependence of the field strength according to Eq. (1) can be partly compensated by the choice of the applied pulse length. In considering the cell radius, the critical field strength is inversely proportional while the relaxation time is directly proportional. For fusing cells of different radii, it may be favorable to apply field pulses with durations which are not long enough to reach the steady-state value.

In designing experiments, field strength and pulse length must be calculated [Eqs. (1) and (2)] to achieve reversible membrane perturbations. Recent results have shown that under some circumstances Eq. (1) has to be extended. Equation (1) represents the generated potential, V_g, which is superimposed on the intrinsic membrane potential, V_m, or, more precisely, on the intrinsic electric field.[21,22] If we assume linear superposition of the external and the intrinsic membrane field (which may be only a rough assumption), then the sum of both potential differences controls the breakdown of the membrane (see Fig. 2).

High intrinsic electric fields lead to an asymmetric electric breakdown. The external field vector is in one hemisphere in parallel to the intrinsic one, whereas in the other hemisphere the vectors are antiparallel. With increasing strength of the external field, the breakdown will be reached in that hemisphere where the vectors are parallel (Fig. 3). After breakdown, the magnitude of current through the cell is restricted in one hemisphere by the resistance of the intact hemisphere. The current is much smaller compared to the case when both hemispheres exhibit breakdown. Such conditions may be favorable both for fusion and for gene (or protein) incorporation. Only at higher field strengths is the breakdown voltage also exceeded in the hemisphere in which the field vectors are antiparallel. At the same time, the angle dependence of the breakdown voltage leads to defects in larger membrane areas of the hemisphere with parallel vectors.

The intrinsic electric field (or membrane potential) rises from electrogenic pumps, diffusion potential, and surface charges. Therefore, the intrinsic electric field changes by the concentration of potential determining ions, surface charges, and the energy supply. These parameters influence the breakdown voltage and the breakdown pattern. Changes in the amount

[21] D. L. Farkas, R. Korenstein, and S. Malkin, *FEBS Lett.* **120,** 236 (1980).
[22] W. Mehrle, U. Zimmermann, and R. Hampp, *FEBS Lett.* **185,** 89 (1985).

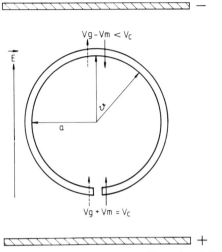

FIG. 2. Schematic diagram of a cell exposed to an uniform electric field pulse between two flat electrodes. a, radius of the cell; θ, angle between a given membrane site and the field direction; V_m, intrinsic membrane potential difference (assume negative on inside); V_g, generated superimposed membrane potential difference; V_c, breakdown voltage; E, field strength.

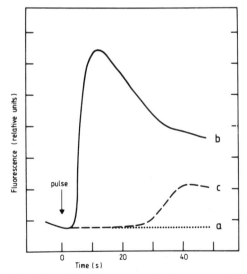

FIG. 3. Microfluorimetric uptake of fluorescent dye by oat mesophyll protoplasts: (a) no field pulse control, (b) change in fluorescent intensity in the cell hemisphere oriented toward the anode, and (c) change in fluorescent intensity in the cell hemisphere oriented toward the cathode.

of surface charge can be achieved by pretreatment of the cells with digestive enzymes (such as dispase, Pronase, and, to a lesser extent, trypsin).[3-17] Surface charge is also influenced by the ionic strength of the external solution. In solutions of low ionic strength, the surface charge concentration is high. In media of high ionic strength (e.g., physiological solutions containing about 150 mM univalent salts), the surface charge concentration and corresponding surface potential are diminished. Addition of small amounts of calcium and/or magnesium ions to the external medium can also change the surface charge concentration significantly. Both procedures of enzyme pretreatment and variations of the ionic strength are used in protocols for electrogene transfer and electrofusion.[5,15,20]

In electrogene transfer, field pulse application is carried out in a solution containing 30 mM KCl and the corresponding amount of nonelectrolyte (inositol or other sugars) to establish isoosmolarity. Using fluorescent dyes as probe molecules and the above conditions, the uptake of membrane impermeable substances in asymmetric in response to the field pulse (Fig. 3 and Ref. 22).

In standard electrofusion technique (combination of dielectrophoresis and field pulse) the media consist mainly of sugars with small additives of ions. Single cell suspensions should show asymmetric breakdown followed by uptake. However, it is not clear if asymmetric breakdown occurs in dielectrophoretically aligned cells. Equations (1) and (2) may no longer be valid due to field line constriction in the contact zone between attached cells (see Fig. 4; see color insert). A large inhomogeneity of the field in this area often results. The field line constriction depends significantly on the number of cell chains and chain length. There are some indications by Mehrle[23] that asymmetric breakdown also occurs in cell chains. Aerotactic bacteria were used for detection of asymmetric breakdown in attached plant protoplasts.[24] Such aerotactic bacteria only accumulate close to the membrane surface of protoplasts if they are intact and produce oxygen.

More detailed study on asymmetric breakdown is required for a complete quantitative prediction of the field pulse parameters involved in electrogene transfer and electrofusion. At the present time, Eqs. (1) and (2) are sufficient for rough calculations of the field conditions. Parallel fluorescent dye studies can provide additional information on the membrane breakdown events in selected cellular systems. The considerations here show that great caution is necessary if protocols for electrofusion and electrogene transfer are changed (e.g., for the design of a new experiment with new cell strains). Small changes in the medium composition can

[23] W. Mehrle, Ph.D. thesis. Universität Tübingen, Tübingen, Federal Republic of Germany, 1986.

[24] R. Hampp, W. Mehrle, and U. Zimmermann, *Plant Physiol.* **81,** 854 (1986).

greatly effect the breakdown conditions and the subsequent processes of uptake or fusion.

Adjustment of the Resealing Parameters

Resealing of field-induced membrane perturbations is a prerequisite for entrapment of membrane impermeable substances (genes, protein, drugs, etc.) within living cells.[1] Resealing also plays an important role in electrofusion because of the membrane breakdown at the chain termini. The resealing properties and the degree of restoration of the original membrane impermeability depend on various factors.

First, the restorative kinetics depends on the temperature.[1-4] At physiological temperatures (37° for mammalian cells, 30° for yeast protoplasts, and about 20° for plant protoplasts) the resealing is very rapid (on the order of a few minutes). At low temperatures (4 to 10°, used for incorporation of drugs and macromolecules), the restoration process is very slow (Fig. 5). After 30 min, cells do not regain their original impermeability. Below 4°, breakdown and the associated permeability changes become irreversible.[5] Uptake of substance requires at least a few minutes because of the time dependent diffusion process. Otherwise large concentrations of a given compound have to be added to the external solution in order to get a sufficient uptake into the cell.

Second, the resealing properties depend on the pulse length and on the field strength.[3,5,15] As a rule, it is more appropriate to increase the field strength than the pulse duration (described above). Breakdown of the cell membrane occurs in nanoseconds.[3] If the cell is exposed to the field pulse for too long, current will flow through the permeabilized membrane. Current flow induces local heating and microelectrophoresis of membrane and cellular components. Transient secondary effects on gene regulation and expression can also occur. The resealing process, also at elevated temperature, is very slow or completely suppressed resulting in the irreversible destruction of the cell.

In order to overcome the nonuniform permeability pattern in the membrane (due to the angle dependence of the generated potential difference), it is important to apply several consecutive pulses. The field strength should be approximately a factor of 2 higher than that calculated for $\cos \theta = 1$ [Eq. (1)]. The time interval between two consecutive pulses should be adjusted up to several seconds.[24a] This time is sufficient to reseal the lipid bilayer structure of the biological membrane (because of the very

---

[24a] H. Stopper, H. Jones, and U. Zimmermann, *Biochim. Biophys. Acta,* in press.

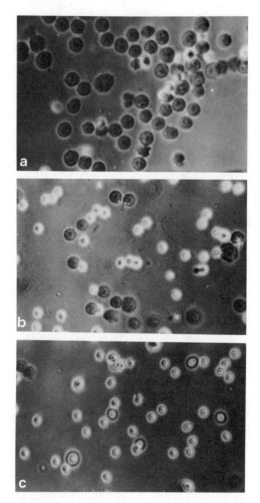

FIG. 5. Uptake of membrane-impermeable dye into mouse lymphocytes after electroper-meabilization at 4° [field strengths: (a) 6kV/cm; (b) 12 kV/cm; (c) 18 kV/cm, pulse duration 500 nsec]. Dye was added to isotonic conductive solution before pulse application. Note, increasing the field strengths leads to increasing dye uptake (phase contrast photography).

rapid resealing time of lipid bilayers).[25] The time is still sufficiently short to avoid complete restoration of the lipid:protein mismatches (at 4°). The restoration of the lipid bilayer again builds up the critical membrane potential difference resulting in breakdown. Breakdown will now occur in different membrane sites because of reorientation of the cell due to Brown-

[25] R. Benz and U. Zimmermann, *Biochim. Biophys. Acta* **640,** 169 (1981).

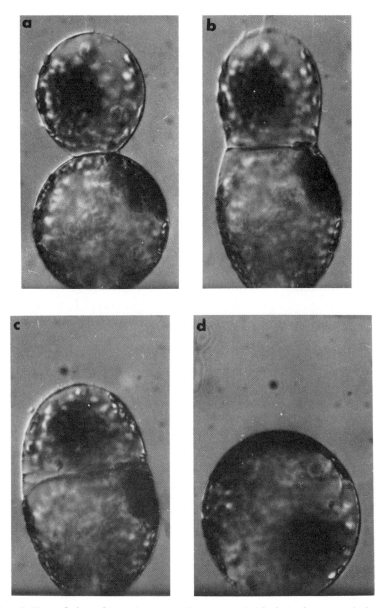

FIG. 6. Electrofusion of two plant protoplasts suspended in isotonic sugar solution. (a) Cells were aligned dielectrophoretically (see Fig. 4, 10 V/cm, 2 MHz). Fusion was induced by a breakdown pulse of 500 V/cm field strength, 20 μsec pulse duration. Photographs (interference contrast) were taken (b) 10 sec, (c) 30 sec, and (d) 120 sec after field strength application.

ian movement. In this way, it is possible to induce uptake of significant amounts of nucleic acids, proteins, or other compounds into cells without dramatic release of intracellular substances or deterioration of the cell membrane. Under these gentle conditions, restoration of the original membrane impermeability is observed when the temperature is raised to its physiological value about 2 to 5 min after field application.

For electrofusion of cells, field application should be performed at room temperature or slightly elevated temperatures. The more rapid resealing of the electropermeabilized membrane at these temperatures is required for two reasons. First, as mentioned above, the resealing of the hemisphere of the terminal cell, facing the external medium, is necessary to avoid substantial release of intracellular substances into the nonelectrolyte environment. Second, the intermingling of the attached membranes is much faster (because of the temperature dependent diffusion coefficient of the membrane molecules). In this context, resealing of the individual adhered membranes in the contact zone does not occur (if the temperature is correct) because of the establishment of a cytoplasmic continuum between the two cells.

Requirements for Membrane Contact

Application of breakdown pulses leads to fusion if the membranes of at least two cells are in close contact (Figs. 6–10; see color insert for Figs. 8

FIG. 4. Positive dielectrophoretic alignment of oat mesophyll photoplasts between two cylindrical electrodes glued in parallel 200 μm apart on a microslide (field strength 50 V/cm, frequency 2 MHz). Field line constriction around aligned protoplasts were visualized by the addition of bacteria. (Photography W. Mehrle, University of Tübingen.)

FIG. 8. Fusion of yeast protoplasts using a rotational chamber (125 μm distance between electrodes, see Fig. 17). (a) Yeast protoplasts were aligned dielectrophoretically (field strength 250 V/cm, 2 MHz frequency). Two minutes after the application of the ac field, 2 consecutive breakdown pulses spaced 1 min apart were applied (10 kV/cm, 10 μsec pulse duration each). (b) Photograph taken 30 sec after breakdown voltage (for media conditions, see Ref. 34).

FIG. 9. Formation of giant red blood cells by electrofusion. High cell suspension density of neuraminidase-pretreated human red blood cells was aligned dielectrophoretically (not shown) and 2 breakdown pulses applied (field strength 6 kV/cm, 7 μsec pulse duration). Photograph taken 3 min after breakdown pulse application.

FIG. 19. Prototype large-capacity high-voltage electropermeabilization chamber for electroinjection of macromolecules. A solution containing cells and material for electroinjection is drawn up into the chamber. The chamber is mounted behind the electronics and the entire chamber cooled by a Peltier element. After pulse application, the solution is ejected from the chamber into a new receptacle.

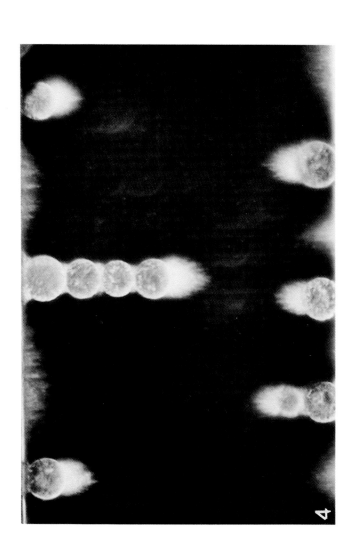

8 a

8 b

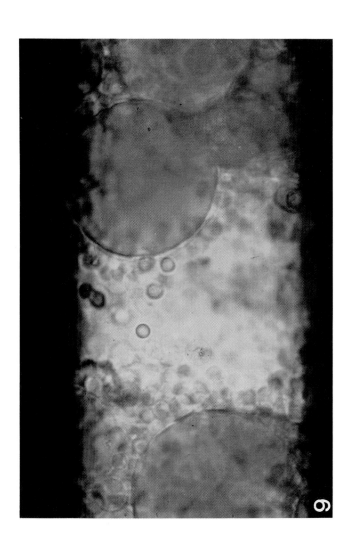

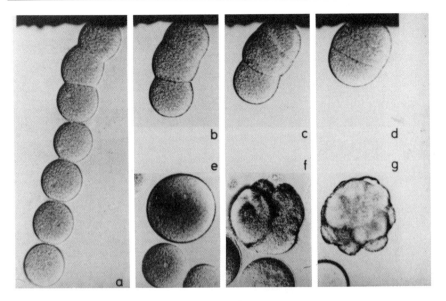

FIG. 7. Time course of sea urchin egg fusion aligned dielectrophoretically after Pronase removal of the vitelline layer. Fusion was induced by 2 breakdown pulses of 400 V/cm field strength, 50 μsec duration. Photographs were taken (a) 1 min, (b) 2 min, (c) 3 min, (d) 7 min, and (e) 15 min after field pulse application. Photographs (f) and (g) show irregular cleavage patterns of fused 2-egg stages after transferring the eggs into artificial sea water and subsequent sperm fertilization.

and 9). Without such contact, only solute exchange between the cell interior and the external medium occurs. This exchange can be used either for gene and protein incorporation into living cells or for release of intracellular substances.

Given a high enough cell density, fusion between freely suspended cells may nevertheless occur. The release of intracellular substances into the external medium is minimized with high cell densities,[7] an advantage for electropermeabilization and in electroinjection of membrane-impermeable substances into cells. Interestingly, high cell densities led to the discovery of electrofusion 9 years ago[26] while using human red blood cells for field-induced drug encapsulation.[7]

At first glance, the fusion of cells under these conditions cannot be understood. Even at very high suspension densities, cells are still some micrometers apart with a homogeneous field between parallel plate electrodes.

[26] U. Zimmermann and G. Pilwat, *Symp. Pap. – Int. Biophys. Congr., 6th* p. 140 (1978).

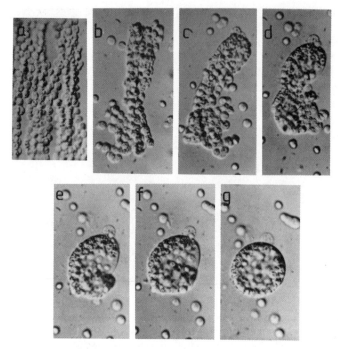

FIG. 10. Formation of giant nucleated cells. High cell suspension density of mouse Friend leukemia cells was dielectrophoretically aligned (200 V/cm, 1 MHz frequency). Photographs of giant cell formation taken after breakdown pulse application (2 kV/cm, 20 μsec duration); (b) 20 sec, (c) 40 sec, (d) 60 sec, (e) 90 sec, (f) 100 sec, and (g) 150 sec.

Closer inspection reveals an inhomogeneous field around the cells during pulse application.[5] A high field strength induces rapid cell movement. Simultaneously, close membrane contact and the build-up of the breakdown voltage is established in the first part of the breakdown pulse. We now know that electrofusion requires the membranes of two adjacent cells be no more than 20 nm apart from each other (quoted in Ref. 17). Calculations show that this distance can be achieved by the breakdown pulse in high suspension densities or pellets.[14] Direct contact and intermingling between the membranes are generated by the disruption of the water layer in the 20 nm gap and by the generation of defects in the membranes.

High suspension densities have been used for electrofusion of cells of different origins. Modifications of this procedure are also described in the literature.[14] Cells are subjected to the breakdown pulses at low suspension density followed by centrifugation. Fusion is observed and hybrids can be obtained by subsequent selection. The yield is naturally lower under these conditions (the defects generated in the membranes are not always in the appropriate positions to each other). In any case, supercritical field strengths are required to permeabilize larger parts of the membrane.

Electrofusion can also be achieved by the addition of fusogenic substances like polyethylene glycol (PEG) and subsequent application of the breakdown pulse (quoted in Refs. 5 and 15). The disadvantages of chemical and viral mediated fusions are the lack of (1) control of cell fusion events and (2) direct visualization of the events.

In the standard electrofusion technique, cell movement and membrane contact can be visualized before and during field pulse application by a separate step (Figs. 4 and 11). A nonuniform ac electric field of low strength and, therefore, of long duration is applied. The use of a dc field is not appropriate because movement of charged cells or particles in a non-

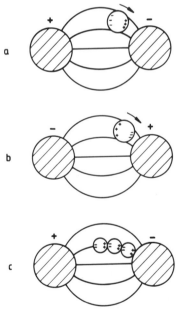

FIG. 11. Schematic representation of dielectrophoresis. This figure represents a cross-section of the parallel electrodes used in Figs. 6–10. External fields induce charge separation in a cell. In an inhomogeneous field, the field strength on the two hemispheres is unequal and therefore pulls the cell toward the region of highest field strength. (a) Cell migration toward an electrode. (b) Similar cell migration even when the external field is reversed. (c) Pearl chain formation of dielectrophoretically charged cells at higher cell densities. Alternating field frequencies between 10 kHz and 1 MHz are sufficient for inducing close membrane contact.

uniform field (also called dielectrophoresis[27]), is masked by electrophoresis. In an ac field of kHz to MHz frequency, electrophoresis is absent and only dielectrophoresis can occur (Fig. 11).

The field strength required for dielectrophoresis depends on the cell volume and the divergence (inhomogeneity) of the field. The dielectrophoretic force is, to a first approximation, proportional to the volume and to the inhomogeneity of the field. For large cells (diameter about 40 μm) only a few volts/cm are needed. Smaller cells require field strengths on the order of 100 V/cm.

The field strength can be adjusted by different electrode geometries and suspension densities which lead to different inhomogeneities of the electric field. In any case, the appropriate parameter choice depends on the critical field strength leading to breakdown. The dielectrophoretic field strength must always be significantly below the breakdown field strength. Otherwise, nonviable hybrids may occur (for reasons, see Refs. 5–17). Dependent on the field parameters, an alternating, nonuniform field leads to cell alignment and close membrane contact within minutes.

To ensure viability of the hybrids, the field pulse application should be as short as possible (e.g., fusion of myeloma cells with lymphocytes 30 sec, fusion of yeast protoplasts 1 to 2 min).

Dipole forces between neighboring cells induce cell alignment (or formation of cell chains) according to the field lines. The dipole force (responsible for the build-up of membrane potential difference) arises because of the charge separation in the cell. These dipole forces are generally much larger than the repulsion forces, i.e., net surface charge, Brownian movement, hydration, or gravitational forces. Recent experiments during a TEXUS (Technologic Experiments Under Microgravity) ballistic flight suggest that under a microgravity environment higher yields of hybrids can be achieved.[28]

The field inhomogeneity and the ratio of the dielectric constants of cells to medium dictate the region in the electric field where cell alignment occurs. If the ratio of dielectric constants (cells to medium) is larger than 1, cell movement and alignment will always occur in that field region where the field strength has its maximum value (i.e., positive dielectrophoresis).

If two parallel electrode wires are arranged on a slide a distance of 100 μm to 1 mm apart (depending on cell size), cell alignment will occur at the electrodes (the site of maximum field strength, Figs. 4, 6, and 8). However, due to interactions of the charged slide surface with cell move-

---

[27] S. Takashima and H. P. Schwan, *Biophys. J.* **47**, 513 (1985).

[28] U. Zimmermann, K.-H. Büchner, and R. Schnettler, "TEXUS 11/12 Final Campaign Report," p. 39, DFVLR, Köln, Federal Republic of Germany (1985).

ment, many cell chains will be formed in the middle of the electrode gap (Fig. 12). When using electrode wires which are 1 mm apart, the inhomogeneity of the field is more uniformly distributed and predominantly generated by the cells introduced into the field. As a result, the field appears more homogeneous and cell alignment will occur across the entire gap.

On the other hand, if the dielectric constant ratio is less than 1, dielectrophoresis and alignment will always occur in the weakest field strength region (i.e., negative dielectrophoresis).

The occurrence of negative or positive dielectrophoresis depends on the frequency of the ac field and the medium composition. Cell and medium dielectric constants are complicated parameters influenced by the frequency of the field and the external medium conductivity. A more detailed account of this subject can be found in the literature cited.[16]

For each unique system, the optimization of alignment parameters is best achieved by manipulation of each variable (e.g., frequency, field strength, and field inhomogeneity) at constant cell density and medium composition. Extrapolation from related systems provided feasible parameter ranges.

The investigator should note that the ac field frequency must never exceed the 10-MHz range since the membrane impedance drastically decreases. Under these conditions, the field lines will go through the cell

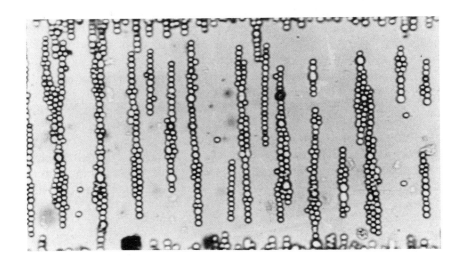

FIG. 12. Cell chain formation in an electrode gap. Friend cells were aligned dielectrophoretically (100 V/cm, 2 MHz).

interior and may interact adversely with the intracellular components. Frequencies lower than 1 kHz should also be avoided because of the occurrence of electrolysis.

Cell alignment has to be carried out in low conductivity media because of ac field induced heating at higher conductivity. This type of media has been utilized extensively with cellular (free flow) electrophoresis with limited damage to most cells.[29] In particular, the membrane permeability of red blood cells to various ions after breakdown has been shown to increase at higher ionic strengths.

Low conductivity media favor high yield of hybrids because of the decrease in membrane resistance following breakdown. The cell interior remains protected from the field which lasts for some time (in microseconds) after the relatively rapid breakdown (in nanoseconds).

The use of dielectrophoresis in combination with breakdown pulses (conventional electrofusion) leads to the production of large numbers of viable hybrids. This method has been used most by investigators.

For special purposes, alternative methods may be used for the establishment of membrane contact in conductive solutions. For example, Kramer et al.[30] were able to produce magnetically coated erythrocytes by the use of Fe_3O_4 particles. By means of two perpendicular magnetic fields the coated cells could be attracted into a very small volume. The force directed toward the center of this region was sufficient (the fields had strong gradients) to give contact between the cells. Application of two field pulses produced fusion products. The possible incorporation of Fe_3O_4 particles into the fusion products may be used as an advantage in subsequent magnetic isolation of fused cells. This procedure has important applications where cells do not display suitable genetic markers for selection (e.g., commercially important yeast cells).

In a sonic field,[31] forces are exerted on cells in a somewhat analogous manner to those exerted during dielectrophoresis (Figs. 13 and 14). The force exerted in the sonic field depends on the density difference between particles and medium (cf. in dielectrophoresis, the difference in dielectric constants is important). Sound wavelengths may be used that are much smaller than the fusion chamber. This permits not only the production of pearl chains (in a purely propagating wave) but also the concentration of cells at standing wave pressure maxima. The effect is so powerful that cells

[29] K. Hannig, *Electrophoresis* **3**, 235 (1982).

[30] I. Kramer, K. Vienken, J. Vienken, and U. Zimmermann, *Biochim. Biophys. Acta* **772**, 407 (1984).

[31] J. Vienken, U. Zimmermann, H. P. Zenner, W. T. Coakley, and R. K. Gould, *Biochim. Biophys. Acta* **820**, 259 (1985).

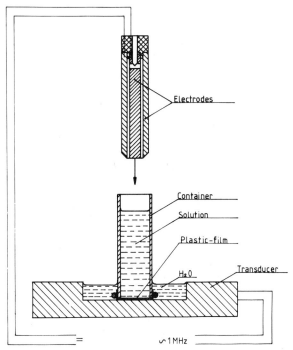

FIG. 13. Schematic diagram of electroacoustic apparatus for erythrocyte and myeloma cell fusions.

are often completely absent from other regions of the standing wave pattern. Vienken et al.[31] were able to use 1 MHz ultrasound (1 mm wavelength) to establish contact between red blood cells, myeloma cells, or myeloma and lymphocyte cells (Fig. 14). The standing wave maxima of 0.5 mm periodicity were developed between two electrodes. Application of breakdown pulses subsequently led to fusion. By variation of the cell density, preferential formation of two cell hybrids was possible.

Recently, we have shown (unpublished data) that a brief application of an alternative electric field (about 1 sec dielectrophoresis) after sonic concentration and prior to pulse application results in very high yields for hybridoma cells.

Standard electrofusion, electroacoustic and magnetoelectrofusion have the advantage that the process can be viewed under the microscope. The fusion process is nearly synchronous and fusion products can be identified in relation to their parental cells. With the appropriate equipment, media, and cell density, it is possible to fuse either two or three aligned cells (which

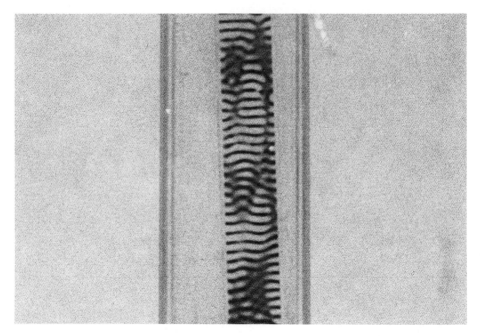

FIG. 14. Banding of erythrocytes as a result of electroacoustic field application. The planes of erythrocytes are separated by one-half a wavelength and at right angles to the direction of sound propagation.

leads to high yields of hybrids, Figs. 6–8) or thousands of cells (which leads to the formation of giant cells, Figs. 9 and 10).

The entire fusion process can often be controlled with the addition of specific agents that enhance chemically mediated membrane contact. An interesting chemical approach for electric field pulse-mediated fusion was recently reported by Lo et al.[32] These authors were able to establish contact between lymphocytes and myeloma cells by synthesizing a selective cross-linking agent. Immunogen–avidin complexes were used to bridge the specific lymphocytes bearing immunoreactive surface immunoglobin to biotinylated myeloma cells. Bridged rosettes were then fused selectively by pulse application. However, this method, valuable for hybridoma production, has been criticized for several reasons.[15,33]

A considerable improvement to this avidin–biotin method was pub-

[32] M. M. S. Lo, T. Y. Tsong, M. K. Conrad, S. M. Strittmatter, and L. D. Hester, *Nature (London)* **310**, 792 (1984).
[33] D. M. Wojchowski and A. J. Sytkowski, *J. Immunol. Meth.* **90**, 173 (1986).

lished recently by Wojchowski and Sytkowski.[33] In the first step of their approach, biotin was attached both to immunogen and to the surface of murine myeloma cells (via N-hydroxysuccinimidobiotin). Streptavidin cross-linked biotinylated myeloma cells with immune lymphocytes bearing biotinylated immunogen. The cellular–molecular complex was fused by electric fields resulting in a high frequency of specific immunoglobulin producing hybridomas. Although presently restricted to hybridoma production, this approach may prove to be very powerful.

The experiments have been interpreted as suggesting increased frequency of contact and physical approximation led to increased number of hybrids. On the other hand, current experimentation in our laboratories suggests secondary cellular physiologic events may be activated during various physical or chemical modifications. Activation of key cellular events appears to have an important role of hybridoma survival postelectrofusion.

Media Constraints

For electropermeabilization (electroinjection) of membrane-impermeable substances, experiment shows that media containing about 30 mM electrolytes are most suitable.[20] KCl, rather than NaCl, results in higher yields of transformants (most likely due to minimization of K$^+$ release from the cell). To achieve isoosmolarity, equivalent amounts of sugars have to be added. As with electrofusion, inositol, sorbitol, or sucrose should be used (mannitol is not recommended). These sugars minimally change the water structure and, in turn, enzyme activities. Although relatively expensive, inositol appears to work best in both electrogene transfer and electrofusion.

Small amounts of Ca^{2+} and Mg^{2+} have to be added to the media during field pulse application both in electrofusion and electropermeabilization. The optimum concentration ranges for Ca^{2+} and Mg^{2+} are very narrow. For example, in electrofusion of yeast protoplasts[34] or lymphocyte : myeloma cells,[35] the Ca^{2+} concentration is adjusted between 0.1 and 0.3 mM and the Mg^{2+} concentration between 0.3 and 0.5 mM. In electrogene transfer, Ca^{2+} and Mg^{2+} should be completely omitted.[20] In electroprotein injections, 1 mM Ca^{2+} should be added.[2,18]

In some cases, electrofusion and electroinjection can be performed in commercially available media. However, the pH indicator dyes are toxic in the electropermeabilized state of the membrane and must be omitted.[11] The dye related toxicity can also be apparent when cells are added immedi-

[34] R. Schnettler and U. Zimmermann, *FEMS Microbiol. Lett.* **27**, 195 (1985).
[35] J. Vienken and U. Zimmermann, *FEBS Lett.* **180**, 278 (1985).

ately to nutrient media postpulse application. Membranes have to be completely resealed in special resealing media as been previously described[2,35] or in the sugar solutions described above (unpublished results).

A very important point which has been overlooked is contamination by trace amounts of heavy metal ions.[11] Heavy metal ions are always present as contaminants in the ppm range in the pulse application media. In the electropermeabilized state, significant uptake of these substances can occur resulting in decreased cell viability. However, as shown in electrorotation studies,[36] small amounts of heavy metal ions (ppb range or nanomolar concentrations) can also interfere with the intact membrane inducing adverse effects on membrane and cellular properties.

Heavy metal ions can be effectively removed by preincubating media with ion selection beads. Chelex beads (Bio-Rad, CA), for example, strongly bind bivalent and multivalent ions. Ca^{2+} and Mg^{2+} are added after the Chelex beads have been removed.

The pH of the pulse media must be adjusted to physiological values (around pH 7) by use of small concentrations of phosphate buffer or histidine. Lower or higher pH values can drastically reduce the yield of transformants or hybrids.

Electronics and Pulse Chambers

Various electronics and chambers for electropermeabilization (electroinjection) and electrofusion are described in literature[3-17] and are commercially available. Electronics and chambers are utilized according to the specific application (Figs. 15–19; see color insert for Fig. 19). The principles for electronic design are shown in Fig. 18. For electropermeabilization (electroinjection), a high-voltage source should generate pulses, on the order of 10 kV, over an electrode distance of 1 cm. For electrofusion, a lower voltage source must be utilized over a smaller distance (100 to 500 μm). In addition, a frequency generator must be present for dielectrophoretic induced membrane contact.

Figure 19 shows a typical pulse generator for electropermeabilization (electroinjection). The reactant solution is aspirated (carefully avoiding bubbles) into the chamber mounted behind the electronics. The appropriate pulse is generated across the electrodes (separated by 1 cm). The solution is then discharged into a new vessel.

Other chamber configurations are shown in Fig. 15. The mäander

[36] W. M. Arnold, B. Geier, B. Wendt, and U. Zimmermann, *Biochim. Biophys. Acta*, **889**, 351 (1986).

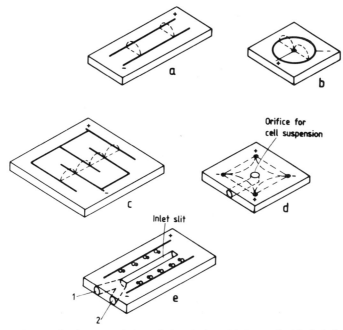

FIG. 15. Schematic diagrams of electrofusion devices. (a) Two cylindrical platinum wires mounted in parallel on a glass slide. (b) Central wire (or cylinder) surrounded by an outer cylindrical electrode. This arrangement allows achievement of a stronger divergence of the electrical field. (c) Mäander chamber with interdigitating platinum electrode configuration for fusion of large number of cells. (d) Flow chamber system where cells are introduced into the central chamber region between 4 opposing electrodes. (e) Flow chamber system for successive cell injection; the two different cell suspensions are aspirated, in turn, through the inlet slit, favoring the formation of mixed cell partners.

chamber (Fig. 15c) can be used both for electroinjection and electrofusion. Mäander chambers have an advantage over plate condensors (when used for electroinjection) because the distance between the electrodes can be much smaller at constant "electrode volume." Therefore, lower voltages can be used to establish the required field strength.

For electrofusion, the helical and rotational chambers (Figs 16 and 17) have been developed to be very efficient in order to fuse large amounts of cells. In both chambers, the electrode distance is between 100 and 500 μm (depending on cells size, see above), whereby, as in the case of the mäander chamber, the electrode volume is very large.

In principle, the helical chamber consists of a hollow Perspex tube around which two cylindrical platinum electrodes are wound at a given

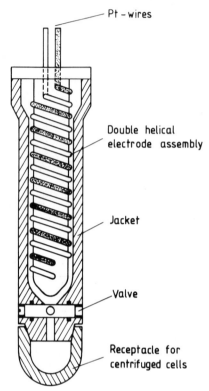

FIG. 16. Schematic diagram of a helical electrofusion chamber. Two parallel cylindrical platinum wires are wound spirally around a hollow tapered cylindrical Perspex tube. The tube is enclosed in a hollow Perspex jacket filled with the cell suspension. On insertion of the electrode into the jacket, the cells are displaced upward into the space between the wires and subjected to electric field application. After fusion, the chamber may be centrifuged and cells recovered from the receptacle.

distance (total length of the wires about 1 m). The Perspex tube can be filled with cooling medium in order to avoid temperature increase during the application of the ac field. The Perspex tube with the platinum wires is introduced into a cylindrical jacket which already contains the cell suspension. The cell suspension occupies the intervening space. Application of the ac field leads to the formation of cell chains between the parallel electrode wires. After application of a breakdown pulse and subsequent fusion, the cells can be collected by centrifugation into a receptacle by way of a valve. The receptacle is disposable and can easily be removed for cloning of the fused hybrids. Instead of centrifugation, the Perspex tube can also be gently removed from the jacket and the suspension is decanted

FIG. 17. Sputtered chamber made of a glass disk containing a bounded metal conducting surface. The disk is sputtered or photoetched with acid to remove a continuous 200 μm wide strip of metal. The remaining conducting surface (on either side of the sputtered surface) serves as electrodes for current applied via metal clips to the center and periphery of the disk.

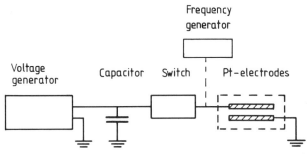

FIG. 18. Schematic diagram for electronic devices used for electropermeabilization and electrofusions. Parallel platinum electrode plates are connected to a capacitor by a switch. When the switch is closed, the capacitor is exponentially discharged. For electrofusion, a source of high frequency voltage is additionally connected to the electrodes for dielectrophoresis.

and diluted by nutrient media. The fusion process can be performed under sterile conditions.

The rotational fusion chamber consists of a cylindrically shaped casing which may be made to act as a centrifuge by spinning it around its axis of symmetry. The floor of the chamber consists of a Perspex carrier plate onto which electrodes have been previously vacuum evaporated. Of the various electrode configurations, Fig. 17 shows a configuration which has proved suitable for mass production of electrically fused yeast hybrids and hybridoma cells. The electrodes cover the circular area of the carrier disk, with electrode pairs (two electrodes running in parallel at a distance of 100 to 500 μm depending on cell size) radiating from the center. The thickness of the vacuum evaporated electrodes is about 1 μm. The circular electrode plate (or disk) is surrounded by a lower annular channel into which the fused cells can be centrifuged after the fusion process has been completed. The channel is filled with special resealing or nutrient medium.

The rotational chamber seen in Fig. 17 can be mounted on a standard centrifugal platform and spun accordingly. After centrifugation an outlet system (valve) allows the cells and the medium to be collected directly from the annular channel into culture vessels. During the fusion process, kinetic events of fusion can be monitored through the carrier plate by means of an inverted microscope.

This new chamber not only permits the large-scale production of hybrids, but also allows rapid screening of optimal media components and other system parameters (described above). For this purpose droplets of cell suspension are applied to different points on the disk area. The droplets do not spread out (because of the cohesive forces of water) and it is therefore possible to rapidly test the influence of different ions in the fusion medium on hybrid production.

Discussion

Over the last few years, a body of literature has accumulated indicating the advantage of field pulse methods in hybridoma production and macromolecule insertion. The purpose of this review is to identify the major contributing factors in electrofusions and electropermeabilization (electroinjection of genes, proteins, and other membrane-impermeable substances). The foregoing considerations should assist in determining the optimum conditions as they apply to each system. The authors of this chapter believe that many pitfalls can be avoided if the described principles are followed with respect to each targeted cell type.

Due to the inherent variation of biological systems, a specific method for electrofusion and electrotransfection is not possible to standardize.

However, once optimal conditions have been achieved for a given cell system, we have found excellent reproducibility. Current experiments are focused on methods for predicting pulse application outcomes as a function of cellular physiology and morphology. The advent of new instruments for detecting cellular and subcellular events during the entire procedure will undoubtedly contribute valuable insight into system control. It is hoped that optimization of these procedures will allow future automated large scale production of transfected cells and new hybrid cell lines.

Acknowledgments

This research was supported in part by the DFG, Sonderforschungsbereich 176 and the BMFT (DFVLR, West German Space Agency, 01 QV 354) for U.Z. We also wish to thank Dr. Garry A. Neil, Department of Immunology, Research Institute of Scripps Clinic, La Jolla, California, for critical review and constructive comments during the writing of this review.

[17] Cell Enucleation, Cybrids, Reconstituted Cells, and Nuclear Hybrids

By JERRY W. SHAY

The use of enucleation procedures results in the production of nuclear and cytoplasmic parts termed karyoplasts and cytoplasts, respectively.[1] By fusing a cytoplast with a karyoplast, a viable "reconstituted" cell can be produced (Fig. 1). Alternatively, by fusing a cytoplast from one cell to another whole cell, a cytoplasmic hybrid or cybrid is formed (Fig. 1). Finally, by fusing a karyoplast to another whole cell it is possible to form a nuclear hybrid (Fig. 1). The various protocols for the production and isolation of these cellular constructs will be described in this chapter.

Enucleation Protocols

Most enucleation procedures use a combination of treatment of cells with cytochalasin B (CB) and exposure to a centrifugal field.[2,3] Cytochalasin B is soluble in dimethyl sulfoxide (DMSO) or ethanol (95% or absolute). Concentrated stock solutions ($> 1/mg/ml$) can be stored for long

[1] J. W. Shay, K. R. Porter, and D. M. Prescott, *J. Cell Biol.* **59,** 311a (1973).
[2] D. M. Prescott, D. Meyerson, and J. Wallace, *Exp. Cell Res.* **71,** 480 (1972).
[3] W. E. Wright and L. Hayflick, *Exp. Cell Res.* **74,** 187 (1972).

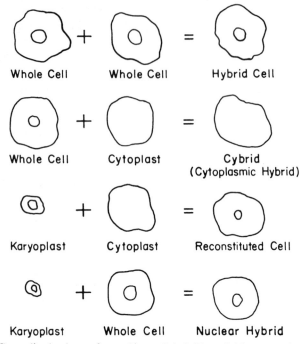

FIG. 1. Generalized scheme for making cell hybrids, cybrids, reconstituted cells, and nuclear hybrids. From Shay and Cram, *in* "Molecular Cell Genetics" (M. M. Gottesman, ed.), p. 161. Wiley, New York, 1985.

periods without losing potency. Diluted stock solutions in culture medium (5–10 μg/ml) are more stable if serum is not added.

Two general techniques are available for cell enucleation: monolayer and gradient.

Monolayer Techniques

Cells that grow as monolayers in culture can be enucleated by two procedures: (1) on coverslips punched out from tissue culture dishes or (2) in tissue culture flasks.

Coverslip Procedure. The original techniques were described by Prescott and Kirkpatrick.[4] Plastic coverslips are prepared from the bottoms of tissue culture dishes using a metal punch which is first heated and then used to "punch out" round disks which can then be spun in centrifuge tubes in the presence of CB to effect cell enucleation. The punched out

---

[4] D. M. Prescott and J. B. Kirkpatrick, *Methods Cell Biol.* **7**, 189 (1973).

disks should be ethanol or UV sterilized and then used for cell growth. Since only one surface of the plastic has been treated to permit cell growth, care must be exercised to identify that surface. An alternative to this is to punch out disks already containing cells.[5] For enucleation the coverslip disks are placed into a sterile, round-bottom centrifuge tube (cells facing bottom), containing 7–10 ml of prewarmed CB, and spun in a prewarmed rotor. Since the efficiency is highly temperature dependent, it is important to centrifuge cells as close to 37° as possible. Coating plastic culture dishes with collagen, poly(L-lysine), or concanavalin A[6] can often improve the monolayer techniques if attachment is poor.

For most murine, well adherent, fibroblast cell lines, a prewarmed (37°) Sorvall SS-34 rotor is spun at 15,000 rpm for 20–30 min at 35–37° in the presence of 10 μg/ml CB. However, the precise conditions to obtain greater than 95% enucleation must be empirically determined for each cell line. The disadvantage of this method is that the number of cytoplasts is somewhat limiting. For a typical experiment, eight disks can be centrifuged at once, which is not sufficient for most fusion experiments; so multiple enucleations or centrifuges are required.

Flask Enucleation. As is illustrated in Fig. 2, entire 25 cm² tissue culture flasks can be filled with medium containing CB, placed in acrylic inserts containing prewarmed distilled water, and centrifuged in the Sorvall GSA rotor. The obvious advantages of this technique are that six flasks can be centrifuged at one time (thus resulting in a greater yield of cytoplasts and karyoplasts) and extra steps of punching out disks are eliminated. The disadvantages of this technique are occasionally the flasks break during centrifugation so lower speeds are used, resulting in poorer enucleation. This can be circumvented by allowing the enucleated cells to recover for a short time and centrifuging a second time. Another disadvantage is that the construction of the inserts is costly. However, this procedure has been used without the acrylic inserts. Warm water can be added directly to the rotor, then flasks filled with CB, inserted, and centrifuged as described in a review by Veomett.[7]

Gradient Techniques

Mammalian cells that do not grow in monolayer or which adhere poorly in monolayer may be enucleated in a density gradient made of Ficoll. As originally reported by Wigler and Weinstein[8] and reviewed by

[5] R. Ber and F. Wiener, *Cytogenet. Cell Genet.* **21,** 304 (1978).
[6] T. V. Gopalarkrishnan and E. B. Thompson, *Exp. Cell Res.* **96,** 435 (1975).
[7] G. E. Veomett, *Tech. Somatic Cell Genet., 1982* p. 67.
[8] M. H. Wigler and I. B. Weinstein, *Biochem. Biophys. Res. Commun.* **63,** 669 (1975).

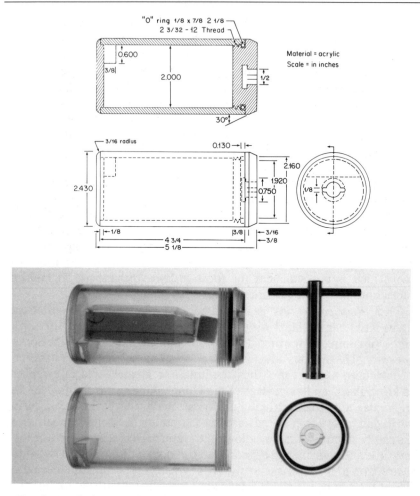

FIG. 2. Acrylic insert design for use in flask enucleation techniques. From Shay and Clark.[14]

Veomett,[7] a discontinuous gradient is prepared at the following Ficoll concentrations: 25, 17, 16, 15, and 12.5%. Solutions are prepared in medium without serum but containing 10 μg/ml CB. The step gradient can be prepared in sterile cellulose nitrate centrifuge tubes and is best allowed to equilibrate for several hours prior to enucleation. At the time of enucleation approximately $1-5 \times 10^7$ cells are suspended in the 12.5% Ficoll step and centrifuged at 25,000 rpm in the SW41 rotor for 60 min in a prewarmed ultracentrifuge. Care must be taken to avoid clumping of cells which inhibits enucleation. After enucleation the cytoplasts are usually located between the 15 and 17% Ficoll layers while the karyoplasts are at the 17–25% interface. The main disadvantage of this procedure is that it requires more time setting up the gradients and all cells do not enucleate well. The advantage is that a large number of cytoplasts and karyoplasts are generated and alternative techniques for suspension cell lines are usually not successful.

Cell Biology of Enucleation

As observed in the scanning electron microscope (Fig. 3a) logarithmically growing Chinese hamster ovary (CHO) cells have a variety of cell morphologies reflecting variations in the cell cycle. Upon 1–2 min centrifugation in the presence of 10 μg/ml CB most cells round up (Fig. 3b), but remain attached to the substrate by cytoplasmic connections. After 5 min enucleation (Fig. 3c and d) the protruding nucleus of each cell becomes distinct from the attached cytoplasm. With increasing time of centrifugation (Fig. 3e, 10 min; Fig. 3f, 20 min), the stalk of cytoplasm becomes thinner until the nucleus and small amount of surrounding cytoplasm are severed and the resulting karyoplast is centrifuged to the bottom of the tube or opposite side of the flask. The stalk of cytoplasm supporting the karyoplast may be in excess of 100 μm in length while only 0.5 μm in diameter. Figure 3g is a whole CHO cell, Fig. 3h is a karyoplast from such a cell, and Fig. 3i is a recovered cytoplast. Within 12–24 hr the cytoplast degenerates (Fig. 3j) and, as seen in a whole mount high voltage electron microscope image (Fig. 4B), cytoplasts contain all the cellular components except the nucleus.

The karyoplasts, though endowed with an intact nucleus, 5–10% of the cytoplasm containing some cytoplasmic organelles, and a continuous plasma membrane (Fig. 4A) also degenerate. As can be seen in Fig. 5, uridine (Fig. 5A) and leucine (Fig. 5B) incorporation decrease rapidly during the first 12 hr postenucleation. In addition, the percentage trypan blue incorporation increases with time (Fig. 5C) indicating decreased karyoplast viability. It has been suggested[9] that some karyoplasts regenerate, but it is much more likely that occasional whole cells detach during centrifugation or that karyoplasts containing a large amount of cytoplasm may be present along with the nonregenerating karyoplasts. Since the karyoplast fraction is heterogeneous, it is often necessary to further purify them. By plating the karyoplast fraction in a new culture dish for 90 min, the contaminating whole cells and larger karyoplasts will adhere while the

[9] G. A. Zorn, J. J. Lucas, and J. C. Kates, *Cell* **18**, 659 (1979).

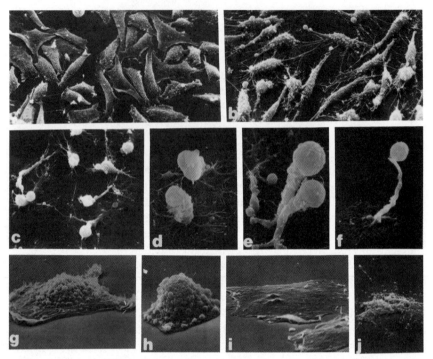

FIG. 3. Scanning electron microscope images of steps in cell enucleation (see text for description).

smaller (nonregenerating) karyoplasts will not. This differential adhesion technique allows one to obtain a greatly purified karyoplast fraction.

Alternatively, a fluorescence-activated cell sorter can separate the smallest karyoplasts on the basis of light scatter. In addition, it is possible to stain cells with the nontoxic mitochondrial dye (R-123) and separate out the smallest karyoplasts with minimal fluorescence and thus without mitochondria (Fig. 6). Finally, another method takes advantage of the cells' ability to ingest nontoxic heavy tantalum (Ta) particles. The Ta-containing cells are centrifuged in CB and the karyoplast fraction is then placed on a discontinuous Ficoll gradient. The whole cells and larger karyoplasts have greatly increased density because of the ingested Ta particles. Thus, the smallest karyoplasts, containing few or no Ta particles, are easily separated from the whole cells and larger karyoplasts.[10]

Cytoplasts can be further purified by several techniques. After the first

[10] M. A. Clark, J. W. Shay, and L. Goldstein, *Somatic Cell Genet.* **6**, 455 (1980).

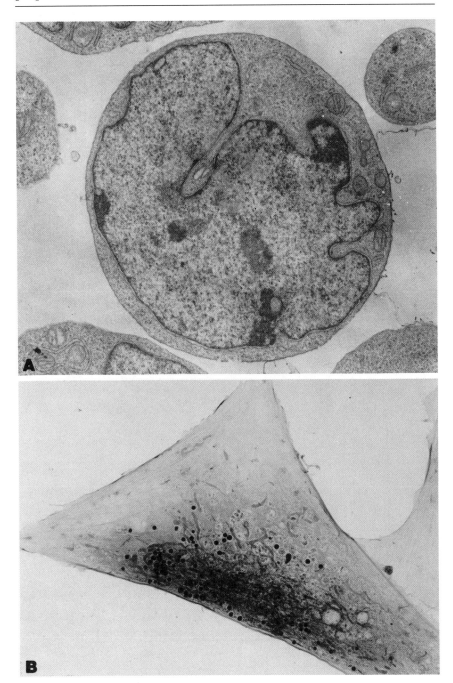

FIG. 4. (A) Transmission electron micrograph of a karyoplast; (B) high-voltage electron micrograph of a whole mount of a cytoplast.

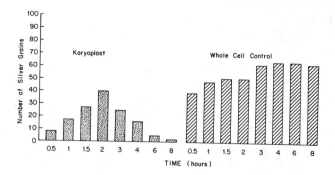

A

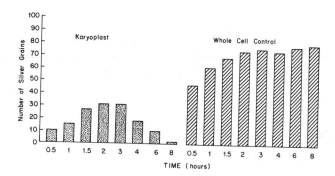

B

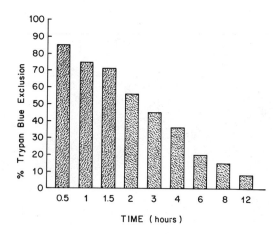

C

Fig. 5. Uridine (A) and leucine (B) incorporation and trypan blue exclusion (C) of karyoplasts. From Clark et al.[10]

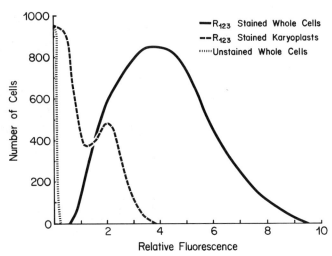

FIG. 6. Fluorescence scan obtained by using rhodamine-123 and the fluorescence-activated cell sorter for isolation of mitochondria-free karyoplasts.

round of enucleation, the cytoplasts and whole cells that did not enucleate can be allowed to recover in fresh medium and the percentage enucleation determined in the inverted microscope. If the percentage enucleation is poorer than desired then the flasks or disks can be centrifuged a second or third time. In addition, the cytoplasts and whole cells can be treated for a brief period with mitomycin C, a DNA cross-linking agent, or physically separated on a Ficoll gradient.[11,12]

Reconstituted Cells

There are several approaches to fusing cytoplasts with karyoplasts and isolating reconstituted cells. Reconstituted cells are the viable proliferating fusion products of a cytoplast with a karyoplast. There are physical isolation and genetic isolation protocols.

Physical Isolation of Reconstituted Cells

As is illustrated in Fig. 7, cells will ingest latex beads of various diameters. In this protocol the cells which will become the karyoplast donor are allowed to ingest 0.5 μm diameter latex beads while the cells which will

[11] W. E. Wright and L. Hayflick, *Exp. Cell Res.* **96**, 1131 (1975).
[12] G. Poste, *Exp. Cell Res.* **73**, 273 (1972).

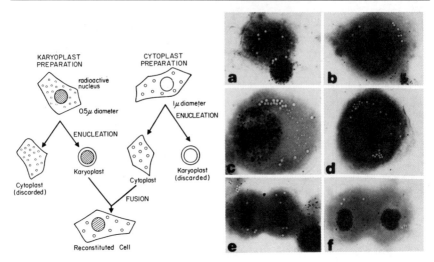

FIG. 7. Schematic representation and photomicrographs (a–f) of physical method for identifying reconstituted cells. From Veomett *et al.*[13]

become the cytoplast donor ingest 1.0 μm diameter latex beads. This serves to mark, temporarily, the cytoplasm of each parental cell. The nucleus of each parental cell can be identified by the karyotype (chromosome number) or, in the case of Fig. 7, by labeling the karyoplast donor with [^{3}H]thymidine (Fig. 7a). After enucleation and fusion, only those cells that have large latex beads (Fig. 7b, d, e, and f) and a labeled nucleus are considered reconstituted cells. Cells that contain both large and small beads are hybrids or cybrids (Fig. 7c), while contaminating whole cells contain only small latex beads. The various images in Fig. 7a–f illustrate some of the products identified from one experiment. Several reconstituted cells are shown in various stages of mitosis (Fig. 7d, e, and f) indicating that they are viable and proliferating. Hybrid cells can be separated from reconstituted cells, since they will initially be heterokaryons (e.g., binucleate cells) and, if missed during the initial screening, can be identified based on karyotypic differences (see Veomett *et al.*[13]).

In a typical experiment one could label the cells with different sizes of latex beads, enucleate each parental cell type, purify the karyoplasts via differential adhesion, fuse with polyethylene glycol, and plate the reconstituted cells on broken glass coverslip fragments at dilutions such that only

[13] G. Veomett, D. M. Prescott, J. W. Shay, and K. R. Porter. *Proc. Natl. Acad. Sci. U.S.A.* **71,** 1999 (1974).

one reconstituted cell attaches per fragment. These fragments are transferred to 48- or 96-well dishes and screened microscopically during the first 24 hr for the presence of a single nucleus and large latex beads. Contaminating whole cells from the karyoplast preparation have small latex beads. Pretreatment of the cytoplast preparation with mitomycin C eliminates any contaminating whole cells that did not enucleate. Usually karyotypic differences between the parental cells makes this step unnecessary. All viable proliferating constructs, that originally had only large latex beads, are karyotyped to ensure presence of the nucleus of the karyoplast donor.

Latex beads can be purchased form Duke Standards (Palo Alto, CA) or Polysciences, Inc. (Warrington, PA). With 1.0 μm diameter latex spheres we make up a solution of approximately 10^6/ml in regular medium and incubate the cells overnight. Generally 5–10 spheres are taken up per cell. Approximately 6% of the cells take up no spheres and 6% take up greater than 20 spheres. Addition of 100 μg/ml of DEAE may enhance bead uptake for some cell types.

Genetic Isolation of Reconstituted Cells

The development of selectable cytoplasmic genetic markers such as resistance to the antibiotic chloramphenicol, which is due to a mutation in a mitochondrial ribosomal RNA gene, has allowed easier isolation of reconstituted cells. As is illustrated in Fig. 8, the cytoplasmic donor is chloramphenicol resistant and is TK$^-$ and thus will die in medium containing HAT but not CAP. The other parental cell type (the nuclear donor) is wild type (e.g., chloramphenicol sensitive and TK$^+$ and thus will grow in HAT medium but die in CAP). Both parental cells are enucleated, fused, and plated in medium containing HAT and chloramphenicol. The parental cells will die in this medium. Cybrids can survive in this medium so latex spheres can help identify which viable proliferating fusion products are cybrids and which are reconstituted cells. The cybrids will grow but initially contain latex spheres while the reconstituted cells will not contain latex spheres (see Shay and Clark[14]).

HAT medium is 10^{-4} hypoxanthine, 10^{-5} aminopterin, 4×10^{-5} thymidine. Glycine is in DMEM medium. HAT is commercially available from Sigma in 50X lyophilized, sterile format. Chloramphenicol (CAP) is used at 50–100 μg/ml. Generally a concentrated stock solution in ethanol (95%) is filter sterilized and added to cells as needed. A 50–200X stock solution is stable at 4° for months.

[14] J. W. Shay and M. A. Clark, *J. Supramol. Struct.* **11**, 33 (1979).

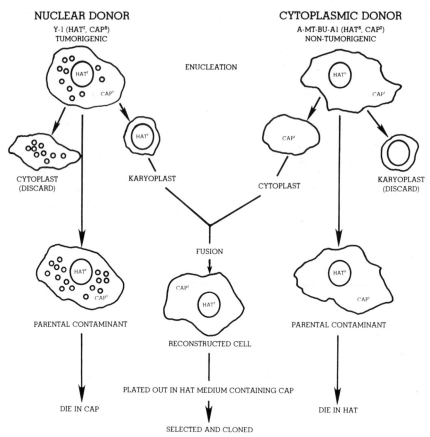

FIG. 8. Schematic representation of the genetic selection protocol for isolating reconstituted cells. From Shay and Clark.[14]

Cybrids

Many of the same techniques developed for isolating reconstituted cells can be adapted to isolate cybrids.

Cybrid Selection Using Fluorescence-Activated Cell Sorting

Even though genetic selection protocols give greater reliability, unfortunately weeks of cell culture must be undertaken before the colonies are sufficiently large enough to be analyzed. In some instances early analysis is desirable, since cytoplasmic regulatory factors may influence gene expression for only a short period. Fluorescence-activated cell sorting can be used for the rapid isolation of cybrids. Cytoplasts can be fluorescently labeled using the mitochondrial vital stain R-123. Rhodamine 123 (R-123) is purchased from Eastman Kodak. A stock solution in distilled water (1 mg/ml) is stable at 4°. R-123 is used diluted 1:100 (v/v) in growth medium 15 min prior to use. Since it is generally not toxic some cells can be continually grown in its prescence. After enucleation the cytoplasts are rinsed several times and fused to unlabeled whole cells or karyoplasts and analyzed immediately on the cell sorter.

In a typical experiment (Fig. 9, top) unstained Y-1 cells have only background levels of fluorescence in comparison with R-123 labeled AMT cytoplasts. Cybrids (Y-1 cells fused to R-123 stained AMT cytoplasts) are isolated by setting the sorter to "gate out" greater than 99.9% of all unstained cells. The sorted cells can be microscopically examined to ensure purity.[15] In a similar experiment (Fig. 9, bottom) R-123 stained AMT cells are fused to unstained T984-15 cells.[16] The dot plots dramatically illustrate the potential utility of the cell sorter in isolating cybrids: (A) enucleated AMT cytoplasts labeled with R-123, (B) unlabeled T984-15 cells, (C) unfused control mixture of labeled AMT cytoplasts and unlabeled T984-15 whole cells (note the small amount of low-level, nonspecific dye uptake of unfused T984-15 cells), and (D) labeled AMT cytoplasts fused to T984-15 cells. Cybrids are larger (increased scatter intensity) and more fluorescent than unfused T984-15 cells in the control mixture. By setting the sorter "gate" it is possible to isolate cybrids with varying amounts of mitochondria. Thus, analysis of cytoplasmic dosages is possible.

Cybrid Selection Using Toxin–Antitoxin Protocols

Genetic selection of cybrids almost always requires mutated cell lines so that the study of somatic cells hybrids or cybrids involving normal diploid cells is difficult. Recently Wright[17] (see also chapter by Wright [18], this volume) has developed a toxin–antitoxin technique for isolating heterokaryons. I have utilized this technique (as illustrated in Fig. 10) to isolate cybrids formed by fusing normal diploid IMR90 cytoplasts to highly tumorigenic HT1080 cells to determine if nontumorigenic cytoplasm can modulate the tumorigenic phenotype. After IMR90 cells are enucleated, they are exposed to mitomycin C for 2 hr in order to kill any cells that did not enucleate. The cytoplasts are exposed to a monoclonal antibody made against ricin and introduced into the cells similar to the Okada and Rech-

[15] M. A. Clark and J. W. Shay, *Proc. Natl. Acad. Sci. U.S.A.* **79**, 1144 (1982).
[16] C. Walker and J. W. Shay, *Somatic Cell Genet.* **9**, 469 (1983).
[17] W. E. Wright, *Proc. Natl. Acad. Sci. U.S.A.* **81**, 7822 (1984).

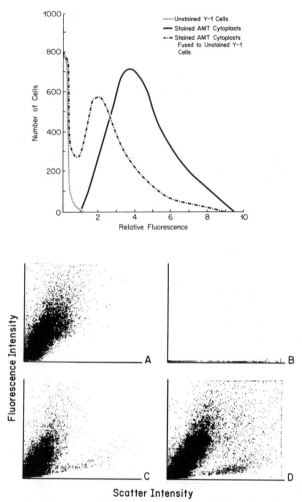

FIG. 9. Fluorescence scan obtained by using rhodamine-123 and the fluorescence-activated cell sorter for cybrid selection (top). From Clark and Shay.[15] Dot plot illustrating cells analyzed using the fluorescence-activated cell sorter (see text) (bottom). From Walker and Shay.[16]

steiner[18] procedure for introducing macromolecules into cells. The cytoplasts are rinsed well, fused to HT1080 cells, and plated in medium containing ricin. The unfused HT1080 cells die in the ricin and the unfused IMR90 cytoplasts degenerate. The mitomycin C step can be omitted if chromosomal analysis is performed to eliminate hybrids.

[18] C. Y. Okada and M. Rechsteiner, *Cell* **29,** 33 (1982).

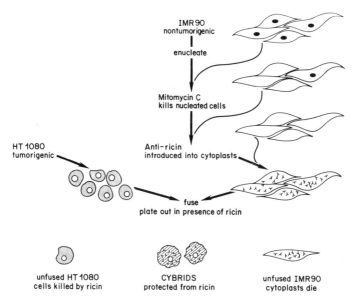

IMR 90
nontumorigenic

enucleate

Mitomycin C
kills nucleated cells

HT 1080
tumorigenic

Anti-ricin
introduced into cytoplasts

fuse
plate out in presence of ricin

unfused HT 1080
cells killed by ricin

CYBRIDS
protected from ricin

unfused IMR90
cytoplasts die

FIG. 10. Schematic representation of the toxin–antitoxin protocol for isolating cybrids.

Nuclear Hybrids

Fusion of nonregenerating karyoplasts to whole cells results in the production of nuclear hybrids that in some instances may be different from whole cell hybrids as previously reported by Weide et al.[19] Whole cell intraspecific hybrids, in addition to containing the nuclear genome of the two parental cells, often have the mitochondrial genome of both cells. Nuclear hybrids have the nuclear genome of both cells but mostly the mitochondrial genome of one cell. As illustrated in Fig. 11, a genetic isolation technique to isolate nuclear hybrids utilizes the resistance to CAP to identify contaminating whole-cell hybrids. In this experiment the karyoplasts are obtained from the TL-1 CAP-resistant cell line. The smallest karyoplasts can be isolated by sorting for absence of mitochondrial R-123 staining and then fused to PCC_4 cells. Both hybrids and nuclear hybrids survive in medium containing HAT while unfused cells die. After clones are of sufficient size they are split and half of each clone is placed in medium containing HAT and the other half in HAT and CAP. The clones that die in HAT and CAP are true nuclear hybrids and the surviving half of the clone that was not tested in CAP can be further analyzed. Clones that survived in HAT + CAP contain enough mitochondria to confer antibiotic resistance and are thus considered hybrids. The procedure allows the char-

[19] L. G. Weide, M. A. Clark, C. S. Rupert, and J. W. Shay, Somatic Cell Genet. **8,** 15 (1982).

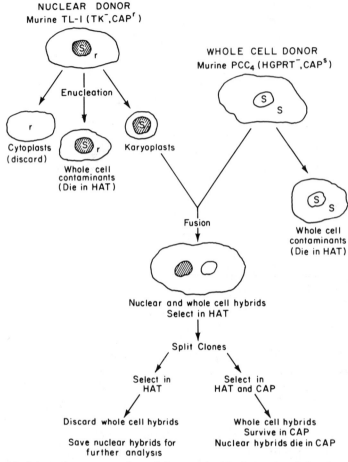

FIG. 11. Schematic representation of the genetic selection protocol for characterizing nuclear hybrids. From Weide et al.[19]

acterization of nuclear–nuclear interactions in the presence of predominantly one cytoplasm and may be useful in understanding the role of mitochondria in hybrids.

Conclusions

Cell enucleation and fusion provide a technology to determine if the cell cytoplasm contains stable regulatory substances that might modulate nuclear gene expression. It has been reported that added cytoplasm can have either a permanent, long-lived (2–8 weeks), short-lived (1–2 days), or no effect on the phenotype of the nuclear recipient. The protocols described in this chapter may be useful for helping to dissect the nature of these cytoplasmic regulatory substances.

Acknowledgments

This research was supported in part by the National Cancer Institute (CA40065), the American Cancer Society (CD-347), and The Council for Tobacco Research-U.S.A., Inc. I thank my former students, M. A. Clark, C. Walker, and L. Weide for their help in this work.

[18] Toxin–Antitoxin Selection for Isolating Heterokaryons and Cell Hybrids

By Woodring E. Wright

Most studies of the behavior of cell hybrids have employed genetic selection techniques, such as the HAT system,[1] to isolate cell fusion products. Although a number of different systems based on this general approach have been extremely useful, there are several practical and theoretical problems associated with their use. For example, only special cell lines that lack the required enzymes can be used. Although in some cases this problem can be overcome by mutagenizing and selecting for cells that carry the necessary defect, this process is often difficult, time consuming, and only conveniently done with immortal cell lines. The cultured proliferative capacity of normal diploid cells[2] is limited, so that by the time mutant clones are isolated they only have a few cell divisions left in their lifespan. Some "half-selective" strategies have been devised where an established immortal cell line is fused to a normal diploid cell type which does not divide in culture, in which it is unnecessary to select against the nondividing normal diploid parental cell type. However, the inability to use genetic selection techniques with normal cells frequently represents a major limitation for many types of studies.

A second limitation that has not been widely appreciated is that a substantial bias is introduced by the fact that these selective systems depend on the ability of cells to *divide* in the selective medium. Fewer than 1% of the cells that actually fuse give rise to growing hybrid colonies in most reports, and in some cases it is as few as 0.001%.[3] Although this may be unimportant in experimental systems where cell division and the phenotype of interest are unlinked, there are many situations in which the two processes are intimately associated. For example, most differentiated cells

---

[1] J. Littlefield, *Science* **145**, 709 (1964).
[2] L. Hayflick, *Exp. Cell Res.* **37**, 614 (1965).
[3] R. A. Miller and F. H. Ruddle, *Cell* **9**, 45 (1976).

are either slowly dividing or postmitotic. The phenotype of the fusion products will clearly be biased against the expression of differentiated products if the analysis is limited to the rare hybrid capable of extensive cell division. The same considerations apply to cell fusion investigations of the control of tumorigenesis.

The initial fusion product between two cell types, when the different nuclei reside within a common cytoplasm, is called a heterokaryon. When heterokaryons divide, they become cell hybrids. The cell hybrids are generally mononucleated, so that both sets of chromosomes reside within a single nucleus. Since most genetic selection systems depend on the selective ability of the fusion products to *divide,* they are (by definition) incapable of isolating heterokaryons. The behavior of heterokaryons has thus remained largely unexplored. It is now becoming clear that the initial regulatory interactions that occur in the heterokaryons before the onset of DNA synthesis can be very different from the ultimate phenotype observed in dividing cell hybrids.[4-6]

I have developed two alternative systems for isolating heterokaryons and cell hybrids. The first one exploits the different specificities of two irreversible biochemical inhibitors. Iodoacetamide is directed primarily toward sulfhydryl groups, while diethyl pyrocarbonate reacts mainly with histidine residues. One population of cells is treated with a lethal concentration of iodoacetamide while the other is exposed to diethyl pyrocarbonate. Both treatments are carried out for 30 min at 0°. The short treatment and cold temperature temporarily prevent the cells from expressing the secondary metabolic effects of the lethal lesion and dying. After washing the unreacted inhibitors away, the two cell types are mixed together and fused with polyethylene glycol. Because the two inhibitors have different specificities, a different spectrum of molecules is inactivated in each cell type. Iodoacetamide-treated cells thus contain active molecules that were inactivated in the diethyl pyrocarbonate-treated cells and vice versa. Complementation can occur and the heterokaryons are viable whereas the unfused parental cells die as a consequence of their lethal biochemical treatments.[7,8] Most of the results that I will discuss at the end of this chapter were obtained using this selective system. However, this method has several important drawbacks. Iodoacetamide and diethyl pyrocarbonate are sufficiently nonspecific that at high doses many molecules are

[4] W. E. Wright, *J. Cell. Biol.* **98,** 427 (1984).
[5] W. E. Wright, *Exp. Cell Res.* **151,** 55 (1984).
[6] H. M. Blau, C.-P. Chin, and C. Webster, *Cell* **32,** 1171 (1983).
[7] W. E. Wright, *Exp. Cell Res.* **112,** 395 (1978).
[8] W. E. Wright, *Tech. Somatic Cell Genet., 1982* p. 48.

inactivated by both agents. If a higher than necessary dose is used, so much overlap is produced that complementation no longer occurs and the heterokaryons die. It is thus necessary to adjust the conditions for each cell type and use a precise and narrow window of concentrations that is just sufficient to kill the parental cells while still permitting complementation in the heterokaryons. The technique is thus difficult to use. In addition, even though many different controls can be performed to show that the biochemical treatments are not producing long-term changes in the parameter being studied, lingering doubts remain as to the consequences of these horrendous treatments on the ultimate phenotype. I thus devised a second selective system that is much simpler to use and avoids these difficulties.

Toxin – Antitoxin System

Okada and Rechsteiner[9] have devised an elegant technique for introducing macromolecules into cells. Cells are first exposed to a hypertonic sucrose solution containing the molecule. Normally, the pinocytic vesicles that are formed during this period would be transported to the lysosomes, where their contents would be degraded. However, if the cells are exposed to a brief hypotonic shock before the hypertonic pinosomes have a chance to reach the lysosomes, they swell and burst, releasing their contents into the cytoplasm. This method permits almost any macromolecule to be "osmotically injected" into mass populations of cells, and thus allows the strategy illustrated in Fig. 1 to be performed. Ricin and abrin are plain toxins that inactivate the 60 S ribosomal subunit. Since both toxins act catalytically, it can take as few as one molecule per cell to inhibit protein synthesis and kill the cell. Our antibodies against these two toxins do not cross-react. Antiricin antibodies are injected into one cell type, while antiabrin antibodies are osmotically lysed into the other. Now the two cell types are fused together and plated in the simultaneous presence of both toxins. The antiricin-injected cells are killed by the abrin, the antiabrin-injected cells are killed by the ricin, but the heterokaryons contain both antibodies and are thus protected from both toxins. Using this approach, we have been able to obtain populations of normal diploid human fibroblasts in which 99% of the nuclei are present in heterokaryons and where virtually 100% of the heterokaryons that were formed survived.[10]

The basic selection protocol is to osmotically inject each cell type, wash away all of the noninternalized antibody, fuse the cells with polyethylene glycol (PEG), plate overnight in various toxin concentrations, then feed

[9] C. Y. Okada and M. Rechsteiner, *Cell* **29**, 33 (1982).
[10] W. E. Wright, *Proc. Natl. Acad. Sci. U.S.A.* **81**, 7822 (1984).

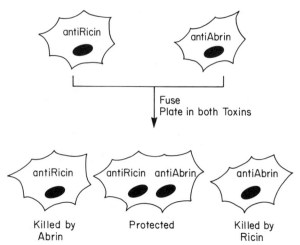

FIG. 1. Strategy for the toxin–antitoxin selection of heterokaryons. Cells osmotically injected with different antitoxin antibodies are fused together and plated in the presence of both toxins. Only heterokaryons containing both antitoxins survive.

toxin-free medium the following day. The parental cells die and detach from the dish by the third day following cell fusion, so that aliquots of the surviving cells can then be fixed, stained, and analyzed for the purity of the population (percentage of nuclei in cells containing more than one nucleus). A detailed consideration of each of these steps will be presented in the following sections.

Osmotic Injection

The original osmotic lysis procedure of Okada and Rechsteiner[9] used a hypertonic solution consisting of 0.5 *M* sucrose and 10% (v/v) PEG (MW 1000) in serum-free medium containing the macromolecule. Cells were first exposed to this solution for 10 min at 37°, either as a suspension of detached cells or as a cell monolayer. The advantage of using cells in suspension is that a much greater number of cells can be treated in a much smaller volume. The cells were then hypotonically shocked by diluting 0.1–0.2 ml of hypertonic solution with 10 ml of 60% medium : 40% distilled water. After 2 min, the cells were washed several times (or centrifuged and resuspended in the case of the detached cells) with isotonic medium and then fed maintenance medium. The important parameters in this protocol include the use of PEG, the temperature, the concentration of sucrose, the time of treatment, and the time and degree of the hypotonic treatment. Adequate assay systems are needed in order to optimize each variable for a particular cell type.

Systems for Assaying Osmotic Injection

Horseradish Peroxidase. Horseradish peroxidase (HRP) is an extremely sensitive tracer for measuring the amount of bulk phase pinocytosis.[11] Total uptake of HRP can easily be measured after lysing the cells (2% Triton X-100 in distilled water) by using a soluble substrate such as ABTS [2,2'-azino-di-(3-ethylbenzthiazolinesulfonic acid), 0.3 mg/ml in pH 4 citrate buffer to which hydrogen peroxide (0.01%) is added just before use]. After stopping the reaction by the addition of one quarter volume of 1.25% sodium fluoride, the absorbance at 417 nm is measured and compared to a standard curve. Measurements of total uptake do not distinguish between enzyme contained within pinosomes and that which is free in the cytoplasm. Okada and Rechsteiner[9] fractionated cells to separate soluble and particulate compartments and then quantitated the free cytoplasmic peroxidase resulting from osmotic injection. This type of cell fractionation has not been sufficiently clean in our hands to permit an adequate optimization of the different variables. Although we have used horseradish peroxidase to quantitate total uptake (and thus the maximum amount that would be released if all of the pinosomes were lysed), it has not been satisfactory for quantitating the amount actually released. Horseradish peroxidase histochemistry, however, is extremely valuable for determining if one is near or far from optimal conditions. In this approach the cells are incubated for 3 hr after osmotic injection in order to permit nonlysed pinosomes to become localized in lysosomes. They are then fixed (2% glutaraldehyde in saline for 3 min at room temperature) and stained with a substrate that gives an insoluble product [diaminobenzidine, 0.5 mg/ml in 0.1 M Tris buffer, pH 7–7.5, to which hydrogen peroxide (0.03%) is added just before use]. The relative amount of diffuse versus particulate staining (Fig. 2) then indicates whether one is lysing most or only a few of the pinosomes.

Antibody Protection. A rough measure for determining either the effective titer of an antibody or the efficiency of osmotic injection is to measure the amount of protection obtained after injecting an antitoxin antibody. Control cells (noninjected, mock-injected, or injected with an irrelevant antibody) and injected cells are plated in serial 3-fold dilutions of the toxin, then fed toxin-free medium the next day. Most of the cells that were killed by exposure to the toxin will have died and detached from the dish by the

[11] R. M. Steinman, J. M. Silver, and Z. A. Cohn, *J. Cell Biol.* **63**, 949 (1974).

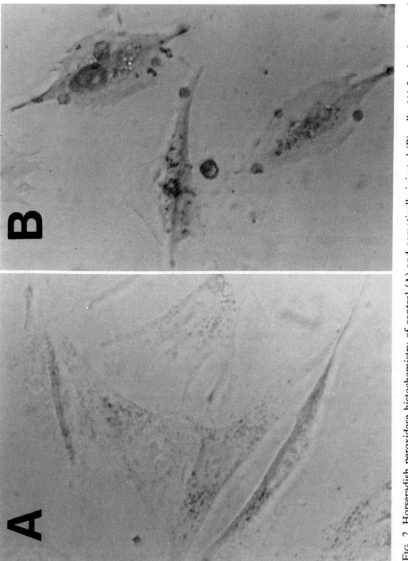

FIG. 2. Horseradish peroxidase histochemistry of control (A) and osmotically injected (B) cells. (A) In the absence of hypertonic treatments, horseradish peroxidase activity is granular (presumably in lysozomes). (B) Following osmotic lysis of the pinosomes, the peroxidase activity is diffusely distributed throughout the nucleus and cytoplasm.

third day after injection, so the cell viability in each dose can be determined by simply estimating the percentage of cells still attached with respect to a no-toxin control. Figure 3 illustrates the results of a typical experiment. The amount of protection is defined as the increase in the LD_{50} for the injected as compared to the control cells. The main disadvantage of using antitoxin protection to quantitate the efficiency of osmotic injection is its lack of sensitivity. Dilution experiments indicate that it takes about a 10-fold increase in antibody concentration in order to produce a 3-fold increase in the amount of protection conferred. Coupled with the fact that the estimation of the LD_{50} is only accurate within about a factor of 2, this assay becomes totally inadequate for determining if one has actually optimized the injection protocol (e.g., it would not distinguish between methods in which 30% versus 60% of the pinosomes had ruptured).

Ricin A Chain. Ricin is composed of two subunits, the cell-binding B chain and the catalytic A chain. The purified A chain is nontoxic, since it is unable to bind or enter cells by itself. It can thus be used as a very sensitive probe for the injection of macromolecules following osmotic lysis. Ricin A chain released into the cytoplasm by the rupture of pinosomes will inactivate the 60 S ribosomal subunit and inhibit protein synthesis, whereas ricin A chain still contained within unbroken pinosomes will travel to the lysosome and be degraded. This approach provides a much more sensitive measure of the efficiency of the osmotic injection technique than the degree of protection afforded by antitoxin antibodies, since the rate of inhibition of protein synthesis is directly proportional to the amount of ricin A chain liberated in the cytoplasm. A typical protocol would be to

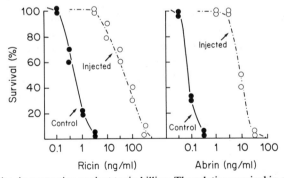

FIG. 3. Antitoxin protection against toxin killing. The relative survival in each aliquot was estimated from the number of human fibroblast cells remaining attached 2 days after an overnight toxin treatment for (A) antiricin toxin-injected cells and (B) antiabrin-injected cells.

inject cells with a constant amount of ricin A chain (e.g., 15 ng/ml) using different osmotic lysis conditions (see below), then wash away the free ricin A chain and put the cells back in culture. After 3 hr the cells would be labeled for 30 min with [³⁵S]methionine, lysed, TCA precipitated, and scintillation counted to determine the rate of protein synthesis. The relative amount of inhibition produced under various conditions would then indicate which condition gave the greatest amount of osmotic lysis.

The following general sequence was followed for optimizing the conditions for human diploid lung fibroblasts described below. Initially, horseradish peroxidase was injected and histochemistry performed 3 hr later. This demonstrated that the starting conditions were suboptimal, since most of the horseradish peroxidase was remaining in pinosomes and going to lysosomes. A variety of parameters were then investigated using the ricin A chain approach. Finally, the revised protocol was retested with horseradish peroxidase histochemistry, which showed that the modifications had been effective and now most of the peroxidase activity was being released into the cytoplasm.

Optimization of Osmotic Lysis Conditions

PEG Concentration. Okada and Rechsteiner[9] reported that 10% PEG 1000 had to be included during the hypertonic sucrose treatment in order to rupture the pinosomes in mouse L cells. We could detect no change in the efficiency of ricin A chain release from hypertonic pinosomes in human diploid fibroblasts with or without PEG. Since 10% PEG in 0.5 M sucrose was somewhat toxic and decreased the viability resulting from a 10 min treatment from 80 to 50%, we have eliminated it from our hypertonic protocol.

Temperature. The rate of pinocytosis varies significantly with temperature, with a Q_{10} of 2.7.[11] The highest possible temperature would thus give the highest rate of pinocytosis. Unfortunately, the toxicity of hypertonic sucrose proved to be directly proportional to temperature. Although 80% of the cells survived a 10 min exposure to 0.5 M sucrose at 37°, only 10% survived a similar treatment at 40°. Great care obviously needs to be taken to maintain a reproducible temperature, since even a 3° temperature change can make such a dramatic difference in cell viability. Using lower temperatures resulted in greater viability but lower rates of pinocytosis and lower efficiencies of lysis of the pinosomes. We selected the physiologic temperature of 37° for optimizing the remaining conditions.

Sucrose Concentration. The efficiency with which pinosomes were lysed was directly proportional to the concentration of sucrose in the hypertonic medium. Increasing the sucrose concentration from 0.5 to 1 M

resulted in approximately three times as much ricin A chain being released into the cytoplasm. Unfortunately, higher concentrations of sucrose produced unacceptably high cell death. Therefore 1 M sucrose was chosen as a compromise between efficiency and toxicity.

Time. Experiments in which cells were exposed to horseradish peroxidase for various intervals during a 10-min hypertonic treatment indicated that 50% of the total was taken up within the first 3 min, while 70% was pinocytosed within the first 5 min. Since the toxicity of a 5-min treatment was significantly less than that of the 10 min one, we selected 5 min as our standard treatment time.

Hypotonic Treatment. Using the original hypertonic protocol (0.5 M sucrose with 10% PEG for 10 min), we found that increased pinosome rupture was obtained when the cells were shocked with increasingly hypotonic solutions. However, the toxicity of the treatments went up in parallel with their increased efficiency. Since horseradish peroxidase histochemistry showed that the modifications described above (1 M sucrose without PEG for 5 min) already resulted in the lysis of most of the pinosomes and little would be gained from further modifications, we did not pursue the use of more hypotonic treatments. In addition, since cell viability was maintained for at least 15 min in 40% distilled water: 60% medium, it seemed prudent to minimize the number of critical variables and maintain the hypotonic exposure within conditions that exhibited a large tolerance for variation.

Calcium Concentration. The fact that toxicity and the efficiency of pinosome rupture paralleled each other in so many of the above experiments suggested that some of the toxicity might in fact result from the lysis of pinosomes. One candidate mechanism for this might be changes in intracellular calcium concentration. The concentration of extracellular calcium is about 1000 times greater than intracellular calcium. Quantitation of uptake using horseradish peroxidase indicated that the cells were pinocytosing about 10 fl of solution during a 10-min exposure at 37°. This represents about 1/1000th of the cytoplasmic volume of a "typical" cell 30 μm in diameter in which the nucleus occupies one-third of the total volume. Even if it were uniformly distributed throughout the cytoplasm, the 10 fl of uptake would be enough to double the intracellular calcium concentration, and might wreak havoc on a multitude of calcium regulated processes and cause cell death. The toxicity of the hypotonic treatments was substantially lowered if the calcium concentration in the hypertonic solution was reduced by adding EGTA or using Ham's F12 (which has about one-sixth as much calcium in it as Dulbecco's MEM). However, these manipulations also resulted in a 50% decrease in the amount of ricin A chain released into the cytoplasm. We have not yet determined if the lower calcium concentrations were directly affecting the rate of pinosome rupture or reducing the rate of pinocytosis. In either case, since reducing the calcium concentration involved a trade-off between cell viability and the efficiency of osmotic injection, the choice of an optimal calcium concentration would depend on the particular system being studied. If toxicity were a problem, the gain in viability achieved by reducing the calcium concentration to 0.3 mM would probably more than compensate for the accompanying decrease in the efficiency of osmotic lysis. Even if cell viability is not a problem, it is important to recognize calcium as a critical variable, since its concentration in serum (e.g., a rabbit antitoxin serum) is probably about 3–4 mM, as compared to only 1.8 mM in Dulbecco's MEM. For example, if one makes a 1 M hypertonic solution by mixing equal volumes of 2 M sucrose (dissolved in Dulbecco's MEM) and the antitoxin serum, the concentration of calcium in the actual experimental solution would be greater than that in test solutions made using Dulbecco's MEM alone. Furthermore, if one subsequently purified the antibodies and used them instead of the crude serum, the calcium concentration would again be different than in the initial tests using the crude serum. Although we have not yet chosen an optimal calcium concentration, its role is clearly important and further studies into its effects on osmotic lysis are indicated.

PEG Fusion

Dimethyl sulfoxide (DMSO) significantly broadens the concentration range over which polyethylene glycol is effective in producing membrane fusion.[12] Although most fusion protocols call for the use of 50% PEG, it is our experience that a solution of 10% DMSO and 35% PEG 1000 in serum-free medium results in a dramatically reduced level of toxicity, while maintaining very effective fusion conditions. Cells to be fused are centrifuged, all of the supernatant is removed to leave a dry pellet, then the cells are resuspended by trituration for 15 sec using approximately 0.1 ml of PEG solution per 5–10 million fibroblasts. The suspended cells are then mixed gently by rotation for an additional 45 sec, and the reaction stopped by the addition of 2 ml of complete medium. The cells can then be gently distributed into their final plating medium, taking care to minimize shear forces that might disrupt newly forming membrane fusions. We have not found it necessary to remove the PEG from the plating medium as long as the PEG solution has been diluted more than 40-fold and the cells are fed the following day.

[12] T. H. Norwood, C. J. Zeigler, and G. M. Martin, *Somatic Cell Genet.* **2,** 263 (1976).

Current Toxin–Antitoxin Selection Protocol

The following sample protocol calls for repeated centrifugations and washes. For reasons that are unclear to us, approximately 20% of the cells are lost during each spin, regardless of the speed and time of centrifugation. Consequently, after four spins, only about 40% of the initial cells are left $(0.8 \times 0.8 \times 0.8 \times 0.8)$. It is thus important to minimize the number of centrifugations by making each wash as efficient as possible. If cells are centrifuged and the supernatant is aspirated, $10–20$ μl of liquid usually adheres to the walls of the tube, which substantially reduces the efficiency of the wash. The most effective procedure is to pellet the cells (e.g., 700 *g* for 2 min, which is about 2000 rpm in a typical benchtop centrifuge), pour off the supernatant, recentrifuge briefly by accelerating until 1000 rpm is reached, and then completely aspirate off the remaining fluid. This procedure will be abbreviated "centrifuge up and down" (for centrifuge, pour off, bring up to speed and down, then aspirate the remaining liquid) and should be used during each centrifugation in the following protocol. "Complete medium" refers to whichever serum supplemented medium is normally used for growth of the cells to be used, while "medium" alone refers to serum-free medium. The cell numbers given are appropriate for cells which grow in monolayer and attain a confluent cell density of about 100,000 cells/cm², and should be adjusted as necessary. Our recommended approach consists of three phases.

Phase 1. Establish the lethal toxin doses for the cell types to be used.

1. Make serial 3-fold dilutions of ricin and abrin in complete medium, with concentrations initially ranging from 0.01 to 100 ng/ml. These solutions are stable for months at 4°.

2. Plate the cells at 100,000 cells/cm² in the various toxin dilutions, including a no-toxin control. The most convenient configuration for this is to use a 48-well dish (1 cm²/well, 0.5 ml/well), although 24-well dishes are also satisfactory.

3. Incubate overnight at 37°.

4. Feed toxin-free complete medium. It is not necessary to wash the cells: simply replacing the toxin-containing medium provides a sufficient dilution of the toxin.

5. Two days later, visually estimate the percentage survival at each toxin dose.

6. Repeat 2–5 until reproducible killing curves are obtained. It is our experience that this usually requires two or three experiments.

Phase 2. Verify that adequate protection is being obtained in injected cells.

1. Make hypertonic antitoxin solutions by diluting one volume of sterile 2 M sucrose in medium with an equal volume of antitoxin serum or ascites.

2. Centrifuge up and down 2.5×10^6 cells of each cell type, each in a different 15-ml sterile conical tube.

3. Vigorously resuspend each cell pellet in 0.2 ml of the appropriate hypertonic antitoxin solution.

4. Incubate for 5 min at 37°.

5. Add 10 ml of hypotonic medium (40% distilled water: 60% complete medium) and mix well.

6. Incubate for 2 min at 37°

7. Centrifuge up and down.

8. Resuspend in 10 ml of normal (isotonic) complete medium.

9. Centrifuge up and down.

10. Resuspend in 10 ml of complete medium.

11. Centrifuge up and down.

12. Plate cells at 100,000 cells/cm² in serial 3-fold toxin dilutions as in phase 1 to determine the amounts of protection obtained with respect to noninjected or mock-injected control cells. The LD_{50} should be increased by 30- 100-fold in the antitoxin injected cells (see Fig. 3).

Phase 3. Perform a toxin–antitoxin selection.

1. Repeat steps 1–10 of phase 2 using 5×10^6 cells of each type to be fused.

2. Remove one-quarter of the cells and use to verify the amount of protection obtained.

3. Combine the remaining cells of the two injected populations.

4. Centrifuge up and down.

5. Resuspend the pellet by aspirating vigorously for 15 sec in 0.15 ml of 35% PEG 1000/10% DMSO in medium using a 1 ml pipet.

6. Gently agitate for an additional 30 sec by rolling the tube back and forth between your fingers.

7. Terminate the fusion by the addition of 10 ml of complete medium 15 sec later.

8. Incubate at room temperature for about 30 min to allow the nascent fusion products to coalesce.

9. Centrifuge the cells.

10. Plate overnight at 100,000 cells/cm² in various toxin concentrations. Initially, this might be a 3×3 matrix (e.g., 1, 3, and 10 ng/ml ricin plus 0.1, 0.3, and 1 ng/ml abrin). The general range should be based upon the dosage curves of phase 1. Later a single dosage combination usually suffices.

11. Feed toxin-free complete medium the next day.

12. Two days later, fix and Giemsa stain aliquots to determine the percentage of all surviving nuclei present in cells with more than one nucleus.

Applications

The complementation analysis of human genetic syndromes permits one to determine the number of alleles involved, and by implication the number of different steps in a metabolic pathway that can lead to a given phenotype. It thus provides an important first step in identifying the specific defects and defining the biochemical basis of a particular syndrome. Complementation analysis has been applied to only a few syndromes due to the inability to isolate fusion products between normal diploid cells. By and large, studies have been limited to syndromes in which single cells can be analyzed, so that the problem of contaminating parental cells is circumvented by only analyzing cells containing more than one nucleus. As a demonstration of the usefulness of the toxin – antitoxin system for complementation analysis, we showed that the known complementation groups of the syndrome xeroderma pigmentosa produced the predicted complementation patterns in heterokaryons purified using antiricin and antidiphtheria toxin antibodies.[10] The toxin – antitoxin selective system is sufficiently easy to use that it should make complementation analysis a routine approach for analyzing human genetic diseases in which an abnormal phenotype has been identified in cultured cells.

The ability to isolate heterokaryons also provides a powerful tool for the study of cell differentiation. Studies using cell hybrids have led to the general conclusion that differentiated functions are suppressed in hybrids between cells of different developmental lineages.[13,14] We[4,5] and others[6] have now shown that different interactions occur at the level of the initial fusion product (the heterokaryon). For example, a differentiated function (rat skeletal myosin light chain synthesis) is induced in a rat glial cell following its fusion to a differentiated chick skeletal myocyte. In contrast to this heterokaryon result, myogenesis is suppressed when the same rat glial cell line is used to form dividing hybrids with a line of rat myoblasts.[4] The simplest interpretation of these superficially contradictory observations is that in heterokaryons one is studying the ability of regulatory molecules already present in a differentiated myocyte to act on the glial genome. On

[13] N. R. Ringertz and R. E. Savage, "Cell Hybrids," pp. 118, 147, 162, and 180. Academic Press, New York, 1976.
[14] R. C. Davidson, in "Somatic Cell Hybridization" (R. L. Davidson and F. de la Cruz, eds.). Raven Press, New York, 1974.

the other hand, in cell hybrids, one is examining the ability of the myogenic precursor component to influence the decision of the entire hybrid cell to terminally differentiate. This decision is subject to a series of steps: the cells must have the appropriate receptors for and be able to respond to the approrpiate environmental signals that stimulate myogenesis, and must successfully complete all the intermediate steps leading to the final expression of the myogenic phenotype. In other words, in cell hybrids the myogenic phenotype must be dominant at every step in the entire sequence if myogenesis is to be expressed, whereas in heterokaryons involving an already differentiated myocyte it may be only the final regulatory stage(s) that are being examined. The ability to examine specific stages in development makes heterokaryons an important new tool for studying the mechanisms regulating cell differentiation.

The isolation of heterokaryons is the most straightforward use for the toxin–antitoxin selection technique. We can routinely produce populations in which greater than 95% of the nuclei are present in heterokaryons, which are eminently suitable for a variety of biochemical analyses. Its use for the isolation of dividing cell hybrids is more problematical and requires very careful controls. Many of the heterokaryons contain more than two nuclei and are thus unlikely to form hybrids, so that if one restricts the analysis to mono- versus binucleates and counts cells rather than nuclei, the purity of the population can drop from 95 to 60%. Since the cloning efficiency of a contaminating parental cell is probably much greater than that of a binucleated heterokaryon, most of the colonies that form may in fact contain parental cells rather than hybrids. It is thus essential to have an independent measure (karyotype, isozyme pattern, etc.) to verify the hybrid status of a particular colony, since the simple ability to form a colony following toxin–antitoxin selection does not provide sufficient evidence to conclude that it is a hybrid.

A good illustration of the absolute necessity for verifying the hybrid nature of growing colonies is the use of the toxin–antitoxin system for producing hybridomas. In our preliminary experiments, normal spleen cells were first injected with antiricin antibodies. The free antitoxin was washed away, then half of the cells were mixed with myeloma cells and fused with polyethylene glycol. For the control, the other half was first fused to itself, and only then mixed with self-fused myeloma cells. Both groups of cells were than plated overnight in 3 ng/ml ricin and washed and distributed at various dilutions among 96-well dishes the next day. The number of colonies in the two groups was determined 1 week later. Whereas the control contained only 5000 colonies, the actual spleen cell–myeloma fusion contained 50,000 colonies, suggesting that 90% of those colonies should represent true hybrids. However, when individual colonies were picked and karyotyped, only about 10% of the colonies exhibited more chromosomes than the parental myoloma. We have repeated this experiment in a variety of ways, and consistently find that the amount of survival in the control does not reflect the amount of parental cell contamination of the hybridized cells.

Our working hypothesis (for which we have no evidence) is that during the process of polyethylene glycol-mediated cell fusion, multiple points of nascent membrane continuity are initiated. These many points of contact may coalesce to provide permanent cytoplasmic continuity between the two cells, so that actual cell fusion results. However, there are probably many situations of abortive fusion in which an insufficient number of points of membrane fusion are established, and the cells then separate back to being two individual cells. If some cytoplasm is exchanged during this process, then some antiricin antibodies from the spleen cell could be transferred to a myeloma cell without heterokaryon formation occurring. These myeloma cells would then be slightly more resistant to ricin killing than the control myeloma cells, which could give the observed result. A consideration of the yields obtained will provide a good illustration of the different dynamics involved in isolating hybrids versus heterokaryons. Fifty thousand colonies represents 0.1% of the input myeloma cells, and thus an extremely small fraction of the initial material. In the human fibroblast experiments described above, approximately 20% of the input cells were recovered as heterokaryons following toxin–antitoxin selection. Under these circumstances, a 0.1% increase in contaminating parental cells as compared to that predicted by the controls would have been totally undetectable. However, such small changes become important if only one in a thousand of the heterokaryons is capable of successfully making the transition and becoming a dividing hybrid cell.

Cells can be enucleated using the combination of high g forces and cytochalasin B.[15,16] Anucleate cytoplasms remain viable for up to a day or so, and maintain the ability to perform a variety of cellular processes, including pinocytosis. One can osmotically inject antitoxin antibodies into anucleate cytoplasts, then fuse them to whole cells. These heteroplasmons can then be isolated following toxin treatment. The toxin–antitoxin system thus provides a flexible approach for the isolation of fusion products between a variety of cell types or cell fragments and should provide a valuable addition to the repertoire of selection techniques. Although in many cases it can be used to isolate cell hybrids, we feel that its major use and importance should be in providing purified populations of heterokar-

[15] W. E. Wright, *Methods Cell Biol.* **7**, 203 (1973).
[16] G. Veomett, *Tech. Somatic Cell Genet.*, p. 67, 1982.

yons for the detailed analysis of regulatory interactions. Because of the previous technical limitations, most scientists have thought of cell fusion only in terms of cell hybrids. It is our hope that the availability of the toxin–antitoxin selection system for isolating heterokaryons will now stimulate a new conceptual approach to the use of somatic cell fusion studies for investigating biological phenomenon.

[19] Chromosome Sorting by Flow Cytometry

By Marty Bartholdi, Julie Meyne, Kevin Albright, Mary Luedemann, Evelyn Campbell, Douglas Chritton, Larry L. Deaven, and L. Scott Cram

The 24 human chromosome types can be sorted by flow cytometry with 90% purity in quantities sufficient for chromosome specific DNA library construction[1] or direct hybridization on filters.[2] The techniques that need to be brought together to sort one million chromosomes of a single type, or 50 thousand of each type, on a commercially available flow cytometer are cell culture, large-scale chromosome preparation, resolution of single chromosome types, and long-term sorting at optimum rates and purity.

The cell types most commonly used as sources of chromosomes for sorting are diploid human fibroblasts, lymphoblasts, and Chinese hamster–human hybrids retaining one or a few human chromosomes. Human chromosomes purified from hybrid cells by flow cytometry have an advantage over the direct use of hybrids because the cross species background is removed. Chromosome sorting by flow cytometry can also be applied to cell strains with abnormal chromosomes or those with important phenotypic traits.

Flow cytometers operate on the principle of rapid analysis of single cells or chromosomes. A suspension of fluorescently stained chromosomes flows at a rate between 1000 to 2000/sec through a finely focused laser beam. Individual chromosome types are resolved on the basis of DNA content and base composition as determined by the cytochemistry of the fluorescent dyes. The chromosomes flow in a small stream that is broken

[1] K. E. Davies, B. D. Young, R. G. Ellis, M. E. Hill, and R. Williamson, *Nature (London)* **293**, 376 (1981).
[2] R. V. Lebo, D. R. Tolan, B. D. Bruce, M. C. Cheung, and Y. W. Kan, *Cytometry* **6**, 478 (1985).

METHODS IN ENZYMOLOGY, VOL. 151

into uniform droplets. The droplet containing the desired chromosome type, as indicated by the fluorescence measurement, can be deflected from the mainstream. Optimal sorting rates are 40 to 50 chromosomes/sec. Sorting can be done separately in two directions for production of one million copies of each of two chromosome types in 6 hr.

Purity is determined by the resolution of the desired chromosome type, the fraction of chromosome debris in the preparation, and instrumental factors. Purity is judged to be near 90% and preliminary characterization of flow sorted chromosomes confirms this. Both cytogenetic analysis and hybridization of chromosome specific probes to flow sorted chromosomes indicate that contamination is about 10%.[3]

In addition to the operation of the flow cytometer, the culturing of cells and chromosome preparation techniques must be highly efficient. A chromosome preparation should provide uniformly stained chromosomes, low debris levels, a high yield of free chromosomes from mitotic cells, and for library construction, chromosomal DNA of very high molecular weight. The yield of free chromosomes is crucial because low yields decrease the chromosome number concentration while increasing the debris fraction, making the chromosome sorting rates much slower than optimum. Techniques for chromosome preparation, resolution of the normal human chromosomes for flow sorting, and the way these interact to produce sorted chromosomes of high purity at optimum rates are described.

Chromosome Preparation

We have routinely prepared chromosomes from diploid human fibroblasts (foreskin strain HSF-7 developed by D. Chen, Los Alamos), diploid human lymphoblasts (GM 130A, Mutant Cell Repository, Camden, NJ), and a number of Chinese hamster–human hybrids. Each cell strain has its own characteristics that requires adjustments to each step in the preparation protocol. The basic steps in chromosome preparation are (1) culture of a sufficient number of cells, (2) blocking cells at metaphase with colcemid and harvest of mitotic cells, (3) cell swelling in hypotonic buffer, (4) dispersal of chromosomes in the presence of isolation buffer for stabilization of free chromosomes, and (5) staining of chromosomes with DNA specific dyes, Hoechst 33258 and chromomycin A_3.

Cell Culture

Chromosome preparation for flow sorting requires from 10 to 100 times the number of mitotics considered necessary to produce the desired

---

[3] B. Moyzis (Los Alamos National Laboratory), personal communication.

number of sorted chromosomes because the yield of isolated chromosomes is usually between 1 and 10% of the expected number. For a large scale chromosome preparation from cells grown in monolayer (fibroblasts or somatic cell hybrids), 10 T150 flasks are seeded with approximately one million cells each. The culture media is α minimal essential media plus 10% fetal calf serum and the cells are grown at 37° in a 5% CO_2 incubator. Monolayer cells are maintained in exponential growth for at least two generations prior to starting the mitotic block. Beyond a certain passage number (about 20, but different for each culture) most human cell strains have an increasing number of senescent cells and become unsatisfactory for high quality chromosome preparations.

For chromosome preparation from lymphoblasts grown in suspension culture, four T75 flasks are seeded with 16 million cells each at a concentration of $3-4 \times 10^5$ cells/ml. The cell culture media is RPMI plus 10% fetal calf serum, and the flasks are placed in 5% CO_2 incubators at 37°. A single generation time is allowed prior to mitotic arrest. The extent of cell culture required for lymphoblasts is significantly less than that for monolayer cells. Also, chromosome yields of 10% and higher are routine, and yields of 20% have been obtained.

Mitotic Arrest and Harvest

Colcemid (0.1 μg/ml) is added to exponentially growing monolayer or suspension cell cultures. The optimal duration of the mitotic block varies considerably for each cell culture, but is typically 12 to 14 hr for diploid human fibroblasts, lymphoblasts, and hamster–human hybrids. Mitotic blocks of shorter duration do not allow as many cells to reach metaphase, and result in preparations of high resolution, low debris, but low chromosome number concentration. Prolonged Colcemid blocks increase debris levels by formation of micronuclei and colcemid toxicity. Increasing the colcemid concentration to increase disruption of the metaphase has been unsuccessful.

Mitotic cells are selectively detached from monolayer cultures by shake off. Usually hybrid mitotic cells are attached less firmly than human fibroblasts and require less vigorous shaking. The flasks are rapped with the palm and examined for the number of cells floating and attached. About 5 to 7 million cells can be harvested from each of the 10 flasks, and the fraction of mitotic cells in the harvest should be between 50 and 95%.

Because the mitotic index of a lymphoblast suspension culture is typically 30%, of the 20 million cells harvested from each of the four flasks seeded, about 6 million are mitotic cells. The mitotic index in the suspension, or fraction of mitotics harvested by shake off, can be determined by swelling cells in 1.0 ml hypotonic sucrose (34 g/liter) after centrifugation of an 8-ml sample of cell culture for 5 min at room temperature. The sample is again centrifuged and suspended in 5 ml of methanol, acetic acid (3 : 1) fix. Mitotic and interphase cells can then be scored under the microscope after staining with propidium iodide.

Cell Swelling

The harvested mitotic cells are allowed to swell in hypotonic buffer to loosen the metaphase chromosomes but not to the point of cell lysis. Careful modification of the hypotonic buffer used, and duration of cell swelling time, can significantly improve the yield of free chromosomes. The failure of the cells to swell uniformly or premature lysis are major factors in the loss of yield.

The harvested cells are pooled and counted in a Coulter counter to determine the number concentration. The cells are distributed at about six million mitotic cells each in 50-ml centrifuge tubes (account for mitotic fraction). In a large-scale preparation of fibroblasts or hybrids, 10 tubes are produced and for lymphoblasts, 4 tubes. The cell harvest is centrifuged (250 g for 10 min) and the cell culture media is aspirated. Fibroblasts or hybrid cells are resuspended in 5.0 ml, 75 mM KCl for 25 min at room temperature. Lymphoblasts are resuspended in 55 mM Ohnuki's buffer (equimolar solution of KCl, NaNO$_3$, and NaC$_2$H$_3$O$_2$ in a ratio of 10 : 5 : 2) for 60 to 100 min, but good yields have also been obtained using 75 mM KCl for 20 min. Leaving the cells too briefly in hypotonic buffer does not swell the cells sufficiently. Prolonged swelling may lyse cells and release chromosomes (usually in clumps) prematurely. A second centrifugation and resuspension in a "wash" buffer (see below) sometimes improves the resolution and lowers debris levels, but can reduce yield and therefore is not routinely done in large-scale preparations.

Chromosome Dispersal

The swollen mitotic cells are centrifuged and resuspended in isolation buffer for physical dispersal of the metaphase chromosomes. Three isolation buffers have been used to stabilize the morphology of the free chromosomes in suspension. The final concentrations of the three buffers are listed here. Enzyme grade reagents are used, when available.

Buffer I (hexylene glycol),[4] final concentrations:
    750 mM hexylene glycol

---

[4] W. Wray and E. Stubblefield, *Exp. Cell Res.* **59,** 469 (1970).

    25 mM Tris base
    1 mM MgCl$_2$
    0.5 mM CaCl$_2$

Adjust pH to 7.5 using HCl prior to addition of hexylene glycol, filter through 0.2-μm filter, store cold or freeze. This buffer can be prepared concentrated and diluted just prior to use.

Buffer II (MgSO$_4$),[5] final concentrations:
    9.8 mM MgSO$_4$
    48 mM KCl
    4.8 mM HEPES
    2.9 mM dithiothreitol

For 10 ml buffer add 1.0 ml of 100 mM MgSO$_4$, 9 ml of 55 mM KCl, and 5.5 mM HEPES, and 0.25 ml dithiothreitol (120 mM in H$_2$O), filter through 0.2-μm filter. Adjust pH to 8.0.

Buffer III (polyamine),[6,7] final concentrations:
    15 mM Tris base
    2 mM EDTA
    0.5 mM EGTA
    80 mM KCl
    20 mM NaCl
    0.2 mM spermine
    0.5 mM spermidine
    14 mM β-mercaptoethanol
    0.1% digitonin (approximately)

Each of the first five components are prepared separately at 10 times the concentration given above, filtered through a 0.2-μm filter, and stored at room temperature. Spermine and spermidine stocks are prepared in distilled water at 0.4 and 1.0 M, respectively, and frozen in 50 μl aliquots. The isolation buffer is prepared on the day of use by combining 10 ml each of the first five stock buffer components (10$\times$) plus 50 ml of distilled H$_2$O. The pH is adjusted to 7.2 with HCl or NaOH. β-Mercaptoethanol is added to a final concentration of 14 mM. (The buffer as prepared to this point can be used as a chromosome "wash" buffer.) To 25 ml of the wash buffer 30 mg of digitonin is added and the solution heated at 37° for 45 min. The undissolved digitonin is removed by filtration. Because not all of the digitonin goes into solution, the final concentration is not 0.12% but probably closer to 0.1%. Add 12.5 μl of each stock solution of spermine and spermidine. Keep the chromosome isolation buffer cold.

[5] G. van den Engh, B. Trask, L. S. Cram, and M. Bartholdi, *Cytometry* **5**, 108 (1984).

[6] A. B. Blumenthal, J. O. Dieden, L. N. Kapp, and J. W. Sedat, *J. Cell Biol.* **81**, 255 (1979).

[7] R. Sillar and B. D. Young, *J. Histochem. Cytochem.* **29**, 74 (1981).

When dispersing chromosomes in Buffer I (hexylene glycol), the mitotic cells from the cell swelling step are centrifuged and resuspended in 0.5 ml buffer and incubated at 37° for 10 to 20 min and then placed on ice for 60 min. The chromosomes are dispersed by forcing the swollen cells through a 1.5 in., 22-gauge syringe needle. The needle is bent slightly to allow the beveled tip to lie flat against the side of the centrifuge tube wall and the syringed suspension should fan out down the wall with little foaming. The suspension is syringed one to three times and the chromosomes are monitored after each pass. Undersyringing results in undispersed chromosomes and oversyringing in chromosome fragmentation, both with significant loss of yield. Usually both conditions are present together and the optimal number of passes must be evaluated by flow cytometry. While microscopic observation is important at this step, it is not a reliable predictor of flow karyotype quality.

Chromosomes from human fibroblasts prepared in Buffer I retain their morphology and can be banded after sorting. This method has been used to confirm the chromosome identity of the peaks in the flow histogram. Purity can also be judged during sorting by identifying each of the sorted chromosomes to determine the number of contaminants. Variable quality of banding, however, makes this practice difficult to implement routinely. The major disadvantage of Buffer I has been high levels of debris, particularly under the smaller chromosomes.

When using Buffer II (MgSO$_4$), the mitotic cells in each tube are resuspended in 1.0 ml buffer, and allowed to stand for 10 min at room temperature. The detergent Triton X-100 is then added to a final concentration of 0.25% for fibroblasts and 1% for hybrids. The suspension stands for an additional 10 min before syringing as described above. Chromosomes from human fibroblasts and hamster–human hybrids prepared in Buffer II usually have excellent uniformity of dye staining, moderate debris levels, and good yields.[5] A disadvantage is difficulty in identification of sorted chromosomes due to a high degree of condensation.

In our laboratory the most consistent results have been obtained with Buffer III (polyamine). The polyamines condense the chromosomes rendering identification difficult but banding has been reported.[2] The resolution, debris levels, and yields are usually superior, especially with the lymphoblast cell line GM 130A.

For preparation of chromosomes in Buffer III, after cell swelling in hypotonic buffer and centrifugation, the cell pellet is resuspended in 1.0 ml cold polyamine isolation buffer. The chromosomes are then dispersed by placing the centrifuge tube on a vortex mixer for up to 60 sec. The time should be adjusted for each cell type. After 60 sec, if more vortexing is indicated, cool the tube before continuing. The presence of chelating agents and cold inhibits nuclease activity, and dispersal by vortexing in the presence of detergent gives more uniform yields.[7] A small portion (10 μl) of the chromosome preparation can be stained with an equal volume of propidium iodide (50 μg/ml) and examined microscopically. An acceptable preparation contains mostly free chromosomes with few clumps or fragments.

Chromosome preparations in Buffer III (polyamine) can be stored unstained at 4° and have long shelf life. Some preparations have been stored for over 18 months with no loss of resolution or chromosomes. Preparations in Buffer I (hexylene glycol) are stored frozen, unstained, and preparations in Buffer II (MgSO$_4$) are stained immediately and should be used within 1 week.

A typical chromosome preparation from fibroblasts and hybrids will produce 10 tubes containing 1.2 ml each with a free chromosome number concentration of about 2.5 million/ml. This corresponds to 3 million chromosomes per tube in which about six million mitotics were swollen for a yield of 1%. About 50% of the preparation is chromosome debris. From the 10 tubes, about one million copies of a single chromosome type can be sorted. Two million chromosomes, one million of two chromosome types, can be sorted with two way sorting.

An optimal preparation from diploid lymphoblasts produces four tubes containing 1.2 ml each and a concentration of free chromosomes of 25 million/ml. This corresponds to 30 million chromosomes per tube or a yield of 10%. Yields of up to 20% have been obtained. The debris is mostly interphase nuclei that can be removed by slow speed centrifugation (75 g, 3 min). Typically, 4 million chromosomes of a single type can be sorted from the total preparation of 4 tubes (two types for two way sorting).

Each step in the isolation protocol, especially metaphase block time, hypotonic treatment, and optimum mitotic cell number concentration, must be adjusted for each cell strain and the results evaluated by flow cytometry. Improvements in yields while maintaining chromosome integrity are needed. The major factor affecting yield appears to be the cell type, with the lymphoblasts routinely more productive than fibroblasts, and with some hybrids working better than others. Also, even with the same cell strain, the yield may be variable from preparation to preparation.

Chromosome Staining

The dyes, Hoechst 33258 and chromomycin A$_3$, provide the optimum resolution of human chromosomes from diploid human cells and Chinese hamster–human hybrids.[8] Hoescht 33258 binds to AT base pairs and

[8] R. G. Langlois, L. C. Yu, J. W. Gray, and A. V. Carrano, *Proc. Natl. Acad. Sci. U.S.A.* **79**, 7876 (1982).

chromomycin A_3 to GC base pairs. The chromosome types are resolved by relative DNA content and base composition.

The stock staining solutions are chromomycin A_3, 0.5 mg/ml in Mc-Ilvane's buffer pH 7, 5 mM $MgCl_2$ (1 : 1), and, Hoescht 33258, 0.5 mg/ml in distilled H_2O. Add 300 μl stock solution of chromomycin A_3 to a 1.0 ml chromosome sample prepared in Buffer III at least 3 hr prior to sorting to allow the stain to reach equilibrium (final concentration, 125 μM) and add 11.2 μl stock solution of Hoescht 33258 (final concentration, 4.8 μM) just before sorting. The chromosome preparations degrade quickly after staining and should be used within 1 week. The stain concentrations given here are optimal for analysis on the EPICS V flow cytometer. For chromosomes prepared in Buffers I and II, lower stain concentrations of Hoescht 33258 (3 μM) and chromomycin A_3 (30 μM) may be used. Uniformity of staining is excellent with chromosome buffers II ($MgSO_4$) and III (polyamine).

Chromosome Sorting

Chromosome sorting with high productivity and high purity requires clear resolution of the chromosome type, fast sorting rates, high yields, and low debris levels in the chromosome preparations. The characteristics of the preparations such as chromosome number concentration, uniformity of staining, and debris levels interact in a complex manner with the capabilities of the flow cytometer. The technique to be described applies to a commercial flow cytometer (EPICS V, Coulter Electronics, Inc., Hialeah, FL).

Resolution

The clear resolution of individual chromosome types gives flow cytometry the capability to separate chromosomes with high purity. Resolution depends partly on the intrinsic properties of the chromosomes and the cytochemistry of staining required to distinguish such differences in size and composition, and partly on the capability of the flow cytometer to measure small differences in fluorescence intensity among the chromosomes types with high precision. Resolution of chromosomes stained with Hoechst 33258 and chromomycin A_3 is done with two argon ion lasers. Hoechst is excited in the ultraviolet (UV) and chromomycin A_3 at 457.9 nm. Laser powers of 200 mW in the UV and 350 mW at 457.9 nm are sufficient for good precision, but lower powers should be used if feasible to prolong laser tube life (Innova 90-6 argon ion lasers, Coherent Inc., Palo Alto, CA). The two laser beams are focused to two separate illumination spots of 16 by 40 μm elliptical cross sections by a single pair of crossed cylindrical lenses of 80 and 40 mm focal lengths (confocal lens assembly).

Fluorescence measurements of higher precision may be possible with higher illumination intensity, but this requires special optics.[2,9]

The sample stream containing the chromosomes is hydrodynamically focused by an outer concentric sheath stream to a diameter of 3 μm. The focusing occurs within the 76 μm orifice of the flow nozzle. The square quartz nozzle (Coulter Electronics, Inc.) is easier to sort with routinely than stream-in-air because the process of breaking the stream into droplets sometimes perturbs the outer surface of the flow stream at the laser intersection. The illumination spot is aligned with its long dimension horizontal and the sample stream flows vertically through its center to provide uniform illumination to each chromosome.

The only special feature used in the EPICS V for chromosome analysis is the measurement of the two fluorescence signals on a single photomultiplier tube. Because the two illumination spots are separate, the two signals occur sequentially in time. A pair of optical filters, KV408 (placed closest to the flow stream) and GG495 block the stray illumination light and light scatter from the chromosomes at the UV and 457.9 nm wavelengths. The fluorescence signal produced by UV excitation is proportional to Hoechst binding. It consists of the red edge of Hoechst fluorescence and chromomycin A_3 fluorescence excited through energy transfer. Because energy transfer acts over a short range, only chromomycin A_3 molecules adjacent to Hoechst molecules are excited to fluoresce by this mechanism. The fluorescence signal produced by 457.9 nm excitation is proportional to the total chromomycin A_3 binding.

The two signals are processed with the electronics provided with EPICS V (Fig. 1). The fluorescence signals occur sequentially at the output of the photomultiplier tube and are integrated to provide a pulse that will produce a single window in the gated amplifier module. The sequential pulses are sent to two signal inputs of the gated amplifier module where they are separated. One set of signals is delayed to align the UV pulse in the gate window and the other set is undelayed to align the 457.9 nm pulse in the window. The action of the window serves to eliminate the pulse outside it. The fluorescence signals now can be amplified and integrated separately and are acquired in a bivariate histogram with 128 channels on each axis.

The fluorescence signals from Hoechst 33258 and chromomycin A_3 stained chromosomes must be within a factor of 10 in relative intensities to be measured with a single photomultiplier tube. The stain concentrations and laser powers can be adjusted to accomplish this. A bivariate histogram of HSF-7 fibroblast chromosomes prepared in Buffer III (polyamine) and stained with Hoechst 33258 and chromomycin A_3 is shown in Fig. 2. Each

[9] M. Bartholdi, D. Sinclair, and L. S. Cram, *Cytometry* **3**, 395 (1983).

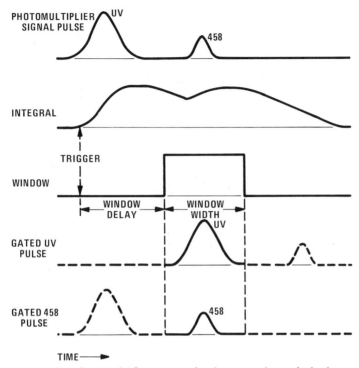

FIG. 1. Separation of sequential fluorescence signals measured on a single photomultiplier tube.

of the human chromosomes except the group 9–12 is resolved. The coefficients of variations, *CV*, of the chromosome peaks are 2%. Resolution in this range is sufficient to resolve chromosomes differing by an amount equivalent to a single band.[8] To achieve this precision the flow cytometer is aligned with 1.6 μm polystyrene beads for a *CV* of 1.8% in the UV and 1.2% at 457.9 nm.

Before analysis the samples are filtered through 62 μm nylon mesh to remove large particles that could clog the flow nozzle. The sheath fluid should match components of the chromosome isolation buffer used (at least for Buffers II and III) to reduce drift of the dye fluorescence during analysis. When Buffer III is used the components in the "wash" buffer are used in the sheath fluid. The temperature of the sheath fluid, sample, and collection tubes is maintained at or near 4°. We have found better long term (6 hr) stability of resolution under these conditions.

A bivariate histogram of a Chinese hamster–human hybrid cell strain

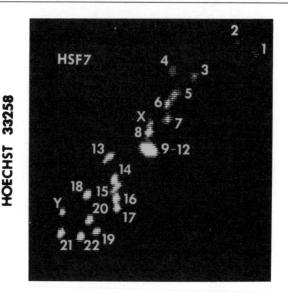

CHROMOMYCIN A₃

Fig. 2. Bivariate histogram of chromosomes isolated from human diploid fibroblast strain HSF-7. About 200,000 counts are displayed as a contour plot.

retaining human chromosomes X, 12, and 15 is shown in Fig. 3. Using hybrids, the chromosomes 9, 10, 11, 12, can be sorted separately. Also, the human chromosomes 1–8 and X are more clearly resolved in hybrids in which the neighboring human chromosomes have segregated. The larger human chromosomes are also well separated from the hamster background due to significant differences in AT to GC content.

The peaks corresponding to the human chromosomes in a hybrid are identified by comparison with those of the hamster parent. If more than one human chromosome is present, then these peaks should fall in the same relative positions as they do in a flow karyotype from normal diploid cells. Also, the normal hamster chromosomes present in the parent can provide benchmarks to locate the peaks from human chromosomes. One difficulty in using hamster–human hybrids is loss of the human chromosome during culture.

Sorting

Chromosomes are sorted by bounding the peak containing the desired chromosome with a rectangular sort window (Fig. 4). The gains of the fluorescent signals are increased to place the peaks to be sorted near the top

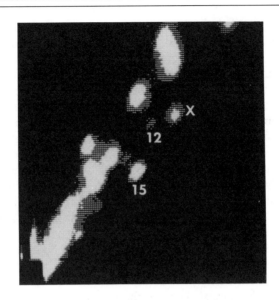

CHROMOMYCIN A₃

HOECHST 33258

FIG. 3. Bivariate histogram of chromosomes isolated from Chinese hamster–human cell strain retaining the human X, 12, and 15 chromosomes.

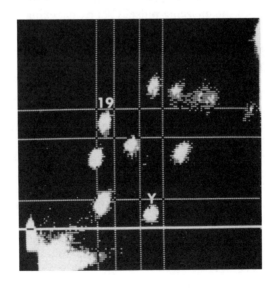

CHROMOMYCIN A₃

HOECHST 33258

FIG. 4. Placement of sort windows around chromosome peaks 19 and Y from diploid human cells. Expanded gain places the larger chromosomes in the uppermost channels.

of the histogram with the larger signals going offscale. The offscale signals in the upper most channels of the histogram are still processed normally.

All particles with fluorescent intensities within the window trigger a "sort" chain of events. The flow stream (sheath and sample) is broken into uniform droplets by a piezoelectric oscillator at 32 kHz. The frequency is adjusted for a minimum droplet breakoff distance from the nozzle. The flow stream velocity is 7 m/sec and diameter is 76 μm forming a single droplet from a 200 μm length of the stream. The droplets form individual spheres of about 150 μm diameter about 18 droplet periods or 600 μsec downstream. After this delay, the "sort" signal generated by the particle within the window causes a charging pulse to be applied to the flow stream. The droplet just breaking off and containing the desired chromosome carries the applied charge and is deflected from the mainstream when passing through a static electric field. Either a positive or negative charge can be applied and two chromosome types can be separately sorted from the same preparation. In practice, more than one droplet is deflected at a time, so that the length of stream that is sure to contain the desired chromosome is charged. The chromosomes flow through the laser beam at random times and therefore may occur at the edge of the droplet cycle and be missed in single drop sorting. We use two drop sorting with a charge pulse of 62 μsec acting on 400 μm of stream.

Before sorting chromosomes, the timing between the fluorescence signal and delay of the charging pulse is set by checking that the recovery of sorted microspheres is 100%. During the sorting procedure the position of the droplet breakoff must be carefully monitored to ensure that the recovery stays at 100%.

The key factor in chromosome sorting by flow cytometry is the rate at which chromosomes can be sorted at high purity. When sorting from diploid human cells, one of every 23 chromosomes is to be sorted. Because a 400 μm length of the stream is deflected, the analysis rate must not be so fast that extensive coincidence of a second different chromosome occurs within this length of stream with the desired chromosome. When the analysis rate is 1250 chromosomes/sec the average spacing between chromosomes is 8000 μm, and due to the random spacing of chromosomes in the flow stream, coincidence within 400 μm occurs for about 7% of the chromosomes.[10] With a chromosome type flow rate of 54/sec, about 50/sec can be sorted with high purity by aborting the sort signal when coincidence is detected. If the sort rate of 50/sec can be maintained, then one million chromosomes can be sorted in 6 hr (one million chromosomes of each of two types with two-way sorting).

[10] M. J. McCutcheon and R. G. Miller, *Cytometry* **2**, 219 (1982).

A total chromosome analysis rate of 1250/sec can be achieved with chromosome preparations of about 2.5×10^7 chromosomes/ml. The sample stream must be maintained at a 3 μm diameter to prevent loss of precision from nonuniform illumination. With a flow velocity of 7 m/sec, about 0.2 ml of sample can be used in a stream this size in one hour, or 0.05 μl/sec. For sufficiently high number concentration, this results in an analysis rate of 1250/sec. For chromosome preparations of lower chromosome number concentration the flow stream is widened as much as possible to optimize resolution and analysis rate. Our experience has indicated that most preparations from human fibroblasts and many from hamster–human hybrid cells must be sorted at less than optimal rates because low yield in the chromosome isolation procedure has reduced the chromosome number concentration. The preparations can be concentrated most easily by overnight storage at 4° during which the chromosomes sediment to the lower third of the tubes. In practice, however, only a finite volume of preparation is available and the total number actually sorted cannot be increased.

Chromosomes are sorted into 1.5 ml polypropylene microcentrifuge tubes at 4°. Five hundred thousand sorted chromosomes occupy a volume of 1.2 ml. The sorted material can be frozen for storage.

Long-term sorting requires stability of both the resolution of the chromosome type and the sort recovery. Resolution is monitored every 15–30 min and sort windows readjusted one or two channels if needed. When the resolution degrades, the flow cytometer must be realigned with microspheres. Recovery of the sorted chromosomes is maintained at 100% by counting chromosomes sorted onto a BSA-coated microscope slide placed above the collection tube and seeing that the number of chromosomes (about 20) coincides with the number of sort pulses generated.

Recovery of 100% does not imply that every chromosome of a given type in a preparation is actually sorted, just that the drop delay is stable and the right portion of the stream is being charged. The efficiency of the flow cytometer at conventional rates is about 80% with about 10% of the preparation used for preliminary evaluation and 10% of the sort signal aborted for coincidence.

Purity

A purity of 90% is used as a nominal figure of what is generally practicable in sorting chromosomes by flow cytometry. To achieve this purity, the chromosomes must be clearly and stably resolved, coincidence aborted, and debris levels low. Maintaining high purity often requires lowering the sort rate and overall efficiency. If the chromosome peak is not clearly resolved the sort windows must be stringently set and may not include all of the desired chromosome type. Chinese hamster–human hybrids are advantageous here because the neighboring human chromosomes have segregated.

Purity of the sorted chromosomes prepared in Buffer I was checked by sorting chromosomes directly onto BSA coated slides. Typically, 85% of the chromosomes were identifiable by morphology or banding as the desired chromosome, and 3 to 4% were positively identified as other chromosomes. The abort circuitry has a "blind" time before it can act allowing a small fraction of coincidence events to be sorted.[10] Around 12% could not be identified, and were either debris or the desired chromosome. Variability in this unidentifiable fraction made it difficult to precisely assess purity directly.

Debris can account for up to 50% of the total analysis rate. Debris includes undispersed metaphase chromosomes, interphase nuclei, clumps of chromosomes, fragments of chromosomes, and even single chromatids. Typically, about 5% of the events within a sort window are considered debris because of the adjacent background level. Other debris is distinguished by its fluorescence intensity and is rejected. But a high debris fraction significantly decreases the sorted events from 1 in 23 to about 1 in 50 with a corresponding increase in coincidence that must be aborted. Thus, the sorting rates can be decreased substantially to about 20 or 30 chromosomes/sec, even with an analysis rate of up to 4000 events/sec.

Clumps of smaller human chromosomes occur as a background continuum lying under the larger human chromosomes in diploid fibroblasts and lymphoblasts, prompting the efforts to sort the larger human chromosomes from Chinese hamster–human hybrids. The fraction of background counts can be as high as 20%. Also, very large debris and nuclei should be removed by slow speed sedimentation because each contributes a significant amount of contamination if sorted.

Contamination by bacteria that can grow in the sheath fluid is avoided by sterile filtering and cleaning. Contamination by "cryptic" genetic rearrangement not observable by banding could occur in virally transformed cells such as lymphoblasts or in very karyotypically unstable cells such as most somatic cell hybrids. Avoiding cryptic contamination is one rationale for sorting chromosomes from diploid fibroblasts despite their lower yields.

Conclusion

Each of 24 normal chromosome types has been sorted with a purity of 90% by flow cytometry. Human diploid fibroblasts, Chinese hamster–human hybrid cell strains, and human diploid lymphoblasts have been used as sources of chromosomes. Improvements in chromosome yield, resolution from new staining protocols,[11] and flow cytometers with higher sorter speeds,[12] will increase the rate at which chromosomes can be sorted.

The techniques described are suitable for application on a commercial flow cytometer with a productivity of one million chromosomes per day. Potential applications lie in the sorting of abnormal chromosomes, evaluation of genetic rearrangement during neoplastic progression, gene mapping, and analysis of chromosome specific proteins. Chromosome analysis and sorting by flow cytometry are becoming an important technique in cellular and molecular genetics.

Acknowledgments

The U.S. Department of Energy provided support for the National Laboratory Gene Library Project. The National Flow Cytometry Resource at Los Alamos sponsored by the Division of Research Resources of NIH also provided support (RR01315). This manuscript was expertly typed by Pat Elder.

[11] B. Trask, G. van der Engh, J. Landegent, N. J. in de Wal, M. van der Ploeg, *Science* **230**, 1401 (1985).
[12] D. Peters, E. Branscomb, P. Dean, T. Merrill, D. Pinkel, M. Van Dilla, and J. W. Gray, *Cytometry* **6**, 290 (1985).

[20] Methods for Chromosome Banding of Human and Experimental Tumors in Vitro

By JEFFREY M. TRENT and FLOYD H. THOMPSON

Chromosomal analysis is increasingly becoming a necessity in studies of somatic cell genetics. The major emphasis of this chapter will be to describe several common cytogenetic procedures which can be utilized to identify the chromosomal constitution of human and experimental tumors grown *in vitro*. Emphasis has been placed on discussion of methods for banding analysis of established cell lines, although a brief description of procedures for analysis of primary tumor material has also been included. Additionally, the discussion has been principally centered around banding analysis of human tumor cells in culture, although the techniques described are amenable to normal and abnormal cells from other mammalian species.

Chromosome Characterization: General Comments

The following procedures are commonly used for chromosomal characterization of mammalian cells grown *in vitro*. Additional procedures for chromosome banding may be found in texts devoted solely to methodology.[1,2] Investigators who are considering using cytogenetic techniques in order to identify cell line cross-culture contamination are referred to the excellent recent article of Pathak and Hsu[3] which details procedure and criteria for identification of interspecies cell line contamination.

In order to fully characterize the chromosomal complement of a cell line or primary tumor specimen, it is necessary to employ one or more of several "chromosome banding" procedures. The most informative and commonly used banding techniques include G-, Q-, or R-banding (described below); all are procedures which induce a pattern of longitudinal differentiation (banding) along the length of each individual chromosome. With sufficient experience in band pattern recognition, it is possible to unequivocally identify all individual normal chromosomes. However, inevitably in cell lines derived from tumors as well as nontransformed cells grown in long-term culture, numeric and structural chromosome alterations will be found. Some of these changes result in the production of "marker" chromosomes, chromosomes which have undergone structural rearrangement to a point where they cannot be fully identified by G-, Q-, or R-banding. In these instances, use of techniques which stain a restricted number of specific bands or structures may be of benefit [e.g., banding procedures for constitutive heterochromatin (C-banding) and nucleolus organizer regions (NORs)]. Finally, although the methods to induce chromosome banding in most cases are only of modest technical difficulty, the correct identification of banded chromosomes (karyotyping) remains a tedious and very labor-intensive "art." Detailed discussion of the accepted criteria for identification of human chromosomes following banding analysis is beyond the scope of this chapter. Instead, the reader is referred to the recent International System for Cytogenetic Nomenclature (ISCN),[4] a volume dedicated entirely to providing an international standard for identification and nomenclature of human chromosomes (similar standardized nomenclature committees have been formed and reports issued for other mammalian species).

[1] J. Yunis, "Human Chromosome Methodology." Academic Press, New York, 1974.
[2] H. Schwarzacher, "Methods in Human Cytogenetics." Springer-Verlag, New York, 1974.
[3] S. Pathak and T. C. Hsu, *Cytobios* **43**, 101 (1985).
[4] "International System for Cytogenetic Nomenclature" (ISCN), p. 21. Karger, Basel, Switzerland, 1985.

Chromosome Banding Techniques

Standard Giemsa Staining

Conventional Giemsa staining may be of considerable use in general descriptive studies of chromosomes from cell cultures. This simple staining procedure may be sufficient if the investigator is concerned only with simple documentation of chromosome number, or screening of metaphases for double minutes (DMs).

Materials
  Giemsa stain: (Gurr's R-66)
  Phosphate buffer: 0.06 M KH_2PO_4 (49 ml)/0.06 M Na_2HPO_4 (51 ml)
Staining Procedure
  1. Prepare a 4% Giemsa stain (in phosphate buffer pH to 6.8).
  2. Stain slides 5–7 min.
  3. Rinse slides briefly with distilled water and air dry.

Giemsa (G-)Banding

G-banding is one of the most useful and commonly employed banding procedures for identification of mammalian chromosomes. Advantages of this method over Q- or R-banding include the use of bright-field (rather than fluorescent) microscopy, and the semipermanent staining induced (allowing extended time to analyze or photograph a metaphase under the microscope). Several methods have been described for inducing G-banding, with most making use of either one or more proteolytic enzyme (e.g., trypsin), or mild heating of slides in a neutral buffer. In addition to identification of all normal chromosomes, G-banding is also very useful in identification of homogeneously staining regions (HSRs). However, G-banding is much less useful in identifying DMs because DMs stain lightly by G-banding and often can be overlooked. Three methods are described below.

Giemsa–Trypsin Method[5]
Materials
  Phosphate buffer: (see Standard Giemsa Staining) (pH 6.8)
  Absolute methanol
  Giemsa stock stain: (Fisher, G-146 1 g, methanol 66 ml, and glycerin 66 ml)
  Trypsin–EDTA 10✕: (GIBCO)

---

[5] N. Sun, E. Chu, and C. Chang, *Mamm. Chromosome Newsl.* **14,** 26 (1973).

Staining procedure

1. Prepare a combined stain–enzyme solution as follows: (a) phosphate buffer 36.5 ml, (b) methanol (absolute) 12.5 ml, (c) Giemsa stain 0.9 ml, and (d) trypsin–EDTA (10×) 0.25 ml.
2. Preincubate slides for 10 min in a covered Coplin jar containing phosphate buffer prewarmed to 57°.
3. Remove slides from buffer and place horizontally on a staining rack. Gently cover the working surface of the slide with the stain–enzyme solution (~3 ml). (The viscosity of the stain–enzyme solution is sufficient to prevent drainage.) Incubate slides for 10–12 min at room temperature.
4. Rinse slides gently but thoroughly with distilled water and allow to air dry.

ASG Method[6]
Materials
Giemsa stock stain: [see Giemsa (G-)Banding]
Phosphate buffer: (see Standard Giemsa Staining) (pH 6.8)
Staining procedure

1. Preincubate slides in phosphate buffer at 56–60° for 8–12 min.
2. Rinse in distilled H_2O and air dry.
3. Stain with Giemsa (4% in phosphate buffer) for 4 min.
4. Rinse in distilled H_2O and air dry.
   (If banding is not well differentiated, destain in 95% ethanol for 2–4 min, air dry, and repeat steps 3 and 4 using only 2–4 min staining.)

Wright Staining Method[7]
Materials
Phosphate buffer: (see Standard Giemsa Staining) (pH 6.8)
Wright stock stain: (0.25% in absolute methanol)
Staining procedure

1. Prepare Wright stain "working solution" as follows: (a) phosphate buffer, 3.0 ml, (b) Wright stock stain, 1.0 ml.
2. *Make up fresh working solution for each slide.*
3. Place slides on staining rack, flood slides with working solution, and stain for 1.5 to 2 min.
4. Rinse briefly with distilled H_2O.
5. If banding is not well differentiated, destain (as described below) and repeat steps 1–4.

[6] S. Schnedl, *Chromosoma* **34,** 448 (1971).
[7] J. Yunis, J. Sawyer, and D. Ball, *Chromosoma* **67,** 293 (1978).

Destaining procedure: (a) preincubate slides in 95% ethanol for 1.5 min (with agitation) and air dry, (b) 95% ethanol + 1% HCl for 30 sec with agitation and air dry, and (c) absolute methanol for 1.5 min and air dry.

Quinacrine (Q-)Banding[8]

Q-banding makes use of the DNA staining fluorochrome quinacrine mustard, or quinacrine dihydrochloride, to provide (like G-banding), unequivocal identification of all normal chromosomes (Fig. 1B). Q-banding is in some regards more labor intensive than G-banding, due to the requirement of fluorescent (rather than bright-field microscopy) and the necessity to immediately photograph mitoses (due to quenching of the fluorochrome dye). Nevertheless, this technique has several advantages over G-banding, including (1) its ability to recognize more readily than G-banding polymorphisms of individual bands, (2) the overall "success rate" of Q-banding is superior to G-banding [especially in cases where metaphases are of marginal quality (e.g., tumor material)], and (3) most importantly, in cases where a minimum number of mitoses are available, Q-banding can be followed by a second banding technique (e.g., Q → G; Q → C) while the reverse order ordinarily cannot be followed. This ability to make use of sequential staining may be very important in identifying chromosomes which have undergone significant structural rearrangement.

Materials
Staining solution: Quinacrine mustard 2 mg/ml (in phosphate buffer)
Phosphate buffer: (see Standard Giemsa Staining) (pH to 5.5)
Ethanol
Mounting media: 10 g sucrose in 15 ml phosphate buffer (saturated solution)

Staining Procedure
1. Preincubate slides for 2 min each in 100% and then 50% ethanol.
2. Stain slides for 7–10 min.
3. Rinse slides for 2 min in phosphate buffer.
4. Mount slides using a #1 thickness coverslip (gently squeezing out excess mounting solution from under coverslip).

[Examination requires microscope with UV source and an excitation filter of 390–490 nm. Photography of fluorescence of Q-banded metaphases can be performed using Technical Pan film (Kodak) using manual exposures at 30 or 60 sec, and development in developer D-19 (Kodak) at 68° for 4 min.]

---

[8] J. Caspersson, L. Zech, C. Johansson, and E. Modest, *Chromosoma* **30,** 315 (1970).

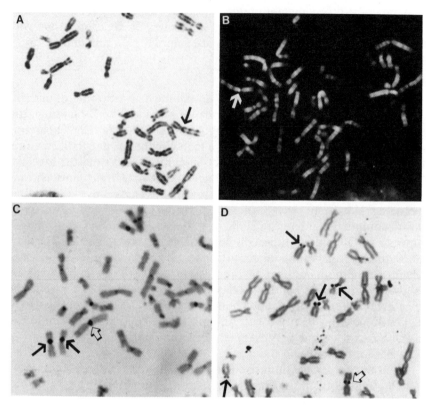

FIG. 1. G-, Q-, C-, or NOR-banded human chromosomes. (A) G-banded metaphase from cultured cell of a human ovarian adenocarcinoma. The arrow points to an inverted chromosome 3 [a consistent (clonal) change in this tumor]. (B) Q-banded metaphase from the same patient as A, with arrow pointing to the inverted chromosome 3. Note that the area under the centromere of chromosome 3 stains very brightly by Q- (but not G-) banding, providing evidence that this altered chromosome is in fact derived from chromosome 3. (C) C-banded cell from a patient with ovarian carcinoma. This cell demonstrates intense staining of constitutive heterochromatin on a normal chromosome 1 (open arrow), as well as intense staining of two structurally aberrant chromosomes 1 (solid arrows). (D) NOR-banded cell from the same ovarian patient shown in A and B. Solid arrows point to the sites of positive silver staining on acrocentric (D- and G-group) chromosomes. The open arrow illustrates an interstitial NOR, a location not found in normal human cells. In this tumor, a translocation between a NOR-bearing chromosome (#13) and a second chromosome (#11) resulted in the retention of ribosomal cistrons normally located on the short arm of chromosome 13.

Reverse (R-)Banding[9]

R-banding provides a "reverse" image of banding recognized by G- or Q-banding techniques. Specifically, the band pattern is opposite in staining intensity to G- or Q-bands with those chromosomal regions displaying very light Giemsa staining (or weak fluorescence) by G- or Q-banding, respectively, staining darkly by R-banding (and vice versa). This banding procedure has particular merit when a chromosomal alteration involves the terminal end of a chromosome (telomeric staining). However, the advantages of using R-banding over G- or Q-banding are minimal and this procedure is less often used than either G- or Q-banding techniques.

As with G-banding, numerous different procedures can be used to successfully induce R-bands in mammalian chromosomes. The following procedure is the one most often successful in our laboratory using established cell lines.

Materials

Staining solution: 0.04% Acridine Orange dye (in phosphate buffer pH 6.8)

Phosphate buffer: (see Standard Giemsa Staining) (pH to 6.8)

Mounting media: [see Quinacrine (Q-)Banding] (pH 6.8)

5-Bromodeoxyuridine (BrdUrd)

Stock solution: 100 μg/ml (in sterile distilled water) (wrap bottle in tin foil to protect from light)

Colcemid: 10 μg/ml in HBSS with phenol red (GIBCO)

Ethanol

Staining Procedure

Note: This method requires preincubation of cells with 5-BrdUrd *in vitro* prior to harvesting and staining of metaphases

1. Four to six hours prior to harvest of metaphases add BrdUrd to a final concentration of 10 μg/ml (reincubate and protect from light).
2. One and one-half hours prior to harvest of metaphases, add colcemid to a final concentration of 0.05 μg/ml (reincubate).
3. Harvest and prepare air-dried slides (see Procedures for Chromosome Procurement).
4. Place slides for 2 min each in 100% and then 50% ethanol.
5. Stain slides with Acridine Orange for 7–10 min.
6. Put slides through 2 rinses of 2 min each (with gentle agitation) in phosphate buffer.
7. Mount with #1 thickness coverslip, gently squeezing out excess working solution [see Quinacrine (Q-)Banding].

[9] M. Bobrow, P. Collacot, and K. Madan, *Lancet* 2, 1311 (1972).

[Examination requires excitation filter of 390–490 nm and addition of a barrier filter at 530 nm. Suggestions for photography of R-banded metaphases include manual exposure time of 0.5, 1, or 2 sec using Tech Pan Film (Kodak) developed with D-19 developer (Kodak) at 68° for 4 min.]

Constitutive Heterochromatin (C-)Banding[10]

C-banding specifically stains areas of constitutive heterochromatin. While C-bands occur in all mammalian species, the amount and distribution may vary significantly between species. As is also true for NOR-banding, this pattern may be used effectively as a cytogenetic marker for species identification.[3] In human cells, C-banding results in some pericentriomeric staining of all chromosomes, as well as staining of "secondary constrictions" found on chromosomes 1, 9, and 16 (Fig. 1C).

As with G-banding, the age of the slide should be taken into account in regards to incubation time. The minimum times presented represent the suggested time for fresh slides; maximum times are given for staining of slides up to 8 months old. This procedure is the most destructive to chromosome morphology of methods described in this report and should be used last in any attempts of sequential staining.

Materials

Barium hydroxide

$Ba(OH)_2$ solution: Prepare a saturated solution of $Ba(OH)_2$ and pour the resulting supernatant into a clean coplin jar prior to staining in order to reduce residue on slides.

Phosphate buffer: (see Standard Giemsa Staining) (pH to 6.8)

Giemsa stock stain: Gurr's R-66

Staining Procedure

1. Slides should be dipped twice in 95% ethanol, immediately dipped five to eight times in phosphate buffer, followed by placement in a saturated solution of $Ba(OH)_2$ for 8–11 min at room temperature.
2. Following exposure to $Ba(OH)_2$, slides are dipped three times in 70% ethanol, three times in a second solution of phosphate buffer, and incubated in phosphate buffer at 60° for 4 hr.
3. Following incubation, slides are rinsed gently in tap water and stained for 5 min in 4% Giemsa (in phosphate buffer).

Silver Staining of Nucleolus Organizer Regions (NOR-)Banding[11]

Staining of NORs recognizes transcriptionally active ribosomal cistrons in mammalian cells, with NOR's occurring in a species-specific pattern on

[10] D. Miller, R. Tantranahi, V. Den, and O. J. Miller, *Musculus Genet.* **88**, 67 (1976).
[11] C. Goodpasture and S. Bloom, *Chromosoma* **53**, 37 (1975).

mammalian chromosomes. In most species, NORs are found on several different chromosome pairs, with NORs in normal human cells restricted to the acrocentric (D and G group) chromosomes (Fig. 1D). The silver staining method for NORs is particularly useful in examination of interspecies cross-culture contamination. The reader is referred to the manuscript of Pathak and Hsu[3] for further discussion of interspecies patterns of NORs.

Materials

Silver nitrate solution: (50%) 1 g $AgNO_3$ in 2 ml distilled deionized water

0.45-μm filter syringe: (Millipore)

Giemsa stain: Gurr's R-66

Staining Procedure

It is essential that the same source of distilled deionized water be used for all steps in this procedure as this will greatly decrease the occurrence of nonspecific silver staining.

1. Prepare a 50% solution of silver nitrate and allow to stand at room temperature 15 min before use.
2. Incubate slides in distilled deionized water for 15 min at room temperature.
3. Prepare a moist chamber for silver incubation. (Use of square petri dish or plastic slide mailer containing moistened filter paper in the base is satisfactory.)
4. Remove slides from distilled water and allow to air dry.
5. Add 3–4 drops of the 50% silver nitrate solution onto each slide using a filter syringe (0.45 μm) and cover with a 22 × 40 mm coverglass.
6. Place slides into moist chamber, cover, and incubate at 56° for 8–18 hr (the length of time may depend on the age of slides, with increased time for older preparations).
7. At the end of 18 hr, examine each slide for the presence of silver staining. You should observe a golden brown tint to the nuclear area and dark brown-black coloration of the nucleolar area. Dots of stained material should appear on the short arms of most of the acrocentric chromosomes (in human) at the conclusion of the incubation.
8. Following incubation, rinse slides with distilled deionized water and counterstain in 1% Giemsa stain for 7 sec.

(An alternate method to greatly decrease the time of incubation is to add three drops of 3% buffered formalin to the 50% $AgNO_3$ solution prior to staining. Slides may then be incubated for only 1–4 hr at 65°.[12]

---

[12] S. Pathak and F. Elder, *Hum. Genet.* **54,** 171 (1980).

Procedures for Chromosome Procurement (Harvesting)

Harvesting of chromosomes for banding analysis has become routine, although some slight modifications must be made for monolayer versus suspension, or agar cultures. The general harvesting procedure will first be described, with modifications for monolayer suspension or agar cultures presented at the end of this section.

Materials

    Colcemid: $10\mu g/ml$ (GIBCO)
    Hypotonic solution: 0.075 M KCl
    Fixative: methanol : glacial acetic acid (3 : 1)

Harvesting Procedure

1. Add Colcemid at a final concentration of 0.05 $\mu g/ml$ and reincubate 1.5 hr. (The optimal time of Colcemid exposure may vary for each cell line, although ordinarily longer colcemid exposures result in contracted chromosomes.)

2. After colcemid exposure, cells should be transferred into a 15 ml conical centrifuge tube and centrifuged at 600–800 rpm for 8–10 min.

3. Remove supernatant (keeping 0.2–0.5 ml) and resuspend cells with gentle aspiration using a Pasteur pipet (being careful not to aspirate cell suspension into the pipet as cell loss will occur onto the glass surface).

4. Add hypotonic 0.075 M KCl (prewarmed to 37°) and resuspend cell–hypotonic solution. Incubate cells at 37° for 15–20 min (optimal time will depend on cells being studied).

5. Following hypotonic treatment, centrifuge cells at 600–800 rpm for 5 min. Remove supernatant (being careful not to disturb cell pellet) and resuspend cells in 5–10 ml of fresh cold fixative (3 : 1 methanol : glacial acetic acid) (suspend thoroughly but do not overpipet cells). Place cells at −20° for a minimum of 30 min.

6. Following Step 5, cells should be centrifuged and one additional change of fix should be made. After removing supernatant from the second fixative, cells should be resuspended in a small volume of fresh fixative (0.2–0.5 ml) for slide preparation.

7. Several procedures exist for air-dried slide preparation, the method described below is routinely utilized in our laboratory. Modifications to this procedure will often need to be made to account for changes in relative humidity and other factors related to individual laboratories.

    a. Slides should be precleaned by placement for 30 min in absolute ethanol at −20°. After removing slides from the ethanol and draining off excess ethanol, slides should be allowed to air dry.

  b. Three to five drops of cell suspension (from Step 6) should be dropped from a Pasteur pipet onto the cleaned microscope slide which is held at an approximate 45° angle (angling the slide will permit the suspension to run down the entire length of the slide, while increasing the distance from the pipet to the slide may assist in spreading of metaphases). A sharp burst of air can also be delivered to the slide to further enhance spreading.
  c. Allow slides to air dry.

Cell Synchronization Procedure

Use of cell synchronization can be of great use in cytogenetic studies of human cell cultures.[7] The advantages of cell synchronization are 2-fold: (1) the mitotic index of most cultures can be improved, and (2) harvesting of chromosomes can be optimized to capture cells closer to prophase than metaphase (resulting in greatly extended chromosomes). This later aspect in optimal preparations may increase the resolution of banding from around 450 total bands in metaphase chromosomes to the recognition of 850 or more bands in very uncondensed prometaphase chromosomes. Although decidedly more information is available in mitoses displaying 850 bands, the complexity of analysis of "high-resolution" banding also increases significantly. Accordingly, most of the work on high-resolution banding has been performed to date on human peripheral blood lymphocytes, where the kinetics of cell division have been thoroughly worked out. The nomenclature for high-resolution preparations of prophase and prometaphase human chromosomes is detailed within the ISCN.[4]

In our laboratory, the use of cell synchronization has been very helpful in obtaining metaphases from cultures with a very long generation time (e.g., early passage epithelial cell cultures). Although this procedure is not capable of providing an increase in the number or quality of metaphases from all cell cultures (particularly tumor cultures), it often is helpful in obtaining large numbers of mitoses of acceptable quality in cases where more standard techniques have failed.

Materials
  Methotrexate stock solution: 10^{-5} (MW 454.46) (in sterile distilled H_2O)
  Thymidine stock solution: 10^{-3} M (MW 242.2) (in sterile distilled H_2O)

Procedure

1. To block cells, add 10^{-7} M (final concentration) methotrexate to the culture medium and reincubate cell cultures for 16–18 hr.

2. To release cell blockage, remove the methotrexate containing medium, allow cells to stand for 5 min, and add fresh medium (without methotrexate), supplemented with 10^{-5} M (final concentration) thymidine, and reincubate for 5–7 hr.

3. Add Colcemid and harvest chromosomes as described in Procedures for Chromosome Procurement.

[The variable times of cell blockage and release must be established for each cell line. The ranges provided will work with a majority of established cell lines. Use of this method for primary tumor specimens is much more variable (due in part to the heterogeneous nature of primary tumor populations).]

Monolayer Cultures

Investigators should examine flasks before chromosome harvesting to insure adequate mitotic activity (evidenced by multiple rounded cells after mitotic arrest). In the event that mitoses are limited (even after prolonged colchicine exposure), cultures should be prepared using cell synchronization as described previously. Cells following mitotic arrest can be obtained mechanically by mitotic shake off or by using mild trypsinization.

Suspension Cultures

Harvesting procedure is identical to that described for monolayer cultures except there is no need to detach cells from plastic substrate.

Agar/Methylcellulose Cultures

Colony forming cells grown in agar/methylcellulose can be harvested for cytogenetic analysis with minor modifications of the procedure described for monolayer cultures.[13,14] For examination of the entire plating layer (containing all colonies or clusters), 2 ml of media containing colchicine (10^{-7} M final concentration) is added, with reincubation for up to 16 hr (for human primary tumors) and, after removing agar layer containing the desired colonies, chromosome harvesting is accomplished as described previously. Alternately, single colonies can be plucked from the agar/methylcellulose layer and harvested using micropipetting techniques.[15]

[13] J. Trent, in "Cloning of Human Tumor Stem Cells," p. 345. Liss, New York, 1980.
[14] J. Trent, Cancer Surv. 3, 395 (1985).
[15] I. O. Dube, C. J. Eaves, D. K. Kalousek, and A. C. Eaves, Cancer Genet. Cytogenet. 4, 157 (1981).

Summary

Technological advances in the study of chromosomes from human and experimental cancers are occurring rapidly. Molecular cytogenetic techniques for *in situ* hybridization, as well as chromosome sorting and even karyotyping via flow cytometry (both described elsewhere within this volume), are important developing areas receiving considerable study. However, there currently remains a significant need for routine karyotyping of mammalian cells. It is hoped that the methods provided in this chapter will be of help in assisting somatic cell geneticists to identify chromosome changes in mammalian cell cultures.

Acknowledgments

Our thanks to Patricia Haight for excellent secretarial assistance. Dr. Trent is a scholar of the Leukemia Society of America. Sponsored in part by PHHS Grants CA-29476 and CA-17094.

[21] *In Situ* Hybridization of Metaphase and Prometaphase Chromosomes

By SUSAN L. NAYLOR, JOHN R. MCGILL, and BERNHARD U. ZABEL

Principle of the Method

In situ hybridization is the direct hybridization of a nucleic acid probe to metaphase chromosomes on a slide. Figure 1 depicts the overall scheme for *in situ* hybridization. Basically, metaphase chromosomes are immobilized on a slide and denatured. A tritiated labeled probe is also denatured and hybridized to the chromosomal DNA. After washing, slides are coated with autoradiographic emulsion. The sites of hybridization are revealed as silver grains produced as a result of the tritium acting on the autoradiographic emulsion. After developing the emulsion-coated slide, the chromosomes are banded, stained, and analyzed. Because of the structure of a metaphase chromosome, accessibility of chromosomal sites complementary to probe can be limiting, that is, not every complementary site will hybridize. Consequently, distribution of the grains over the metaphase chromosomes is analyzed statistically to distinguish specific from nonspecific background hybridization.

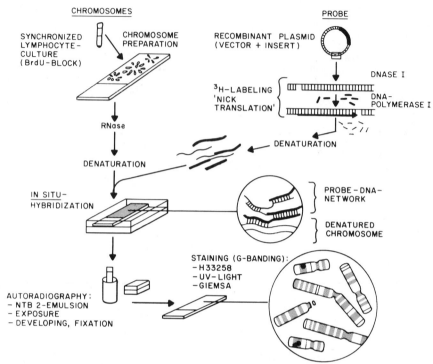

FIG. 1. Schematic representation of *in situ* hybridization to elongated chromosomes.

Materials

Materials for Preparation of Metaphase Spreads

Syringe and needles
Sodium heparin
Chromosome medium 1A with phytohemagglutinin (GIBCO)
Culture tubes (Falcon 3033)
Bromodeoxyuridine (BrdUrd) (8mg/ml; Sigma)
Thymidine (2.5 mg/ml; Sigma)
Amethopterin (methotrexate; Lederle)
Colcemid (10 μg/ml; GIBCO)
RPMI 1640 (with antibiotics and glutamine) (GIBCO)
Phytohemagglutinin (M form; GIBCO)
Fetal bovine serum
Hypotonic KCl (0.075M; GIBCO)

Carnoy's fixative (methanol : glacial acetic acid, 3 : 1)
Prepared slides (cleaned with 7X detergent and alcohol)

Probe Preparation

[3]H-labeled nucleoside triphosphates (New England Nuclear)
   dCTP (specific activity 40–60 Ci/mmol)
   dTTP (specific activity 90–110 Ci/mmol)
   dATP (specific activity 40–60 Ci/mmol)
dGTP (unlabeled) (PL Biochemicals)
DNase I (1 mg/ml; Worthington DPFF)
DNA polymerase I
Nick translation buffer (10X)
   Tris–HCl, pH 7.8, 0.5 M
   $MgCl_2$, 75 mM
   BSA, 150 μg/ml (nuclease free)
   2-Mercaptoethanol, 0.1 M
0.5 M EDTA
10% Trichloroacetic acid (TCA)
Sephadex G-50 medium hydrated in column buffer (10 mM Tris, pH 7.6, 1 mM EDTA, and 0.1% SDS)
Oligo labeling buffer
   LS: 1 M HEPES, pH 6.6/DTM/OL in ratio 25/25/7, store at −20°
   DTM: 100 μM dGTP, 250 mM Tris–HCl, pH 8.0, 25 mM $MgCl_2$, 50 mM 2-mercaptoethanol
   OL: 1 mM Tris–HCl, pH 7.5, 1 mM EDTA, 90 OD units/ml oligo primers (hexamers from PL Biochemicals)

Hybridization

RNase I (2 mg/ml; Sigma) place stock in boiling water bath for 10 min
20X SSC (standard saline citrate)
   3 M NaCl
   0.3 M sodium citrate
   pH to 7.0 with HCl and autoclave
Dextran sulfate (50% in water, autoclaved)
Salmon sperm DNA (10 mg/ml in water, sheared through an 18-gauge needle, autoclaved)
formamide (BDH or AnalR) (should be ~pH 7.0, after opening bottle, store remainder −20°in 50 ml conical tubes)
1 M NaOH

1 M sodium phosphate buffer pH 6.0 (made from 1 M stocks of monobasic and dibasic phosphate buffers in the ratio 2 : 1)
coverslips, 24 × 60 mm (siliconized)

Photography

Kodak NTB-2 nuclear track emulsion
Kodak D-19 developer, diluted 1 : 1 with water
Safelight, dark red filter (Kodak No. 2, Cat. 152 1723)
Slide boxes (light tight)

Staining

Wright stain stock (EM Sciences: 0.25% in methanol, acetone free)
Hoeschst H33258 (1 μg/ml in 2× SSC; Sigma)
Giemsa stain (7% in 0.2 M phosphate buffer, pH 6.8)

Two alternative methods for chromosome preparation and banding are presented. One method produces elongated prometaphase chromosomes (approximately 1000 bands) for precise gene localization. The other method yields metaphase chromosomes of 400 bands, and are easier to recognize for the inexperienced eye. It is of the utmost importance to use the best chromosome spreads possible for *in situ* hybridization.

Methods

Chromosome Preparation (Method 1, Elongated Chromosomes; Adapted from Dutrillaux and Viegas-Pequignot[1])

Five milliliters of blood is drawn into a syringe containing 500 units heparin. The blood should be stored at room temperature and used while fresh. Twelve drops of heparinized blood are added to 4 ml chromosome medium 1A with PHA in 12.5 ml Falcon culture tubes and cultured for 72 hr. BrdUrd is then added to 200 μg/ml. After 16–17 hr, the cells are washed twice with medium and then incubated in new culture tubes with 4 ml chromosome medium 1A with PHA and 10^{-5} M thymidine. After 7 hr cells are harvested by centrifugation (5 min at 1000 rpm) and resuspended in 6 ml warm (37°) 0.075 M KCl. The cells are incubated in the hypotonic solution for 8 min at 37°. The cells are pelleted by centrifugation (10 min, 1000 rpm) and fixed by resuspending them in Carnoy's fixative for 20 min. After three more washes with Carnoy's fixative, the cell suspension is

[1] B. Dutrillaux and E. Viegas-Pequignot, *Hum. Genet.* **57**, 93 (1981).

dropped onto cold slides (0°). The slides are air dried and stored at room temperature until used for hybridization.

Chromosome Preparation (Method 2; Modified from Yunis and Chandler[2])

Three milliliters of peripheral blood is drawn into a heparinized (0.1 ml sodium heparin, 1000 U/ml). The blood is expelled into 42.4 ml RPMI 1640 containing fetal bovine serum, antibiotics, and PHA (1.6 ml of reconstituted PHA from GIBCO). The mixture is divided into eight T-25 flasks which are incubated at 37°. Seventy-two hours later sterile methotrexate (60 μl of 10^{-5} *M* stock) is added to each flask and incubated at 37° for 17 hr. The contents of 2 flasks are combined into each of four 15 ml conical centrifuge tubes. The tubes are spun for 8 min at 400 *g*. Aspirate supernatant to just above pellet (note: lymphocyte layer is on top of pellet). Ten milliliters of warm (37°) RPMI 1640 without supplements is added to the cells and the cells are resuspended. The cells are pelleted and the wash repeated. Add 10 ml of RPMI 1640 with fetal bovine serum and 10^{-5} *M* thymidine. Place tubes into a 37° water bath for 5 hr (2 tubes) and 5 hr and 10 min (2 tubes). After incubation times are complete add 60 μl of colcemid (10 μg/ml) to tubes. Place tubes back into 37° waterbath for exactly 10 min. Centrifuge 8 min at 400 *g*. Aspirate supernatant; carefully add 10 ml warm (37°) hypotonic (0.075 *M* KCl) solution to tube and gently but thoroughly resuspend pellet. Incubate 10–20 min at 37° (time varies depending on blood sample). Centrifuge tubes 5 min at 400 *g*. Aspirate supernatant except for 0.5 ml above pellet. Gently resuspend pellet. Add freshly prepared Carnoy's fixative dropwise to 6–8 ml. Let tubes sit 1 hr. Wash cell pellet 4–5 times in fresh fixative; prepare slides.

Radiolabeling of DNA

DNA can be labeled by the same methods used in Southern filter hybridization. Both nick translation and random priming with oligonucleotides using tritiated nucleotides have been successful in our laboratory. The nick translation protocol is essentially that of Rigby *et al.*[3] and the random priming method is that of Feinberg and Vogelstein.[4]

Nick Translation. To obtain the best hybridization, the nick translated probe should average 400 bases and have a specific activity of 10^7 cpm/μg DNA.

[2] J. J. Yunis and M. E. Chandler, *Prog. Clin. Pathol.* **7**, 267 (1977).
[3] W. J. Rigby, C. Dieckmann, C. Rhodes, and P. Berg, *J. Mol. Biol.* **113**, 237 (1977).
[4] A. Feinberg and B. Vogelstein, *Anal. Biochem.* **132**, 6 (1983).

1. Whole plasmid (0.25 to 1 μg) is added on ice to a mixture of 6 μM labeled nucleotides ([^{3}H]dATP, [^{3}H]dTTP, and [^{3}H]dCTP), 60 μM cold dGTP, 1X nick translation buffer, and sterile water. The total volume of the reaction after addition of DNase I and DNA polymerase I should be between 20 and 30 μl. If the nucleotides are supplied in ethanol, first remove the ethanol by vacuum.

2. Dilute the stock of DNase I to 1 × 10^{-5} mg/ml with sterile water. Add DNase I at a final concentration of 10^{-7} mg/ml in the reaction and mix. Let sit on ice for 30 min. Then heat for 10 min at 65°.

3. Add 5 units *Escherichia coli* polymerase I and incubate at 15° for 1–5 hr until at least 10^7 cpm/μg DNA is obtained. The reaction is monitored by TCA precipitation of the DNA onto glass fiber filters and counting in a scintillation counter. Stop the reaction by adding 5 μl of 0.5 M EDTA, pH 8.0.

4. Remove the unincorporated label by column chromatography with Sephadex G-50 hydrated in column buffer. Apply the contents of the reaction tube to a column made in a siliconized Pasteur pipet. Elute with column buffer. Collect 100 μl fractions into each of 15 tubes. Count a 2-μl aliquot of each fraction in aqueous scintillation fluid. Combine the tubes from the first peak and freeze at −20° until needed. The tritium-labeled plasmid is stable and can be used for up to 1 year.

Random Priming with Oligonucleotides; Adapted from Lin et al.[5]

1. Add the following reagents in order into a microcentrifuge tube: 0.75 nmol each of [^{3}H]dATP, [^{3}H]dCTP, and [^{3}H]dTTP, water to a volume of 36.5 μl, 10 μl of oligo-labeling buffer containing 96.8 μM dGTP, 0.5 μl bovine serum albumin (50 mg/ml), and 2 μl DNA (100 ng/μl). Denature in a boiling waterbath for 5 min and immediately quench in an ethanol–dry ice bath. Add 1 μl (4 units) Klenow fragment of DNA polymerase I and incubate at room temperture for 1.5 to 4 hr.

2. The reaction is stopped by adding 5 μl of 0.5 M EDTA, pH 8.0.

3. The unincorporated label is removed by column chromatography through Sephadex G-50 as described above.

In Situ Hybridization

The *in situ* hybridization procedure is that of Harper and Saunders[6] with some modifications.

1. Incubate slides with RNase A (100 μg/ml in 2X SSC) for 1 hr at 37°. Rinse slides 4 times in 2X SSC (2 min per rinse) and then dehydrate with an ethanol series (2 min each in 70, 80, and 95% ethanol).

[5] C. C. Lin, P. N. Draper, and M. DeBraekeleer, *Cytogenet. Cell Genet.* **39**, 269 (1985).

[6] M. E. Harper and G. F. Saunders, *Chromosoma* **83**, 431 (1981).

2. Two methods are described for denaturing the chromosomal DNA on a slide. The sodium hydroxide/ethanol appears to prevent the loss of DNA from the slide.

Formamide Method. Place coplin jar containing 70% formamide/2× SSC in a waterbath and heat to 70°. Incubate slides 2–6 min. Dehydrate slides in an ethanol series and air dry.

Sodium Hydroxide/Ethanol Method. Warm coplin jar containing 1 *M* NaOH to 37°. Incubate slides 2–6 min and dehydrate in an ethanol series. Air dry.

3. The amount of probe needed is calculated by considering the concentration to be used (usually 50–100 ng/ml), the number of slides needed, and the amount of probe per slide (80–100 μl). The final concentration of the hybridization mixture (pH 7.0 to 7.2) is 50% formamide, 2× SSC, 0.04 *M* sodium phosphate, pH 6.0, carrier salmon sperm DNA (10 μg/ml) in 500 to 1000-fold excess over the probe DNA, and tritiated probe. The hybridization mixture is placed in a boiling waterbath for 5 min to denature the probe immediately before applying to the slides. Cool quickly in a dry ice–ethanol bath.

4. Hybridization mixture (80–100 μl) is added to each slide and covered with a 24 × 60 mm siliconized coverslip. The slides are incubated 18–48 hr at 37° in a humidified chamber saturated with 2× SSC.

5. The washing procedure is initiated by dipping the slides in 50% formamide/2× SSC (pH 7.0) at 39° to remove coverslips. The slides are then washed in Wheaton dishes: 4 washes of 50% formamide/2× SSC for 5 min at 39°, 4 washes in 2× SSC for 5 min at 39°, 3 washes of 2× SSC at room temperature for 5 min. Dehydrate the slides by an ethanol series (70, 80, and 95%) for 2 min each. Air dry slides.

Autoradiography

Kodak nuclear track emulsion NTB2 is melted at 43–45° and diluted 1:1 with water using a safelight with a Kodak dark red filter (No. 2, cat. 152 1723). The emulsion is aliquoted into clean new 20-ml glass scintillation vials which are stored in light-proof containers at 4° until needed. Approximately 12 aliquots of 20 ml can be made from one bottle of diluted NTB2 emulsion.

To coat the slide warm an aliquot of emulsion to 43°. Pour warmed emulsion into a clean slide mailer box and place in a coplin jar. Dip slides one at a time and place slides in a slide rack. Place the racks into a light-tight, humid drying box. After drying (usually overnight), load coated slides into light-tight boxes and wrap with aluminum foil. Store slide boxes at 4° for 10–21 days.

While using the safelight with a dark red filter for illumination, develop slides in D-19 diluted 1 : 1 with water for 2 min at 15°. The slides are placed in stop bath for 30 sec, fixed in Kodak fixer for 5 min, and rinsed in distilled water for 15 min. Scrape emulsion from the back of the slide with a razor blade and allow the slides to air dry.

Staining Chromosomes

Chromosome Preparation Method 1: Giemsa/Hoeschst 33258 Staining of Elongated Chromosomes. First stain the slides in Hoeschst fluorescent dye H 33258 (1 μg/ml in 2× SSC) at 15°, then rinse with distilled water and expose to long-wavelength UV light for 1 hr. The slides are exposed to the UV light covered with 2× SSC at a distance of 20 cm. After rinsing and air drying stain the slides in 7% Giemsa in 0.2 M phosphate buffer, pH 6.8, rinse in distilled water and air dry (see Fig. 2).

Chromosome Preparation Method 2: Wright Stain. Combine 1 ml Wright stain with 3 ml phosphate buffer (0.06 M), mix, and pour onto slide. Stain 5–10 min and then rinse with water. Air dry. Insufficient rinsing will result in particulate matter that has the appearance of silver

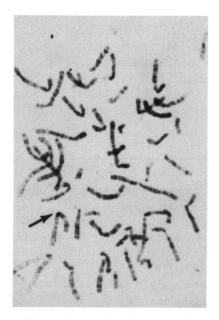 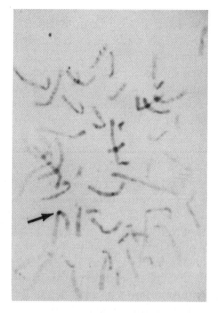

Fig. 2. Metaphase spread of elongated chromosomes (approximately 1000 bands) hybridized to a single copy probe located at chromosome 3q12. The chromosomes were prepared using chromosome method 1 and the Giemsa/Hoescht 33258 staining method. Right, metaphase field focused on chromosome bands; left, focused on silver grains.

grains. If the staining is too light or there is a lack of chromosome banding, destain and restain again. Destain is as follows: place slide in 95% ethanol for 1.5 min with continuous agitation, dry with a stream of air, 95% ethanol/1% HCl for 35 sec with continuous agitation, dry with a stream of air, and 100% methanol for 1.5 min with continuous agitation and dry. Restain as above.

Alternatively, slides can be stained with Wright stain and sodium borate solution. Place coplin jar in 37° waterbath and add 50 ml of sodium borate/sodium sulfate buffer (2.5 mM sodium borate, 50 mM sodium sulfate, pH 9.0) and let equilibrate. Incubate slides in the sodium borate/sodium sulfate buffer for 20–30 min. Combine 1 ml Wright stain with 3 ml sodium borate/sodium sulfate buffer. Mix well and pour onto slide. Stain for 2–4 min and rinse with water. Air dry. If slides are too dark destain by placing the slide into the 37° sodium borate/sodium sulfate buffer (see Fig. 3).

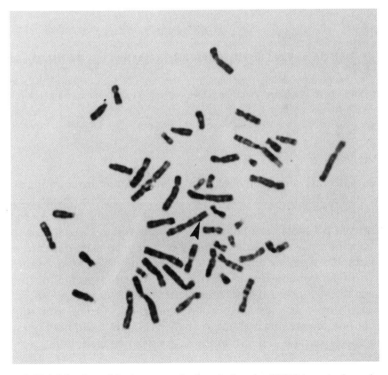

FIG. 3. Hybridization of single copy probe (ceruloplasmin cDNA) to metaphase chromosomes (about 400 bands). Chromosome method 2 and the Wright stain in sodium borate was used.

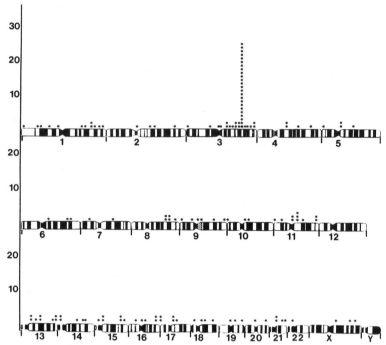

FIG. 4. An example of the presentation of data in a histogram format. The human chromosomes are end to end, p to q. The cDNA probe is for ceruloplasmin and is located on chromosome 3q25.

Data Analysis

Genes that are single copy or low copy number require counting at least 100 metaphase spreads to determine the exact chromosomal location. Generally 10–15% of grains will be located at a specific band; 3–5 grains are observed per metaphase spread. Very rarely will one metaphase spread have both homologs labeled for a single copy gene. The data are represented graphically using a histogram of the chromosomes. An example of the type of data obtained is given in Fig. 4. Visual inspection of the data indicates the significant hybridizing site(s), but the data can also be analyzed statistically as illustrated by Morton et al.[7] Genes with multiple copies at one site are much easier to analyze. Not only will both homologs have grains at the site, but also multiple grains often are found.

[7] C. C. Morton, R. Taub, A. Diamond, M. A. Lane, G. M. Cooper, and P. Leder, *Science* **223,** 173 (1984).

Problem	Possible reason	Corrective action
Nick translation/random priming reaction		
Specific activity too low	Improper nucleotides used	Tritiated nucleotides from New England Nuclear routinely have the highest specific activity and are of very high purity
	Reaction volume too large	Keep reaction volumes as small as possible; 20 μl reactions are optimal
	Input plasmid DNA too high concentration	Use 0.10–0.25 μg of DNA per reaction to increase specific activity
	Enzymes no longer active	Do not freeze and thaw DNase I. Try new enzymes
DNA fragments too small	DNase I concentration is not correct	Titrate DNase I for optimal conditions
Synchronization		
Few mitoses	Improper amethopterin (method 2)	Use Lederle methotrexate 2 mg/ml vial only
	Improper incubation times	Timing is critical on the day of harvest. The reaction should be timed exactly. The wave of synchronized cells could be missed if you are 10 or 20 min off schedule
	Insufficient removal of blocking agent (method 2)	After centrifugation resuspend cell pellets completely. We use warm (37°) wash media. Aspirate to just above the cell pellet layer
	Thymidine decomposition	Thymidine solutions should be prepared immediately prior to use. Refrigerated or frozen stock solutions *do not* work
	Improper incubation temperature	Place tubes in a water bath after release from the block rather than using a dry incubator
Poor chromosome spreading	Improperly fixed cells	You must repeatedly wash the cell pellets in fresh Carnoy's fixative 4–6 times. The cell pellet should be very white in color. Light brown to tan color indicates more fixative changes are necessary

(*Continued*)

TABLE I   *(Continued)*

Problem	Possible reason	Corrective action
	Time in hypotonic solution improper	Use longer hypotonic time or add Quabain when hypotonic times exceed 15–20 min (see Ref. 2)
	Cell pellet too viscous	Dilute with fresh fixative
	Too many drops of pellet added to slide	Add fewer drops (2–4)
	Evaporation rate of fixative improper	Try using cold or wet slides. This slows the evaporation time. The use of a drying oven or hot plate can be optimized by using a variety of temperature setting; avoid using excess heat for drying. Blowing across the slide with a few short puffs of air tends to spread the cells and allows better morphology than dropping chromosome pellets from a 2′–3′ height
Hybridization Unlabeled chromosomes	Improper probe concentration	Increase probe amount
	Probe not denatured	Boil or heat probe to 70° prior to hybridization
	Improper slide making technique	Avoid using heat to prepare chromosome slide preparations. Color of chromosomes on slide prior to hybridization should be light gray. Highly refractile chromosomes hybridize poorly
	Too high stringency washes	Decrease stringency conditions
	May actually be significant	Do a statistical analysis
Excessive chromosome labeling	Improper probe concentration	Decrease probe concentration; 50 ng/ml of probe is sufficient
	Too low stringency washes	Increase stringency of washes, i.e., increase temperature or decrease salts
	Insufficient amount of carrier DNA	Add 500–1000 × carrier salmon sperm DNA
	Autoradiography improper	If background is highly labeled as well as chromosomes this suggests either light leaks in the darkroom or emulsion was outdated or exposed to radiation

TABLE I   *(Continued)*

Problem	Possible reason	Corrective action
Autoradiography		
High background	Light leaks	Darkroom must be absolutely light tight
	Improper safety light	Use only Kodak dark red No. 2 filter from a distance of 4 feet
	Drying time too fast	Slow drying time by putting wet sponge in drying box
	Emulsion exposed or heated previously	Dip test slides prior to use to determine emulsion condition
Wrinkled emulsion layer	Improper development and fixation temperatures	All solutions in processing slides must be 15–16°. Rinse with room temperature tap water after fixation however
Staining and banding		
Too lightly stained	Stain too old	Keep spare recently prepared (at least 2 weeks in advance) stain. Discard old stain bottle
	Buffer solutions improper	Be sure solutions are recently prepared and of proper pH
	Staining time too short	Increase staining time
Heavy staining	Improperly or insufficient rinsing after staining	Rinse slide well with tap water to remove all excess stain. With sodium borate buffer stain place back in 37° solution to destain
Banding insufficient	Emulsion may be too thick or chromosomes exposed to excessive heat when preparing slides	Destain/restain once or twice more

Remarks

We have used *in situ* hybridization with probes cloned into plasmids and phage, genomic sequences and cDNA. The longer the probe, the more labeling is seen at a specific site. However, a longer probe is more likely to contain repetitive sequences. *In situ* hybridization is often done in conjunction with somatic cell hybrid studies. To ensure an unbiased assay, grains are scored by a person who does not know the chromosome location.

Genes with multiple copies are hybridized in a manner similar to single copy probes. Statistical analysis of the data will reveal the number of sites. However, the identification of a specific member of a gene family is often verified with Southern filter analysis of somatic cell hybrids.

The importance of having excellent metaphase chromosomes before beginning *in situ* hybridization cannot be overemphasized. All the steps involved in this procedure only exaggerate the problems with poorly prepared slides. Table I lists some of the problems that are encountered in *in situ* hybridization. Despite the difficulties encountered with *in situ* hybridization, the data obtained yield localization to single chromosome bands. More precise localization can be obtained by using *in situ* hybridization with cells containing defined chromosomal aberrations.

Acknowledgments

Support for these studies has come from the Council for Tobacco Research and the National Foundation March of Dimes. B.U.Z. acknowledges Thomas B. Shows for support during the period in which these techniques were developed. The authors are grateful to A. Y. Sakaguchi and J. McCombs for their suggestions and E. Fischer for typing the manuscript.

[22] Gene Mapping with Sorted Chromosomes

By ROGER V. LEBO and BARRY D. BRUCE

Sorted chromosomal DNAs isolated by single laser chromosome sorting and analyzed by standard restriction enzyme analysis were used to map the β-, γ-, and δ-globin[1] and insulin genes to the short arm of chromosome 11. Hybridization of radiolabeled probe to chromosomes sorted onto spots with a single laser sorter determined the Y-chromosome containing histogram peak[2] and further sublocalized the *myc* gene.[3] Spot-blot analysis of chromosomes separated with a high-resolution dual-laser chromosome sorter provided a straightforward, rapid method to map genes to whole chromosomes.[4] Several developments contributed to the efficacy of the spot-blot method: (1) development of spermine buffer[5] and improved lymphocyte chromosome suspension preparation,[4] (2) identification of

[1] R. V. Lebo, A. V. Carrano, K. Burkhart-Schultz, A. M. Dozy, L.-C. Yu, and Y. W. Kan, *Proc. Natl. Acad. Sci. U.S.A.* **76,** 5804 (1979).

[2] A. Bernheim, P. Metezeau, G. Guellaen, M. Fellous, M. E. Goldberg, and R. Berger, *Proc. Natl. Acad. Aci. U.S.A.* **80,** 7571 (1983).

[3] J. G. Collard, P. A. J. de Boer, J. W. G. Janssen, J. F. Schijven, and J. W. Oosterhuis, *Cytometry* **6,** 179 (1985).

[4] R. V. Lebo, F. Gorin, R. J. Fletterick, F.-T. Kao, M.-C. Cheung, B. D. Bruce, and Y. W. Kan, *Science* **225,** 57 (1984).

[5] R. Sillar and B. D. Young, *J. Histochem. Cytochem.* **29,** 74 (1981).

chromosomes sorted in spermine buffer,[4] (3) sorting chromosomes directly onto filters,[2,4] (4) dual laser chromosome sorting,[6] (5) improved sorter design and performance,[7,8] and (6) characterization of an additional chromosome stain pair.[4] To date we have used spot-blot analysis to confirm the location of 14 cloned genes. Five of these 14 assignments helped confirm the designation of the putative gene clone. In addition, we have mapped 43 recently cloned genes by spot blot analysis. Eighteen of these 43 assignments have been confirmed by another mapping method. Thirty-one of the 43 probes hybridized to homologous gene sequences on more than one chromosome. Eight of the 31 probes hybridized to more than one chromosome spot at high stringency. Often the same genomic DNA sequence as the clone can be distinguished from homologous sequences by increasing the stringency of hybridization and washing. In a few cases like the ferritin-L gene, increasing hybridization stringency failed to distinguish the chromosome carrying the most homologous gene sequence. Then miniaturized restriction enzyme analysis of sorted chromosomal DNA was required to determine which restriction fragments are on each positive chromosome.[9]

Chromosome Suspension Preparation

Lymphocytes

The normal male GM130[4] and female GM131 spontaneously transformed lymphocyte cultures (≈ 26 hr doubling time) were purchased from the NIGMS Human Genetic Mutant Cell Repository, Camden, NJ. These two cell lines were started from the peripheral blood of two normal donors without the addition of Epstein–Barr (EB) virus (A. Bloom, personal communication). Since about 1 in 20 permanent B cell cultures transformed with Epstein–Barr virus shed active virus particles into the medium,[10] we chose spontaneously transformed cultures GM130 and GM131 to minimize this likelihood. All GM130 and GM131 culture manipulation, chromosome preparation, and sorting are conducted at a P-2 biosafety level. All EB-transformed cultures we have analyzed by flow analysis had multiple chromosome rearrangements. In contrast GM130 and

[6] J. W. Gray, R. G. Langlois, A. V. Carrano, K. Burkhart-Schultz, and M. A. Van Dilla, *Chromosoma* **73**, 9 (1979).

[7] R. V. Lebo and A. M. Bastian, *Cytometry* **3**, 213 (1982).

[8] R. V. Lebo, B. D. Bruce, P. Dazin, and D. Payan, *Cytometry* **8**, 71 (1987).

[9] R. V. Lebo, Y. W. Kan, M.-C. Cheung, S. K. Jain, and J. Drysdale, *Hum. Genet.* **71**, 325 (1985).

[10] P. K. Pattengale, R. W. Smith, and P. Gerber, *J. Natl. Cancer Inst.* **52**, 1081 (1974).

GM131 have had normal flow karyotypes for 5 years. When the cultured GM130 cells recently developed an abnormal chromosome histogram peak to the right of the normal chromosome 14 peak, we thawed another frozen aliquot with a normal flow cytogenetic histogram.

These cells are grown to resting phase to synchronize the culture in RPMI-1640 medium which is supplemented with 25 mM HEPES-NaOH (pH 7.3), 2.0 g/liter NaHCO$_3$, and 18% fetal calf serum.[4] The HEPES buffer maintains the pH without CO$_2$ in the incubator and protects the cells from pH changes during manipulation. The cells are diluted 4-fold in this medium and cultured 16 hr. Then the cell suspension is enriched in mitotic cells by adding 0.04 μg/ml colcemid for 5 hr.[4,11] The cells are harvested without mitotic cell enrichment by centrifuging at 280 g for 5 min in 50 ml Corning tubes (Cat. no. 25330). The pellets are pooled and resuspended gently in hypotonic 75 mM KCl for 30 min at 37° to swell the cells. After recentrifugation at 170 g for 5 min, the cells are resuspended in 1× spermine buffer (15 mM Tris–HCl, 0.2 mM spermine, 0.5 mM spermidine, 2 mM EDTA, 0.5 mM EGTA, 80 mM KCl, 20 mM NaCl, 14 mM 2-mercaptoethanol titrated to pH 7.2),[5] aliquoted to 25 × 10^6 cells/Falcon 2063 polypropylene tube, and recentrifuged at 280 g for 7 min. These cell pellets are resuspended in spermine buffer plus 10% stock digitonin solution by vortexing.[11] Our stock supersaturated digitonin solution (10 mg/ml) is stored at 4°, heated 30 min at 37°, and an aliquot filtered through a Millipore filter, and diluted 1 : 10 with spermine buffer just before use in Falcon 2063 polypropylene tubes. These same tubes are used for all chromosome preparation steps following hypotonic swelling to minimize nonspecific chromosome sticking and maximize recovery. With our protocols an average of 15% of the starting number of lymphocyte chromosomes are delivered to the focused laser beams for analysis and sorting.

Initially hexylene glycol buffer was the method of choice for all chromosome suspension preparations[12] since chromosome suspensions could be frozen indefinitely at −80° before sorting, and banded after sorting with modified Leif buckets (see below) to identify the chromosomes represented in histogram peaks. We now use only spermine buffer since we found that lymphocyte and fibroblast chromosomes prepared in spermine buffer can be identified if sorted without freezing.[4] In addition, using a sterile protocol to prepare our chromosome suspensions in spermine buffer increases the shelf life from 2 weeks to many months at 4°. Finally, sorting with sper-

[11] R. V. Lebo, D. R. Tolan, B. D. Bruce, M.-C. Cheung, and Y. W. Kan, *Cytometry* 6, 478 (1985).

[12] R. V. Lebo, *Cytometry* 3, 145 (1982).

mine buffer yeilds chromosomal DNA over 80 kb in size for optimal cloning of large fragments. (Lebo *et al.*, in preparation).

Chromosomes are separated by vortexing 2-ml aliquots of 12×10^6 colcemid arrested cells for 15 sec to 2 min on a Vortex Geni Mixer (American Scientific Products) at top speed. Hoechst stained aliquots are studied with 0.05 mm microslides (Vitro Dynamics, Rockaway, NJ) to monitor chromosome separations. Vortexing is complete when 90% of chromosomes are separated individually, 5% or less are doublets and triplets, and 5% or less are stretched and broken. Clumps of 4 or more chromosomes are not scored. Excessive vortexing is avoided since whole mitotic chromosomes will split in half to form chromatids that give a duplicate histogram peak pattern half the distance from the origin as the whole chromosome pattern. After the chromosome suspensions are pooled to 4-ml aliquots, the interphase nuclei are pelleted by centrifuging at 280 g for 2 min. The supernatant with the suspended mitotic chromosomes is removed and the pellets are pooled, resuspended, and recentrifuged. All supernatants are stained with 4.0 μg/ml chromomycin A_3 and 0.4 μg/ml Hoechst 33258 or 0.4 or 0.8 μg/ml DIPI at least 12 hr before sorting. Stains are prepared and stored at 4° as 100× Hoechst or DIPI (40 μg/ml in H_2O) [DIPI = 4′,6-bis(2′-imidazolinyl-4′-5′H)-2-phenylindole] (from O. Dann) or Hoechst 33258 (Polysciences, Warrington, PA) and 400× chromomycin A_3 (Sigma, St. Louis, MO) (1.6 mg/ml dissolved at 25° overnight and stored at −20° in ethanol) diluted to a 40× working stock in H_2O and stored up to 1 month at 4°.

Fibroblasts

Good chromosome suspensions are considerably more difficult, time consuming, and expensive to prepare from fibroblasts than from lymphocytes. Lymphocyte suspension cultures of spontaneously transformed cell lines GM130[4] or GM131 can provide enough chromosomes from 1 liter of culture grown in two roller bottles to sort five complete sets of spot-blot filter panels[4] or to test 10 fractions of sorted chromosomal DNA by miniaturized restriction enzyme analysis.[9] In contrast, about 120 roller bottles of cultured fibroblasts are required to provide a similar quantity of sortable chromosomes. Thus we reserve fibroblast or amniocyte cultures for sorting translocated or deleted chromosomes that permit assignment of a gene to a chromosome that is not sorted uniquely or to sublocalize a gene to a chromosome region. We have separated as many normal lymphocyte chromosome types as possible to minimize the need to grow fibroblast cell lines with chromosome translocations (see below). Rearranged chromosomes are usually different in size from the normal chromosome. If a gene moves to a new histogram location with the derivative chromosome, the gene may be assigned to the region carried by the rearranged chromosome. Even though lymphocyte cultures are easier to use, we have avoided Epstein–Barr transformed lymphocyte cultures with known chromosome translocations since all cultures we studied had additional chromosome rearrangements. Most available cell lines with known translocations can be purchased from the Human Genetic Mutant Cell Repository, Camden, NJ. A list of fibroblast lines with translocations that can be used to separate each chromosome type with very poor chromosome sorting resolution has been published.[12]

Human fibroblast suspensions are prepared from cells that are partially synchronized by passing confluent cultures to Corning plastic roller bottles. Forty-eight hours later the cultured cells should be about 80% confluent when adding 0.03 μg/ml colcemid. After 10–14 hrs mitotic cells were harvested by swirling the roller bottles until about 80% of the rounded cells were removed. This cell supension is filtered through autoclaved 37 μm mesh (Small Parts Inc, Miami, FL) to remove nonmitotic cell clumps. Following swelling in hypotonic 75 mM KCl with 0.03 μg/ml Colcemid for 30 min at 37° and washing in spermine buffer without digitonin, 7×10^6 cells are resuspended in 1 ml spermine buffer with digitonin by vortexing. After 3 hr on ice the mitotic chromosomes are separated by homogenizing 1-ml aliquots of cells in a Virtis homogenizer microcup with 8 sec bursts at 8000 rpm with the microblade mounted in reverse fashion. Separations are monitored and the separation stopped as when vortexing lymphocytes. Mitotic fibroblast chromosomes break in half much more easily with the higher homogenization shear forces required to separate these chromosomes.

Chromosome Sorting

Improved Optical Bench

We have analyzed and sorted chromosomes with a custom dual laser Becton-Dickinson FACS IV sorter equipped with two 18-W Spectra Physics argon ion lasers[7] and with a triple laser sorter with improved optics[8]. We find that the following modifications each contribute to improved chromosome resolution: (1) expanding high-power laser beams to protect optical coatings and obtain a sharper focal spot, (2) redirecting each beam independently with microadjusters to position prisms that are optimally coated and configured to deliver the maximum laser light energy, (3) focusing the different wavelength beams independently to correct for different wavelength focal distances with simple optics, and (4) improving the light signal collection.[7,8] In a blind study of cells carrying rearranged chromosomes these instruments have resolved nearly all chromosome abnormalities.[13] The triple laser sorter resolves similar chromosomes at well at a flow rate of 8000 chromosomes/sec as the dual laser sorter at 2000 chromosomes/sec.[8] We also added a rapid sheath switchover to change between sterile cell and sterile chromosome sorting without interrupting to change sheath buffers and resterilize the system.[8] The characteristics of a commercially available instrument are described elsewhere in this volume.[14] We use the following sorter protocols for spot-blot analysis.

1. The 18-W Spectra Physics lasers are cooled with water entering at $80-85$ lb/in^2 at 95°F which exists at $130-140$°F to protect and extend glass plasma tube life to an average of 14 months.

2. Chromosome suspensions are introduced through a siliconized, glass-lined flow restrictor and then floppy siliconized tubing sample line to minimize nonspecific sticking and maximize recovery. In order to prevent nonspecific hybridization of DNA probes to chromosome spots, the system must be aseptic as well as sterile. Thus the sorter sheath reservoir is washed with 0.1% sodium hypochlorite (bleach) then 70% ethanol and autoclaved H_2O. The sample and sheath lines are cleaned with 0.1% sodium hypochlorite for 20 min, then 70% ethanol for 20 min, and finally autoclaved H_2O for 20 min before sorting. All $1\times$ spermine sheath buffer stock solutions are autoclaved or filter sterilized.

3. The fluorescent chromosome signals are collected through three common Schott KV 389 barrier filters to exclude the scattered ultraviolet laser light. Optochrome LWP 457, Schott OG 475, and Corning 3-71 filters are placed in front of the first photomultiplier tube to measure chromomycin A_3 fluorescence and exclude the 458 nm scattered laser light and Hoechst fluorescence. Four Corning holmium oxide filters are placed before the second photomultiplier tube to exclude the 458 nm scattered laser light and pass the Hoechst fluorescence.[7]

4. The signals are recorded on a FACS IV Nuclear Data 64×128 channel histogram without gating at a rate of 2700 chromosomes/sec.[4,8]

5. The sheath exiting the 80 μm ceramic nozzle tip (Becton-Dickinson FACS Systems) is collected by a centrally located vacuum tube attached to a peristaltic pump (Buchler) to transport the unsorted droplets to a collection reservoir (Fig. 1). The waste reservoir liquid is decontaminated with an iodine-based detergent or autoclaved before disposal, consistent with P-2 biosafety requirements for animal cells.

[13] R. V. Lebo, M. S. Golbus, and M.-C. Cheung, *Am. J. Med. Genet.* **25**, 519 (1986).

[14] M. Bartholdi, J. Meyne, K. Albright, M. Luedemann, E. Campbell, D. Chritton, L. L. Deaven, M. Van Dilla, and L. S. Cram [19], this volume.

FIG. 1. Sorting onto filter disks. The chromosome-carrying stream exits the 80 μm ceramic nozzle tip orifice. The laser beams intersect the stream 200 and 400 μm below the nozzle. The piezoelectric crystal vibrates the nozzle to produce 26,000 droplets/sec. The uncharged droplets are collected by the central tube connected by a peristaltic pump to the waste reservoir. Charged droplets are deflected to either side onto the nitrocellulose filter on spots delineated by the circles. Reprinted from Lebo *et al.*[11]

6. The desired chromosome fractions are deflected to either side onto a 25-mm-diameter nitrocellulose filter disk. To avoid confusion, pairs of chromosome peaks are sorted left and right according to the relative positions on the flow histogram. Futhermore, the same filter panel pattern is repeated for any one cell line and stain pair.

7. Each filter disk lot number must be tested since general nonspecific binding of radiolabeled probe can be quite high. The filter disk is placed on a scintered glass filter holder (Millipore) mounted in a 6.4 × 8.7 × 14.6 cm wooden block (Fig. 1). This wooden block is held in place in a Becton-Dickinson FACS sample box which is closed during operation. The scintered glass filter is cleaned by soaking for 20 min in Nochromix (Godax Laboratories, Inc., New York, NY) dissolved in H_2SO_4, then washed thoroughly in autoclaved, distilled H_2O.

8. Building vacuum (20 in. Hg) is applied to the cleaned, mounted Millipore filter unit and a nitrocellulose filter added. A protective wax-coated filter paper, supplied by the manufacturer to separate the individual nitrocellulose filters, is placed on top and test droplet deflection used to confirm that the holder and mount are centered.

9. The protective paper is removed, the test droplets deflected for less than 1 sec, and the wet spots on the nitrocellulose filter circled with a VWR labmarker pen (VWR, San Francisco, CA) for reference when studying the autoradiographs. (Eight other pen inks tested hybridized nonspecifically to radiolabeled gene probes.)

10. The two fractions of 30,000 chromosomes each are sorted with three droplets deflected for each chromosome. The side streams must be set and monitored carefully to assure that each deflected chromosome is deposited on the same spot less than 1 mm in diameter.

Identification of Sorted Chromosomes

1. The chromosome content of the histogram peaks is determined by quinacrine-banding analysis of the chromosomes that are prepared in spermine buffer and sorted with spermine buffer sheath into alternate silicone rubber wells of a modified Leif bucket[15] (Fig. 2) at room temperature.

2. The sorted chromosome suspensions are fixed immediately by adding 100 μl of 3:1 methanol:glacial acetic acid (mixed fresh) to each well. The fix is added to the well rapidly with a pipetman in order to mix the fix and sorted suspension immediately.

[15] R. C. Leif, J. R. Easter, R. L. Warters, R. A. Thomas, L. A. Dunlap, and M. F. Austin, *J. Histochem. Cytochem.* **19,** 203 (1971).

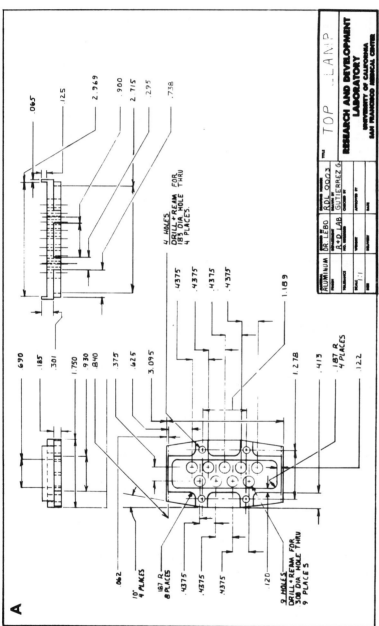

Fig. 2. Machine drawings of modified leif buckets. These Leif buckets,[15] adapted from the modified buckets developed by Anthony V. Carrano, are machined from aluminum jigplate to the sizes labeled in inches. Steel (SS) pins support the buckets in the centrifuge head. A clean, labeled slide is placed in the base. A silicone rubber insert $1 \times 2\frac{7}{8} \times \frac{3}{4}$ in. with 9 holes $\frac{3}{16}$ in. in diameter is stored in 10% detergent and scrubbed, rinsed, and dried with compressed air just prior to use. The cleaned insert is placed on top of the glass slide, the top added, and the unit clamped tightly with four screws through the top (A) into the base (B). Chromosomes are sorted directly into alternate holes. A $2\frac{1}{4}$ in. bolt is added to the bottom to keep the bottom of the silicon insert away from the axis of the Sorvall GLC-3 centrifuge when spinning the sorted chromosomes onto the glass slide. Pairs of buckets are spun at 1000 g for 15 min.

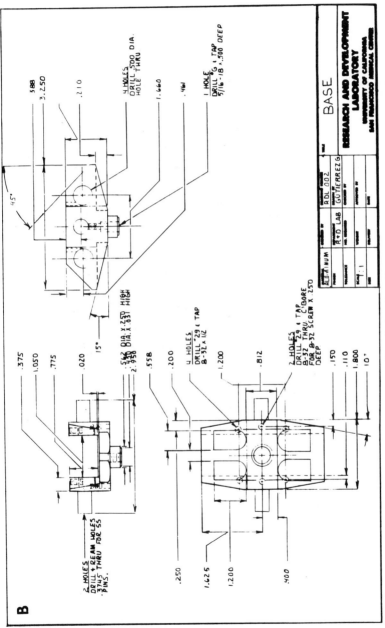

Fig. 2. (continued)

3. After 15 min the Leif buckets are centrifuged at 1000 g for 15 min.

4. The slide is removed by bending away the silicone rubber insert and dipped in the first fresh fix for 30 min and in a second fix overnight in a sealed 50 ml Corning centrifuge tube (Cat. no. 25330).

5. After 1 week at room temperature to improve banding, the slides are stained for 3 min with 0.15% quinacrine dihydrochloride (Sigma), washed 4 min in running deionized water, and air dried for 30 min.

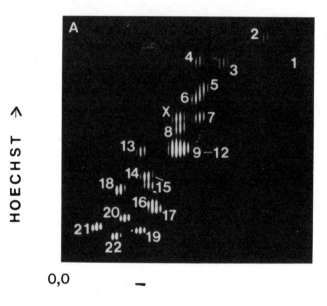

FIG. 3. (A) Hoechst–chromomycin-stained chromosomes from female cell line GM131. This three-dimensional 64 × 128 channel Nuclear Data histogram of flow analyzed lympho-cyte chromosomes is viewed from above. The dual-laser excited fluorescence signals were measured separately and correspond to chromomycin fluorescence from the first laser on the abscissa and to DIPI or Hoechst fluorescence from the second laser on the ordinate. (A) illustrates the separation of each chromosome type except chromosomes 9–12. The chromo-some 14 peak also carries some chromosome 15. (B) Standard DIPI–chromomycin-stained small chromosomes from GM131. Note the separation of chromosome 9 from chromosomes 10–12 with 0.4 μg/ml DIPI. (C) Decreased DIPI staining. Lowering the DIPI concentration to about 0.08 μg/ml moves chromosome 9 toward the 10–12 peak, but moves chromosome 12 away from chromosomes 10 and 11. Thus chromosome 12 can be enriched without chromosomes 10 and 11. In this fashion genes may be assigned to each whole normal chromosome except chromosomes 10 and 11. Furthermore, spot-blots of translocations involving either chromosome 10 or 11 can distinguish which of these chromosomes carry an unknown cloned gene. The data in these histograms were acquired without gating or mathe-matical manipulation.

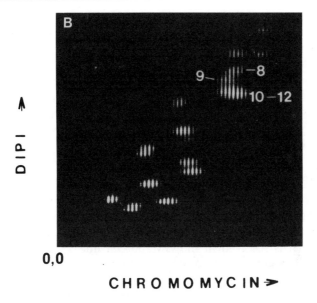

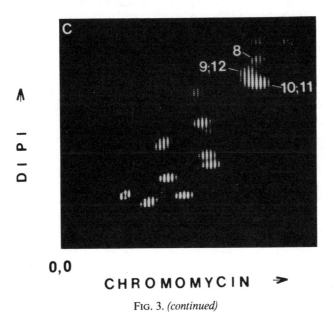

FIG. 3. *(continued)*

6. The dry slides are mounted with 50% sucrose in 0.2 M sodium phosphate buffer (pH 5.5) using No. 0 thickness coverslips (Clay Adams) to minimize glare. The coverslips are sealed with rubber cement (Carter's).

7. Slides are studied using a high-power fluorescence microscope like the Leitz orthomat equipped with a 100× NPL fluotar objective with iris and 15× Nikon eyepieces. The fraction of chromosomes recognizable by banding has been considered to represent a random sample without scoring bias. Typical separations of most normal lymphocyte chromosome peaks have been 98–99% pure.

Unique Chromosome Separations

Hoechst–chromomycin-stained chromosomes can resolve 20 types of human chromosomes with chromosomes 9–12 remaining in the same histogram peak (Fig. 3A). Chromosome 14 is not resolved uniquely in GM131 but it is resolved in the GM130 cell line.[4] Staining with DIPI-chromomycin resolves chromosome 9 from chromosomes 10–12 in both GM131 (Fig. 3B) and GM 130.[4] Lowering the DIPI concentration with GM131 resolves chromosomes 9 and 12 from chromosomes 10 and 11 (Fig. 3C). The exact concentration of DIPI is dependent upon the density of the chromosome suspension preparation and age of the DIPI stock solution. The relative purity of the 9–12 chromosomes sorted from these peaks was sufficient to unambiguously assign previously mapped genes to each peak as illustrated (Fig. 5A–C).

Rearranged Chromosome Separations

Usually chromosome deletions, translocations, or insertions are sufficiently different in length to be sorted separately from the normal remaining homologous chromosomes.[13] For example cell line GM201 carries a translocation between chromosomes 1 and 17 that has been sorted and tested to sublocalize the c-src-2 oncogene (Fig. 4). The terminal portion of the short arm of chromosome 1 is translocated to the short arm of 17 to give a larger derivative chromosome 17. Note that this translocated segment carries more Giemsa-negative, chromomycin-positive (white) bands than Giemsa-positive, Hoechst-positive (dark) bands. These idiograms correctly predict that the derivative chromosome 17 [der(17)] will move farther from the original chromosome 17 position in the chromomycin direction than in the Hoechst direction. We have reported that measuring the band widths of derivative chromosomes will predict the new histogram peak positions.[13] This approximation permits one to pick the most optimal available derivative chromosomes for sorting. To date all predictions of histogram peak positions agreed with the identified sorted chromosomes.

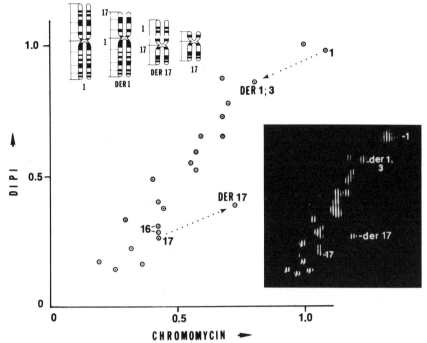

FIG. 4. Sorting analysis of GM201. Upper Left: Idiograms of the large normal chromosome 1, the small chromosome 17, and the intermediate size derivative chromosomes emphasize the difference in size. Each is well-separated from the others as illustrated in the right histogram. The peaks of the largest chromosomes in this histogram are exposed longer for clarity. The graph (left) illustrates the electronic histogram peak positions relative to the chromosome 2 peak position defined as coordinate (1,1). Note that the derivative chromosomes move in approximately equal and opposite parallel directions as expected for a reciprocal chromosome translocation with no change in total DNA content.

Initial tests of two entire spot-blot filter sets showed that the c-*src*-2 gene probe hybridized only to chromosome 1.[16] When derivative chromosome 17 was sorted from derivative chromosome 1 in the normal chromosome 3 peak and tested with c-*src*-2 gene probe, derivative chromosome 17 (Fig. 4) sequences hybridized to the probe (Fig. 5D). This result sublocalized the c-*src*-2 oncogene to the terminal portion of the short arm of chromosome 1.

[16] R. C. Parker, G. M. Mardon, R. V. Lebo, H. E. Varmus, and J. M. Bishop, *Mol. Cell. Biol.* **5,** 831 (1985).

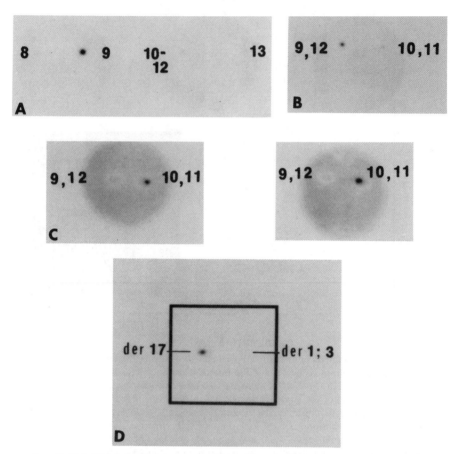

FIG. 5. Hybridization of chromosome 9, 10, 11, and 12 probes to sorted chromosome fractions. (A) Results from GM130 chromosomes stained and sorted as in Fig. 3B and hybridized to aldolase B gene probe map aldolase B to chromosome 9.[11] The light hybridization to the 10, 11, 12 chromosome-containing spot may represent contamination from the adjacent chromosome 9 fraction or cross-hybridization of the aldolase B probe to the homologous ψA aldolase gene on chromosome 10. (B) Chromosomes sorted from histogram B hybridized to Von Willebrands' factor on chromosome 12. (C) β-Globin probe (left) (on chromosome 11) and aldolase ψA probe (right) (on chromosome 10) (D. R. Tolan, J. Niclas, B. D. Bruce, and R. V. Lebo, *Am. J. Hum. Genet.*, in press) hybridize to the spot with both chromosomes. Aldolase A was mapped to chromosome 9 by spot blot analysis of normal chromosomes.[11] The Von Willebrand's factor gene was mapped to chromosome 12 by *in situ* hybridization.[22] Aldolase ψA was mapped to the long arm of chromosome 10 (D. R. Tolan, J. Niclas, B. D. Bruce, and R. V. Lebo, *Am. J. Hum. Genet.*, in press) and γ-globin to 11p15 by spot blot analysis of rearranged chromosomes [R. V. Lebo, M.-C. Cheung, B. D. Bruce, V. M. Riccardi, F. T. Kao, and Y. W. Kan, *Hum. Genet.* **69**, 316 (1985)]. (D) c-*src*-2 oncogene probe[16] hybridized to derivative chromosome 17 DNA to localize the gene to the terminal portion of chromosome 1.

Spot-Blot Hybridization

Since chromosomes are sorted and denatured directly onto nitrocellulose filters, the chromosomal DNA binds quantitatively without loss during extraction. This protocol reduces chromosome sorting time considerably.

1. A spot-blot panel is constructed from chromosome suspensions prepared from 4×10^7 cells grown in 200 ml of lymphocyte suspension culture.

2. Two fractions of 30,000 chromosomes of each type representing 3–15 ng of enriched DNA are sorted directly onto filter disks to provide enough nitrocellulose-bound DNA for an average of seven probe tests. Only flat tipped forceps are used to handle filters to maximize filter life.

3. The sorted chromosomal DNA is denatured on the nitrocellulose filter paper by placing the filter with the chromosome side up on 3MM Whatman filter paper saturated with 0.5 N NaOH, 1.5 M NaCl for 3 min.

4. The filters are neutralized by three transfers to 0.5 M Tris–HCl (pH 7.5), 3.0 M NaCl saturated Whatman paper for 5 min each.

5. Then the filters are placed between dry Whatman filter papers and baked for 90 min in an oven at 80° with a vacuum of 20 in. Hg.

6. The filters are prehybridized at 41° overnight in 5 ml prehybridization solution (50% v/v deionized boiled formamide, 50 mM Na-HEPES (pH 7.0), 3× SSC, 200 μg/ml denatured salmon sperm DNA, 150 μg/ml yeast RNA, and 1× Denhardt's solution) in a 60 mm petri dish with all spots exposed. The dish is sealed with Parafilm to prevent evaporation. Denhardt's solution: (100×) [2 g/100 ml polyvinylpyrrolidone 360, 2 g/100 ml Ficoll 400, 2 g/100 ml BSA (Sigma A-4378)].

7. The prehybridization solution is removed and the filters are then hybridized at 41° for 1 day to 2×10^6 cpm/ml radiolabeled probe in hybridization solution (1.25× prehybridization solution diluted 4:5 with a 50% dextran sulfate stock solution). All radiolabeled probes are tested with 2–3 lanes of total restriction enzyme digested human DNA that has been electrophoresed and transferred to nitrocellulose filter paper. Test Southerns are first tested with known, good-quality radiolabeled probes. Autoradiographs of a Southern blot tested with probe to be mapped must show unique bands with no spots from randomly bound radiolabeled probe debris, and minimal nonspecific gray background confined to the DNA-containing tracks. Since we are sorting total chromosomal DNA on each spot, nonspecific hybridization of very small radiolabeled fragments or a highly repetitive probe sequence will give significant signal in each sorted chromosome spot. In these cases the positive spot will not be more intense than the negative chromosome spots. Centrifuging the probe in an Eppen-

dorf centrifuge 10 min at 4° immediately before removing a supernatant aliquot for hybridization removes most radiolabeled debris and minimizes nonspecific spots in the autoradiograph. (See below for radiolabeling protocol.)

8. The filters are washed out of the petri dish with a pipet and the next washing solution. The filters are washed twice for 5 min in 2× SSC, 1× Denhardt's at 25°.

9. Typically the filters are then washed twice for 1.5 hr in 0.1× SSC, 0.1% SDS at 51° with gentle shaking and one solution change.

10. Two brief washes in 2× SSC, 0.1× SSC at 25° and 4 more washes in 2× SSC at 25° remove additional unbound radiolabeled probe and SDS.

11. The filters are suspended in the last washing solution, transferred to Whatman filter paper and dried at 25°, placed between Saran Wrap, and autoradiographed between two Dupont Cronex Lightning-Plus intensifying screens in a light-tight cassette at −40 to −80°.

Whenever possible the hybridization and washing conditions are chosen to assure that the primary gene sequence gives a considerably darker signal than homologous gene sequences. Our standard conditions for testing human probes against human DNA are given. Stringencies can be increased by raising hybridization temperatures up to 50° or washing temperatures up to 60°. A gene probe that gives a good specific signal on a standard size Southern blot of 7 μg of total human DNA per slot will normally provide fine probe to map genes by spot blot analysis. When using probes from another laboratory, the precise hybridization conditions that were successful for human Southern blot analysis are substituted. Each nitrocellulose spot blot filter panel can be hybridized about 7 times for gene localization.

We have not switched to nylon filters in spite of the improved durability. Thus far nylon filter panels have given random extra chromosome spots of equal intensity in addition to the single correct chromosome spot observed in duplicate nitrocellulose filters. The extra spots have been observed with three of seven probes that gave unique spots on duplicate nitrocellulose filter panels.

Radiolabeling Gene Probe

The majority of our successful gene probes have been excised from the cloned fragment and nick-translated.[17] A smaller percentage of probes labeled by other methods have given high quality test Southerns and spot blots. Our nick-translation protocol follows.

---

[17] P. W. J. Rigby, M. Dieckmann, C. Rhodes, and P. Berg, *J. Mol. Biol.* **113,** 237 (1977).

1. DNase (Boehringer-Mannheim, grade 1) is prepared by dissolving 1 mg/ml in 10 mM HCl and activated by placing the solution on ice for 2 hr. Then 50 μl aliquots are stored at $-20°$. At the time of use the 50 μl DNase aloquot is diluted to about 2.5 ml in 10 mM Tris–HCl (pH 7.5), 5 mM MgCl$_2$. (This concentration and the subsequent volume are determined experimentally to give the proper concentration of DNase.)

2. The nick-translation mixture consists of 150 ng DNA to be nick-translated, 5 mM Tris–HCl (pH 7.8), 50 μg/ml BSA, 5 mM MgCl$_2$, 10 mM 2-mercaptoethanol, 1 mM dGTP, 1 mM dTTP (Boehringer-Mannheim), 10 μl [^{32}P] dATP (3000/Ci/mmol) (Cat. no. 10204), and 10 μl [^{32}P] dCTP (3000/Ci/mmol) (Cat. no. 10205, Amersham, Arlington Heights IL) in a final volume of 25 to 50 μl, depending on the volume of the DNA sample to be nick-translated. The reaction is begun by mixing 2 μl diluted DNase solution and placing the solution at 37° for 5 min.

3. Then 15 units of DNA polymerase I (Bethesda Research Laboratories, Cat. no. 8010) is added and the solution placed at 15° for 2 hrs in a waterbath prepared from tap water in an insulated container.

4. The reaction is stopped by adding 3 μl of 0.2 M EDTA and 10 μg salmon sperm DNA and heating at 65° for 5 min. The protein is removed from the reaction mix by adding 50 μl phenol, mixing gently, centrifuging 1 min in an Eppendorf centrifuge, and removing the DNA-containing supernatant. Bromphenol blue dye is added as a marker to follow the subsequent column separation.

5. Sephadex G-50 fine resin is swollen in 10 mM Tris–HCl, pH 7.5, 1 mM EDTA equilibration buffer overnight at 25° and stored at 4°. Just before pouring, the Sephadex G-50 suspension is warmed to room temperature and added to a closed column to give a resin bed of 0.7 $\times$ 14 cm. After the resin has settled, the column is washed with 2 resin bed volumes of equilibration buffer.

6. The labeled DNA is separated from the unincorporated nucleotides by adding the reaction mix to the Sephadex column and eluting with equilibration buffer behind a protective Plexiglas shield. The radiolabeled DNA is excluded from the Sephadex G-50 and comes through in the void volume. The bromphenol blue dye marker is retarded along with the unincorporated nucleotides. The eluted fractions with radiolabeled probe are identified by Cerenkov counting (i.e., placing the tubes in empty, capped scintillation vials and counting the disintegrations in the ^{3}H channel with an efficiency of approximately 30%). Then the first peak fractions containing the radiolabeled DNA are pooled and an aliquot counted in PCS (Phase Combining System, Amersham) in the ^{32}P channel. The specific activity of useful probe has been between 1×10^8 and 2×10^9 cpm/μg.

7. Any double-stranded, radiolabeled polynucleotide probe is denatured for 2 min at 98° in a heat block (or held in boiling water) and then plunged on ice to cool. The probe is centrifuged 5 min in an Eppendorf centrifuge at 4° and the supernatant added to hybridization mix at 25°.

8. Every probe is tested on a control Southern blot of total human DNA that has been digested completely with a restriction enzyme electrophoresed to separate the fragments by size, and transferred. If gene-specific bands can be visualized without significant hybridization to the remaining human DNA and nonspecific binding or background is not observed, then the radiolabeled DNA is hybridized to a spot-blot filter panel.

A low specific activity of nick-translated probe can result if the DNA has not been nicked enough to allow the polymerase to label the DNA thoroughly. Another cause of the same problem is the presence of salts in the DNA sample which inhibit the nick-translation enzymes. If the DNase has nicked the sample too thoroughly, the small radiolabeled fragments will hybridize nonspecifically to all the DNA fragments on a Southern blot of restriction enzyme-digested total human DNA.

Spot-blot analysis has been a rapid, effective gene mapping method. The chromosome preparation, instrumentation, and hybridization of good quality probe have been routine. The most significant problem has been to obtain good quality probe from numerous laboratories with the problems associated with shipping ^{32}P-labeled nucleotides including the radiation danger and short half-life. Probes are routinely tested according to the conditions that give good quality Southern analysis. Probes are rejected (1) if precipitate not removed by Eppendorf centrifugation remains to give nonspecific random spots, (2) if probe nonspecifically binds to prehybridized nitrocellulose filter paper, or (3) if significant specific gray hybridization backgrounds are observed in the DNA slots between the Southern bands. The third reason gives significant signal in each sorted chromosome spot. While this observation confirms that DNA was sorted on each spot, confirmation has been unnecessary. Every nitrocellulose filter lot should be tested since significant nonspecific hybridization with suboptimal probes can obscure positive signals.

Confirming Results

Permanent lymphocyte cultures may carry small, undetected chromosome rearrangements or undetected mosaic rearrangements in a normal chromosome peak. If this happens, two spots will be positive instead of one. In these cases a good control would be to repeat the test with a spot-blot panel from a second cell line or to routinely hybridize complete panels from two different cell lines such as GM130 and GM131.[21] Should only a single spot be positive, confirmation of the assignment with a translocation involving that chromosome will sublocalize the gene as well as confirm the assignment.[18] An entire second spot-blot filter set would confirm that a second chromosome spot did not fail to hybridize in the first filter set. We have yet to discover evidence of chromosome rearrangements in spot-blot filter panels prepared from cell lines GM130 and GM131 after testing with 48 gene probes.

Miniaturized Restriction Enzyme Analysis

Spot-blot analysis may be done directly on filter panels of 3 to 15 ng of enriched DNA per spot representing 30,000 metaphase chromosomes of each type. Miniaturized restriction enzyme analysis can be done readily with 0.03 to 0.15 μg of enriched sorted DNA representing 300,000 chromosomes of each type. These relatively small quantities of DNA are extracted with 0.5 μg total yeast RNA carrier. When mapping the ferritin-L gene,[9] 300,000 chromosomes of each chromosome type positive by spot-blot analysis were sorted, the DNA extracted, restriction enzyme digested, and applied to a 0.8 × 4.0 mm slot. The dye front was electrophoresed 9 cm into the 4.0 mm thick gel in 6 hr (Fig. 6). These conditions were chosen to resolve the multiple ferritin-L restriction fragments.[9] The number of sorted chromosomes may be increased to 600,000 to compensate for suboptimal restriction enzyme analysis conditions.

Mapping Disease Loci with Chromosome Libraries

Chromosome-specific recombinant DNA libraries have been constructed and used to isolate many fragments from a single chromosome.[19-21] This approach has led to the description of a restriction enzyme fragment within the chromosome region deleted in some Duchenne muscular dystrophy patients.[22,23] Whole X chromosomal DNAs from male

[18] A. Joyner, R. V. Lebo, Y. W. Kan. R. Tjian, C. R. Cox, and G. R. Martin, *Nature (London)* **314,** 173 (1985).
[19] K. E. Davies, B. D. Young, R. G. Elles, M. E. Hill, and R. Williamson, *Nature (London)* **293,** 374 (1981).
[20] C. M. Disteche, L. M. Kunkel, A. Lojewski, S. H. Orkin, M. Eisenhard, E. Sahar, B. Travis, and S. A. Latt, *Cytometry* **2,** 282 (1982).
[21] R. V. Lebo, L. A. Anderson, Y.-F. Lau, R. Flandermeyer, and Y. W. Kan, *Cold Spring Harbor Symp.* **51,** 169 (1986).
[22] D. Ginsberg, R. I. Handen, D. T. Bonthron, T. A. Donlon, G. A. P. Bruns, S. A. Latt, and S. H. Orkin, *Science* **228,** 316 (1985).
[23] R. J. Barlett, A. Monaco, L. Kunkel, M. A. Pericak-Vance, J. Lanman, T. Siddique, and A. D. Roses, *Am. J. Hum. Genet.* **37,** A143 (1985).

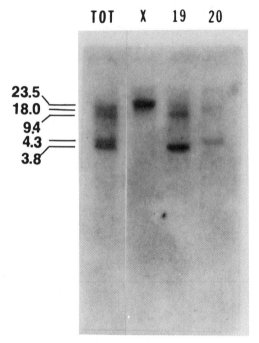

FIG. 6. Miniature restriction enzyme analysis of sorted chromosomes. In the left lane, restriction enzyme analysis of total *Bgl*II digested human DNA (TOT) reveals all five restriction fragment lengths. *Bgl*II digested sorted X, 19, and 20 chromosomal DNA hybridized to ferritin-L probe reveals each of the 5 restriction fragment lengths are carried by unique chromosomes. Reprinted from Lebo *et al.*[9]

patients with one X chromosome can be tested directly for DNA fragment deletions by restriction enzyme analysis. A productive alternative to locate polymorphic restriction enzyme fragments on autosomes (not including the X and Y sex chromosomes) would be to use spot-blot analysis of a series of rearranged chromosomes tested with multiple fragments isolated from a chromosome-specific DNA library. Only the rearrangements in the chromosome region of interest need be sorted to rapidly sublocalize the sequences of immediate interest. In this fashion we are searching for a DNA fragment closely linked to the Charcot-Marie-Tooth disease locus.[21]

Primary Physical Gene Mapping Methods

Somatic cell hybridization has been used to assign the largest number of genes and remains the only method to map genes which can be identified by the gene product but have not been cloned. The method relies on identifying the chromosome content of cells with unstable karyotypes that must be verified. Care must be taken to resolve the rodent restriction fragments from the human with the proper restriction enzyme. However, if homologous gene sequences have different length restriction fragments, the location of each fragment length is determined at the same time as the gene when restriction enzyme analysis of somatic cell hybrid DNA is done with cloned gene probe.

In situ hybridization requires high quality radiolabeled gene probe that is hybridized directly to metaphase human chromosomes on a slide. This method has the advantage of subchromosomally localizing the gene at the same time the chromosome assignment is made. However, less homologous DNA fragments on other chromosomes may confuse the picture. In this case, the chromosome assignments can be confirmed by spot-blot analysis and the restriction fragments located on each positive chromosome can be identified by somatic cell hybrids or miniaturized restriction analysis of sorted chromosomal DNA.

Spot blot panels are very rapid since constructed filter panels can be tested quickly once quality probe is prepared. Although the sorting equipment is expensive and specialized, an existing facility can readily prepare fractions for many gene localizations. Some probes are not suitable for testing by this method although an answer can be obtained by restriction enzyme analysis since repetitive sequences will hybridize to each spot and obscure gene-specific signal. Nevertheless, this method has routinely produced high quality results with 48 of the first 50 probes that produced good results on a test Southern of total human DNA.

[23] Transfer and Selective Retention of Single Specific Human Chromosomes via Microcell-Mediated Chromosome Transfer

By PAUL J. SAXON and ERIC J. STANBRIDGE

Human gene mapping has been greatly facilitated by the development of interspecific somatic cell hybrids, in which human cells are fused to rodent cells, resulting in human–rodent hybrids.[1] These hybrids initially contain a complete set of chromosomes from both parents, but then rapidly segregate out most of their human chromosome complement in a

---

[1] V. A. McKusick and F. H. Ruddle, *Science* **196**, 390 (1977).

random fashion. Following segregation, semistable segregant hybrid cells arise which retain partial human chromosome complements in a complete rodent chromosome background. By generating a panel of hybrid clones, which contain overlapping combinations of human chromosomes (representing the entire human karyotype), genes can be localized to a particular human chromosome by screening each hybrid for the presence or absence of the particular gene sequence or gene product.[2]

An important extension of the interspecific hybrid panel concept is the development of a complete set of human–rodent hybrid clones, each containing a different biochemically selectable human chromosome in the rodent cell background. This type of hybrid panel would provide a permanent source of specific human chromosome DNA for restriction fragment length polymorphism (RFLP) analysis,[3] gene mapping studies, or cloning of specific chromosome libraries. In addition, such a panel would provide a source of intact selectable human chromosomes for chromosome transfer studies which require the presence of a specific intact human chromosome in different cell types.

Transfer of specific chromosomes can be accomplished in a number of ways. Metaphase chromosomes may be separated by fluorescence-activated cell sorting and then dispersed in suspension onto monolayers of recipient cells. The metaphase chromosomes are taken up by the cells but are then rapidly degraded by endogenous nucleases. Therefore, although genetic information is transferred to and expressed in recipient cells by this technique,[4] there is only stable retention of genetic information for which there is positive selection, e.g., expression of hypoxanthine–guanine phosphoribosyltransferase (HGPRT), which allows cells to survive in HAT medium. In one sense, this procedure may be viewed as a variation of DNA transfection, owing to the relatively small amount of genetic information that is stably retained in the recipient cell, presumably by integration into host chromosomal DNA.

This problem of degradation of transferred chromosomes is avoided by the microcell-mediated transfer technique.[5,6] In this technique, cells are micronucleated by prolonged colcemid treatment followed by enucleation, which is accomplished by centrifugation of flasks containing monolayers of the micronucleate cells in medium containing cytochalasin B. The result-

---

[2] M. C. Weiss and H. Green, *Proc. Natl. Acad. Sci. U.S.A.* **58,** 1104 (1967).

[3] D. Botstein, R. White, M. Skolnick, and R. W. Davis, *Am. J. Hum. Genet.* **32,** 314 (1980).

[4] O. W. McBride and H. L. Ozer, *Proc. Natl. Acad. Sci. U.S.A.* **70,** 1258 (1973).

[5] T. Ege, N. R. Ringertz, H. Hamberg, and E. Sidebottom, *Methods Cell Biol.* **15,** 339 (1977).

[6] R. E. K. Fournier and F. H. Ruddle, *Proc. Natl. Acad. Sci. U.S.A.* **70,** 319 (1977).

ing microcells are then fused to recipient whole cells. In this case the transferred chromosome is retained in the recipient cell as a complete structural unit that now becomes a stable heritable component of the cell. The main problem with this technique is a lack of selection for the cell that has acquired a given chromosome via microcell-mediated transfer. This may be accomplished by growth in selective media. However, relatively few chromosomes contain naturally occurring genes whose products allow for growth in a selective medium. Examples of such products include HGPRT (X chromosome), adenine phosphoribosyltransferase (human chromosome 16), and thymidine kinase (human chromosome 17).

In order to overcome this lack of selective retention for most human chromosomes, we have devised strategies whereby selectable genetic markers are integrated into individual chromosomes.[7]

In this chapter (see also Ref. 8) we present an improved procedure for generating microcell hybrids containing single, selectable human chromosomes in both mouse and human recipient cells. The system combines the technique of DNA-mediated transfection, which facilitates integration of a dominant selectable marker into human chromosomes, and the technique of microcell-mediated chromosome transfer to isolate the marked chromosome in rodent or human recipient cells.

General Strategy

A human cell line is transfected with plasmid vectors which contain dominant selectable markers (e.g., pSV2-gpt or pSV2-neo). These plasmids integrate randomly into different chromosomes. Because integration of the plasmid occurs randomly in the genome it should be possible, by isolating a sufficiently large number of transfected clones, to obtain a set of clones with each clone containing the plasmid integrated into a different human chromosome. Transfected clones (transformants) are screened for those containing only single integrations of the plasmid per cell (and thus containing a single marked chromosome). These transformants are micronucleated, yielding microcells which contain one or more chromosomes. The microcells are fused to mouse A9 recipient cells and the resulting human–mouse microcell hybrids, containing the single marked human chromosome, are isolated by growth in appropriate selective medium. The karyotype of the microcell hybrid is examined and the marked chromosome is

[7] E. S. Srivatsan, E. J. Stanbridge, P. J. Saxon, P. J. Stambrook, J. J. Trill, and J. A. Tischfield, Cytogenet, Cell Genet. 38, 227 (1984).

[8] P. J. Saxon, E. S. Srivatsan, G. V. Leipzig, J. H. Sameshima, and E. J. Stanbridge, Mol Cell. Biol. 5, 140 (1985).

then identified by established cytogenetic procedures (G-11 staining, G-banding, isoenzyme assay, and chromosome-specific DNA probes). Examination of sufficient hybrids should result in a library of microcell hybrids containing single human chromosomes that represent the entire human karyotype. The microcell hybrid clones can be used either for subsequent microcell transfer of the identified selectable human chromosome to other human or rodent cells, or genomic DNA can be isolated directly from the microcell hybrids and used for gene mapping studies, preparation of RFLP probes, etc.

Materials and Reagents

Cell Lines and Plasmids

D98/AH-2, an HPRT-deficient derivative of HeLa, was originally described by Szybalski et al.[9] A9, a HPRT-deficient mouse line derived from L cells, was described by Littlefield.[10] HT1080-6TG, a HPRT-deficient human fibrosarcoma,[11] was mutagenized to 6-thioguanine resistance from the HT1080 cell line described by Rasheed et al.[12] The plasmids pSV2-gpt and pSV2-neo were constructed by Mulligan and Berg.[13]

Media and Culture Conditions

Microcell hybrids and rodent cells are maintained on Dulbecco's modified Eagle's medium (DMEM) (Gibco Laboratories) supplemented with 10% fetal calf serum (FCS) (JR Scientific) and 100 IU/ml penicillin/streptomycin/Fungizone (PSF). Hybrid clones are selected in either HAT medium (DMEM, 10% FCS, 1×10^{-4} M hypoxanthine, 4×10^{-5} M methotrexate, 1.6×10^{-5} M thymidine), or G418 medium (DMEM, 10% FCS, 800 μg/ml G418-Geneticin). All cell cultures are incubated in water-jacketed tissue culture incubators at 37° in a 5% CO_2 atmosphere.

Fusion Reagents

Polyethylene glycol (PEG) of molecular weight 1000 (Baker Chemical) is diluted to 48% w/v with DMEM (no serum). The diluted PEG was detoxified by filtering through a layer of mixed-bed resin (20–50 mesh,

[9] W. S. Szybalski, E. H. Szybalski, and G. Ragni, *Natl. Cancer Inst. Monogr.* **7**, 75 (1962).
[10] J. W. Littlefield, *Exp. Cell Res.* **41**, 190 (1966).
[11] C. M. Croce, *Proc. Natl. Acad. Sci. U.S.A.* **73**, 3248 (1976).
[12] S. Rasheed, W. A. Nelson-Rees, E. M. Toth, P. Arnstein, and M. B. Gardner, *Cancer* **33**, 1027 (1974).
[13] C. Mulligan and P. Berg, *Science* **209**, 1422 (1980).

Bio-Rex RG grade, Bio-Rad Laboratories), then filter-sterilized through 0.2-μm bacterial filters.

Phytohemagglutinin P (PHA-P) (Difco) is diluted to 50 μg/ml in DMEM without serum. Cytochalasin B (Sigma Chemical) is diluted to 10 μg/ml in DMEM without serum. Colcemid (Calbiochem-Behring) is diluted to 1 mg/ml stock solution in sterile doubly distilled water.

Procedures

Transfection of Human Cells

The following procedure describes the transfection of two established human tumor cell lines (D98/AH-2 and HT1080-6TG) with a selectable marker plasmid. These particular cell lines were orginally chosen because they readily take up and integrate exogenous DNA via DNA-mediated gene transfer, and because they efficiently yield microcells in quantities sufficient to ensure a relatively high frequency of microcell-mediated chromosome transfer. The HeLa derivative, D98/AH-2, was used for our initial transfections but because this cell line has an aneuploid karyotype, including numerous marker chromosomes and partially deleted chromosomes which might also be sites of plasmid integration, the bulk of the studies were undertaken using the pseudodiploid human fibrosarcoma HT1080-6TG cell line. The pseudodiploid karyotype, and the presence of only two marker chromosomes in HT1080-6TG, increases the probability that individual unrearranged human chromosomes will be the recipients of an integrated plasmid vector.

The marked chromosomes described below are derived from neoplastic cells. For gene mapping studies it will be important to study microcell hybrids that contain single chromosomes derived from a normal human cell background. Therefore, the procedure also describes how normal human diploid fibroblasts can be successfully transfected and single chromosomes containing an integrated plasmid marker transferred to recipient rodent cells.

The transfection of D98/AH-2 and HT1080 cells was described by us in a previous paper[7] and is a modification of the calcium phosphate precipitation procedure of Graham and van der Eb.[14] Briefly, 500 ng of either pSV2-gpt or pSV2-neo DNA, plus 20 ng homologous human carrier DNA are coprecipitated in a calcium phosphate precipitate suspension. Add 1 ml of this suspension to 100 mm dishes containing 5×10^5 of either D98 or HT1080 cells in 10 ml growth medium. After incubation at 37° for 24 hr,

[14] F. L. Graham and A. J. van der Eb, *Virology* **52,** 456 (1973).

replace the growth medium with fresh growth medium, and 24 hr later by either HAT or G418 selective medium. Isolate HAT-resistant or G418-resistant colonies and remove with glass cloning rings, expand, and assay for plasmid integration number using the Southern blot method (see below).[15]

Select transformants, which have single copy integrations of plasmid DNA, for stable integration by growing each transformant clone in nonselective (growth) medium for 8–10 passages (2–3 weeks) followed by a return to selective medium. Unstable transformants are unable to survive the return to selective medium.

While normal human fibroblasts are relatively more difficult to transfect with exogenous DNA recent experiments have reported successful transfection with pSV2-neo DNA. The fibroblast transfection method of Debenham et al.[16] results in transfection frequencies of approximately 1×10^{-4} for human diploid fibroblasts (J. Wilkinson and E. Stanbridge, unpublished observations). Also, human fibroblasts can be transfected with plasmid DNA by protoplast fusion techniques.[17] Transfected human diploid fibroblasts usually have a very limited population doubling capacity and tend to senesce before enough cells can be obtained for subsequent microcell transfer. Also, we have found that human fibroblasts are very difficult to micronucleate and they yield relatively few microcells. Therefore, when a transfected fibroblast clone has reached a population density of 1×10^6 cells, they are then fused with mouse A9 cells in order to "immortalize" the marked human fibroblast chromosome in a mouse A9 background. These human–mouse somatic whole cell hybrids are immortal and easy to micronucleate and thus are used as microcell donors to transfer just the marked human chromosome to either rodent or human recipient cells.

Microcell-Mediated Chromosome Transfer

Seed cells (transformants or human–mouse hybrids) to be used as microcell donors into six 25-cm² tissue culture flasks (Nunc) and incubate until 80–90% confluent. Add colcemid (Calbiochem-Behring) to a final concentration of 0.2 µg/ml for D98-gpt transformants and 0.02 µg/ml for either HT1080-gpt transformants or human–mouse A9 hybrids, and incubate 48 hr at 37°. Other cell types may require different concentrations of colcemid for maximum micronucleation. The optimal concentration and

[15] E. M. Southern, J. Mol. Biol. 98 503 (1975).
[16] P. G. Debenham, M. B. T. Webb, W. K. Masson, and R. Cox, Int. J. Radiat. Biol. 45, 525 (1985).
[17] R. N. Sandri-Goldin, A. L. Goldin, M. Levine, and J. C. Glorioso, Mol. Cell. Biol. 1, 743 (1981).

time of treatment are determined by seeding a number of glass coverslips with the cells and incubating them in different concentrations of colcemid for varying lengths of time. Micronuclei are visualized by fixing cells in 0.2% paraformaldehyde for 1 hr, then staining with DNA-specific fluorescent stains such as Hoechst 33258[18] or DAPI (4,6-diamidino-2-phenylindole).[19]

For enucleation, fill flasks to the neck with DMEM (without serum) containing 10 μg/ml cytochalasin B (Sigma Chemical). Preincubate the flasks at 37° for 30 min, then place (with growth surface facing inward) in a Beckman JA-14 fixed-angle rotor, containing 100 ml of water in each well for a cushion, and centrifuge at 25,000 g (14,000 rpm) for 60 min at 34°.

Resuspend the microcell pellets, pool, and filter through polycarbonate membrane filters (Nucleopore) using 8 and 5 μm pore sizes in series, respectively. Typical yields of 1×10^7 microcells that are approximately 3 μm in diameter or smaller are routinely obtained.

Fuse microcells to 95% confluent monolayers of recipient cells by suspending the filtered microcells in 2 ml of DMEM plus 50 μg/ml phytohemagglutinin P (Difco Laboratories) and then adding the suspension to the recipient cell layer. Allow the microcells to settle by gravity and attach to recipient cells for 15–20 min at room temperature. Fuse the attached microcells to recipient cells with 2 ml of detoxified polyethylene glycol (PEG 1000), 48% in DMEM, for 60 sec. Rinse the cells quickly three times with PBS to remove excess PEG and then incubate for 12 hr in DMEM plus 10% FCS. Remove fused cells by trypsinization and split into 8 or 10 100-mm tissue culture dishes (Falcon) and allow the cells to grow for 2–3 days in growth medium. After 2–3 days expression time change the growth medium to HAT or G418 selection medium, depending on the integrated selectable marker. Pick surviving clones after 2–3 weeks and expand for further characterization. A microcell to recipient cell ratio of 10:1 gives viable microcell hybrid frequencies of approximately 1×10^{-5} of input cell number, but lower ratios will also give reasonable frequencies.

Whole-Cell Fusions

Whole-cell fusions are accomplished using monolayer cultures consisting of a 1:1 mix of both parental cell types. Seed a mixture of 1×10^6 cells of each parental type into 25-cm² tissue culture flasks and allow to attach for 24 hr at 37°. Following incubation, rinse the monolayer once with PBS

[18] T. R. Chen, *Exp. Cell Res.* **104,** 255 (1977).
[19] W. C. Russell, C. Newman, and D. H. Williamson, *Nature (London)* **253,** 461 (1975).

and treat with 2 ml of 48% polyethylene glycol (Baker PEG 1000) in DMEM for exactly 60 sec. Remove excess polyethylene glycol and rinse the monolayer three times with PBS. Incubate the monolayer for 12–24 hr in growth medium (DMEM + 10% FCS) at 37°. Following incubation, transfer the fused cell into 8 or 10 100-mm tissue culture dishes and incubate with growth medium for an additional 2–3 days. When the cells are 50% confluent change the growth medium to selection medium (HAT or G418). Hybrid colonies should appear within 1–3 weeks. Using glass cloning cyclinders remove hybrid colonies from selection dishes. Expand the colonies stepwise in 48-well plates, 24-well plates, 25-cm² flasks, and 75-cm² flasks, respectively. Maintain cell hybrids continuously in selection medium during expansion and freezing down of stocks.

Southern Blot Analysis of Transformants and Microcell Hybrids

The human transformants and resulting microcell hybrids are analyzed for plasmid integration copy number using Southern blot analysis. Isolate high-molecular-weight genomic DNA of cell lines and hybrids according to established procedures.[20] Perform digestions of DNA samples with restriction enzymes, using conditions specified by the manufacturers. For example, in our studies 5–10 units of restriction enzyme were added to 5–10 μg of DNA in 1× universal restriction buffer (URB) [20× URB is 0.66 M Tris (pH 7.8), 1.32 M potassium acetate, 0.2 M MgCl$_2$, 0.01 M dithiothreitol], then incubated at the appropriate temperature overnight. Terminate digestion by incubation at 65° for 10 min. Mix the digested DNA with loading buffer (5% glycerol and 5% bromphenol blue/xylene cyanol) and load onto 0.8% agarose slab gels. Electrophorese the DNA in TAE buffer [40 mM Tris-acetate (pH 7.01), 1 mM EDTA] at 0.25 mA/cm² for 10–15 hr at room temperature. Transfer the electrophoresed DNA from the agarose gel to nitrocellulose filters by the method of Southern.[15] Bake the filters in a vacuum oven at 80° for 2 hr to bind DNA to the nitrocellulose. Place the filters in plastic freezer bags, add the hybridization mixture and seal. Perform the hybridization in 5× SET buffer [150 mM Tris–HCl (pH 8.0), 5 mM EDTA, 750 mM NaCl] containing 1× Denhardt's (0.02% each of bovine serum albumin, Ficoll, and polyvinylpyrrolidone), 0.5% sodium dodecyl sulfate (SDS), 10% dextran sulfate, and 100 μg/ml of heat-denatured salmon sperm DNA as carrier. Heat-denatured ^{32}P-labeled probe DNA (specific activity, 5×10^8 to 1×10^9 cpm/μg of DNA) gives an adequate signal for single copy genes. The DNA probe is labeled with [α-^{32}P]dGTP, using a random-primer labeling procedure.[21] Wash the hybri-

[20] N. Blin and D. W. Stafford, Nucleic Acids Res. 3, 2303 (1976).
[21] A. P. Feinberg and B. Vogelstein, Anal. Biochem. 137, 266 (1984).

dized filters in 2× SET buffer (pH 8.0) containing 0.2% SDS for 1–2 hr at 65°, then in 0.2× SET/0.1% SDS for 30 min at 60°, and finally in 3 mM Tris-base (pH 8.0) for 30 min at room temperature. Air dry the filters and expose them to Kodak XAR-5 film with an intensifying screen at -70° for 2–14 days.

A typical pSV2-gpt integration number analysis of four transformants is seen in Fig. 1. Transformant 5 contains multiple-site integrations of the plasmid while transformants 1, 4, and 7 contain single-site integrations.

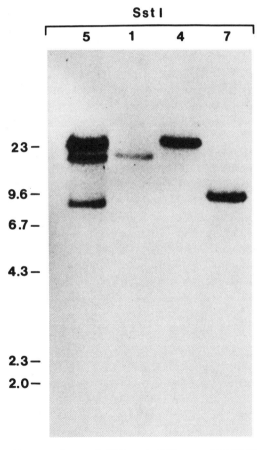

FIG. 1. Nature of integrated state of pSV2-gpt in D98-gpt and HT1080-gpt transformants. The transformant DNAs were digested with SstI, an enzyme that does not cut within the plasmid. Southern blot analysis with ^{32}P-labeled pSV2-gpt as probe showed that D98-gpt transformant 5 contains multiple integrants, whereas HT1080-gpt transformants 1 and 4 and D98 transformant 7 contain single integrants of pSV2-gpt. Reproduced with permission from Saxon et al.[8]

Using the transfection procedure described above, approximately 75% of our transformants contained integrations into single chromosome sites and 25% contained integrations of the plasmid into multiple sites. Further analysis of these transformants has shown that the single-site plasmid integrations usually contain 2–5 copies of the plasmid in tandem (E. Srivatsan and E. Stanbridge, unpublished observations).

Chromosome Analysis

Although there are numerous established procedures for preparing chromosome spreads for karyotype analysis, experience has shown that the choice of a procedure depends on the particular cell line being analyzed. Chromosome spreads of our cell lines were prepared by a modification of the method of Nelson-Rees et al.[22] Incubate a 90% confluent monolayer of each cell type in a T-75 flask with 1 μg/ml colcemid in growth medium for 45 min at 37°. Rinse the monolayer with PBS and trypsinize briefly (less than 60 sec). Harvest the metaphase cells by shaking the flask until approximately 10% of total cells detach from the monolayer and appear in the supernatant. Collect the detached cells and pellet by centrifugation. Resuspend the pellet in 6 ml hypotonic solution (0.075 M KCl, 0.25 M sodium citrate) and incubate at room temperature for 60 min. Pellet the treated cells and resuspend in 5 ml freshly prepared Carnoy's fixative (3:1, methanol:glacial acetic acid) which is added dropwise. After a 15 min fixation, rinse the cells an additional 2× with fresh fixative and finally resuspend in 300 μl Carnoy's fixative. Drop the fixed cells onto clean, wet microscope slides and air dry overnight at 60°. The spreads are then analyzed by two different methods: (1) G-11 staining to differentiate human chromosomes from rodent chromosomes, and (2) trypsin-Giemsa banding to identify specific chromosomes.

1. *G-11 staining.* The G-11 staining procedure of Friend et al.[23] is advocated. After drying spreads overnight at 60°, rinse the slides in doubly distilled H_2O for 3 hr to remove excess traces of cytoplasm that might interfere with staining of chromosomes. Stain the slides 3–4 min in freshly prepared G-11 stain (4% Giemsa stain in phosphate buffer—NaH_2PO_4, pH 11.3). The slides are then rinsed in doubly distilled H_2O, dried, and examined. Rodent chromosomes stain magenta with pale blue centromeres while human chromosomes stain pale blue throughout.

2. *Trypsin-Giemsa banding.* The G-banding procedure of Worton and

[22] W. A. Nelson-Rees, R. R. Flandermeyer, and P. K. Hawthorne, *Science* **184,** 1093 (1974).
[23] K. K. Friend, S. Chen, and F. H. Ruddle, *Somatic Cell Genet.* **2,** 183 (1976).

colleagues[24] has proven adequate. Dip the aged slides in 0.025% trypsin in Hands' buffered saline for 45–60 sec, then quickly dip in Hanks' buffered saline containing 10% calf serum to inactivate the trypsin. Then briefly rinse the slides in running water and stain in 2% Giemsa stain in Gurr's buffer, pH 7.0 for 4 to 6 min. The positions on the slides of good banded spreads are noted and then photographed with high-resolution Kodak Tech Pan 2415 film. Next, destain the slides for 3 min in 4:1 methanol:glacial acetic acid solution, rinse thoroughly in water for 1 hr, and restain with G-11 stain to locate the human chromosome in the previously photographed spreads. It is our experience that a particular spread can be G-banded, destained, and then successfully G-11 stained. However, if a spread is G-11 stained initially, it does not G-band well, presumably because the high pH treatment during G-11 staining reduces the banding potential of chromatin.

Identification of Chromosomes by Isoenzyme Marker Assay

The tentative identification of human chromosomes in microcell hybrids by Giemsa-banding patterns may be confirmed by assaying hybrid cell extracts for marker isoenzyme activity (Table I) using methods outlined by Harrison and Hopkinson.[25] The details of each isoenzyme assay is beyond the scope of this chapter, and the above reference or the chapter by Siciliano and White ([15] in this volume) should be consulted for technical details.

Comments

Using the system described above, we have been able to isolate microcell hybrids containing at least 10 different selectable human chromosomes. We have also isolated a number of hybrids which contained partial or fragmented human chromosomes and translocated chromosomes. While these hybrids represented a minority of all hybrids isolated, it should be noted that fragmentation and rearrangements of the marked human chromosomes can occur and probably do so at the time of micronucleation or microcell transfer. Once a particular chromosome has been transferred to the recipient cell, it appears to remain stable and unaltered even during long-term passage in culture.

In addition, the integration of the selectable plasmid appears to be stable both before and after microcell-mediated chromosome transfer,

[24] R. G. Worton and D. Duff. this series, Vol. 58, P. 322.
[25] H. Harris and D. A. Hopkinson, "Handbook of Enzyme Electrophoresis in Human Genetics." North-Holland Publ. Amsterdam, 1976.

TABLE I
CHROMOSOME-SPECIFIC MARKER ISOENZYMES[a]

Chromosome	Marker	Location
1	PGM-1 (phosphoglucomutase-1)	1p221
2	MDH-1 (malate dehydrogenase-1)	2p23
3	ACY1 (aminoacylase)	3pter-q13
4	PGM-2 (phosphoglucomutase-2)	4p14-q12
5	HEXB (hexosaminidase B)	5q13
6	PGM-3 (phosphoglucomutase-3)	6q11
7	MDH-2 (malate dehydrogenase-m)	7p22-q22
8	GSR (glutathione reductase)	8p21.1
9	ACO-1 (Aconitase, soluble)	9p22-p13
10	GOT-1 (glutamate oxaloacetate transaminase, soluble)	10q25.3-q26.1
11	LDH-A (lactate dehydrogenase A)	11p1203-p1208
12	PEP-B (peptidase B)	12q21
13	ESD (esterase D)	13q14.1
14	NP (nucleoside phosphorylase)	14q13
15	HEXA (hexosaminidase A)	15q22-15q25.1
16	PGP (phosphoglycolate phosphatase)	16p13-p12
17	GALK (galactokinase)	17q21-q22
18	PEPA (peptidase A)	18q23
19	GPI (glucose-phosphate isomerase)	19p13-q13
20	ADA (adenosine deaminase)	20q13-qter
21	SOD1 (superoxide dismutase-s)	21q211
22	ACO2 (aconitase-m)	22q11-q13
X	HPRT (hypoxanthine–guanine phosphoribosyltransferase)	Xq26-q27

[a] From "*Genetic Maps* 1984," Vol. 3. Cold Spring Harbor Lab., Cold Spring Harbor, New York, 1984.

ensuring the retention of the chromosome in the hybrid. A useful feature of this system is that rare hybrid segregants which have lost the pSV2-gpt marked chromosome can be isolated by growth in medium containing 6-thioguanine. Thus, a particular chromosome can be selected for or against in a microcell hybrid, depending on the selection medium used.

The development of this type of system was a natural extension of DNA-mediated gene transfer and microcell-mediated chromosome transfer techniques. Other researchers have independently developed similar approaches to isolating single human chromosomes in microcell hybrids and these procedures[26,27] should be consulted if difficulties are encountered with cell types other than those used in this system.

[26] A. Tunnacliffe, M. Parker, S. Povey, B. O. Bengtsson, K. Stanley, E. Solomon, and P. Goodfellow, *EMBO J.* **2,** 1577 (1983).
[27] R. S. Athwal, M. Smarsh, B. M. Searle, and S. S. Deo, *Somatic Cell Mol. Genet.* **11,** 177 (1985).

Acknowledgments

The studies that led to the development of this experimental procedure were supported by Grant CA19401 from the National Cancer Institute and Grant 1475 from the Council for Tobacco Research-U.S.A. PJS was supported by PHS Training Grant T32-CA 09054 from the National Cancer Institute.

Section IV

Isolation and Detection of Mutant Genes

[24] Strategies for Isolation of Mutant Genes

By Michael M. Gottesman

The following section of this volume will deal with specialized techniques which have been developed for the isolation of mutant genes from cultured mammalian cells. In recent years, several strategies have been developed which make it possible to detect and isolate genes by virtue of their mutant properties. It is also possible to use standard techniques to isolate such genes. The first section of this chapter will briefly summarize standard approaches to gene isolation which have been amply described in other volumes in this series,[1-3] and the second section of this chapter will serve as an introduction to more detailed protocols presented in the following chapters.

Standard Techniques for Gene Isolation

There are several well-defined pathways to the isolation of DNA segments from a gene of interest. If the protein product of that gene is known and antibodies to this protein are available, it is possible to create and screen cDNA expression libraries made from mRNA purified from tissues which express the protein of interest. The current preferred technique utilizes cDNA libraries made in the λ cDNA expression vectors λ gt11 or λ gt10. Large numbers of plaques can be screened for expression of antigenic determinants detectable after transfer to nitrocellulose paper.[4,5] If the amino acid sequence of the protein of interest is known, mixed oligonucleotide probes can be synthesized based on this sequence and used to screen genomic or DNA libraries which carry the cDNA or the genomic DNA of interest. Other, more labor intensive, techniques include (1) purification of polysomes containing mRNA encoding the protein of interest through the use of antibody affinity chromatography or immunoprecipitation, and the use of this purified mRNA to make cDNA libraries; (2) hybrid selection and translation using individual cDNA clones to identify the clone of interest; and (3) differential hybridization techniques in cases where mRNA levels are inducible or repressible. Frequently, combinations

[1] R. Wu (ed.), this series, Vol. 68 (1979).

[2] R. Wu. L. Grossman, and K. Moldave (eds.), this series, Vol. 100 (1983).

[3] R. Wu, L. Grossman, and K. Moldave (eds.), this series, Vol. 101 (1983).

[4] R. A. Young and R. W. Davis, *Proc. Natl. Acad. Sci. U.S.A.* **80**, 1194 (1983).

[5] R. A. Young and R. W. Davis, *Science* **222**, 778 (1983).

of these techniques can be used to isolate specific cDNAs. Many of these techniques and approaches have been recently reviewed.[6]

Once a cloned DNA segment is obtained which represents the gene of interest, this DNA can be used as a probe to isolate a full-length wild-type or mutant gene (including promoter and regulatory sequences) or a full-length cDNA in an expression vector in which the cDNA is downstream from a strong eukaryotic promoter. Techniques for the preparation of cosmid libraries, which can carry up to 40 kb of foreign DNA (enough for most, but not all mammalian genes), are described in the chapter by Fleischmann *et al.* [29] and procedures for screening such libraries using cloned DNA probes are presented in the chapter by Troen [30]. Smaller genes can be completely encoded within inserts in λ phage vectors. Once isolated, the cloned gene or cDNA can be transferred to a wild-type cell by the technique of DNA-mediated gene transfer (see the chapter by Fordis and Howard [27]) or by a specialized technique which allows very high efficiency transfer of cDNA libraries (see the chapter by Okayama [32]). If the cloned gene or cDNA encodes a dominantly acting mutant protein, expression of this gene will result in an altered phenotype of the recipient cell, and the mutant nature of the cloned gene will be established. If the mutant phenotype is recessive, it will be necessary to sequence the isolated clone (in this case, cDNA clones would be preferable to the larger, genomic clones) and compare it to known wild-type sequences.

Innovative Strategies for Isolation of Mutant Genes

In many cases, a DNA probe for the gene under study is not readily available, but mutant cell lines with an altered gene have been isolated. There are several ways in which such cell lines can be utilized to clone the mutant gene of interest.

If it can be established that the mutated gene is dominant, either by formation of somatic cell hybrids (see the chapter by Gottesman [9] on drug resistance) or by DNA-mediated gene transfer techniques (see the chapter by Fordis and Howard [27]), then it should be possible to prepare a cosmid library or phage library containing this gene and transfer the dominant phenotype using this library. In the transferents, the mutant gene will be linked to associated plasmid or phage vector sequences, as well as selectable markers such as G418 resistance, and these sequences can be used to rescue the transferred mutant gene. In the case of the cosmid system, genomic DNA from the primary transferents containing λ *cos*

---

[6] P. J. Doherty, *in* "Molecular Cell Genetics" (M. M. Gottesman, ed.), p. 235. Wiley, New York, 1985.

sequences can be directly packaged into phage heads and infected into bacteria where they are grown as cloned cosmids and can be retested by transfection into appropriate recipient cells (see the chapter by Lau [31] for techniques of cosmid rescue). An alternative to this approach is to prepare a new cosmid or phage library from the transfected recipient cells and use the linked vector sequences to pull out positive clones or phage. Mutant somatic cells can also be used to clone wild-type genes in cases where the mutant phenotype is recessive to the wild-type phenotype. Variations on this technique which reduce the work of screening positive cosmid clones or phage libraries include the introduction of linked DNA sequences such as nonsense suppressors or antibiotic resistance determinants which will be expressed in the bacterial hosts.

Another strategy which utilizes mutant somatic cells to isolate genes depends on the improved ability to isolate full-length cDNA clones for genes of interest.[7] It is theoretically possible to prepare a full-length cDNA library in an expression vector using mRNA from a mutant cell line and use this library to transfect an appropriate donor cell. Improved techniques for transfer of such cDNA libraries (see the chapter by Okayama [32]) increase the efficiency of this transfer to the level where this approach is feasible. Selected recipients expressing the mutant phenotype will carry the cDNA of interest which can be rescued using linked sequences. Selections which increase mRNA levels for the gene of interest, either as a result of regulatory mutations or gene amplification (see the chapter by Schimke *et al.* [7]), make the task of purifying an appropriate cDNA much easier. Genomic segments of amplified genes can also be cloned (see the chapter by Roninson [25] on gel renaturation).

The above approaches utilize the power of selection in tissue culture to purify the gene of interest. Another approach, which in some cases is more practical, is the use of *sib* selection to sequentially narrow down the population containing the gene of interest (see the chapter by McCormick [33] for theoretical and experimental details of this approach). The disadvantage of this approach is that it may involve several cycles of transfection and analysis, depending on the copy number of the gene in the original population. The significant advantage is that the net result is a single cloned gene, obviating the need for rescue of a transferred gene from the morass of a recipient cell's chromosome. In many cases, *sib* selection will be used in combination with selective approaches, since this latter approach might yield several hundred clones of which only one encodes the gene of interest.

There are several techniques under development by molecular geneti-

[7] H. Okayama and P. Berg, *Mol. Cell. Biol.* **2,** 161 (1982)

cists which are worthy of consideration in the cloning of mammalian genes. One approach uses insertional mutagenesis, usually by a retrovirus marked with an independent selectable marker, to mark genes of interest. This approach is described in detail in the chapter by Goff [36]. Another approach uses the technology of pulsed-field gradient gel electrophoresis and the formation of linkage libraries involving stretches of hundreds of kilobases of DNA to isolate genes which are more distantly linked than those which might be carried on a single cosmid or phage clone. This technology is described in the chapter by Smith *et al.* [35]. This approach could be used to isolate a gene known to be linked to another gene for which a probe exists.

The somatic cell geneticist is frequently interested in isolating more than one mutated version of a gene for comparative sequence studies or to examine the phenotype of the expression of different mutant genes. In their chapter, Chasin *et al.* [34] describe a technique for repetitive cloning of mutant genes in cosmid libraries by exploiting the properties of restriction sites which might flank these genes.

The chapters which follow will provide experimental protocols and detailed descriptions of many of the strategies outlined above.

[25] Use of in-Gel DNA Renaturation for Detection and Cloning of Amplified Genes

By Igor B. Roninson

The in-gel DNA renaturation technique[1] has been developed for the analysis of cell populations where gene amplification is suspected, but no probes corresponding to the amplified genes are available. This technique allows one to detect amplified fragments in restriction digests of total cellular DNA, identify those fragments that are likely to correspond either to the essential gene within the long ($10^2 - 10^3$ kb) region of amplified DNA (the amplicon) or to the immediate flanking sequences of such a gene, and to clone the amplified fragments. Since gene amplification is a common genetic mechanism for increasing the expression of various genes in mammalian cells in culture, as well as in tumor cells growing *in vivo*,[2-5] in-gel

[1] I. B. Roninson, *Nucleic Acids Res.* **11,** 5413 (1983).
[2] G. R. Stark and G. M. Wahl, *Annu. Rev. Biochem.* **53,** 447 (1984).
[3] R. T. Schimke, *Cancer Res.* **44,** 1735 (1984).
[4] K. Alitalo, *Med. Biol.* **62,** 304 (1985).
[5] J. L. Hamlin, J. D. Milbrandt, J. D. Heintz, and J. C. Azizkhan, *Int. Rev. Cytol.* **90,** 31 (1984).

DNA renaturation provides a convenient approach for the isolation of genes responsible for almost any cellular phenotype, as long as an appropriate selection protocol is available. This procedure has been used to detect and clone amplified genes associated with multidrug resistance[6-8] and to detect and analyze amplified DNA in various human tumors.[9-12] Alternative strategies for cloning amplified DNA include purification of chromosomal structures that are known to contain amplified DNA, i.e., double minute chromosomes[13] or unusually large chromosomes with homogeneously staining regions,[14] as well as differential cloning of amplified DNA by screening with C_0t-fractionated probes.[15,16] The advantages of in-gel renaturation over the above procedures include the ability to use it as a primary method for identification of cells carrying amplified DNA, in the absence of any information about the chromosomal localization or approximate copy number of amplified DNA sequences, and the ability to compare amplified DNA sequences in different cell populations prior to cloning. Other applications of the in-gel DNA renaturation technique, which are not addressed in this chapter, include identification of evolutionarily conserved DNA sequences among different organisms and characterization of the structure and abundance of repetitive DNA sequences.

Principle of the Method

Restriction endonuclease digestion of an amplicon present in the DNA from a cell line or tumor under study gives rise to a set of repeated

[6] I. B. Roninson, H. T. Abelson, D. E. Housman, N. Howell, and A. Varshavsky, *Nature (London)* **309,** 626 (1984).

[7] A. T. Fojo, J. Whang-Peng, M. M. Gottesman, and I. Pastan, *Proc. Natl. Acad. Sci. U.S.A.* **82,** 7661 (1985).

[8] P. Gros, J. M. Croop, I. B. Roninson, A. Varshavsky, and D. E. Housman, *Proc. Natl. Acad. Sci. U.S.A.* **83,** 337 (1986).

[9] H. Nakatani, E. Tahara, H. Sakamoto, M. Terada, and T. Sugimura, *Biochem. Biophys. Res. Commun.* **130,** 508 (1985).

[10] P. Meltzer, K. Kinzler, B. Vogelstein, and J. M. Trent, *Cancer Genet. Cytogenet.* **19,** 93 (1986).

[11] J. Trent, P. Meltzer, M. Rosenblum, G. Harsh, K. Kinzler, R. Mashal, A. Feinberg, and B. Vogelstein, *Proc. Natl. Acad. Sci. U.S.A.* **83,** 470 (1986).

[12] K. W. Kinzler, B. A. Zehnbauer, G. M. Brodeur, R. C. Seeger, J. M. Trent, P. S. Meltzer, and B. Vogelstein, *Proc. Natl. Acad. Sci. U.S.A.* **83,** 1031 (1986).

[13] D. L. George and V. E. Powers, *Cell* **24,** 117 (1981).

[14] N. Kanda, R. Schreck, F. Alt, G. Bruns, D. Baltimore, and S. Latt, *Proc. Natl. Acad. Sci. U.S.A.* **80,** 4069 (1983).

[15] O. Brison, F. Ardeshir, and G. R. Stark, *Mol. Cell. Biol.* **2,** 578 (1982).

[16] K. T. Montgomery, J. L. Biedler, B. A. Spengler, and P. W. Melera, *Proc. Natl. Acad. Sci. U.S.A.* **80,** 5724 (1983).

DNA fragments which are absent from a restriction digest of control DNA, where the corresponding region is not amplified. When the degree of DNA amplification is 500-fold or higher, the amplified fragments can be detected in ethidium bromide-stained gels as distinct bands against the background smear, which is produced primarily by single-copy fragments. In most cases, however, the degree of amplification is insufficient for direct visualization of amplified fragments. The amount of the repeated and amplified fragments relative to single-copy fragments in a restriction digest of genomic DNA can be selectively increased by in-gel renaturation of restriction fragments, since the higher local concentration of repeated fragments results in their higher efficiency of reassociation.

This is the principle of the in-gel DNA renaturation technique, as illustrated in Fig. 1. Following digestion with a restriction enzyme that cuts at a 6 bp recognition sequence, a portion of each DNA preparation (tracer) is labeled with ^{32}P by replacement synthesis using T4 DNA polymerase.[17] The labeling reaction includes limited exonucleolytic degradation of DNA using the 3'-exonuclease activity of T4 DNA polymerase, followed by the resynthesis of degraded ends in the presence of a labeled precursor. As a result, high specific activity is achieved without significant degradation of tracer DNA fragments. Tracer DNA is then mixed with an excess of unlabeled (driver) DNA digested with the same restriction enzyme and the mixture is electrophoresed in an agarose gel. After electrophoresis, DNA in the gel is denatured with alkali, then the gel is neutralized and DNA fragments are allowed to renature. Following renaturation, single-stranded DNA is selectively degraded *in situ* by diffusing single-strand specific nuclease S1 into the gel. To achieve additional enrichment for repeated and amplified fragments, the whole cycle of denaturation, renaturation, and restriction enzyme digestion is repeated once more. As a result of this treatment, the single-copy fragments are almost completely degraded and eluted from the gel, whereas repeated and amplified fragments that are present at sufficient concentration to efficiently reanneal in the gel can be detected as distinct bands after autoradiography.

Performing denaturation and renaturation in the gel rather than in solution imposes an additional limitation on the nature of the DNA fragments capable of reannealing. In order for different DNA fragments to reanneal after electrophoretic separation in agarose gels, such fragments should not only contain homologous sequences, but also have to be of the same length. The length-dependent reassociation greatly decreases cross-hybridization of otherwise unrelated fragments containing common short interspersed repeated sequences (SINEs), which is the main problem that

[17] M. D. Challberg and P. T. Englund, this series, Vol. 65, p. 39.

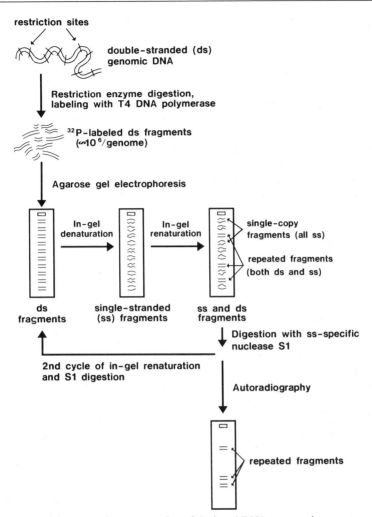

FIG. 1. Diagrammatic representation of the in-gel DNA renaturation process.

prevents the reassociation of large restriction fragments of vertebrate DNA in solution. Another feature of our protocol provides for further decrease in the background resulting from cross-hybridization of different restriction fragments of the same size containing a common SINE element. This decrease is achieved by labeling tracer DNA under the conditions where the label is incorporated only in a short stretch of 50–100 nucleotides at the fragment termini, so that SINE elements are unlikely to acquire the label unless they are immediately adjacent to the termini. As a result, partial duplexes bound only through the common SINE sequences would usually contain the label only in their single-stranded regions and therefore would not be detectable after S1 nuclease digestion.

Once a set of amplified fragments has been detected, it becomes necessary to identify within the set those fragments that are closely linked to the essential gene that provides a selective advantage for the cells in which it is amplified. Since the amplicons are known to vary with regard to the length and the composition of flanking sequences coamplified together with the essential gene,[18-20] identification of a subset of amplified fragments that are commonly amplified in different idnependently derived cell populations allows one to delineate a relatively short region which is likely to contain the gene of interest. Commonly amplified fragments in different DNA preparations can be readily identified by using DNA from one preparation as a tracer and DNA from another preparation as a driver. In such mixtures, the fragments that are amplified only in the tracer but not in the driver would have insufficient concentration to reanneal in the gel, and therefore would be degraded by S1 nuclease. The fragments that are amplified in the driver but not in the tracer would, in fact, reanneal in the gel, but the amount of label incorporated by these fragments in the tracer DNA, where they are present at a single copy, would be insufficient for detection of the corresponding bands. Therefore only those fragments that are amplified both in the tracer and in the driver would be detectable in such mixtures.

After a subset of fragments amplified in common in different independently derived cell populations has been identified, the commonly amplified bands can be cloned using in-gel DNA renaturation as a preparative method. During the cloning procedure, the enrichment for the amplified fragments of interest is achieved in two steps. Since cloning is performed after the size of the fragment of interest has been determined, enrichment at the first step is achieved by size fractionation in agarose gels. Further enrichment for the amplified fragment is obtained after denaturation and renaturation of DNA within the gel strip, followed by cloning of the renatured fragments. These two steps of enrichment, in addition to the initial *in vivo* amplification of the fragment, result in a high proportion of the recombinant clones containing the desired amplified fragment.[6]

Two modifications of the in-gel DNA renaturation procedure do not

[18] C. Tyler-Smith and C. J. Bostock, *J. Mol. Biol.* **153,** 203 (1981).
[19] N. A. Federspiel, S. M. Beverly, J. W. Schilling, and R. T. Schimke, *J. Biol. Chem.* **259,** 9127 (1984).
[20] F. Ardeshir, E. Giulotto, J. Zieg, O. Brison, W. S. L. Lio, and G. R. Stark, *Mol. Cell. Biol.* **3,** 2076 (1983).

require labeling of the tracer DNA. These methods provide for detection of a subset of fragments containing specific DNA sequences within the amplicon. These procedures include the removal of single-copy fragments from unlabeled DNA by two cycles of in-gel renaturation and S1 nuclease digestion, followed by Southern transfer[21] and hybridization of the repeated and amplified fragments remaining in the gel. In the first method,[22] the blots are hybridized with a probe containing a cloned SINE element, which hybridizes to a large number of fragments derived from the amplicon, but only to a few of the repeated fragments found in control DNA and derived from tandemly repeated or long interspersed repeated sequences (LINEs). In the second method, ^{32}P-labeled cDNA, corresponding to the mRNA population from cells containing amplified DNA, is used as a probe for detection of transcriptionally active fragments within the amplicon. The relative merits and specific applications for these methods are discussed later in this chapter.

Factors Influencing the Sensitivity of the in-Gel Renaturation Assay

The sensitivity of in-gel renaturation for detection of amplified DNA fragments depends on the intensity of the signal from the amplified fragments relative to the background. The signal intensity is determined by the extent of in-gel renaturation for the given amplified fragment and by the efficiency of labeling of this fragment in the tracer DNA preparation. The extent of renaturation for the given fragment is proportional to the C_0t value, where C_0 is the initial concentration of the fragment in the gel, and t is the time of the renaturation reaction.[23] However, increasing the time of the reaction does not necessarily increase the efficiency of renaturation, since in the course of incubation a part of DNA is diffused out and lost from the gel. Thus, we have found that the intensity of the bands changes very little between 2 and 24 hr of renaturation. The initial concentration of DNA fragments in the gel (C_0) can be maximized by using the largest amount of driver DNA that can be loaded onto each lane without causing distortion of the electrophoretic pattern, and by performing electrophoresis under the conditions that provide for the maximum sharpness of the bands. Another factor that influences the initial concentration of DNA fragments is the size of genomic DNA prior to restriction enzyme digestion. Degradation of genomic DNA preparations to the average size of less

[21] E. Southern, *J. Mol. Biol.* **98**, 503 (1975).
[22] M. Fukumoto and I. B. Roninson, *Somat. Cell Mol. Genet.* **12**, 611 (1986).
[23] R. J. Britten and D. E. Kohne, *Science* **161**, 529 (1968).

than 50 kb may greatly decrease the concentration and hence the efficiency of renaturation for the large (over 10–15 kb) restriction fragments. Therefore minimizing hydrodynamic shearing of genomic DNA during the extraction is crucial to this procedure. The efficiency of renaturation may also vary depending on the size and GC content of a DNA fragment, as well as on the presence of highly repeated SINE elements. Therefore different restriction fragments present at the same copy number in the genome may differ in their signal intensity.

The efficiency of labeling a given fragment with T4 DNA polymerase may be reduced because of hairpin formation by the sequences near the fragment termini, which become single-stranded after 3′-exonucleolytic digestion. The probability of hairpin formation varies among individual fragments, and it may prevent resynthesis of the termini by T4 DNA polymerase.[24] One can minimize the variations of labeling efficiency among different fragments by limiting the extent of exonucleolytic digestion at the first step of T4 DNA polymerase reaction and therefore reducing the probability of hairpin formation by the single-stranded regions. The extent of exonucleolytic degradation can be minimized without a corresponding reduction in the specific activity of tracer DNA by using larger amounts of the enzyme and shorter times of exonuclease reaction.

The background that hinders detection of amplified fragments by in-gel renaturation is produced by the repeated fragments that are normally present in restriction digests of DNA from the given organism, as well as by incomplete denaturation or residual renaturation of single-copy fragments, or by incomplete S1 nuclease digestion. Among these factors, the presence of a large number of normal repeated fragments appears to be the most frequent source of high background. The number and size distribution of repeated fragments are determined by the distribution of restriction sites within different tandemly repeated sequences and LINEs, which varies for different species and restriction enzymes used. It is advisable therefore to identify for each species a restriction enzyme that would produce the smallest number of bands in the control DNA preparation. An additional consideration in the choice of the restriction enzyme is the average size of the fragments, which should be readily separable under usual electrophoretic conditions. We have found that relatively low background can be achieved by using HindIII with human DNA and BamHI with Chinese hamster DNA, but we have not yet been able to identify an enzyme that would give a similarly low background with mouse or rat DNA.

In our laboratory, the in-gel DNA renaturation assay can detect amplified fragments present at as few as 20–25 copies per haploid genome in

---

[24] K. C. Deen, T. A. Landers, and M. Berninger, *Anal. Biochem.* **135,** 456 (1983).

human or Chinese hamster DNA, but only at about 40 copies in mouse or rat DNA. These estimates have been made on the basis of either indirect experiments, where genomic DNA was mixed with known amounts of HindIII fragments of λ phage DNA[1] or by direct studies, where fragments initially detected by in-gel renaturation were cloned, and the copy number of cloned sequences was estimated by filter hybridization.[6] The combination of in-gel renaturation and Southern hybridization with a SINE probe, as described later in this chapter, can give a higher sensitivity. It should also be noted that the minimum copy number required for detection of an amplified fragment would be lower for nonmammalian systems with smaller genome sizes, e.g., as few as 10–15 copies of an amplified fragment can be detected in the chicken genome, which is approximately three times smaller than the mammalian genome.

Preparation of Restriction Digests

Genomic DNA for in-gel renaturation analysis can be extracted by any of several standard procedures,[25–27] which allow for adequate deproteinization with minimal shearing of DNA. To obtain genomic DNA of high molecular weight, we avoid ethanol precipitation of DNA prior to restriction enzyme digestion and use cut-off pipet tips for handling undigested DNA. The concentration of DNA can be determined either by the diphenylamine reaction,[28] as described in Procedure 1, or by fluorometry.[29] It should be noted that measurement of genomic DNA concentration by UV absorption at 260 nm may result in significant errors, since cellular DNA preparations may have a considerable degree of RNA contamination. The size of undigested DNA is determined relative to intact λ phage DNA (49 kb) by electrophoresis in low-percentage agarose gels. Only those DNA preparations where the bulk of DNA is larger than λ DNA should be used for in-gel renaturation.

Some cellular DNA preparations are contaminated with inhibitors of restriction enzymes. To assure complete digestion, we use 4- to 6-fold excess of the restriction enzyme and do the reactions in the presence of 3 mM spermidine, which suppresses many of the restriction enzyme inhibitors.[30] Spermidine, however, should not be used in low-salt restriction

[25] N. Blin and D. W. Stafford, Nucleic Acids Res. 3, 2303 (1976).
[26] M. Gros-Bellard, P. Oudet, and P. Chambon, Eur. J. Biochem. 36, 32 (1978).
[27] P. Krieg, E. Amtmann, and G. Sauer, Anal. Biochem. 134, 288 (1983).
[28] K. W. Giles and A. Meyers, Nature (London) 206, 93 (1965).
[29] C. Labarca and K. Paigen, Anal. Biochem. 102, 344 (1980).
[30] J. P. Bouche, Anal. Biochem. 115, 42 (1981).

enzyme buffers (10 mM NaCl or less) since some of the DNA may be precipitated. To achieve optimum electrophoretic separation and band sharpness, the restriction digests are purified by phenol extraction and two rounds of ethanol precipitation (Procedure 2). Purification may result in variable losses of DNA (up to 30–50%), and therefore it is necessary to measure the DNA concentration after purification of the restriction digests. The completeness of digestion and the purity of digested DNA are determined by minigel electrophoresis of an aliquot of the DNA digest and ethidium bromide staining. The degree of digestion is determined from the size distribution in the DNA smear and from the intensity of the characteristic bands corresponding to highly repeated fragments. The purity of the DNA preparation is established by the absence of streaks and salt effect in the DNA smear (see Diagnosis and Correction of Problems).

Procedure 1. Measurement of Genomic DNA Concentration by the Diphenylamine Reaction (Modified from Giles and Meyers[28])

Solutions

20% (w/v) perchloric acid (Fisher)

4% (w/v) diphenylamine (Sigma) in glacial acetic acid

0.16% (v/v) acetaldehyde (Fisher): prepare from 100% acetaldehyde immediately before use; acetaldehyde is highly volatile and flammable; it should be kept on ice before opening and closed immediately after use

Salmon sperm DNA (Sigma D-1626); 50 μg/ml standard solution: prepare by diluting 100 mg/ml stock and adjusting the DNA concentration to give an OD of 1 at 260 nm

1. Prepare a set of standard solutions containing 0, 1, 2, 3, 4, and 5 μg of salmon sperm DNA in 1.5 ml Eppendorf tubes. Adjust the volume in each tube to 100 μl with distilled water.

2. Determine approximate concentrations of genomic DNA preparations by measuring OD_{260}. Take an aliquot containing approximately 5 μg of DNA and adjust the volume to 100 μl with distilled water. Between 1 and 10 μg of DNA can be used for measurement.

3. To each tube add 100 μl of 20% perchloric acid, then 200 μl of 4% diphenylamine, and then 20 μl of 0.16% acetaldehyde. Mix by vortexing.

4. Incubate at room temperature overnight (4–5 hr minimum), until blue color develops.

5. Measure OD of each solution at 595 and 700 nm. Calculate $OD_{595} - OD_{700}$.

6. For the standard solutions, plot the amount of DNA (in μg) against the value of $OD_{595} - OD_{700}$. Draw a straight line approximating all the points.

7. Use the straight line to determine the amount of DNA in the genomic DNA samples, and calculate the starting DNA concentration.

Procedure 2. Estimation of the Genomic DNA Size by Minigel Electrophoresis

Solutions

80× electrophoresis buffer: 1.8 M Tris base, 0.8 M sodium acetate, 0.08 M EDTA, adjusted to pH 8.3 with acetic acid

TE buffer: 10 mM Tris–HCl, pH 8.0, 1 mM EDTA

10× loading buffer: 0.4% bromphenol blue, 50% glycerol in TE

10 mg/ml ethidium bromide (store in a light-proof container)

Equipment.

Hoeffer "Minnie" electrophoresis apparatus with 7 × 10 cm tray and 12-well comb

300 nm wavelength UV transilluminator (Photodyne)

1. Prepare 35 ml of 0.3% agarose (Bethesda Research Labs., ultrapure) in 1× electrophoresis buffer, dissolve by heating in a microwave oven or in an autoclave (10 min, liquid cycle).

2. Cool the solution to 50°, add 2 μl of 10 mg/ml ethidium bromide, mix and pour into the gel mold.

3. Transfer the mold to the cold room (4°) and allow agarose to solidify for 1 hr. Remove the well comb.

4. Mix 0.2 μg of genomic DNA sample with 10× loading buffer and TE to the final volume of 8 μl. As a size standard, use 0.05 μg of intact λ phage DNA (Bethesda Research Labs.).

5. Perform electrophoresis in the cold room at 55 V for 1.5 hr. Photograph DNA using 300 nm wavelength UV transilluminator. Estimate the size of genomic DNA relative to λ DNA (49 kb).

Procedure 3. Restriction Endonuclease Digestion and Purification of the Digests

Solutions

10× restriction buffer (as recommended by the supplier)

0.1 M spermidine (Sigma); store at −20°

TE Buffer: 10 mM Tris–HCl, pH 8.0, 1 mM EDTA

3 M sodium acetate, pH 7.0

100% ethanol (ice-cold)
70% ethanol (ice-cold)

1. Combine 50 μg of DNA (in TE buffer) with the appropriate volumes of 10× restriction buffer and water in a 1.5 ml Eppendorf centrifuge. Add 3 μl of 0.1 M spermidine per each 100 μl of the reaction volume. Mix by tapping the tube (do not vortex). Place on ice for 3 min.

2. Add 100–150 units of the restriction enzyme. Mix and centrifuge briefly to collect the liquid at the bottom of the tube. Incubate 1.5 hr at 37°. Add additional 100–150 units of the enzyme and incubate for another 2 hr.

3. Add equal volume of phenol:chloroform:isoamyl alcohol (25:24:1). Mix by inversion until a homogeneous suspension is formed. Centrifuge 3 min in an Eppendorf centrifuge. Collect the aqueous phase.

4. Add equal volume of chloroform:isoamyl alcohol (24:1). Mix, centrifuge 1 min. Remove chloroform, and reextract the aqueous phase one more time with chloroform:isoamyl alcohol. Transfer the aqueous phase into another tube and extract once with two volumes of ether. Remove the ether layer.

5. Add 1/10 volume of 3 M sodium acetate (pH 7.0) and 2.5 volumes of ice-cold ethanol. Mix by inverting the tube 5–6 times. Place for 5 min into dry ice–ethanol bath. Centrifuge 10 min in an Eppendorf centrifuge at 4°. Remove the supernatant with the tip of a 1 ml automatic pipetter.

6. Add 200 μl of TE (pH 7.5) to the pellet. Allow DNA to dissolve for 30–60 min at room temperature, periodically tapping the tube. Centrifuge 5 min at room temperature. In some but not all cases the insoluble particles form a visible pellet. Transfer the solution to another tube, taking care not to transfer any part of the loose pellet.

7. Add 20 μl of 3 M sodium acetate (pH 7.0) and 500 μl of ice-cold ethanol. Mix, place into dry ice–ethanol bath for 5 min, centrifuge 10 min at 4°. Remove the supernatant with a pipet tip. Centrifuge another 30 sec to collect the liquid that may remain on the walls of the tube and remove the liquid completely with a pipet tip.

8. Add 200 μl of ice-cold 70% ethanol. Wash the pellet by tapping the tube. Centrifuge 5 min at 4° and remove the supernatant completely. Dry the pellet briefly in a Speed Vac concentrator (Savant) or a desiccator. Dissolve in 35 μl TE for 30 to 60 min at room temperature.

9. Remove a 2.5-μl aliquot to measure the DNA concentrations by the diphenylamine reaction (Procedure 1).

10. Remove a 0.5-μl aliquot for electrophoresis in a 1% agarose minigel. Electrophoresis is performed as in Procedure 2, except that agarose is allowed to solidify and electrophoresis is carried out at room temperature.

Continue electrophoresis until bromphenol blue reaches the bottom of the gel and photograph the gel using a 300 nm UV transilluminator.

Labeling of Tracer DNA and Gel Electrophoresis

Tracer DNA is labeled by replacement synthesis with T4 DNA polymerase.[31] The labeling reaction consists of three steps. In the first step, when DNA is incubated with the enzyme in the absence of nucleotide precursors, the 3′-exonucleolytic activity of the enzyme removes a stretch of nucleotide residues from each 3′-terminus. In the second step, a mixture of [α-^{32}P]dCTP and unlabeled dATP, dGTP, and dTTP is added, and the degraded ends are repaired by the DNA polymerase activity of the enzyme. In the third step, excess unlabeled dCTP is added to ensure complete resynthesis of the termini. The length of the stretch of labeled nucleotides at the fragment termini is determined by the time of the 3′-exonuclease reaction, which proceeds at a roughly linear rate.[31] As discussed previously, the tracer DNA fragments should be labeled in such a way that the label is confined to a short (50 nucleotides or less) stretch of DNA near the termini, which is achieved by limiting the 3′-exonuclease reaction to 3 min. We have observed that at a fixed time of the 3′-exonuclease reaction, the total incorporation of [α-^{32}P]dCTP is determined primarily by the amount of the enzyme and is almost independent of the amount of DNA in the reaction. For the optimal rate of resynthesis at the second step of the reaction, the rate-limiting concentration of [α-^{32}P]dCTP should be 3–4 μM or larger. In order to achieve this concentration, [α-^{32}P]dCTP, available at 3000 Ci/mmol, should be diluted approximately 3-fold with unlabeled dCTP. The other three deoxynucleotide triphosphates are used in excess (200 μM concentration). The efficiency of label incorporation may vary among different DNA preparations, as a result of partial inhibition of T4 DNA polymerase by contaminants in cellular DNA preparations.

In order to detect the weakest bands after an overnight exposure, each lane is loaded with 5–15 × 10^6 dpm of tracer DNA. The amount of DNA used for labeling depends on whether the tracer will be used only for homologous reactions (same DNA preparation used as a tracer and as a driver) or also for heterologous reactions (different DNA preparations used as a tracer and as a driver). If the tracer is used only for homologous reactions, the specific activity of tracer DNA is not essential, and it is advantageous to use large amounts of DNA (0.5–1.5 μg) with 1.5–2 units

---

[31] P. O'Farrell, *Focus (Bethesda Research Labs.)* **3**, 1 (1981).

of T4 DNA polymerase. These conditions ensure adequate incorporation of the label while minimizing the length of the labeled stretch at the termini. The use of large amounts of DNA also allows one to omit tRNA carrier in ethanol precipitation, resulting in better electrophoretic patterns.

When the tracer is used for heterologous reactions, no more than $0.1-0.15$ μg of tracer DNA should be used in each reaction in order to minimize the renaturation of amplified fragments present only in the tracer but not in the driver DNA. Therefore the specific activity of the tracer should be at least 5×10^7 dpm/μg. In this case, the amount of DNA to be used for labeling can be calculated as $0.15-0.2$ μg times the number of lanes containing the tracer. To achieve the desired specific activity, we use 1.5 units of T4 DNA polymerase for each 0.2 μg of tracer DNA. To avoid salt effects in electrophoresis and to minimize the amount of ^{32}P released into the electrophoretic buffer, labeled tracer DNA is purified by two rounds of ethanol precipitation and rinsing with 70% ethanol.

In order to compare the band patterns among different lanes, it is essential to use the same amount of tracer (in dpm) and the same total amount of DNA (in μg) in each lane. The total amount of DNA in each mixture is determined empirically as the maximum amount that can be loaded onto each lane without distorting the electrophoretic pattern. Using gel wells that are 9 mm long, 1.5 mm wide, and 3 mm deep, we can load 15 μg of DNA per well in $25-30$ μl volume. As size standards, we usually use HindIII fragments of λ phage DNA. The standard mixture is prepared by combining $2-3 \times 10^4$ dpm of λ HindIII tracer and 0.8 μg of the same driver DNA in a convenient volume of sample buffer.

To obtain adequate resolution of most repeated and amplified fragments, it is important to use gel slabs that are at least 30 cm long. In our laboratory we use horizontal gel chambers, where the gels are cast and run on UV-transparent plastic trays of 20×40 cm size with a 17-well comb. The use of relatively long (9 mm) wells results in improved resolution of bands in autoradiograms. We use $1-1.2\%$ agarose gels that are 4 mm thick, which makes them sufficiently sturdy to withstand handling and shaking during in-gel renaturation, while allowing rapid diffusion of ions and S1 nuclease into the gel. We do not recommend using 0.8% or lower agarose gels, since they are likely to break during in-gel renaturation. To ensure even gel thickness, the gel trays are carefully leveled before molding. To maximize the band sharpness, electrophoresis is performed at the voltage gradient of 1.2 V/cm or lower. The electrophoretic buffer contains ethidium bromide so that DNA movement can be monitored during electrophoresis. Electrophoresis is terminated when the 2.0 kb size marker traverses $15-17$ cm from the origin.

Procedure 4. Labeling Tracer DNA with T4 DNA Polymerase (Modified from O'Farrell[31])

Solutions

5× reaction buffer: 165 mM Tris-acetate (pH 7.9), 330 mM sodium acetate, 50 mM magnesium acetate, 2.5 mM DTT, 0.5 mM BSA; store at −20°

Deoxyribonucleotide solutions: 2 mM dATP, 2 mM dGTP, 2 mM dTTP, 2 mM dCTP, 30 μM dCTP; store at −20°

Resynthesis mixture (for 12 reactions): 30 μl of 2 mM dATP, 30 μl of 2 mM dGTP, 30 μl of 2 mM dTTP, 24 μl of 30 μM dCTP, 36 μl of 5× reaction buffer, 30 μl of water; store at −20°

[α-³²P]dCTP (3000 Ci/mmol, 1 mCi/ml in 50% ethanol; Amersham); store at −20°

10 mg/ml tRNA (Boehringer Mannheim); store at −20°

5 M ammonium acetate

3 M sodium acetate, pH 7.0

100% ethanol (ice-cold)

70% ethanol (ice-cold)

10% trichloroacetic acid (TCA; ice-cold)

1. For each reaction, place 80 μl of [α-³²P]dCTP in a 1.5 ml Eppendorf tube and dry using Speed Vac concentrator or a lyophilizer. Add 15 μl of resynthesis mixture and dissolve by vortexing. Chill the tubes in an ice-water bath.

2. In separate tubes, mix 0.15–1.0 μg of digested genomic DNA in TE buffer and 2 μl of 5× reaction buffer; add water to the final volume of 8.5 μl. Mix by tapping, centrifuge briefly, and chill in ice-water for at least 2 min.

3. Add 1.5 μl of T4 DNA polymerase (1 unit/μl; Bethesda Research Labs.) to each tube with DNA. Mix by tapping and place in a 37° water bath for 3 min. Centrifuge briefly to collect the liquid at the bottom of the tube and place into an ice-water bath.

4. Transfer each DNA mixture into a tube containing the resynthesis mixture and [α-³²P]dCTP (step 1). Mix by tapping and centrifuge briefly. Incubate at 37° for 30 min.

5. Add 2.5 μl of 2 mM dCTP to each tube. Mix by tapping and centrifuge briefly. Incubate at 37° for 20 min.

6. Stop the reactions by adding 27 μl of 5 M ammonium acetate, 1 μl of 10 mg/ml tRNA, and 130 μl of ice-cold ethanol. If the amount of DNA in the reaction is >0.5 μg, tRNA can be omitted. Mix by inverting the tube

5–6 times. Place for 5 min into dry ice–ethanol bath. Centrifuge 10 min in an Eppendorf centrifuge at 4°. Remove the supernatant with a pipet tip.

7. Dissolve each pellet in 100 μl TE. Remove 2-μl aliquots into separate tubes for TCA precipitation (step 9). To the rest, add 100 μl of 5 M ammonium acetate and 500 μl of cold ethanol. Mix, place into dry ice–ethanol bath for 5 min, centrifuge 10 min at 4°. Remove the supernatant with a pipet tip. Centrifuge another 30 sec to collect the liquid that may remain on the walls of the tube and remove the liquid completely with a pipet tip.

8. Add 200 μl of ice-cold 70% ethanol. Wash the pellet by vortexing 5 sec. Centrifuge 5 min at 4° and remove the supernatant completely. Dry the pellet briefly in a Speed Vac concentrator or a desiccator. Tracer DNA pellets are stored at −20° up to 24 hr before use.

9. To the tubes containing 2-μl aliquots from step 7, add 20 μl of 10 mg/ml tRNA. Add 1 ml of ice-cold 10% TCA, mix by vortexing, and chill on ice for 5 min. Collect the precipitates by filtering through glass-fiber filters (Whatman 934-AH). Wash the filters six times with 5 ml of 10% TCA, followed by 5 ml of ethanol. It is not necessary to dry the filters before counting in a scintillation counter provided that water-mixable scintillation fluid, such as Scintiverse II (Fisher), is used.

Procedure 5. Preparation of DNA Mixtures and Gel Electrophoresis

Solutions
  TE buffer: see Procedure 3
  80× electrophoresis buffer: see Procedure 2
  10× loading buffer: see Procedure 2
  10 mg/ml ethidium bromide; store in a light-proof container
Equipment
  horizontal gel electrophoresis apparatus (Dan Kar Plastics, Reading, MA) with a "superslab" UV-transparent gel tray, 20 × 40 cm size; 17-well comb, 9 × 1.5 mm wells

1. Calculate the total amount of label in each tracer DNA preparation and divide it by the number of mixtures in which this tracer will be used. Since the amount of tracer DNA (in dpm) should be the same for all lanes of the gel, this amount is determined by the tracer preparation with the lowest incorporation of the label. Dissolve each tracer preparation in the smallest convenient volume of TE (e.g., 5 μl) and calculate for each preparation the volumes of aliquots containing the required amount of label. Calculate the amounts of tracer DNA (in μg) in each aliquot.

2. Mix aliquots of tracer DNA with the appropriate driver DNA preparations, so that the total amount of tracer and driver DNA would be 15 μg

in each mixture. Add TE to equalize the volumes of all mixtures. Add 1/10 volume of 10× loading buffer. For wells of 9 × 1.5 mm size and a 4-mm-thick gel, the final volume of mixtures should not exceed 30 μl. Mix thoroughly by tapping. Centrifuge briefly. Mix and centrifuge one more time.

3. To make a 4-mm-thick gel slab, prepare 320 ml of 1% agarose (Bethesda Research Labs., ultrapure) in 1× electrophoresis buffer, dissolve by autoclaving (15 min, liquid cycle). Cool the solution to 50°, add 16 μl of 10 mg/ml ethidium bromide, mix, and pour into a level gel tray. Remove any bubbles that may be trapped in the gel with a transfer pipet. Allow the gel to solidify at room temperature for at least 40 min before removing the comb and the spacers.

4. Place the gel tray into the electrophoresis chamber, add 1× electrophoresis buffer containing 0.5 μg/ml ethidium bromide to the level of 2–3 mm above the surface of the gel. Load the DNA samples and start electrophoresis at 70 V. Once bromphenol blue moves 7–8 mm from the origin, start recirculating the buffer with a peristaltic pump and decrease the voltage to 40 V. During electrophoresis, monitor DNA movement using a 300 nm UV transilluminator. Continue electrophoresis until the 2.0 kb size marker traverses 15–17 cm from the origin (35–40 hr). After electrophoresis, photograph the gel on the electrophoresis tray using a 300 nm UV transilluminator. When using an electrophoresis apparatus without a UV transparent tray, the gel can be photographed through an unfolded polystyrene sheet protector (available at office supply stores).

In-Gel Renaturation

To ensure adequate exchange of solutions inside the gel, diffusion of S1 nuclease into the gel, and elution of degradation products from the gel, it is necessary to shake the gel continuously during all steps of in-gel renaturation. To avoid gel breakage during shaking, several precautions should be observed. First, the gel should not be loose in the incubation chamber. For this purpose, the gel is trimmed after electrophoresis to the size slightly smaller than the inner surface of the box used for incubation. The boxes used in our laboratory include polyethylene "Freezette" boxes (28.5 × 16.5 cm) and 30 × 19.5 cm Tupperware boxes. Second, the shaker should provide smooth rotation, without jerky movements at a low rotation speed. We use a gyrotory shaker equipped with a slow speed conversion device, which allows smooth rotation at 20–30 rpm. The temperature of the reaction is regulated by keeping the shaker in a variable temperature air incubator. All the solutions are prewarmed to the appropriate temperature prior to use. Finally, used solutions are removed by careful aspiration using a water pump and a trap.

S1 nuclease digestion and elution of degradation products can be monitored with a Geiger counter. Usually elution is complete 2 or 3 hr after the end of incubation with S1 nuclease. For human or hamster DNA, the reading on the Geiger counter in the "hottest" part of the gel is decreased 10- to 20-fold after each round of digestion. At the end of the second cycle, the gel is dried on a gel dryer at 60° and autoradiographed for 6–48 hr, either with or without an intensifying screen (a sharper band pattern is obtained in the latter case). It is recommended to make several autoradiograms of each gel in order to achieve best resolution of both weak and strong bands.

Procedure 6. In-Gel Renaturation and S1 Nuclease Digestion

Stock Solutions

   A: 2.5 M sodium hydroxide, 3.0 M sodium chloride, 0.02% thymol blue
   B: 0.2 M sodium phosphate (pH 7.0), 3.6 M sodium chloride, 2 mM EDTA
   C: 100% formamide (Kodak P565); deionize formamide before use by adding 200 g of ion-exchange resin (Bio-Rad RG501-X8) to 1 gallon of formamide, stirring for 1 hr at 4° and filtering through Whatman 3MM paper
   D: 0.5 M sodium acetate (pH 4.6), 2 M sodium chloride, 0.01 M zinc sulfate
   E: 10% SDS

Working Solutions

   Denaturing buffer (1.5 liters): 300 ml solution A, 1200 ml water
   Renaturing buffer (3.5 liters): 875 ml solution B, 1500 ml solution C, 875 ml water; check pH of the buffer and, if necessary, adjust to pH 7.0 with HCl
   S1 nuclease digestion buffer (3.5 l): 350 ml solution D, 3150 ml water
   Elution buffer (1 liter): 150 ml solution B, 10 ml solution E, 840 ml water

All working buffers are prepared shortly before use and prewarmed to 37 or 45°, as required by the protocol.

Equipment

   Polyethylene "Freezette" box (28.8 × 16.5 cm), available from most department stores
   Gyratory shaker (New Brunswick model G-2) with slow speed conversin and rubber-covered utility platform

Variable-temperature incubator (Bellco model 7728-06005)
Slab gel dryer (Hoeffer SE-1150)

1. Trim the gel on an electrophoresis tray or a polystyrene sheet to 28.5 × 16.3 cm size. Place the polyethylene box on top of the gel and flip over the gel tray together with the box. With the polyethylene box resting on the bench top, use a spatula to detach the gel from the tray and let the gel fall into the box. Immediately add 300–400 ml of denaturing buffer.

2. Place the box on the gyratory shaker in an air incubator set at 45° and incubate for 35 min with shaking at 30 rpm. Remove the buffer by aspiration and add 350 ml of fresh denaturing buffer. Continue incubation for 35 min.

3. Remove the denaturing buffer and neutralize the gel by washing with 350 ml of renaturing buffer, 5 times for 20 min at 45°, with shaking at 30 rpm. Once the color of thymol blue in the gel changes from blue to yellow, check pH of the buffer with a pH stick. pH of the buffer should be below 8.0 in the last wash; additional washes with renaturing buffer may be done if necessary.

4. Reduce the speed of the shaker to 10–20 rpm and incubate the gel in the renaturing buffer for 2 hr at 45°.

5. Remove the renaturing buffer, change the temperature of the incubator to 37°, increase the speed of shaking to 30 rpm, and wash the gel five times for 15 min with 350 ml of S1 nuclease digestion buffer. After the second buffer change, thoroughly rinse the lid of the incubation box with distilled water to remove traces of previously used solutions.

6. Remove the buffer after the last wash. Scan the gel surface with a Geiger counter and note the reading in the "hottest" area of the gel. Add 20,000 units of S1 nuclease (Sigma N-7385) to a flask containing 250 ml of S1 nuclease digestion buffer. Mix and add to the gel box. Incubate for 2 hr at 37°, shaking at 30–40 rpm.

7. Remove the buffer. Scan the gel with a Geiger counter; the reading in the "hottest" area should be at least 2–3 times lower than before the addition of S1 nuclease. Change the incubator temperature to 45°. Denature DNA in the gel by two washes with 350 ml of the denaturing buffer, 35 min in each wash. Continue monitoring the removal of degraded tracer DNA with a Geiger counter.

8. Neutralize the gel with the renaturing buffer as described in step 3.

9. Decrease the speed of shaking to 10–20 rpm and incubate the gel in the renaturing buffer overnight at 45°.

10. Wash the gel with S1 nuclease digestion buffer as described in step 5.

11. Incubate the gel with S1 nuclease as described in step 6.

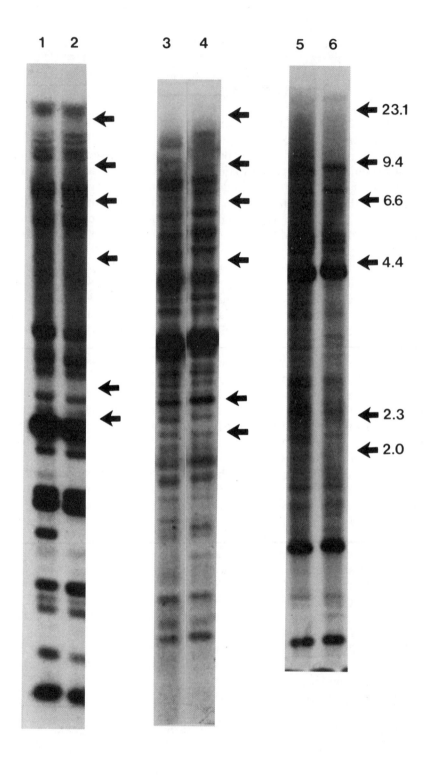

12. Wash the gel three times for 50 min with 350 ml of elution buffer, shaking at 30 rpm. Monitor the removal of degraded tracer DNA with a Geiger counter.

13. Remove the buffer after the last wash. Place an unfolded polystyrene sheet protector on the bench top. Turn the box with the gel upside down, holding it over the polystyrene sheet. Carefully detach the gel from the bottom of the box with a spatula and let the gel drop onto polystyrene. Cut a piece of Whatman 3MM paper slightly larger than the gel. Carefully cover the gel with 3MM paper, avoiding air bubbles. Wearing gloves, lift the "sandwich" consisting of the polystyrene sheet, the gel, and the sheet of 3MM paper, and place it on the gel drier with 3MM paper at the bottom. Remove the polystyrene sheet and cover the gel with Saran wrap, avoiding folds and air bubbles. Dry the gel using a vacuum pump and two refrigerated traps at 60°. After drying, turn off the heat and let the gel cool for 5–10 min before disconnecting the vacuum.

14. The dried gel is autoradiographed overnight at −70° using Kodak XAR-5 film either with or without an intensifying screen.

Analysis of the Results of In-Gel Renaturation Experiments

In order to identify amplified bands in the DNA preparation under study, it is necessary to make sure that the lane containing the control DNA digests shows all the repeated fragments that are usually detected in DNA from the given organism. Since detection of the weakest bands varies depending on the quality of the DNA preparation, some of the normal repeated fragments may be visible in the experiment but not in the control DNA lane, leading to their erroneous identification as amplified bands. The patterns of normal repeated fragments in the most commonly used mammalian systems are shown in Fig. 2. They include *Hin*dIII-digested human DNA (lanes 1,2), *Bam*HI-digested Chinese hamster DNA (lanes 3,4), and *Bam*HI-digested mouse DNA (lanes 5,6). Note that there are

---

FIG. 2. Repeated fragment patterns in mammalian DNA. The same DNA digests were used both as a tracer and as a driver in each lane. Each lane contains a total of 15 μg of DNA and 7–12 × 10⁶ dpm of ³²P-labeled tracer. Lanes 1 and 2 contain human DNA from peripheral blood cells of two leukemia patients (gift of Dr. M. Minden, Ontario Cancer Institute), digested with *Hin*dIII. Lanes 3 and 4 contain Chinese hamster DNA (gift of Dr. H. Abelson, Dana Farber Cancer Institute), digested with *Bam*HI. DNA in lane 3 is from CHO Aux B1 cell line, and DNA in lane 4 is from V79 cells. Lanes 5 and 6 contain *Bam*HI-digested mouse DNA. DNA in lane 5 is from FM3A mammary carcinoma cell line, and DNA in lane 6 is from B16 melanoma cells. Positions of λ *Hin*dIII fragments are indicated with arrows. The time of autoradiography was 14–20 hr for lanes 1–4 and 4–5 hr for lanes 5–6.

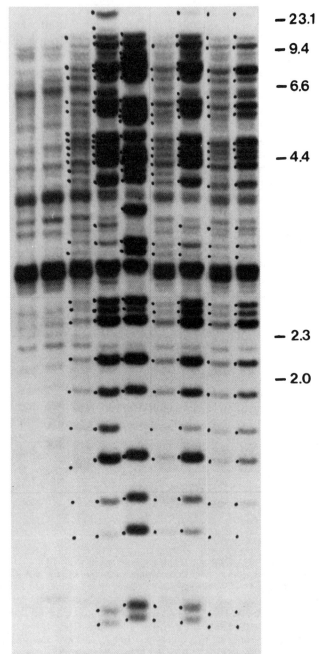

some polymorphic variations in the repeated fragment patterns among different individuals or different cell lines. The polymorphism is apparently of germline rather than somatic origin, since we have not observed such variations among several different tissues of the same individual (unpublished data). Therefore, it is preferable to use a control DNA preparation from the same individual or animal strain from which the experimental DNA sample was obtained. If such controls are not available, it may still be possible to draw a conclusion about DNA amplification in the sample. Since polymorphic variations usually involve a relatively small number of brands, and since the polymorphic bands may be either stronger or weaker in one individual than in the other, the presence of more than 7–10 bands specific for the experimental but not the control DNA preparation is a reliable indication of gene amplification, even if the control DNA was derived from a different individual.

For a sample analysis, we will consider an experiment shown in Fig. 3. In this experiment, we have used DNA preparations from a series of Chinese hamster ovary (CHO) cell lines, corresponding to different steps of selection for resistance to methotrexate (MTX; DNA preparations were a gift of Dr. Joyce Hamlin, University of Virginia). These cells contain an amplicon which includes the gene for dihydrofolate reductase (DHFR).[32] All DNA samples were digested with BamHI. In lanes A–E, the same DNA preparations were used both as a tracer and as a driver (homologous experiments), whereas lanes F–I represent heterologous experiments. Lane A contains DNA from the parental CHO cells, and it shows a characteristic pattern of repeated BamHI fragments. Lane B contains DNA from a cell line CHOC 0.01, which is resistant to 0.01 μg/ml MTX. No amplified bands are detectable in lane B, suggesting that the degree of amplification in these cells is too low for detection by in-gel renaturation. Lane C

---

[32] J. D. Milbrandt, N. H. Heintz, W. C. White, S. M. Rothman, and J. L. Hamlin, *Proc. Natl. Acad. Sci. U.S.A.* **75**, 5553 (1981).

FIG. 3. DNA amplification in methotrexate-resistant Chinese hamster ovary cells. Amplified bands are indicated with dots. Each lane contains DNA samples from the following cell lines: A, CHO parental (tracer and driver); B, CHOC 0.01 (tracer and driver); C, CHOC 0.2 (tracer and driver); D, CHOC 5.0 (tracer and driver); E, CHOC 400 (tracer and driver); F, CHOC 0.2 (tracer) and CHOC 400 (driver); G, CHOC 5.0 (tracer) and CHOC 400 (driver); H, CHOC 0.2 (tracer) and CHOC 5.0 (driver); I, CHOC 5.0 (tracer) and CHOC 0.2 (driver). A total of 14 μg of DNA and 2.5 × 10⁶ dpm of tracer were loaded into each lane. Positions of λ HindIII fragments are indicated. The heterologous experiments contained 0.15–0.2 μg of tracer DNA. Autoradiography was done for 16 hr.

contains DNA from a cell line CHOC 0.2, resistant to 0.2 μg/ml MTX, and in this lane a set of amplified bands, absent in the parental DNA, is readily detectable (indicated with dots). The same bands are much more intense in lane D, which contains DNA from a cell line CHOC 5.0, resistant to 5 μg/ml MTX, indicating further increase in the degree of amplification. Lane D also contains several amplified bands that are not visible in lane C. DNA in lane E, derived from CHOC 400 cells resistant to 400 μg/ml MTX, does not show a significant increase in band intensity relative to lane D, suggesting no great increase in the degree of amplification. This lane, however, contains several bands not found in lane D. In addition, some of the amplified bands detectable in lane D are missing or have a decreased intensity in lane E. This result suggests that a rearrangement in the structure of the amplicon has occurred at this step of selection.

These conclusions are confirmed by the heterologous experiments. In lane H, DNA from CHOC 0.2 was used as a tracer and DNA from CHOC 5.0 was used as a driver, whereas lane I contains a reciprocal combination. Comparison of these lanes with lanes C and D indicates that the structure of the amplicon present in CHOC 0.2 cells remains practically unchanged in CHOC 5.0. In lane F, DNA from CHOC 0.2 was used as a tracer and DNA from CHOC 400 was used as a driver, and in lane G DNA from CHOC 5.0 was the tracer and DNA from CHOC 400 was the driver. These experiments confirm that CHOC 400 cells contain several amplified fragments absent in either CHOC 0.2 or CHOC 5.0. The experiment in lane G, where DNA from CHOC 5.0 was used as a tracer and CHOC 400 DNA was used as a driver, indicates that those bands that are found in CHOC 5.0 (lane D) but not in CHOC 400 (lane E) remain amplified in CHOC 400 (though to a lower degree), since these bands are detectable in this heterologous experiment. However, these weak bands might also be a result of self-reassociation of tracer DNA. This possibility could be excluded by a control experiment where the same amount of CHOC 5.0 tracer DNA were mixed with the parental CHO driver DNA.

Some of the differences between band patterns in different lanes may result from minor variations in the quality of DNA mixtures. For example, it would be premature to conclude that the uppermost band detectable in lane D but not in lane E is present at a higher copy number in CHOC 5.0 relative to CHOC 400 DNA. This fragment is very large (>23 kb), and therefore even a small difference between the average sizes of undigested DNA preparations may cause considerable variations in the yield of this restriction fragment and the efficiency of its renaturation. Also note that some of the amplified fragments have slightly different mobility in different lanes (for example, the lowest bands in lanes C, D, and E), but the heterologous experiments prove that such fragments are in fact identical.

Diagnosis and Correction of Problems

Since the efficiency of in-gel renaturation depends on a number of parameters that may vary depending on the manual skills and the equipment used, it is recommended that new users start by estimating the sensitivity of the procedure in their hands. This can be done by determining the minimum amount of HindIII-digested λ phage DNA that can be detected in the mixtures with genomic DNA from the organism under study, expressed as the number of copies per haploid genome.[1] The size of λ DNA is 49 kb, and the size of haploid mammalian genome is approximately 2.9×10^6 kb; therefore a mixture of 100 μg of mammalian DNA and 170 ng of λ DNA would correspond to approximately 100 copies of λ per mammalian genome. If HindIII fragments of λ DNA are detectable only at a copy number which is significantly higher than 20–30 for human or hamster or 40–50 for mouse DNA (see Sensitivity of In-Gel Renaturation), the causes of low sensitivity should be diagnosed and corrected. Since even minor distortions of the electrophoretic pattern may have a severe detrimental effect on the efficiency of in-gel renaturation, the most common causes can be usually found at the early steps of the procedure: DNA extraction, purification of restriction digests, and gel electrophoresis. The best source for the diagnosis of specific problems is a photograph of the ethidium bromide-stained gel, taken immediately after electrophoresis. An autoradiogram obtained at the end of the experiment is less informative, since it would generally indicate only one of the two symptoms: a small number of detectable bands with little or no background, or a high nonspecific background smear obscuring individual bands. Below specific symptoms are enumerated with suggestions about their underlying causes along with the means for their correction.

Poor Restriction Enzyme Digestion or T4 DNA Polymerase Labeling of Genomic DNA Preparation

These problems stem from the presence of inhibitors of enzymatic reactions in genomic DNA preparations. Inhibition of restriction enzymes can be usually overcome by using spermidine in the reactions, as described in Procedure 3. If this does not help, additional phenol extraction and dialysis may alleviate these problems. In many cases, however, DNA extraction has to be repeated, using fresh preparations of phenol and longer times for phenol extraction.

Few Bands Visible in the Autoradiogram, Low Background

1. The ethidium bromide pattern of digested DNA shows a shift of density toward the lower part of the gel. Electrophoresis of undigested

DNA in 0.3% gels shows that the bulk of DNA migrates faster than the band of intact λ phage DNA, either as a compact band or as a smear.

This problem is caused by degradation during DNA extraction, most likely as a result of excessive hydrodynamic shearing either during phenol extraction or after ethanol precipitation followed by centrifugation. To avoid this problem, use gentler shaking during phenol extractions and use dialysis against TE instead of ethanol precipitation of DNA. Alternatively, DNA may be precipitated by ethanol, but instead of centrifugation it should be recovered by spooling it out of solution with a bent glass rod. The problem may also be caused by nuclease contamination of solutions used for DNA extraction.

2. Similar to (1) above, but the ethidium bromide pattern of digested DNA reveals a ladder of diffuse bands at the bottom part of the lane. Undigested DNA gives some smearing in a 0.3% gel, even though the bulk of DNA may migrate as a compact band at or above the position of the λ DNA band.

The diffuse bands correspond to a nucleosome ladder, indicating that DNA degradation occurred within chromatin, as a result of action of endogenous nucleases either before or at the very beginning of the extraction procedure. This problem is most often encountered with DNA prepared from tissue samples. To avoid this problem, make sure to extract DNA from tissues that are either fresh or frozen within 15–30 min after dissection. If the tissue sample contains necrotic elements, try to remove them as completely as possible prior to DNA extraction. Use a high concentration of EDTA (0.1–0.2 M) during the first steps of DNA extraction.

3. The ethidium bromide pattern of digested DNA shows streaks, as illustrated in Fig. 4a.

This artifact results from the presence of solid particles or incompletely dissolved DNA in the restriction digests of either the tracer or the driver DNA. To correct this problem, dilute the DNA preparation to a 100 μl volume with TE, mix, leave for 30 min at room temperature, and centrifuge for 5 min in an Eppendorf centrifuge. Transfer the solution into another tube without disturbing the pellet that may be visible at the bottom. Precipitate DNA with ethanol, as described in Procedure 3, rinse the pellet with 70% ethanol, dry, and redissolve in a convenient volume of TE. Measure the DNA concentration with diphenylamine (Procedure 1) and electrophorese an aliquot in a minigel. If necessary, repeat this procedure one more time.

4. Ethidium bromide staining shows higher density of DNA at the edges relative to the central part of the lane (Fig. 4b).

This artifact reflects the presence of salt in the tracer or the driver DNA

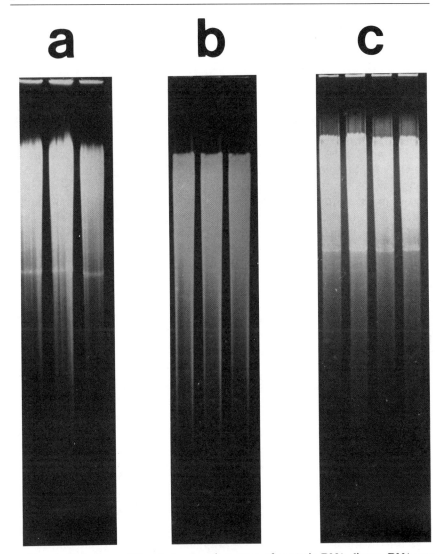

FIG. 4. Distortions of the electrophoretic pattern of genomic DNA digests. DNA was electrophoresed in 1% agarose gels, stained with ethidium bromide, and photographed using 35 mm Kodak Panatomic X film. (a) Streaking due to the presence of solid particles or incompletely dissolved DNA (*Bam*HI-digested mouse DNA). (b) Salt effect resulting in edge-specific distribution of DNA in the lanes (*Eco*RI-digested chicken DNA). (c) Normal electrophoretic pattern (*Bam*HI-digested mouse DNA).

digests. To avoid the salt effect, reprecipitate DNA with ethanol (Procedure 3), taking care to remove the supernatant completely after precipitation. Wash the pellet by brief (5 sec) vortexing with 70% ethanol. Remove ethanol, dry the pellet, and redissolve in TE. Measure the DNA concentration with diphenylamine before use.

5. The bands visible in the autoradiogram, as well as the bands of highly repeated DNA detectable with ethidium bromide, are convex rather than straight.

The most likely cause for this artifact is the presence of a large amount of RNA contaminating the DNA preparation, resulting in overloading of the gel. RNA can be detected by ethidium bromide staining after electrophoresing an aliquot of undigested DNA in 1.5% agarose, until the bromphenol blue moves about 1 cm. To correct this problem, add RNase A (DNase-free) to the final concentration of 20 μg/ml, either in TE or in a restriction buffer and incubate 30 min at 37°. Do not use tRNA carrier in ethanol precipitation of labeled tracer DNA.

High Background in the Autoradiogram

1. The background is observed with some but not all tracer DNA preparations used in the experiment.

The exact source of this infrequently occurring artifact is unclear, but it may reflect either nicking of tracer DNA, resulting in labeling of interspersed repeated sequences, or the presence of tightly bound proteins or DNA crosslinks which prevent complete denaturation of the tracer. When encountered in our laboratory, this problem could not be readily corrected, other than by repeating the DNA extraction. It should be noted, however, that a DNA preparation which gives high background when used as a tracer, is usually quite suitable for use as a driver in heterologous experiments.

2. High background in all lanes of the gel.

This may represent either the same problem as above, occurring with all the tracer DNA preparations used in the experiment, or inefficient digestion of single-stranded DNA with S1 nuclease. These possibilities can be distinguished by using a tracer DNA preparation previously found to give a low background. If such preparation is unavailable, control experiments can be done using *Hind*III-digested λ phage DNA as a tracer and increasing amounts of the same DNA as a driver.[1] Poor digestion with S1 nuclease may result from decreased enzyme activity (replace the enzyme) or from inhibition of the enzyme by contaminants in the enzyme buffer. To avoid contamination, make sure that all the glassware and plasticware which are used for the enzyme buffer are carefully rinsed several times with distilled water prior to use, and that the lid of the box used for incubation, which may contain traces of the denaturing buffer, is rinsed with distilled water before S1 nuclease is added. Incomplete S1 nuclease digestion may also be caused by poor diffusion of the enzyme into the gel. To ensure adequate diffusion, prepare the gel no thicker than 4–5 mm and shake the gel continuously at 30–40 rpm during the incubation with S1 nuclease.

Detection of Amplified Fragments Containing Short Interspersed Repeated Sequences

As discussed above (see Principle of the Method), the background of in-gel renaturation results mostly from repeated fragments that are normally present in the DNA from the organisms under study. This background is especially high in mouse and rat DNA, decreasing the sensitivity of detection of amplified fragments in these organisms to 40–50 copies per haploid genome. The background bands are produced primarily by tandemly repeated and long interspersed repeated sequences (LINEs), and therefore are not expected to hybridize to short interspersed repeated sequences (SINEs) unless they share a common evolutionary origin with the SINE element. On the other hand, the amplicons are known to include unique DNA sequences interspersed with SINE elements, as is the usual pattern throughout the bulk of mammalian DNA.[33] Therefore, a cloned SINE element used as a hybridization probe on a Southern blot would be expected to hybridize to a subset of restriction fragments derived from almost any amplicon, but only to a few of the normal repeated fragments, resulting in a lower background and increased sensitivity.

It should be noted, however, that not every cloned SINE element can be used as a probe for detection of amplified DNA. On the one hand, such an element should be present in the genome at sufficiently high frequency to occur one or more times in the amplicons that are at least 100 kb long. On the other hand, if the SINE element occurs at a very high frequency, the concentration of SINE DNA sequences would be high enough for them to reanneal in the gel, resulting in a high background after Southern hybridization. This is the case, for example, with human *Alu* sequences, which are found on average every 3–6 kb in the human genome,[34] and therefore give a high background when used as a probe after in-gel renaturation.

In the case of mouse DNA, we have found that a SINE element B2,

[33] E. H. Davidson, B. R. Hough, C. S. Amenson, and R. J. Britten, *J. Mol. Biol.* **77,** 1 (1973).
[34] F. P. Rinehart, T. G. Ritch, P. L. Deininger, and C. W. Schmid, *Biochemistry* **20,** 3003 (1981).

which occurs about 10^5 times in the mouse genome[35] corresponding to the average frequency of once per 30 kb, provides a convenient probe for detection of amplified sequences after in-gel renaturation.[22] Using mixtures of normal mouse DNA with DNA from cell lines containing amplified DHFR or *c-Ki-ras* gene,[36,37] we were able to detect amplified fragments hybridizing with the B2 probe in the mixtures containing as few as 10–15 copies of these genes. Comparison of the DNA patterns obtained by in-gel renaturation with labeled tracer DNA and by Southern hybridization using the B2 probe clearly indicates a higher sensitivity for the latter protocol.[22] The in-gel renaturation/SINE hybridization procedure also simplifies the screening of multiple DNA preparations for the presence of amplified DNA, since this technique does not require the labeling of each individual tracer DNA preparation. It should be noted, however, that SINE hybridization cannot discriminate between identical and incidentally comigrating amplified fragments, and therefore labeling genomic tracer DNA is still necessary for identification of commonly amplified sequences in different DNA preparations.

Either nitrocellulose or nylon membranes can be used for Southern transfer after in-gel renaturation. For labeling the B2 probe, we use the "oligolabeling" technique of Feinberg and Vogelstein,[38] which gives specific activity in excess of 10^9 dpm/μg. Hybridization is done under conditions of intermediate stringency in order to account for the divergence among different B2 sequences. When using this protocol for detection of amplified sequences in mouse DNA, it is essential to use DNA from the same mouse strain in the experimental and control lanes, since there are polymorphic variations among B2-containing repeated fragments in different mouse strains.[22]

Detection of Amplified Fragments Containing Transcriptionally Active Sequences

When amplified genes express high levels of mRNA, it is sometimes possible to identify transcriptionally active fragments within the amplicon prior to cloning. This can be done by hybridizing repeated and amplified fragments, obtained after in-gel renaturation, with ^{32}P-labeled cDNA representing the total mRNA population from cells containing amplified

[35] A. S. Krayev, T. V. Markusheva, D. A. Kramerov, A. P. Ryskov, K. G. Skryabin, A. A. Bayev, and G. P. Georgiev, *Nucleic Acids Res.* **10**, 7461 (1982).
[36] P. C. Brown, S. M. Beverly, and R. T. Schimke, *Mol. Cell. Biol.* **1**, 1077 (1981).
[37] M. Schwab, K. Alitalo, H. E. Varmus, J. M. Bishop, and D. George, *Nature (London)* **303**, 497 (1983).
[38] A. P. Feinberg and B. Vogelstein, *Anal. Biochem.* **132**, 6 (1983).

DNA. A typical experiment is shown in Fig. 5. In this experiment, done in collaboration with E. Rose and H. Munro (M.I.T., unpublished data), this approach was used to detect transcriptionally active amplified DNA sequences in a human neuroblastoma cell line IMR-32, which contains an amplicon that includes the *n-myc* gene.[39,40] Genomic DNA from normal human blood (lane 1) or IMR-32 (lane 2) was digested with *Hind*III, and the mixture of labeled tracer and unlabeled driver DNA (A) or driver DNA alone (B) was electrophoresed in an agarose gel and subjected to two cycles of in-gel renaturation. Lanes containing tracer DNA (A) were dried and autoradiographed, revealing multiple amplified bands specific to IMR-32. Lanes containing unlabeled DNA (B) were transferred onto a nitrocellulose filter and hybridized to ^{32}P-labeled cDNA prepared by reverse transcription of poly(A)$^+$ cytoplasmic RNA from IMR-32 cells using an oligo(dT) primer. The use of the oligo(dT) primer for cDNA synthesis, resulting in preferential labeling of 3'-terminal sequences, was chosen over random priming in order to simplify the hybridization pattern and to avoid labeling the ribosomal and tRNA sequences, that are usually present as contaminants in poly(A)$^+$ RNA preparations. The cDNA probe hybridized to a subset of amplified fragments in IMR-32, as well as to the mitochondrial DNA bands (mt). Hybridization of the cDNA probe to blots containing genomic DNA that was transferred either directly after electrophoresis or after one cycle of in-gel renaturation and S1 digestion did not allow detection of specific bands because of a high background (not shown).

Caution should be exercised, however, in the interpretation of the experiments of the type shown in Fig. 5, since poly(A)$^+$ RNA from human cells includes *Alu*[41] and possibly other repeated sequences, and therefore a cDNA probe may hybridize to repeat-containing amplified fragments, even if such fragments are not transcribed. When amplified DNA is derived from cells that have been selected *in vitro* for a specific phenotype, e.g., drug resistance, rehybridization of the blots with cDNA corresponding to poly(A)$^+$ RNA from the parental cell line provides a good control for nonspecific hybridization of amplified fragments. However, in the case of tumor-derived cells, it is much more difficult to select a control cell population. It may be possible to decrease the hybridization of repeated sequences in the cDNA probe by prereassociation of the probe with total genomic DNA from a control cell line.[42]

[39] M. Schwab, K. Alitalo, K.-H. Klempnauer, H. E. Varmus, J. M. Bishop, F. Gilbert, G. Brodeur, M. Goldstein, and J. Trent, *Nature (London)* **305**, 245 (1983).
[40] N. E. Kohl, N. Kanda, R. R. Schreck, G. Bruns, S. A. Latt, F. Gilbert, and F. W. Alt, *Cell* **35**, 359 (1983).
[41] W. R. Jelinek, R. Evans, M. Wilson, M. Salditt-Georgieff, and J. E. Darnell, *Biochemistry* **17**, 2776 (1978).
[42] P. G. Sealey, P. A. Whittaker, and E. M. Southern, *Nucleic Acids Res.* **13**, 1905 (1985).

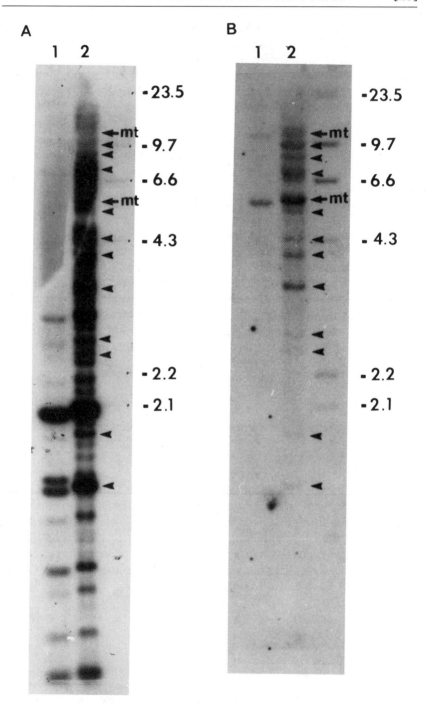

Identification of transcriptionally active fragments within the amplicon does not necessarily indicate that such fragments are derived from the essential gene, since transcriptionally active but apparently nonessential genes were found within flanking sequences of several amplicons.[43-45] Therefore, as a primary approach to selection of amplified DNA fragments for cloning and further analysis we recommend, whenever possible, identifying those fragments that are amplified in common in different independently derived cell populations. Hybridization with a cDNA probe, on the other hand, can be used as a secondary criterion for the selection of important fragments.

Procedure 7. Preparation of DNA Probes by Oligolabeling (Modified from Feinberg and Vogelstein[38])

Stock Solutions

Deoxyribonucleotide solutions: 0.1 M dATP, 0.1 M dGTP, 0.1 M dTTP, each dissolved in 3 mM Tris–HCl (pH 7.0), 0.2 mM EDTA; store at $-20°$

O: 1.25 M Tris–HCl (pH 8.0), 0.125 M MgCl$_2$

A: 1 ml solution O, 18 μl 2-mercaptoethanol, 5 μl each of dATP, dGTP, and dTTP; store at $-20°$

B: 2 M HEPES, adjusted to pH 6.6 with NaOH

C: Hexadeoxyribonucleotide mixture (Pharmacia 27-2166-01), suspended at 90 OD units/ml in 3 mM Tris–HCl (pH 7.0), 0.2 mM EDTA; store at $-20°$

[43] G. M. Wahl, L. Vitto, and J. Rubnitz, *Mol. Cell. Biol.* **3**, 2066 (1983).

[44] M. Debatisse, B. R. De Saint Vincent, and G. Buttin, *EMBO J.* **3**, 3123 (1984).

[45] A. M. Van der Bliek, T. Van der Velde-Koerts, V. Ling, and P. Borst, *Mol. Cell. Biol.* **6**, 1671 (1986).

FIG. 5. Transcription of amplified DNA sequences in IMR-32 neuroblastoma. DNA from peripheral blood cells of a healthy individual (A, B, lane 1) and from IMR-32 cells (A, B, lane 2) was digested with *Hin*dIII. Each lane contains 15 μg of DNA. DNA in A was mixed with 10^7 dpm of homologous tracer DNA. No tracer DNA was added in B. After two cycles of in-gel renaturation, lanes containing tracer DNA (A) were dried and autoradiographed, and unlabeled DNA (B) was transferred onto a nitrocellulose filter. The filter was hybridized with 5×10^7 dpm of ^{32}P-labeled cDNA prepared from poly(A)$^+$ cytoplasmic RNA of IMR-32 using an oligo(dT) primer. The amplified fragments hybridizing to the cDNA probe are indicated with arrowheads in both A and B. The position of mitochondrial DNA bands (mt) was determined from the restriction map of human mitochondrial DNA [J. Drouin, *J. Mol. Biol.* **140**, (1980)] and by the comparison of repeated fragment patterns in human DNA preparations extracted from either total cells or from purified nuclei (data not shown).

Working Solutions

OLB buffer (0.5 ml): 100 μl solution A, 250 μl solution B, 150 μl solution C; store at $-20°$

10 mg/ml bovine serum albumin (Sigma A-7906); store at $-20°$

0.5 M EDTA (pH 8.0)

2.5% blue dextran (Sigma D-5751) in water

10 mg/ml denatured salmon sperm DNA (Sigma D-1626); store at $-20°$

[α-^{32}P]dCTP (3000 Ci/mmol, 10 mCi/ml in stabilized aqueous solution; Amersham PB.10205); store at $-20°$

1. Place 50–100 ng of DNA in TE buffer into a 1.5 ml screw-cap polypropylene centrifuge tube. Add water to 32 μl. Place into a boiling water bath for 3–4 min to denature the DNA. Chill on ice for 2 min.

2. Add 10 μl of OLB buffer, 2 μl of 10 mg/ml bovine serum albumin, 5 μl of [α-^{32}P]dCTP, and 1 μl of large fragment of DNA polymerase I (4 units/μl, Amersham). Mix by tapping and centrifuge briefly. Incubate 3 hr at room temperature. Stop the reaction by adding 10 μl of 0.5 M EDTA and 40 μl of 2.5% blue dextran.

3. Prepare a Sephadex G-100 column in a 5 ml serological pipet plugged with sterile glass wool. Equilibrate the column in TE buffer. Pass 500 μg of denatured salmon sperm DNA through the column to decrease nonspecific binding of the probe. Apply the reaction mixture to the column and elute the DNA at a flow rate determined by gravity. Collect the exclusion volume (approximately 1.5 ml), containing the blue dextran, into a 1.5 ml screw-cap polypropylene centrifuge tube.

4. Mix the liquid in the tube. Remove a 10-μl aliquot and add it to a scintillation vial containing 5–7 ml of water-mixable scintillation fluid, such as Scintiverse II (Fisher). Count the vial in a scintillation counter and calculate the concentration of the probe.

5. Store the probe at $-70°$ up to 1 week. Prior to use, heat the probe for 10 min in a boiling water bath and chill on ice.

Procedure 8. Preparation of cDNA Probes

Solutions

5× reaction buffer: 0.25 M Tris–HCl (pH 8.3), 0.25 M potassium chloride, 50 mM magnesium chloride, 10 mM dithiothreitol; store at $-20°$

dNTP mixture: 1 mM dATP, 1 mM dGTP, 1 mM dCTP, 1 mM dTTP; store at $-20°$

1 mg/ml oligo(dT)$_{12-18}$ (Pharmacia 27-7858-01); store at $-20°$

[α-^{32}P]dATP (3000 Ci/mmol, 1 mCi/ml in 50% ethanol; Amersham PB.204); store at $-20°$

[α-^{32}P]dCTP (3000 Ci/mmol, 1 mCi/ml in 50% ethanol; Amersham PB.205); store at $-20°$

[α-^{32}P]dGTP (3000 Ci/mmol, 1 mCi/ml in 50% ethanol; Amersham PB.206); store at $-20°$

[α-^{32}P]dTTP (3000 Ci/mmol, 1 mCi/ml in 50% ethanol; Amersham PB.207); store at $-20°$

1. Dry 250 μCi of each of the four [α-^{32}P]dNTP in the same Eppendorf tube, using Speed Vac concentrator or a lyophilizer. Add 5 μl of 1 mg/ml oligo(dT)$_{12-18}$, 5 μl of 5$\times$ reaction buffer, and 1 μl of 1 mM unlabeled dNTP mixture. Mix by tapping and place on ice.

2. Mix 5 μg of poly(A)$^+$ cytoplasmic RNA with 11 μl of water in a 1.5 ml screw-cap polypropylene tube. Heat in a boiling water bath for 5 min and chill on ice. Centrifuge briefly to collect the liquid at the bottom of tube.

3. Transfer RNA into the tube containing the reaction mixture (step 1). Add 3 μl of AMV reverse transcriptase (20 units/μl, Bio-Rad 170-3110). Mix and incubate 1 hr at 42°.

4. Purify the cDNA probe on a Sephadex G-100 column, determine the label incorporation, and use as described in Procedure 7, steps 3–5.

Procedure 9. Southern Transfer and Hybridization of Repeated and Amplified Fragments

Stock Solutions

20$\times$ SSPE: 0.2 M sodium phosphate (pH 7.0), 3 M sodium chloride, 25 mM EDTA

20$\times$ SSC: 0.3 M sodium citrate (pH 7.0), 3 M sodium chloride

10% SDS

100$\times$ Denhardt's: 2% Ficoll, 2% polyvinylpyrrolidone, 2% bovine serum albumin

10 mg/ml denatured salmon sperm DNA

Working Solutions

Denaturing buffer: see Procedure 6

Neutralizing buffer: 3 M sodium acetate (pH 5.5)

3$\times$ SSPE

Hybridization buffer: 5$\times$ SSPE, 5$\times$ Denhardt's, 0.2% SDS, 500 μg/ml denatured salmon sperm DNA (Sigma D-1626)

2$\times$ SSC, 0.5% SDS

1$\times$ SSC, 0.5% SDS

Equipment

Gyratory shaker (New Brunswick model G-2)

Gyratory water bath shaker (New Brunswick model G-76)

Bag sealer (Sears Seal-N-Save) and sealing pouches (10 in. × 20 ft rolls)

1. Load 15 μg of unlabeled DNA, digested and purified as described in Procedure 3, into each lane of a 1% agarose gel. After electrophoresis (Procedure 5, steps 3–4), perform two rounds of in-gel renaturation and S1 nuclease digestion, as described in Procedure 6, steps 1–11.

2. After the second S1 nuclease digestion, wash the gel two times for 45 min with 500 ml of 3× SSPE at room temperature, shaking at 30–40 rpm.

3. Denature DNA in the gel by washing it with 350 ml of denaturing buffer, two times for 15 min, shaking at 20–30 rpm.

4. Neutralize the gel by washing it with 350 ml of neutralizing buffer for 30 min, shaking at 20–30 rpm.

5. Transfer DNA from the gel onto Biodyne (Pall) nylon membrane, cut to the size of the gel, using capillary transfer in 20× SSC, essentially as described by Southern.[21]

6. Bake the membrane for 1.5 hr at 80° in a vacuum oven.

7. Seal the membrane in a plastic bag containing 200 μl of hybridization buffer per 1 cm² of the membrane. Prehybridize 1 hr at 65° in a shaking water bath.

8. Open the bag and replace the buffer solution with fresh hybridization buffer containing 6–10 × 10⁷ dpm of denatured probe, at a volume of 30 μl per 1 cm² of the membrane. Seal and incubate overnight in a shaking water bath at 65°.

9. Remove the membrane from the bag and place it into a tray containing 500 ml of 2× SSC, 0.5% SDS. Wash at room temperature two times for 5 min, shaking at 50–60 rpm.

10. Seal the membrane in a bag containing 100 ml of 1× SSC, 0.5% SDS. Wash two times for 60 min in a shaking water bath at 65°

11. Cover the membrane on both sides with Saran wrap and autoradiograph overnight at −70°, using Kodak XAR-5 film and an intensifying screen.

Cloning of Amplified DNA

After identifying a subset of amplified fragments which are likely to contain the essential gene or to be located in the immediate vicinity of such a gene, either on the basis of amplification of these fragments in different independently isolated cell lines or by their hybridization to cDNA, the region of DNA containing the essential gene can be cloned. The simplest cloning strategy consists of isolating one of the fragments from this region and using it as a starting point for chromosome walking. For initial cloning, it would be reasonable to select a fragment that is best separated from other repeated or amplified fragments in the gel. In most cases, this would be the smallest of the amplified fragments. To achieve maximal enrichment for the fragment of interest, it is important to establish precisely the position of this fragment in the gel and to cut out a narrow strip of the gel containing the fragment. The most reliable way to locate the fragment of interest in the gel is by performing the complete in-gel renaturation assay with labeled tracer DNA using several lanes of the same gel. Further enrichment is obtained by denaturation and renaturation of DNA in the gel strip. Since only precisely reannealed double-stranded molecules can be ligated into the appropriate restriction site of a plasmid or phage vector, it is possible to avoid digestion with S1 nuclease, which would otherwise result in degradation of cohesive ends of the restriction fragments and loss of some double-stranded DNA. Even though in this procedure single-stranded DNA is not degraded after the first cycle of renaturation, we have observed that two cycles of in-gel renaturation still give better enrichment for amplified fragments than a single cycle. This may be due to a higher diffusion rate of single-stranded relative to double-stranded DNA fragments, resulting in partial loss or a decrease in concentration for those fragments that do not reanneal during the first cycle.

After in-gel renaturation, the DNA preparation enriched for the desired amplified fragment is recovered from the gel and ligated into an appropriate restriction site of the cloning vector. The choice of a plasmid or a bacteriophage vector for cloning depends on the starting amount of genomic DNA, the apparent degree of amplification of the desired fragment (as estimated from the band intensity) and the size of the fragment. While phage vectors have a much higher cloning efficiency, they require a more laborious cloning procedure, and the size of the desired fragment has to fit into the upper and lower limits for cloning into the particular restriction site of a phage vector. It is helpful to estimate the approximate amount of the amplified fragment that would be recovered after purification, assuming that the recovery of the fragment after in-gel renaturation may vary between 0.1 and 5%, depending on the degree of amplification. If the estimated amount is higher than 50–100 pg, plasmid vectors can be used, while lower amounts of DNA would require phage vectors.

The resulting clones are analyzed by digestion with restriction enzymes. Once several clones producing identical restriction digests are found, such clones should be tested for their amplification in the genome by hybridizing them with genomic DNA from both the control cells and the cells containing amplified DNA. A clone containing the amplified fragment can then be used for isolation of adjacent genomic sequences by chromosome walking.[46,47] In the course of the latter procedure, it is desirable to know if the phage or cosmid clones containing amplified DNA accurately represent the corresponding genomic sequences. This can be done by in-gel renaturation, using genomic DNA from both the control and the experimental cells as tracers, and DNA from individual clones, digested with the same restriction enzyme, as a driver.[8] The presence of SINE repeats within the clones has practically no effect on this hybridization procedure, whereas in Southern hybridization it constitutes a serious problem. The cloned segments of amplified DNA can then be analyzed for the presence of transcriptionally active sequences, or used for biological assays until the essential gene within the amplicon is identified.

Procedure 10. Cloning of Amplified DNA Fragments

Solutions and Equipment
See Procedures 3–9

1. Digest 25–30 µg of DNA from control cells and 150–200 µg of DNA from cells containing amplified genes with an appropriate restriction enzyme. Purify the digests as described in Procedure 3.

2. Label 0.5–1.0 µg of the control and experimental DNA preparations, as well as *Hin*dIII-digested λ phage DNA as described in Procedure 4.

3. Prepare electrophoresis mixtures as described in Procedure 5. Load the gel lanes in the following order: lane 1, control DNA, tracer and driver (15 µg total); lane 2, experimental DNA, tracer and driver (15 µg total); lane 3, λ *Hin*dIII DNA, tracer and driver (1 µg total); lane 4, 1 µg of unlabeled λ *Hin*dIII DNA. In each of the other lanes load 15 µg of unlabeled experimental DNA, the number of lanes being determined by the amount of digested DNA. In the last lane, load 1 µg of unlabeled λ *Hin*dIII DNA.

4. After electrophoresis, photograph the gel and cut off the part of the gel containing lanes 1–3. Wrap the remainder of the gel containing unlabeled DNA in Saran wrap and store it at 4° in the dark.

5. Place the gel containing lanes 1–3 in a polyethylene box and perform two rounds of in-gel renaturation and S1 nuclease digestion, as described in Procedure 6. After autoradiography, identify in lane 2 the band corresponding to the amplified fragment selected for cloning. Cut the

[46] C. L. Smith, S. K. Lawrance, G. A. Gillespie, C. R. Cantor, S. M. Weissman, and F. S. Collin, this series [35].
[47] T. Maniatis, E. F. Fritsch, and J. Sambrook, "Molecular Cloning: A Laboratory Manual." Cold Spring Harbor Lab., Cold Spring Harbor, New York, 1982.

autoradiogram so that the bands of the λ HindIII digest in lane 3 are on the edge of the X-ray film.

6. Place the autoradiogram on a 300 nm UV transilluminator, next to the untreated portion of the original gel. Align the λ HindIII bands in lane 3 of the autoradiogram with the ethidium bromide stained bands on the edge of the untreated gel. Mark the position of the amplified fragment of interest relative to λ HindIII fragments on both sides of the gel. Using a razor blade and a ruler, cut out a 2–3 mm thick gel strip containing the fragment of interest.

7. Place the gel strip into a 15 ml screw-cap polypropylene tube filled with the denaturing buffer. If necessary, cut the gel strip into two or more pieces. Incubate at 45° for 30 min, shaking at 50–60 rpm. Wearing gloves, carefully pour the liquid out of the tube, holding the gel strip in the tube with a finger. Add fresh denaturing buffer and continue incubation for another 30 min.

8. Neutralize the gel by three 15 min washes with the renaturing buffer, shaking at 50–60 rpm.

9. Incubate the gel in the renaturing buffer for 2 hr at 45° without shaking.

10. Denature DNA with the denaturing buffer as described in step 7.

11. Neutralize the gel with the renaturing buffer as described in step 8.

12. Incubate the gel in the renaturing buffer overnight at 45° without shaking.

13. Wash the gel at room temperature with three 20 min changes of 1/2× electrophoresis buffer.

14. Recover DNA from the gel by electroelution.[48]

15. Adjust salt concentration in the electrophoresis buffer to 0.2 M NaCl and purify the eluted DNA using an Elutip-d column (Schleicher and Schuell), under conditions recommended by the manufacturer. Elute DNA in 0.4 ml volume of high salt buffer [20 mM Tris (pH 7.5), 1 M sodium chloride, 1 mM EDTA]. Add 10 μg of tRNA (Boehringer-Mannheim) and 2.5 volumes of ice-cold ethanol. Mix, place into a dry ice–ethanol bath for 5 min, centrifuge 10 min at 4°. Remove the supernatant and wash the pellet with 200 μl of ice-cold 70% ethanol. Centrifuge 5 min at 4° and remove the supernatant completely. Dry the pellet briefly in a Speed Vac concentrator or a desiccator. Dissolve in 4 μl of 2× ligation buffer [1× ligation buffer is 50 mM Tris (pH 7.5), 10 mM MgCl$_2$, 20 mM DTT, 1 mM ATP, and 5 μg/ml bovine serum albumin].

16. Digest 1 μg of pBR322 or any other appropriate plasmid vector

---

[48] D. Hanahan, in "DNA Cloning Techniques: A Practical Approach" (D. Glover, ed.), p. 109. IRL Press, Oxford, England, 1985.

with 10 units of the same restriction enzyme as used for genomic DNA digestion, for 1 hr at 37°, in a 10 μl reaction volume. Add 2/3 units of calf intestinal phosphatase (Boehringer-Mannheim) and continue incubation for another 30 min. Add 40 μl of 10 mM Tris (pH 7.5), 0.1 M NaCl, 1 mM EDTA. Extract DNA once with 50 μl of phenol:chloroform:isoamyl alcohol (25:24:1), followed by three extractions with 50 μl of chloroform:isoamyl alcohol (24:1) and one extraction with 100 μl of ether. To the aqueous phase, add 5 μl of 3 M sodium acetate and 125 μl of ice-cold ethanol. Mix, place into a dry ice–ethanol bath for 5 min, centrifuge 10 min at 4°. Remove the supernatant and wash the pellet with 200 μl of ice-cold 70% ethanol. Centrifuge 5 min at 4° and remove the supernatant completely. Dry the pellet briefly and dissolve in 10 μl of TE.

17. Mix 2 μl of insert DNA from step 15 and 1 μl of vector DNA from step 16 (<100 ng). Incubate 5 min at 55°, chill on ice. Add 1 μl of T4 DNA ligase (2 units/μl), mix, and incubate overnight at 16°.

18. Use one-half of the ligated mixture to transform competent *E. coli* cells as described.[48] Plate transformed bacteria on agar plates containing an appropriate antibiotic.

19. Identify transformants containing recombinant plasmids, e.g., ampicillin-resistant/tetracycline-sensitive colonies if the insert is cloned into *Bam*HI site of pBR322. Extract miniscale DNA preparations[48] from 30–50 recombinant colonies. Digest plasmid DNA with the same restriction enzyme as used for cloning and determine if the clones contain inserts of the expected size.

20. Digest plasmid DNA preparations with a second restriction enzyme recognizing 4 or 5 bp restriction sites, selected so as to generate fewer than 10–12 fragments in vector DNA, e.g., *Hinf*I in the case of pBR322. Separate the digests in an agarose gel and compare the fragment patterns among different clones. Identify serial patterns occurring in more than one clone.

21. Pick one clone from each series and test them for the presence of amplified DNA. This can be done by in-gel hybridization between the clones and genomic DNA, using 0.1–0.2 μg of genomic DNA preparations from the control and the experimental cells as tracers, and 1–10 ng of each clone DNA preparation as a driver. All DNA preparations should be digested with the same restriction enzyme that was used for cloning. Electrophoresis and in-gel renaturation are performed as described in Procedures 5 and 6, except that in this case a smaller gel can be used. For example, it is possible to use an electrophoresis apparatus from Bethesda Research Laboratories (model H4) with a 20 × 25 cm tray and a 20-well comb. The size of the box for in-gel renaturation and the buffer volumes for all washes should be reduced accordingly. The in-gel hybridization

assay allows one to identify the clones that contain single-copy DNA fragments (no bands detectable with either the control or the experimental genomic tracer DNA), the clones that contain normal repeated fragments (bands visible with both tracers), and the clones containing amplified fragments (a band is visible only with the experimental but not with the control tracer DNA).

22. Once a clone containing an amplified DNA fragment has been identified, it can be used as a probe for Southern[21] hybridization with genomic DNA. If the clone contains an interspersed repeated sequence, hybridization will produce a smeared pattern. In this case, isolate smaller subfragments of cloned DNA and identify those that give no background in Southern hybridization. The repeat-free subfragments can then be used to screen phage or cosmid libraries of genomic DNA in order to isolate the adjacent genomic sequences.

Acknowledgments

I would like to thank Drs. Joyce Hamlin and Elise Rose for the permission to use unpublished data from our collaborative studies, Drs. Mark Minden and Herbert Abelson for the gifts of DNA preparations used in the experiments shown in Fig. 2, Drs. Manabu Fukumoto and Kevin Noonan for their comments on the manuscript, and Drs. Alexander Varshavsky and Vernon Ingram for their support during the development of the in-gel DNA renaturation methodology. This work was supported by Grants CA 39365 and CA 40333 from the National Cancer Institute.

[26] Detection and Characterization of Specific mRNA by Microinjection and Complementation of Mutant Cells

By PIN-FANG LIN, DAVID B. BROWN, PAT MURPHY, MASARU YAMAIZUMI, and FRANK H. RUDDLE

Various methods have been developed for the transfer of RNA into eukaryotic cells. These include free uptake of exogenously supplied polyadenylated RNA,[1] CaCl$_2$ or DEAE-dextran facilitated transfer,[2] red cell-mediated transfer,[3] and the glass capillary microinjection technique that was

[1] B. Mroczkowski, H. P. Dym, E. J. Siegel, and S. M. Heywood, J. Cell Biol. 87, 65 (1980).
[2] J. J. Greene, C. W. Dieffenbach, and P. O. P. Ts'o, this series, Vol. 79, p. 104.
[3] C. Boogaard and G. Dixon, Exp. Cell Res. 143, 175 (1983).
[4] C. Boogaard and G. Dixon, Exp. Cell Res. 143, 191 (1983).

pioneered by Chambers,[5] Diacumakos,[6] and Graessmann and Graessmann.[7] Of these methods direct capillary injection is the most efficient.

Microinjection of mRNA into intact mammalian cells offers several distinct advantages. It provides greatly enhanced sensitivity in detecting translation products from small amounts of mRNA and can also be applied to almost every cell type. The recipient cell can be subjected to a variety of manipulations and its response after translation of the injected RNA can be studied directly. Moreover, if a mutant cell is used as the recipient, the microinjection of wild-type mRNA and transient restoration of the mutant phenotype can serve as a powerful technique for characterizing the complementing mRNA and the nature of the mutation.

Several groups of investigators have successfully used RNA microinjection to study the translation and function of specific messages or viral RNA.[8-11] Liu et al.[12] have utilized similar techniques to introduce total human mRNA into mutant cells. The biological activity of one specific mRNA in the pool was detected by complementation of the mutant phenotype after in vivo translation of the injected mRNA. Lin et al.[13] were able to complement thymidine kinase (TK) and hypoxanthine–guanine phosphoribosyltranferase (HPRT) mutants by injecting mutant cells with specific mRNA fractions isolated from wild-type cells. Since the mRNAs were separated on the basis of size, these investigators were able to determine the molecular weight of the biologically active TK and HPRT messages. A similar approach was later used to partially purify and characterize the mRNA complementing the defect in a temperature-sensitive S phase cell cycle mutant,[14] and in cells from individuals with xeroderma pigmentosum group A and G.[15]

In the following report, we describe the procedures used in our laboratory to characterize TK and HPRT mRNA in order to illustrate the details

[5] R. Chambers, *Anat. Rec.* **24,** 1 (1922).

[6] E. G. Diacumakos, *Methods Cell Biol.* **7,** 287 (1973).

[7] M. Graessmann and A. Graessman, this series, Vol. 101, p. 482.

[8] M. Graessmann and A. Graessmann, *Proc. Natl. Acad. Sci. U.S.A.* **73,** 366 (1976).

[9] D. W. Stacey and V. G. Allfrey, *Cell* **9,** 725 (1976).

[10] D. Stacey, *Cell* **21,** 811 (1980).

[11] D. W. Stacey, this series, Vol. 79, p. 76.

[12] C. P. Liu, D. L. Slate, R. Gravel and F. H. Ruddle, *Proc. Natl. Acad. Sci. U.S.A.* **76,** 4503 (1979).

[13] P. F. Lin, M. Yamaizumi, P. D. Murphy, A. Egg, and F. H. Ruddle, *Proc. Natl. Acad. Sci. U.S.A.* **79,** 4290 (1982).

[14] A. Fainsod, M. Marcus, P. F. Lin, and F. H. Ruddle, *Proc. Natl. Acad. Sci. U.S.A.* **81,** 2393 (1984).

[15] R. J. Legerski, D. B. Brown, C. A. Peterson, and D. L. Robberson, *Proc. Natl. Acad. Sci. U.S.A.* **81,** 5676 (1984).

of the techniques employed. This will be followed by a presentation of two other examples which used similar methods to study mutant genes. We conclude the report with a discussion of strategies that can be used to isolate the gene corresponding to the mRNA responsible for the complementation event.

Procedure

Microinjection

Microinjection of RNA solutions into the cytoplasm of individual cells was performed, essentially, by the method of Graessmann and Graessmann.[7] The recipient cells were grown on coverslips (25 × 25 mm) containing photoengraved microscopic grid patterns[16] which facilitate the localization of injected cells after photoradiography. The injection pipettes were prepared from glass capillaries with fine glass filaments attached to their inner walls (1.2 mm o.d., American Glass, Bargaintown, NJ) to facilitate sample loading. The vertical pipet puller (model 700C) was from David Kopf Instruments, Tujunga, CA. Injection was carried out on an inverted phase-contrast microscope (Leitz Diavert ×200). Movement of the micropipets was controlled with a Leitz micromanipulator. As recipients, HT1080 HPRT$^-$ cells were injected individually. Since LTK$^-$ cells are small, they were fused with polyethylene glycol (1540 or 3000) for 30–45 sec to form polykaryons containing three or four cells to facilitate the injection process. RNA samples of 5–10 μl were centrifuged at 60,000 g for 45 min in glass capillaries to remove debris that might clog the injection pipette. The glass capillaries were spun in a Beckman ultracentrifuge (SW 40 rotor) with homemade adaptors to hold the capillaries. After centrifugation, samples were withdrawn with glass needles and loaded into injection micropipets. The volume injected into each single cell was approximately 5×10^{-11} ml (average value).

Preparation of Coverslips with Microgrid Pattern.[16]

Equipment and Material. A slide projector (Leitz); a photoresist spinner (from Headway Research, Inc., Garland, TX); positive photoresist (AZ-1370) and developer (AZ-351) (from Shipley Co., Newton, MA); glass coverslip no. 2, 25 mm² (from Corning, NY); hydrofluoric acid (2.4% v/v in deionized H$_2$O); acetone.

Method of Photoengraving. Mask making. An original grid pattern is

[16] P. F. Lin and F. H. Ruddle, *Exp. Cell Res.* **134,** 485 (1981).

A   **PFL**

O	I	2	3	4	5	6	7	8	9	a	b	c	d	e	f	g	h	i	j
OA	IA	2A	3A	4A	5A	6A	7A	8A	9A	aA	bA	cA	dA	eA	fA	gA	hA	iA	jA
OB	IB	2B	3B	4B	5B	6B	7B	8B	9B	aB	bB	cB	dB	eB	fB	gB	hB	iB	jB
OC	IC	2C	3C	4C	5C	6C	7C	8C	9C	aC	bC	cC	dC	eC	fC	gC	hC	iC	jC
OD	ID	2D	3D	4D	5D	6D	7D	8D	9D	aD	bD	cD	dD	eD	fD	gD	hD	iD	jD
OE	IE	2E	3E	4E	5E	6E	7E	8E	9E	aE	bE	cE	dE	eE	fE	gE	hE	iE	jE
OF	IF	2F	3F	4F	5F	6F	7F	8F	9F	aF	bF	cF	dF	eF	fF	gF	hF	iF	jF
OG	IG	2G	3G	4G	5G	6G	7G	8G	9G	aG	bG	cG	dG	eG	fG	gG	hG	iG	jG
OH	IH	2H	3H	4H	5H	6H	7H	8H	9H	aH	bH	cH	dH	eH	fH	gH	hH	iH	jH
OI	II	2I	3I	4I	5I	6I	7I	8I	9I	aI	bI	cI	dI	eI	fI	gI	hI	iI	jI
OJ	IJ	2J	3J	4J	5J	6J	7J	8J	9J	aJ	bJ	cJ	dJ	eJ	fJ	gJ	hJ	iJ	jJ
OK	IK	2K	3K	4K	5K	6K	7K	8K	9K	aK	bK	cK	dK	eK	fK	gK	hK	iK	jK
OL	IL	2L	3L	4L	5L	6L	7L	8L	9L	aL	bL	cL	dL	eL	fL	gL	hL	iL	jL
OM	IM	2M	3M	4M	5M	6M	7M	8M	9M	aM	bM	cM	dM	eM	fM	gM	hM	iM	jM
ON	IN	2N	3N	4N	5N	6N	7N	8N	9N	aN	bN	cN	dN	eN	fN	gN	hN	iN	jN
OO	IO	2O	3O	4O	5O	6O	7O	8O	9O	aO	bO	cO	dO	eO	fO	gO	hO	iO	jO
OP	IP	2P	3P	4P	5P	6P	7P	8P	9P	aP	bP	cP	dP	eP	fP	gP	hP	iP	jP
OQ	IQ	2Q	3Q	4Q	5Q	6Q	7Q	8Q	9Q	aQ	bQ	cQ	dQ	eQ	fQ	gQ	hQ	iQ	jQ
OR	IR	2R	3R	4R	5R	6R	7R	8R	9R	aR	bR	cR	dR	eR	fR	gR	hR	iR	jR
OS	IS	2S	3S	4S	5S	6S	7S	8S	9S	aS	bS	cS	dS	eS	fS	gS	hS	iS	jS
OT	IT	2T	3T	4T	5T	6T	7T	8T	9T	aT	bT	cT	dT	eT	fT	gT	hT	iT	jT
OU	IU	2U	3U	4U	5U	6U	7U	8U	9U	aU	bU	cU	dU	eU	fU	gU	hU	iU	jU
OV	IV	2V	3V	4V	5V	6V	7V	8V	9V	aV	bV	cV	dV	eV	fV	gV	hV	iV	jV
OW	IW	2W	3W	4W	5W	6W	7W	8W	9W	aW	bW	cW	dW	eW	fW	gW	hW	iW	jW
OX	IX	2X	3X	4X	5X	6X	7X	8X	9X	aX	bX	cX	dX	eX	fX	gX	hX	iX	jX
OY	IY	2Y	3Y	4Y	**5Y**	6Y	7Y	8Y	9Y	aY	bY	cY	dY	eY	fY	gY	hY	iY	jY
OZ	IZ	2Z	3Z	4Z	5Z	6Z	7Z	8Z	9Z	aZ	bZ	cZ	dZ	eZ	fZ	gZ	hZ	iZ	jZ

B

C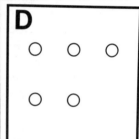

D

FIG. 1. (A) Original grid pattern. (B,C) The microscopic grid pattern. The area shown corresponds to the circled region in A. The grid is in focus in B with an arrow delineating out of focus murine LMTK⁻ cells. In C, the murine LMTK⁻ cells are in focus (arrow delineates two cells) with the image of the grid still apparent. (D) Coverslip with five 2 mm circles scribed in an asymmetric pattern.

shown in Fig. 1A. The letters PFL above the grid are used to facilitate coverslip orientation. The mask is the negative image of the reduced grid pattern. The final dimension of each grid square is 0.56 mm. The mask is then taped onto a glass slide (6 × 6 cm) for ease of handling during subsequent steps.

A photolithographic process, as described below, is then used to transfer the grid pattern onto coverslips. This process can be repeated as many times as the number of coverslips required, permitting mass production of engraved coverslips.

The following steps are performed under a yellow safe light.

Photoresist coating. (Coverslips are uniformly coated approximately to a 1 μm thickness with positive photoresist. This is done by putting a few drops of photoresist on the coverslip, followed by spinning on a photoresist spinner at 3000 rpm for one minute. If a photoresist spinner is unavailable, one may use a cytocentrifuge (Shandon Southern Instrument, Sewickley, PA) to accomplish the same end.

Preexposure baking and exposure. The coated coverslips are baked at 90° for 15 min to harden the photoresist. Then the coverslip is sandwiched between the mask and an opaque backing slide, with the coated side of the coverslip facing the mask. The mask, coverslip, and the black-backing slide are held together by two paper clamps, and placed at a distance of 15.5 cm in front of a Leitz slide projector, as shown in Fig. 2. The exposure time is 2 min for an incident light intensity of 3×10^{-2} W/cm². The purpose of the opaque backing slide is to prevent exposure to stray and reflected light.

Developing. The exposed coverslips are developed in diluted AZ-351 developer (1–5 dilution in deionized H_2O) for 40 sec and then rinsed in deionized H_2O for 1 min. The grid pattern now appears in the developed photoresist image.

Postdevelopment baking and etching of coverslips. After baking at 90° for 20 min, the coverslips are subsequently etched in 2.4% HF/H_2O for 10 min and rinsed in deionized H_2O for 1 min. The pattern has now been engraved onto the coverslips.

Photoresist stripping and cleaning. The procedure is completed when the remaining photoresist is washed away with acetone. This step requires about 1 min if an ultrasonic bath is used.

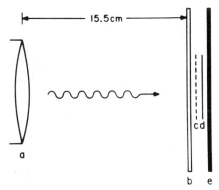

Fig. 2. Photoexposure apparatus. (a) Tungsten light source of a Leitz slide projector; (b) transparent glass slide; (c) photomask; (d) photoresist coated coverslip; (e) glass slide with black backing.

The grid on an etched coverslip is shown in Fig. 1B. The area shown corresponds to the circled region in Fig. 1A. To perform a microinjection using the Graessmann procedure, one full grid can be seen under ×200 magnification. When a Zeiss light microscope is used to scan the injected cells after autoradiography, two full grids can be observed at ×156 magnification. Figure 1C shows that when cells are brought to focus under the microscope, the grid pattern is still visible in the background. The small grid size and the identification number within each grid facilitates the localization of individual micromanipulated cells. Photoengraved coverslips are currently available from Bellco Glass, Inc., Vineland, NJ.

Alternative to the Coverslips with the Microgrid Pattern

For many microinjection studies, it is important to be able to relocate a group of microinjected cells that can then be compared to noninjected (control) cells on the same coverslip. We have found that glass coverslips of number 2 thickness that are scribed using a diamond scriber, in a pattern of 2 mm circles as shown in Fig. 1D, are useful for such studies and easy to prepare. The coverslips are sterilized by dipping in 70% EtOH and then air drying in a tissue culture room or hood flooded with ultraviolet light. Cells to be microinjected are plated onto the coverslip and cells that attach within the circles are microinjected. By scribing the circles in the pattern shown in Fig. 1D, it is easy to keep a record of what was injected into the cells within each of the five circles without having to worry about the orientation of the coverslip. Up to five different aliquots of fractionated mRNA can be tested per coverslip. All cells outside of the circles are the control noninjected cells.

Preparation of Poly(A)+ RNA

HeLa cell cultures containing 5×10^9 cells were pelleted and washed twice with saline. Subsequently, the cells were dispersed and resuspended in 50 ml of ice-cold buffer containing 30 mM Tris–HCl (pH 7.4), 30 mM KCl, 0.1 M NaCl, heparin (1 mg/ml), and 0.15% Nonidet P-40. When the cell membranes were lysed completely, nuclei and mitochondria were removed by centrifugation at 9500 rpm for 10 min with a Sorvall SS34 rotor. The supernatant was made 10 mM in EDTA and immediately mixed with 0.5 volume of buffered phenol. After the addition of sodium dodecyl sulfate to 0.5% and 0.5 volume (of the supernatant) of chloroform/isoamyl alcohol, 24:1 (v/v), the supernatant was thoroughly extracted. The aqueous phase was repeatedly extracted with an equal volume of phenol/chloroform/isoamyl alcohol, 24:24:1 (v/v), until no interface material was apparent, and then was extracted twice with chloroform to remove residual phenol. Total cytoplasmic RNA then was precipitated

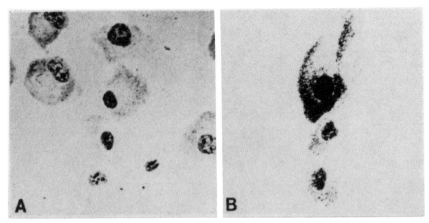

FIG. 3. Autoradiographic analysis for HPRT activity present in HT 1080 (HPRT⁻) cells before (A) and after (B) direct cytoplasmic injection with poly(A)⁺ RNA from HeLa cells. After injection, cells were fed with medium containing [³H]hypoxanthine (3.8 Ci/mmol, 5 μCi/ml) for 16 hr, followed by autoradiography.

with ethanol. Poly(A)⁺ RNA was purified twice by oligo(dT)-cellulose (Collaborative Research) column chromatography, precipitated with ethanol, and dissolved in 1 mM Tris (pH 7.5) for microinjection at 1 mg/ml or in water for gel electrophoresis. Alternatively, mRNA can be prepared by the guanidinium isothiocyanate method[17] followed by oligo(dT)-cellulose column chromatography.

Biological Detection of Human HPRT and TK mRNA by Microinjection and Complementation of Mutant Cells[13]

In order to assay for HPRT mRNA, human HT1080 HPRT⁻ cells were injected with cytoplasmic poly(A)⁺ RNA from HeLa cells at 1 mg/ml. After incubation with [³H]hypoxanthine for 16 hr, the cells were washed, fixed, and autoradiographed. The results are shown in Fig. 3. The injected cells incorporated label into DNA and RNA because of the *in vivo* translation of injected HPRT mRNA. Uninjected cells were always negative. The majority of injected cells incorporated radioactivity into RNA and hence, only the cytoplasms were heavily labeled.

To assay for TK mRNA, polyethylene glycol-fused murine LTK⁻ cells were used as recipients for the injection of HeLa cytoplasmic mRNA. Expression of TK mRNA was scored by the incorporation of [³H]thymidine into nuclear DNA, as illustrated in Fig. 4.

[17] T. Maniatis, E. F. Fritsch, and J. Sambrook, "Molecular Cloning: A Laboratory Manual," p. 196. Cold Spring Harbor Lab., Cold Spring Harbor, New York, 1982.

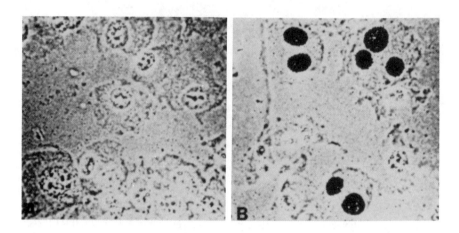

FIG. 4. Autoradiographic analysis for TK activity present in uninjected LTK⁻ cells (A) and cells injected with poly(A)⁺ RNA from HeLa cells. (B). The polykaryons were injected at 3 or 4 hr after fusion with polyethylene glycol. After injection the cells were incubated with [³H]thymidine (10 Ci/mmol, 5 μCi/ml) for 16 hr and subsequently autoradiographed.

These results indicated that microinjection and complementation of mutant phenotypes provide sensitive bioassays for HPRT and TK mRNA and that this system can be extended as a general method for mRNA activity assays.

Analysis of Human HPRT or TK mRNA by Methylmercuric Hydroxide Gel[13]

Methylmercuric hydroxide is known to fully denature RNAs, and biologically active RNA can be recovered after methylmercuric hydroxide gel fractionation—an essential factor for our RNA assay. To determine the molecular size of human HPRT and TK, we have electrophoresed HeLa cytoplasmic poly(A)⁺ RNA through 10 mM Ch₃HgOH/1% agarose gels. The procedures used[18,19] are as follows: HeLa cytoplasmic poly(A)⁺ RNA (100 μg) was electrophoresed through a 1% agarose/10 mM CH₃HgOH (Alfa-Ventron, Beveral, MA) slab gel (13 × 13 × 0.3 cm) in borate buffer (0.1 M boric acid/6 mM sodium borate/1 mM EDTA/10 mM sodium sulfate, pH 8.2) at room temperature. Electrophoresis was for 1 hr at 40 V, followed by 18 hr at 25 V. Unlabeled marker RNAs (200 μg of HeLa 28 S and 18 S rRNA and 200 μg of *E. coli* 28 S and 16 S rRNA) were

---

[18] J. M. Bailey and N. Davidson, *Anal. Biochem.* **70,** 75 (1976).
[19] P. B. Sehgal and A. Sagar, *Nature (London)* **288,** 95 (1980).

electrophoresed in adjacent lanes. Their mobilities were identified by UV absorption. The gel was then sliced manually into 2-mm fractions. Gel fractions were mixed with 0.2 ml of buffer (10 mM Tris–HCl, pH 7.4/0.1 NaCl/10 mM EDTA/0.2% sodium dodecyl sulfate/1 mM2-mercaptoethanol), melted at 95° for 5 min, and frozen at −70°. After thawing and removing the collapsed agarose by centrifugation, we extracted the supernatant once with phenol/CHCl$_3$/isoamyl alcohol, 24:24:1 (v/v), and once with chloroform/isoamyl alcohol, 24:1 (v/v). RNA was precipitated with alcohol in the presence of carrier yeast tRNA. Finally, RNA pellets were washed with 2 M LiCl and then with ethanol and dissolved in 30 μl of 1 mM Tris (pH 7.5).

The localization of HPRT or TK mRNA on gels was evidenced by the expression of their respective enzymatic activities after the injection of the fractionated RNAs into HPRT⁻ or TK⁻ cells. In typical experiments, 150–200 cells were injected for each fraction. The RNA fractions ranging from the sample origin to the approximately 40,000-Da position were checked for HPRT or TK activity. In the positive fractions, greater than 80% of the injected cells showed enzymatic activity. In two experiments, mRNA coding for HPRT and TK activity was obtained in a single peak near 16 S (Fig. 5). The position of the peak is equivalent to 1530 nucleotides in length. Molecular sizes were calibrated by plotting the square root of the marker RNA against logarithmic relative mobilities with a least-squares method.

Application

Our results have demonstrated that mRNA microinjection and complementation of mutant phenotypes provide a general method for detection of specific mRNAs. Moreover, one can study the relevant mRNA without detailed knowledge of the gene product, provided a phenotypic assay is available. The method can be applied to study any specific mRNA sequence for which a biological, immunological, autoradiographic, or selective assay is available.

Combining mRNA bioassays and gel fractionation, one can characterize the size of the relevant mRNA and the nature of a variety of mutations, and also enrich for the species of interest.[13-15] This method allows one to study the structure of eukaryotic mRNAs without being limited to systems that yield large amounts of RNA, or to a reliance on available cDNA probes. In the case of HPRT, our results provided information on transcriptional and translational processing.[13] In the TK system, our results helped to resolve the molecular size of the human cytosolic TK mRNA and provide information about the molecular structure of biologically active enzymes.[13]

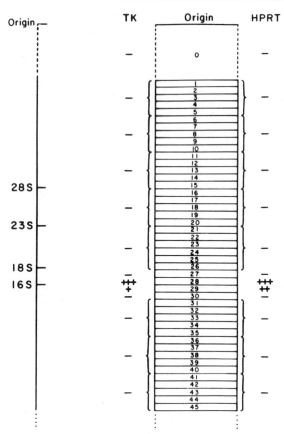

Fig. 5. Methylmercuric hydroxide/agarose gel fractionation of HeLa poly(A)$^+$ RNA for sizing the HPRT and TK mRNAs. HeLa cytoplasmic poly(A)$^+$ RNA was electrophoresed on a 10 mM CH$_3$HgOH/1% agarose slab gel. The gel was sliced into 45 fractions (2 mm per fraction); fractions 6–45 contained mRNA ranging from 4 MDa to 40,000 Da. The positions of HPRT and TK mRNA were recognized by injection of RNA eluted from gel slices into HPRT$^-$ or TK$^-$ cells and expression of enzyme activities. RNA from groups of five fractions were combined and injected as an initial screen, except for fraction 0. Fraction 0, an unsliced 18-mm gel near the origin, was injected as a single sample. Individual samples were then injected in the positive regions to further refine the position of the strongest signal. Signals were judged by the relative number of silver grain counts in autoradiographs; negative (−), weakly positive (+), moderately positive (++), and strongly positive (+++). Ribosomal RNAs from human (28 S, 18 S) and *E. coli* (23 S, 16 S) were used as size markers and run adjacent to the sample. Marker RNAs were identified by UV absorption bands.

In another study done in our laboratory,[14] the defect of E36ts24, a temperature-sensitive cell cycle mutant of Chinese hamster lung cell line E36, was transiently corrected by direct capillary injection of the enriched mRNA ($\simeq$940 nucleotides) that was obtained using methymercuric hydroxide gel fractionation. Microinjected cells that were held in early S phase at the restrictive temperature (40.3°) were found to resume cycling, as demonstrated by the incorporation of [³H]thymidine.[14] This enriched RNA fraction was eventually used in a cloning strategy described below to isolate the gene that could stably transform and correct the mutant cell phenotype.

A similar approach was used by Legerski et al.[15] to partially purify the mRNAs that complement cells from individuals with xeroderma pigmentosum (Groups A and G). In their study, they used poly(A)⁺ RNA fractionated on a 15–30% sucrose gradient. They found that RNA sedimenting at 11 S and 12 S restored DNA repair activity in Group A and G, respectively, following the direct capillary injection into quiescent XP-A and −G cell lines.[15] Transient complementation of the injected cells was determined by measuring unscheduled DNA synthesis.[15]

The previously discussed studies have provided important information regarding the size of the mRNAs of interest. More importantly, the partially purified mRNAs can then be used as a probe (in a differtial screening) or as starting material for molecular cloning of the DNA sequences corresponding to the normal counterpart of the mutant gene.

Recently, Fainsod et al.[20] have used the partially purified mRNA, obtained by the aforementioned method as a probe, to clone the gene that can correct the cell cycle mutation in E36ts24 cells. The cloning strategy that they developed[20] involved screening a λ human genomic library with ³²P-labeled cDNA synthesized from the enriched mRNA fraction. To identify the appropriate plaque, the resulting 845 hybridizing plaques were subdivided into 10 groups, and DNA of the pooled plaques were transfected into the mutant cell line E36ts24 using the calcium phosphate method. One of the groups contained the DNA that could transiently restore the mutant phenotype. Two more screens of further subdivided λ-human genomic DNA were performed before isolating the pure wild-type genomic DNA sequence that could complement the mutant cell line.[20]

Another cloning strategy would be to use the enriched RNA as the starting templates for constructing a partial cDNA library within an expression vector. The resulting positive clone could then be identified by DNA transfection or glass capillary microinjection of the subdivided clones into the mutant cells, and tested for the transient complementation

[20] A. Fainsod, G. Diamond, M. Marcus, and F. H. Ruddle, Mol. Cell Biol. 7, 775 (1987).

of the mutant phenotype. Clones containing only a fragment of the DNA of interest could be identified by hybridizing to and selecting for the specific mRNA that can restore the mutant phenotype following microinjection.

In conclusion, the glass capillary injection of fractionated RNA into mutant cell lines has already proven to be a powerful approach to study the nature of mutations. It is also an important step toward the isolation and characterization of genes capable of complementing mutant phenotypes.

Acknowledgments

We wish to thank Dr. A. Graessmann for instruction in the microinjection procedure, Dr. T. P. Ma for advice concerning photolithography, Drs. H. B. Lieberman, R. J. Legerski, and A. Fainsod for many thoughtful discussions, M. Sinischalchi for typing the manuscript, and S. Pafka for photography. This work was supported by N.I.H. Aging Grant 5R01 AG D1940-02.

[27] Use of the CAT Reporter Gene for Optimization of Gene Transfer into Eukaryotic Cells

By C. MICHAEL FORDIS and BRUCE H. HOWARD

Many questions awaited the development of techniques for the transfer of either cloned or genomic DNA sequences into eukaryotic cells. With their appearance came novel studies of unanticipated diversity. However, certain investigations have been hampered in part by a continued inability to introduce genes into relevant target cells. This in turn has led to a proliferation in the number and types of techniques available for gene transfer. Unfortunately, the application of any particular transfer technique can often be difficult. The gene of interest may be expressed at a low level and assays may be tedious if not time consuming. Without easily detectable expression, it becomes virtually impossible to distinguish low level expression from inefficient gene transfer. In frustration one may prematurely abandon one technique only to confront similar problems with use of another.

The application of easily detectable reporter genes to the optimization and monitoring of gene transfer can avert unnecessary hardship. Here we will consider the use of the gene that confers bacterial resistance to the antibiotic chloramphenicol in a reporter gene system pioneered by Gorman *et al.*[1,2] The gene product is an enzyme, chloramphenicol acetyltransferase (CAT), that is peculiar to prokaryotes. As such, no significant expression is demonstrable in eukaryotic cells that may serve as targets. Rapid and simple assays for CAT activity have been available since the investigations of bacterial resistance by Shaw.[3] Use of an enzymatic assay for the gene product can facilitate detection of even low levels of product. Furthermore, experiments in *Saccharomyces cerevisiae* suggested that CAT enzyme could be expressed in and was apparently nontoxic to eukaryotes.[4] Gorman and colleagues constructed expression vectors in which the CAT gene was fused to the SV40 early promoter (pSV2CAT)[1] or to the 3' long terminal repeat of Rous sarcoma virus (pRSVCAT).[2] In the process a plasmid similar to those above but lacking a known eukaryotic promoter (pSVOCAT)[1] was also constructed. Both pRSVCAT and pSV2CAT expressed at high level in a variety of cell types and were useful in optimizing gene transfer techniques for the selection of stably transformed cells.[5]

Below we have included a sample protocol and illustrations relevant to use of the CAT system in optimization of gene transfer procedures. However, first we would like to present the protocol used to prepare CAT reporter gene DNA for transfer. The method differs from many in that a high concentration of ethidium bromide is used to complex excess protein prior to purification of the plasmid over two cesium chloride–ethidium bromide gradients. Albeit a more lengthy purification procedure, DNA so prepared can be transferred with high efficiency.[5]

Preparation of Plasmid DNA

Materials

Tris	International Biotechnologies Inc., (New Haven, Conn.) and Bethesda Research Laboratories (Gaithersburg, MD)
Sucrose	
Ethylenediaminetetraacetic acid (EDTA)	
Triton X-100	
Cesium chloride	

[1] C. M. Gorman, L. F. Moffat, and B. H. Howard, *Mol. Cell. Biol.* **2,** 1044 (1982).
[2] C. M. Gorman, G. T. Merlino, M. C. Willingham, I. Pastan, and B. H. Howard, *Proc. Natl. Acad. Sci. U.S.A.* **79,** 6777 (1982).
[3] W. V. Shaw, *J. Biol. Chem.* **242,** 687 (1967).
[4] J. D. Cohen, T. R. Eccleshall, R. B. Needleman, H. Federoff, B. A. Buchferer, and J. Marmur, *Proc. Natl. Acad. Sci. U.S.A.* **77,** 1078 (1980).
[5] C. M. Gorman, R. Padmanabhan, and B. H. Howard, *Science* **221,** 551 (1983).

Lysozyme, grade I, Sigma Chemical Co. (St. Louis, MO)
Ethidium bromide, Sigma Chemical Co. (St. Louis, MO)
Tryptone, Difco Laboratories (Detroit, MI)
Yeast extract, Difco Laboratories (Detroit, MI)
Glycerol
Potassium phosphate (dibasic K_2HPO_4)
Potassium phosphate (monobasic KH_2PO_4)
Sodium acetate

Solutions

1. Superbroth (pH 7.2), for 10 liters
   Autoclave solutions A and B separately and combine.
   A. Tryptone, 120 g
      Yeast extract, 240 g
      glycerol, 50 ml
      H_2O, 9000 ml
   B. K_2HPO_4, 125 g
      KH_2PO_4, 38 g
      H_2O, 1000 ml
2. TE, 10 mM Tris hydrochloride (pH 8.0)/1 mM EDTA
3. TS, 50 mM Tris hydrochloride (pH 8.0)/25% (w/v) sucrose
4. TSL, 10 mg/ml lysozyme freshly dissolved in TS
5. 0.25 M EDTA (pH 8.0)
6. 1.0 M Tris hydrochloride (pH 8.0)
7. Triton solution, prepared from stock solutions
      10% Triton X-100, 3 ml
      0.25 M EDTA (pH 8.0), 25 ml
      1 M Tris–HCl (pH 8.0), 5 ml
      H_2O, 67 ml
8. Ethidium bromide solution, 10 mg/ml in TE
9. Cesium chloride–ethidium bromide solutions for diluting plasmid
   DNA: Final volumes of the solutions below approximate 45 to
   46 ml. One can scale the volumes up or down as required to fill
   the requisite number of Quick-Seal centrifugation tubes.
   A. Mix-1: Prepare a solution in a ratio of 33.3 g cesium chlo-
      ride/30.8 ml TE/5.19 ml ethidium bromide solu-
      tion
   B. Mix-2: Prepare a solution in a ratio of 33.3 g cesium chlo-
      ride/34.9 ml TE/1.13 ml ethidium bromide solu-
      tion

Procedure

1. Inoculate 800 ml of Superbroth with an overnight culture of bacteria (5 ml) harboring the CAT plasmid of interest. Place the culture in a shaking incubator (37°) for 36 hr. Amplification of plasmids with chloramphenicol is not required for high yields of pRSVCAT and pSV2CAT, as the ColE1 copy control sequences that code for ROM (RNA I inhibition modulator)[6] in pBR322 have been deleted. Indeed, bacteria harboring pRSVCAT do express the CAT gene product and are resistant to chloramphenicol—a fact that proves useful in the preparation of positive controls for the CAT assay (see below).

2. Collect the bacteria from each 800 ml culture in one 500 or 1000 ml Sorvall bottle. Two centrifugations in a GS3 Sorvall rotor (5000 rpm for 10–15 min at 4° are required to consolidate the bacteria from an 800 ml culture into one 500 ml bottle.

3. Discard the supernatants and place the bottles on ice.

4. Liquify the bacterial pellet using a rubber policeman and then thoroughly resuspend the bacteria in 100 ml of ice-cold TE.

5. Collect the bacteria by centrifugation at 5000 rpm for 10 min at 4°. Discard the supernatant.

6. The bacterial pellet may be stored at $-20°$.

7. Liquify the pellet using a rubber policeman and then resuspend the bacteria in 9 ml ice-cold TS. From step 7 to step 13 work rapidly and maintain the bacteria at 0° where possible. Failure to do so may result in contamination of the preparation with degraded RNA.

8. To the resuspended bacteria add 0.9 ml TSL and mix using a gentle swirling motion of the bottle. With occasional mixing, incubate the bacteria on ice for 5 min.

9. Add 3.7 ml of ice-cold 0.25 M EDTA (pH 8.0) and incubate on ice for 5 min as before.

10. Add 14.5 ml of ice-cold Triton solution, mix, and incubate on ice for 10 min. Check for bacterial lysis by gently swirling the mixture every few minutes. Bacterial lysis will be accompanied by a marked increase in viscosity of the solution. If at the end of 10 min the solution has not become viscous, incubate for an additional 10 to 15 min on ice. If lysis has not occurred with the additional incubation, brief warming to room temperature may be useful. Failure to achieve adequate lysis commonly results from use of inactive lysozyme or from failure to maintain the pH required for enzymatic activity (pH 8.0).

[6] J. Tomizawa and T. Som, *Cell,* **38,** 871 (1984).

11. After bacterial lysis has occurred, remove the cellular debris by spinning the mixture for 30 min at 25,000 rpm at 4° in either a Beckman SW27 rotor (in polyallomer tubes) or in a Beckman 50.2 Ti rotor (in polycarbonate tubes).

12. Decant the supernatant into a tared 50 ml polypropylene tube and adjust the weight of the solution to 30.17 g with TE.

13. Add 28.14 g cesium chloride to the plasmid solution and mix gently by inversion at room temperature until dissolved. The cesium chloride solution may be stored at 4° for extended periods if desired.

14. Add 4.5 ml of ethidium bromide solution and mix gently. Plasmid in the presence of ethidium bromide should be handled in subdued lighting to minimize nicking of the DNA.

15. Divide the DNA cesium chloride–ethidium bromide solution equally between two 40-ml Beckman Quick Seal tubes. (A 20-ml syringe attached to a 19-gauge needle may be used as a funnel.) Fill the tubes to the top with Mix-1 cesium chloride–ethidium bromide solution. Each tube will take approximately 19 to 20 ml of Mix-1 to fill.

16. Balance and seal the tubes. Centrifuge at 34,000 rpm for 72 hr at 15° in a Beckman 60 Ti rotor or equivalent. Use of a vertical rotor for the first centrifugation step should be avoided.

17. After centrifugation visualize the plasmid band with long wave UV light and withdraw the plasmid in a volume of 4 ml or less. Use of a 16-gauge needle is recommended.

18. Combine the two identical plasmid fractions into one 40-ml Quick Seal tube and fill the tube with Mix-2 cesium chloride–ethidium bromide solution. Balance and seal the tubes.

19. Centrifuge at 34,000 rpm for 72 hr at 15° in a Beckman 60 Ti rotor or for 48 hr at 15° in a Beckman VTi50 rotor.

20. After centrifugation remove the plasmid band in a volume of 3 ml or less.

21. In a darkened room remove the ethidium bromide by extracting the plasmid solution four times with equal volumes of 2-propanol equilibrated with cesium chloride-saturated TE. After each extraction, discard the upper organic layer.

22. Remove the cesium chloride by dialysis for 24 hr at 4° against three, 4-liter volumes of TE.

23. Estimate the DNA concentration by spectrophotometry.

24. Precipitate the DNA with two volumes of ice-cold, 100% ethanol and one-tenth volume of 3 M sodium acetate, pH 5.2, at −20° overnight. Centrifuge at 10,000 g for 15 min at 4°, discard the supernatant, wash with 70% ice-cold ethanol, centrifuge, discard the supernatant, dry, and resuspend in TE at 1 $\mu g/\mu l$.

25. Confirm the plasmid identity by restriction endonuclease digestion and agarose electrophoresis. Plasmids not subjected to endonuclease digestion should migrate as (greater than 90%) supercoiled circles. Significant fractions of plasmid migrating as linear or open circular forms may indicate a random nicking of the DNA that can be associated with reduced expression.

Gene Transfer

Any of a number of transfer methods can be used to introduce DNA sequences into target cells. For purposes of illustration in the examples below, cells are transfected using the technique of calcium phosphate–DNA coprecipitation.[7] Forty-eight to 72 hr after transfection, cells are harvested for analysis of CAT enzymatic activity. A calcium phosphate precipitation protocol is included below as well as a laboratory protocol for measurement of CAT enzymatic activity with examples to demonstrate the utility of the CAT reporter gene in the optimization of a gene transfer technique.

Introduction of Plasmid DNA into Cultured Cells via the Technique of Calcium Phosphate–DNA Coprecipitation

Materials

> Calcium chloride, dihydrate, AnalaR from BDH Chemicals Ltd., Poole, England
> Glycerol
> HEPES (N-2-hydroxyethylpiperazine-N'-2-ethanesulfonic acid), Pharmacia P-L Biochemicals, Piscataway, NJ
> Sodium chloride
> Sodium phosphate dibasic heptahydrate ($Na_2HPO_4 \cdot 7H_2O$)

Solutions

> 1. 2 M $CaCl_2$
> 2. 2 × HBS, 280 mM NaCl/50 mM HEPES/1.5 mM $Na_2HPO_4 \cdot 7H_2O$
>   Prepare as follows: NaCl, 8.18 g
> $\qquad\qquad\qquad\qquad$ HEPES, 5.95 g
> $\qquad\qquad\qquad\qquad$ $Na_2HPO_4 \cdot 7H_2O$, 0.20 g
> $\qquad\qquad\qquad\qquad$ Distilled water to 500 ml

---

[7] F. L. Graham, and A. J. van der Eb, *Virology* **52**, 456 (1973).

Adjust the pH to 7.10 with 1 N NaOH and filter sterilize. The solution can be stored at 4° for up to 4 weeks.
3. 50% glycerol/water (weight/volume)
4. 15% glycerol/HBS
Prepare as follows: 50% glycerol, 30 ml
$\qquad\qquad\qquad$ 2 × HBS, 50 ml
$\qquad\qquad\qquad$ H_2O, 20 ml
Filter sterilize and store at 4°.

Procedure

1. By way of introduction it should be mentioned that for some cell lines the efficiency of gene transfer with calcium phosphate appears to be passage dependent. In particular, the transfer efficiency may drop after 10 to 15 passages. For such cell lines it is useful to discard the cells after 10 to 15 passages and thaw a vial of cells frozen at a lower passage. Additionally, the transfer efficiency of certain cells is maintained at a higher level if the cells are not permitted to grow to confluence.

2. On day 1 split the cells and replate. One can start by plating cells at 10^4 cells/cm². However, for each cell line the optimal plating density for gene transfer should be determined. When plating monolayer cells prior to transfection, use the most gentle treatment with trypsin and/or EDTA necessary for separating the cells.

3. On day 2 feed the cells two or three hours prior to the addition of the calcium phosphate precipitate.

4. Bring to room temperature the solutions required for precipitation. Take a 50-ml aliquot of the 2 × HBS, recheck the pH, readjust if necessary, and filter sterilize. (Sterilization units are prefiltered with water prior to filtration of the solution of interest).

5. The precipitate is added in one-tenth the volume of the medium overlying the cells. For a flask containing 5 ml of medium covering a surface area of 25 cm², prepare the precipitate in 0.5 ml as described below. Use 17 × 100 mm polypropylene tubes (2059 Tube from Falcon) or equivalent.
$\qquad$ Tube 1: 2 M CaCl$_2$, 31 μl
$\qquad\qquad$ DNA (5–10 μg) + H_2O, 219 μl
$\qquad\qquad$ Total, 250 μl
$\qquad$ Tube 2: 2 × HBS, 250 μl
For a circular 100 mm (diameter) dish containing 10 ml of medium, double each of the volumes above.

6. Use a clamp to attach Tube 2 to a ringstand and insert a sterile 1-ml pipet into the tube. The tip of the pipet should reside opposite the 3 ml

mark on Tube 2. The pipet is connected to a tank of nitrogen via tubing containing an inline filter. The nitrogen flow is directed to one side of the tube and adjusted to create a gentle stream that indents (no more than halfway) the surface of the 2 × HBS producing a trough at one side of the liquid. Bubbling and splashing the liquid is to be avoided. The nitrogen stream rapidly but gently mixes the components. One can test the mixing action by using colored dyes. Add dropwise the mixture contained in Tube 1 to the top of the trough formed by the nitrogen stream. Upon combining the contents of Tube 1 with that of Tube 2, allow the precipitate to mix under the nitrogen stream for 10 sec and then remove. Add the precipitate to the cells within 10 min and mix gently. (Delays in adding the precipitate to the flask can result in the formation of large aggregates of calcium phosphate which decrease the efficiency of transfer).

7. Return the flasks to a 37° incubator. (It is important to adjust the pCO_2 of the incubator so that medium placed within equilibrates to pH 7.3 to 7.4.)

8. Allow the precipitate to remain on the cells for 3.5 to 4 hr. At this point microscopic examination of the cells should reveal cells peppered with a fine precipitate. For certain cell lines optimal transfers can be achieved by increasing the incubation with precipitate up to 24 hr. However, not all cell lines will tolerate prolonged exposure to calcium phosphate.

9. Remove the calcium phosphate containing medium and wash the cells with medium without serum.

10. At this juncture the cells may be shocked with glycerol. Transfer efficiency for many lines increases with exposure to glycerol. To glycerol shock the cells remove the serum-free wash medium and add 1.5 ml 15% glycerol/HBS per 25 cm^2 flask and return the flask to the 37° incubator. After 30 sec to 3 min (the period of incubation with the glycerol solution must be optimized for each cell line) remove the glycerol solution and wash the cells once with serum-free medium. Prolonged exposure to glycerol solution will produce cell distortion which should be avoided.

11. Feed with complete medium containing serum and after 48 to 72 hr harvest the cells for analysis of CAT activity.

Analysis of CAT Enzymatic Activity

The CAT enzyme acetylates chloramphenicol at the number 1, 3, or 1 and 3 carbons in the side chain. Enzymatic activity can be estimated by monitoring the appearance of acetylated forms. A radiochemical assay first developed by Shaw[3] and later modified by Cohen[4] is based upon the acetylation of [^{14}C]chloramphenicol. The precursor and products are separated by thin-layer chromatography and quantitated by scintillation counting. The following protocol is based upon the adaptation of that assay for the analysis of expression of eucaryotic vectors containing the CAT gene.[1] In particular, the high level expression of pRSVCAT and pSV2CAT in a variety of cell types has proven quite useful in the optimization of gene transfer techniques.[5]

Materials

Acetyl-coenzyme A, lithium (Pharmacia P-L Biochemicals, Piscataway, NJ) (store at −20°)

[14C]Chloramphenicol, 40–60 mCi/mmol (New England Nuclear, Boston, MA) (store at −20°)

Ethyl acetate, Sequanal Grade (Pierce Chemical Co., Inc., Rockford, IL)

Baker Flex Silica Gel 1B TLC plates (J.T. Baker Chemical Co., Phillipsburg, NJ)

Silica Gel 60 plates (EM Science, division of EM Industries, Inc. Cherry Hill, NJ)

Eppendorf Micro Test Tubes (Brinkman Instruments Co., Westbury, N.Y.) Tubes from other manufacturers have performed variably through the ethyl acetate extraction step and may be associated with poor resolution of labeled products by thin-layer chromatography.

Chloroform

Methanol

Tris

Ethylenediaminetetraacetic acid (EDTA)

Sodium chloride

Solutions

PBS (phosphate-buffered saline, without Mg^{2+} or Ca^{2+})

TEN, 0.04 M Tris–HCl, ph 7.4/1 mM EDTA/0.15 M NaCl

0.25 M Tris–HCl, pH 7.8

1 M Tris–HCl, pH 7.8

4 mM acetyl-coenzyme A: prepare in 0.25 M Tris, pH 7.8. A fresh solution is generally prepared for each assay, although storage of the solution at −20° for short periods (1 to 2 weeks) is acceptable.

Procedure

1. Wash the cells three times with PBS. Cells that grow in suspension or loosely attached to plastic can be washed in a centrifuge tube, resuspended

in 1 ml PBS, and transferred to an Eppendorf tube. Some cell lines that grow attached to plastic may be loosened from the substratum by incubation with TEN at room temperature for 5 min. Use 1 ml/100 mm tissue culture dish and transfer the cells to an Eppendorf tube. Other cells will require incubation with TEN and use of a tissue culture scraper. Transfer the cells to an Eppendorf tube.

2. As soon as possible, centrifuge the cells, discard the supernatant, and resuspend the cells in 0.100 ml of 0.25 M Tris–HCl, pH 7.8. The cell suspension can be stored at $-20°$.

3. Disrupt the cells by sonication or by subjecting them to several cycles of freezing and thawing.

4. Pellet the cellular debris by centrifugation at 9000 to 12,000 g at 4° for 5–10 min.

5. Recover the supernatant and place in a new Eppendorf tube. The supernatants may be assayed immediately or stored at $-20°$.

6. A positive control for enzyme activity can be prepared from the HB101 strain of *E. coli* containing pRSVCAT. Because the RSV long terminal repeat contains a bacterial promoter, these bacteria do express the CAT gene product and are resistant to chloramphenicol. To prepare control enzyme grow a 5 ml overnight culture of bacteria. Collect the bacteria in a single Eppendorf tube by repeated centrifugations. Resuspend the cells in 0.100 ml of 0.25 M Tris–HCl, pH 7.8. Lyse the cells with several cycles of freezing and thawing followed by sonication. Pellet the bacterial debris and recover the supernatant for assay. Generally 1 μl of extract contains sufficient activity to convert 80–90% of the [^{14}C]chloramphenicol substrate to acetylated product in less than 60 min. Store the bacterial extract at $-20°$.

7. To perform the radiochemical assay combine the following:

> 1 to 50 μl extract
> <u>124 to 75 μl</u> 0.25 M Tris–HCl, pH 7.8
> 125 μl total volume of extract and buffer
> 5 μl [^{14}C]chloramphenicol
> <u>20 μl 4 mM acetyl-CoA</u>
> 150 μl total reaction volume

The assay can be modified to use less than 5 μl of labeled substrate if so desired.

8. In general the protein concentration of each of the cellular extracts is determined. Assaying identical quantities of protein permits direct comparison of enzymatic activities. Should widely different protein concentrations be used, verify the linearity of the assay with respect to protein concentration. Bear in mind that some cells do contain inhibitory activity (see below).

9. Incubate the reaction mixture at 37° for 60 min. Incubations have been performed for as short as 10 min to as long as 18 hr. Assays with longer incubation times may require increased concentrations of acetyl-CoA to maintain linearity with time.

10. Terminate the reaction by extracting the incubation mix with 1 ml ethyl acetate. Use a glass pipet to dispense the ethyl acetate. Vortex vigorously for 30 sec. Separate the phases by centrifugation for 30 sec.

11. Collect the top organic layer avoiding debris that may be trapped at the interface.

12. Evaporate the ethyl acetate using a Speed Vac Concentrator (Savant) or equivalent. If vacuum devices are unavailable, evaporate the ethyl acetate overnight in a fume hood.

13. Prepare the TLC plates by marking the tops and subjecting them to ascending chromatography in freshly prepared 95 : 5 chloroform : methanol solvent. Dry the plates in a fume hood.

14. Resuspend the dried samples thoroughly in 30 μl of ethyl acetate and spot onto a TLC plate prepared as above. Directing a stream of air through a glass pipet onto the plate will expedite the application of samples by capillary pipet.

15. Separate the acetylated products from unreacted substrate by ascending chromatography in freshly prepared 95 : 5 chloroform : methanol. Chromatography will be completed in 2.5 hr or less depending on the plates used. Poor resolution or compression of the acetylated products on the TLC plate can result from performing chromatography in tanks inadequately equilibrated with solvent. To saturate the atmosphere in the tank place chromatography paper around the inside of the tank. Wet the paper with solvent and place the paper's bottom edge into the solvent. The paper will act as a wick to supply the requisite solvent for saturating the atmosphere within the tank.

16. Dry the plates in a fume hood (5 min) and expose to film.

17. To quantitate the reaction products, use the exposed film as a template to locate the labeled reactant and products. Cut the spots from the plastic-backed plates and place into separate scintillation vials for counting. Calculate the percent acetylated chloramphenicol.

$$\frac{\% \text{ acetylated}}{\text{chloramphenicol}} = 100 \times \frac{\text{counts in acetylated products}}{\text{total counts in acetylated and unreacted chloramphenicol}}$$

By way of illustration we describe here a few experiments in which pRSVCAT proved useful in improving the efficiency of gene transfer. We were interested in studying the expression of certain globin promoters in

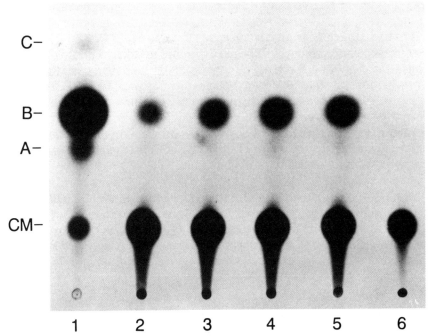

FIG. 1. CAT activity in K562 human erythroleukemia cells transfected with pRSVCAT. Extracts from cells transfected with pRSVCAT and subjected to glycerol shock for 30, 60, 120, or 180 sec were assayed in lanes 2 to 5, respectively. Lane 1 contains an assay on an extract from bacteria expressing CAT activity. Lane 6 contains unreacted [^{14}C]chloramphenicol. The positions after chromatography for unreacted [^{14}C]chloramphenicol (CM), the 1- and 3-monoacetylated products (A and B, respectively), and the 1,3-diacetylated product (C) are shown.

a human erythroleukemia cell line.[8] These particular cells grow in suspension and our attempts to introduce exogenous globin genes were unsuccessful according to our analyses of RNA. We selected a subline of cells that attaches loosely to plastic for use as a target. However, without a signal it became extremely difficult to optimize the transfer conditions for calcium phosphate–DNA coprecipitation.

We attempted to transfer pRSVCAT into the erythroleukemia cells. The results of the first experiment (Fig. 1) demonstrated the successful transfer of the reporter gene and permitted us to choose an optimal period for glycerol shock. In a second experiment we tested a 5-fold range of DNA

[8] C. M. Fordis, N. Nelson, A. Dean, A. N. Schechter, N. P. Anagnou, A. W. Nienhuis, M. McCormick, R. Padmanabhan, and B. H. Howard, *Prog. Clin. Biol. Res.* **191**, 281 (1985).

concentration to be used for gene transfer (Fig. 2) and improved expression. Subsequent experiments demonstrated that the glycerol shock step could be omitted if the precipitate were left on the cells for 18 hr. (Not all cell lines tolerate similarly prolonged exposures to calcium phosphate.) We tested the effect of cell density on expression. Unlike many cell lines that grow in monolayer, we discovered that expression increased if, before gene transfer, the attached erythroleukemia cells were allowed to reach greater than 80% confluence. High cell densities could only be achieved by using Primaria flasks. A typical CAT assay after optimization of gene transfer is shown (Fig. 3). The advantage of using the RSV promoter is quite apparent, since activity of globin promoters fused to CAT genes is less than 5% that of RSV.

The activity of globin promoters here further illustrates the utility of a reporter gene that codes for an enzyme. Levels of gene product that may be difficult to detect directly can be reflected in product formation over time. In Fig. 4 the activity of one globin promoter can be distinguished from the absence of activity from other globin promoters or from pSVOCAT by

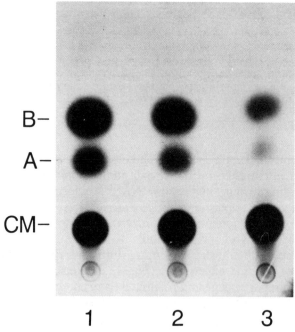

FIG. 2. CAT activity in K562 human erythroleukemia cells transfected with pRSVCAT. Cells transfected with 50, 20, or 10 μg pRSVCAT plasmid DNA and assayed for CAT activity are shown in lanes 1 to 3, respectively.

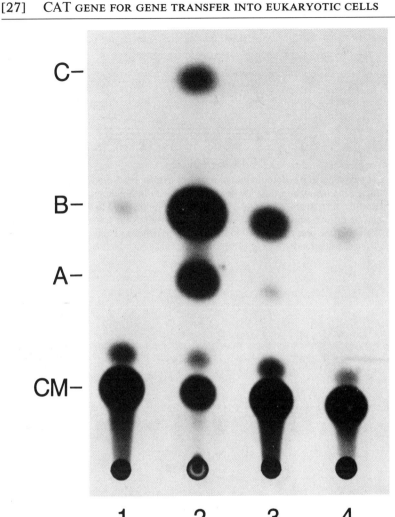

FIG. 3. CAT activity in K562 human erythroleukemia cells. CAT assays are shown for extracts from cells transfected with pSVOCAT (lane 1), pRSVCAT (lane 2), a plasmid containing the human embryonic ($\epsilon-$) globin gene promoter fused to the CAT gene (lane 3), and a plasmid containing the human adult ($\beta-$) globin gene promoter fused to the CAT gene (lane 4). Reproduced in part from Fordis *et al.*[8]

increasing the incubation time of the enzyme assay. However, increasing assay incubation time is no panacea for low level expression. Certain cell lines[8,9] and tissues[10] contain an endogenous activity that interferes with the

[9] M. Mercola, J. Goverman, C. Mirell, and K. Calame, *Science* **227,** 266 (1985).
[10] P. A. Overbeek, A. B. Chepelinsky, J. S. Khillan, J. Piatigorsky, and H. Westphal, *Proc. Natl. Acad. Sci. U.S.A.* **82,** 7815 (1985).

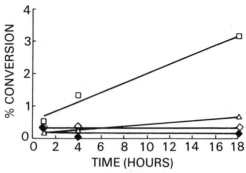

FIG. 4. Enzymatic conversion of chloramphenicol to acetylated product as a function of time. By increasing incubation time low level CAT activity in human erythroleukemia cells transfected with a human embryonic ($\epsilon-$) globin gene promoter fused to the CAT gene (open squares) can be distinguished from the absence of activity in cells transfected with human adult ($\beta-$) globin gene promoters fused to CAT (open triangles and open diamonds) or with pSVOCAT (closed diamonds).

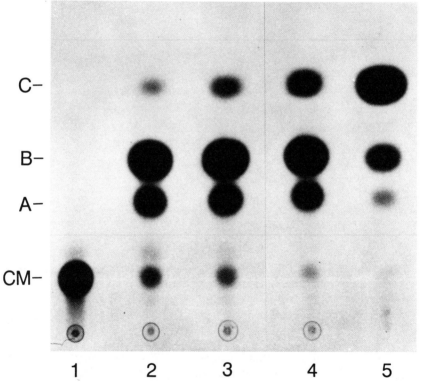

FIG. 5. CAT activity in 293 cells, an adenovirus-transformed human embryonal kidney cell. CAT assays are shown for extracts from cells transfected with no DNA (mock transfected) (lane 1), pSVOCAT (lane 2), a plasmid containing a human ϵ-globin promoter fused to the CAT gene (lane 3), and pRSVCAT (lane 4). Lane 5 contains an assay on an extract of HB101 containing pRSVCAT.

CAT assay. The inhibitory activity can be inactivated by heating the extract (diluted for assay) to 65° for 5 min.[8-10]

A final caveat must be included. If the use of the CAT gene is extended to the study of various promoters, a promoterless but otherwise identical plasmid should be included as a negative control (e.g., pSVOCAT for the pSV2 series of CAT vectors). Significant expression from the control plasmid should cast suspicion on the relevance of expression from other like plasmids. By way of example we include a CAT assay from 293 cells transfected with several plasmids including pSVOCAT (Fig. 5). High level expression was observed with plasmids whether or not a known eukaryotic promoter had been inserted in front of the CAT gene. In such cases expression could represent upstream transcriptional starts within pBR322 vector sequence and should be defined by RNA mapping analyses.

In conclusion the high level expression from pRSVCAT or pSV2CAT coupled with the sensitivity of the CAT enzymatic assay should continue to be of use in the optimization and development of techniques for the transfer of genes into cells.

Acknowledgments

We wish to thank Dr. Alan N. Schechter for his support in the studies on erythroleukemia cells. We are grateful to Dr. Ann Dean for the initial work in selecting a subline of K562 cells that attach to plastic. We appreciate the assistance of Ms. Joyce Sharrar in preparation of the manuscript.

[28] Cosmid Vector Systems for Genomic DNA Cloning

By MARY McCORMICK, MAX E. GOTTESMAN, GEORGE A. GAITANARIS, and BRUCE H. HOWARD

We have developed two cosmid vector systems for the construction of genomic DNA libraries. The first, λSV2cos, is a bacteriophage λ-based vector that can replicate autonomously, or be propagated as an integrated copy in the *Escherichia coli* chromosome. The second, pSV13, contains the pBR322 origin of replication and is maintained as a high copy number plasmid. Both cosmids carry the *E. coli* gene for guanine phosphoribosyltransferase *(gpt)* under the control of the SV40 early region promoter/enhancer.[1] The *gpt* gene serves as a dominant selectable marker when transferred into animal cells. Cosmid vectors bear the λ *cos* site and may be packaged into infectious particles by the λ *in vitro* packaging system after the insertion of DNA fragments of the appropriate size (40–45 kb).[2] For selection in *E. coli,* both λSV2*cos* and pSV13 carry bacterial genes encoding ampicillin (Ap^r) and chloramphenicol resistance (Cm^r). In addition, both vectors contain several unique and convenient restriction sites for cloning.

The λSV2 System

Vector λSV2*cos* was constructed by inserting the 1.7 kb *Bgl*II *cos* containing fragment from pHC79[2] into λSV2 at the *Bam*HI site,[3,4] as shown in Fig. 1. The vector contains the λ attachment site (attP) and the λ rightward promotor/operator (p$_R$o$_R$) which controls *cro* and the λ replication genes *O* and *P*. After transformation of a sensitive host, λSV2*cos* is maintained as an autonomously replicating plasmid. When introduced into an *E. coli* strain containing a specialized λ prophage (N6106 or N6377), λSV2*cos* is propagated as a single copy inserted in the *E. coli* chromosome. Induction of the prophage results in the excision and autonomous replication of the cosmid.

The integration and excision of λSV2*cos* is shown diagrammatically in Fig. 2. The lysogenic strains N6106 and N6377 contain a modified λcI857 prophage. The H1 deletion inactivates *cro* and deletes the lambda replication and late functions. A section of the λ p$_L$ operon including the *kil* gene, which is lethal to the host, has been deleted. The prophage is bordered by the bacterial attachment site *(attB)* and carries *int* and *xis* under the control of the λ leftward promoter/operator (p$_L$o$_L$). Expression of *int* and *xis* occurs when the lysogens are shifted from 32 to 42° denaturing the cI857 repressor. Restoring the culture to 32° after 15 min prevents further expression of these genes and allows the decay of the unstable Xis protein. Int protein, required for phage integration, persists. Introduction of λSV2*cos* into these transiently induced cells, either by transformation or infection, results in the efficient integration of the cosmid into the bacterial chromosome at *attB*. λSV2*cos* remains stably integrated until the culture is shifted

[1] R. C. Mulligan and P. Berg, *Science* **209**, 1422 (1980).
[2] B. Hohn and J. Collins, *Gene* **11**, 291 (1980).
[3] B. H. Howard and M. E. Gottesman, *in* "Eukaryotic Viral Vectors" (Y. Gluzman, ed.), p. 211. Cold Spring Harbor Lab., Cold Spring Harbor, New York, 1982.
[4] G. A. Gaitanaris, M. McCormick, B. H. Howard, and M. E. Gottesmann, *Gene* **46**, 1 (1986).

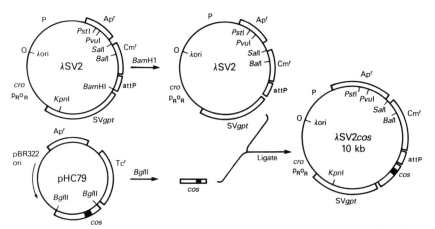

FIG. 1. Construction and map of λSV2*cos*. λSV2*cos* was constructed from λSV2[3] and pHC79[2] as shown schematically in this figure and described in the text. λSV2*cos* has unique *Pst*I and *Pvu*I sites in Ap[r], a *Bal*I site in Cm[r], and an intergenic *Sal*I site.

to 42°. After temperature shift, *int, xis, O,* and *P* are expressed. The cosmid then excises and replicates autonomously.

Test libraries were constructed with λSV2*cos* carrying genomic DNA ligated into the *Pst*I site. The clones were introduced into strains of N6106 and N6377 by infecting with *in vitro* packaged ligation mixtures, and selecting for Cm[r]. (The construction of genomic libraries will be described

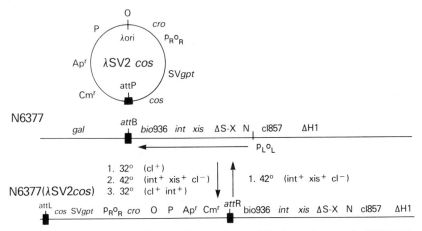

FIG. 2. Integration and excision of λSV2*cos* in specialized prophage strain N6377. The process is identical in strain N6106. As described in the text, a 42° heat pulse, followed by a return to 32° results in a *cI857*[+], Int[+] background suitable for integration of λSV2*cos*. Heating to 42° creates a *cI857*[−], *int*[+], *xis*[+] environment, resulting in excision and replication of λSV2*cos*. AttP is the λ attachment site, and attB is the bacterial attachment site.

by Fleischmann *et al.*[5] in another chapter in this volume.) We have obtained efficiencies in N6377 of up to 1.7×10^5 bacterial colonies per microgram of cosmid vector in the ligation mixture.

The λSV2*cos* system is recommended for the generation and maintainance of large clones (40 kb) as single copy integrates. Such integrates may be more stable than plasmid or phage clones and can be useful for site-specific mutagenesis. In addition, we have used the λSV2 system to reconstitute DNA segments from overlapping fragments.[4]

An *in Vivo* Packaging System for Cosmids

The λSV2*cos* clones can be recovered from the *E. coli* genome by inducing excision and isolating the replicating plasmid DNA (as described above), or by packaging the excised plasmid into λ phage particles *in vivo.* Jacobs *et al.*[6] have also developed a system for *in vivo* packaging. In our system, we lysogenize a λSV2*cos* lysogen at 32° with $\lambda imm^{434} rev - \lambda p_L o_L$.[7] The relevant properties of the phage are (1) it integrates at a site in the bacterial genome which is distinct from the site of λSV2*cos* integration at *attB;* (2) the λ packaging functions are fused to a $\lambda p_L o_L$ insert and are controlled by the *cI857* repressor; (3) the 434 repressor remains active after temperature shift, thus $\lambda imm^{434}rev - \lambda p_L o_L$ does not replicate after thermal induction.

To obtain *in vivo* packaged λSV2*cos* clones, the lysogens are grown at 32° and then shifted to 42°. The denaturation of the *cI857* repressor results in the synthesis of Int, Xis, and the O and P gene products as well as the expression of the λ packaging and lysis functions from the $\lambda imm^{434}rev - \lambda p_L o_L$ prophage. The *in vivo* packaged cosmids are then recovered as λ transducing phage from the cell lysate. A 100 ml lysate usually yields sufficient infectious cosmid particles to permit visualization of a band in a CsCl equilibrium density gradient. For three random λSV2*cos*::HeLa clones, we were able to obtain titers between 2.4×10^6 and 4×10^8 Cm^r transducing particles/ml of induced culture titered on N6106 and N6377 (data not shown). Gottesman *et al.*[8] have used this technique to isolate and propagate clones carrying the *E. coli* genes, *rscB* and *rscC,* that regulate capsular polysaccharide synthesis. The detailed protocols for λSV2*cos* integration, excision and *in vivo* packaging are described below.

---

[5] R. Fleischmann, M. McCormick, and B. H. Howard, this volume [29].

[6] W. R. Jacobs, J. F. Barrett, J. E. Clark-Curtiss, and R. Curtiss III, *Infect. Immun.* **52,** 101 (1986).

[7] M. M. Gottesman, M. E. Gottesman, S. Gottesman, and M. Gellert, *J. Mol. Biol.* **88,** 471 (1974).

[8] S. Gottesman, P. Trisler, and A. Torres-Cabassa, *J. Bacteriol.* **162,** 1111 (1985).

Bacterial Strains, Phage, and Media

N6106 and N6377 are C600 derivatives. Their complete genotypes are as follows: N6106 [F⁻, *thi*-1, *thr*-1, *leu*-B6, *lac*Y1, *ton*A21, *sup*E44, $r_k^- m_k^-$, (*chl*D – *blu*)^{de18}, (λcI857, *bio*936, delS-X delH1)]; N6377 [F⁻ *thi*-1, *leu*-B6, *lac*Y1, *ton*A21, *sup*E44, $r_k^- m_k^+$, *pro*, (λcI857, *bio*936, delS-X, delH1)]. The S-X deletion in the λ prophage removes sequences from the *Sal*I site in *redB* to the *Xho*I site in *c*III. The H1 deletion extends from within λ *cro* to beyond the bacterial *bio* and *uvr*B genes.

The HB101 genotype is [F⁻, *hsd*S20 (r_B^-, m_B^-), *rec*A13, *ara*-14, *pro*A2, *lac*Y1, *gal*K2, *rps*L20 (Sm^r), *xyl*-5, *mtl*-1, *sup*E44, λ⁻]. λ*imm*^{434} *rev*–λ$o_L p_L$ carries the *Sau*3A fragment encoding λ N and λ $p_L o_L$ (λ coordinates 34992 to 35711) inserted upstream of gene *Q* into the *Bam*HI site at position 41732 on the λ map.

LB broth:
Tryptone, 10 g
Yeast extract, 5 g
NaCl, 10 g
Bring to 1 liter with H₂O.
Adjust to pH 7.5 with 1 *N* NaOH
Autoclave to sterilize
Superbroth (pH 7.2):
A. Tryptone, 12 g
   Yeast extract, 24 g
   Glycerol, 5 ml
   H₂O, 900 ml
B. K₂HPO₄, 12.5 g
   KH₂PO₄, 3.8 g
   H₂O, 100 ml
   Autoclave solutions A and B separately and combine.
LB top agar:
LB broth, 1 liter
Agar, 8 g
Autoclave to melt agar and sterilize
Low melt top agarose:
Bring 1 liter of LB broth to a boil while sitrring on a hot plate
Slowly add 8 g of low melt Sea Prep agarose purchased from FMC
   Corp.
Autoclave to sterilize
LB agar plates with antibiotic:
LB broth, 1 liter
Agar, 15 g
Autoclave to melt agar and sterilize

Add chloramphenicol to 30 μg/ml or ampicillin to 50 μg/ml final concentration when LB agar is approximately 45°. Pour 40 ml portions into plastic petri dishes.

Integration of λSV2cos Clones

Inoculate LB broth with a 1/100 volume of a fresh 32° overnight LB culture of N6106 or N6377. Grow the culture in a 32° shaking water bath to midlogarithmic phase ($OD_{650} = 0.3-0.4$).

Shift the culture to a 42° shaking water bath for 15 min.

Shift the culture back to the 32° shaking water bath for 15 min, and then transfer to ice.

Infect these induced cells with a λSV2cos library that has been packaged into λ phage particles (see Fleischmann et al.,[5] this volume), or transform[9] the culture with a λSV2cos DNA construct.

After transformation or infection, incubate cells at 32° for 1 hr without shaking to permit expression of Cm[r] or Ap[r].

Mix cells with LB top agar and spread on antibiotic containing plates. (For libraries, mix cells in low melt top agarose, and allow the agarose to solidify for 1–2 hr at 4° before shifting the plates to 32°. Do not put plates at 42° to melt agarose when collecting cells; otherwise, follow the protocol of Fleischmann et al.[5] for recovery and storage of libraries.) Select on LB agar plates with 30 μg/ml chloramphenicol or 50 μg/ml ampicillin. Incubate plates at 32° for 24–36 hr.

Excision and Replication of a λSV2cos Clone

Inoculate LB broth with a 1/100 volume of a 32° LB broth overnight culture of the λSV2cos lysogen. Grow this culture at 32° to midlogarithmic phase ($OD_{650} = 0.3-0.4$).

Shift the culture to 42° for 2–3 hr.

Isolate plasmid DNA.[9] A crude DNA preparation from 5 ml of induced bacterial culture provides sufficient plasmid for one restriction analysis.

In Vivo Packaging of a λSV2cos Clone

Superinfect a λSV2cos lysogen grown overnight at 32° in LB broth plus 0.2% maltose with λimm[434] rev–λp$_L$o$_L$ (moi = 5).

Incubate at 32° for 15 min to allow phage to adsorb.

[9] T. Maniatis, E. F. Fritsch, J. Sambrook, "Molecular Cloning: A Laboratory Manual," pp. 86–94 and 366–369 for plasmid preparations, pp. 250–253 for bacterial transformation. Cold Spring Harbor Lab., Cold Spring Harbor, New York, 1982.

Select for lysogens by plating on EMBO plates seeded with coimmune phage (10^9 PFU/plate) λimm^{434} b2cI and λimm^{434} h80cI. Lysogens appear as pink staining colonies after 24–36 hr at 32°[10]

Inoculate Superbroth plus 0.003% biotin, and 10 mM MgSO$_4$ with a 1/100 volume of a 32° overnight LB culture of the λSV2cos, $\lambda imm^{434}rev-\lambda$p$_Lo_L$ lysogen. Grow the culture at 32° to midlogarithmic phase (OD$_{650}$ = 0.3–0.4).

Shift the culture to 42° for 2–3 hr.

Add a 1/50 volume of CHCl$_3$ to aid cell lysis.

Remove cellular debris by centrifuging 5000 rpm, 10 min, 4° in an SS34 Sorvall rotor. Store the phage-containing supernatant at 4°.

pSV13 Cosmid System

As shown in Fig. 3, pSV13 was constructed from λSV2cos and pSV2gptM1.[1,11] λSV2cos and pSV2gptM1 were each digested with PstI and KpnI. The λSV2cos PstI/KpnI fragment (5.5 kb) containing the cos site and Cm^r was ligated to the pSV2gptM1 PstI/KpnI fragment (2.2 kb) containing the pBR322 origin of replication. This ligation reconstituted SVgpt and Ap^r, and resulted in the 7.7 kb cosmid vector, pSV13. Similar cosmid vectors, with dominant selectable markers, have been constructed by Lau and Kan.[12] Eukaryotic libraries have been constructed with pSV13 in E. coli strain HB101 as described by Fleischmann et al.[5] The genomic DNA inserts are 40–45 kb in size. We routinely obtain $1-4 \times 10^5$ bacterial colonies per microgram of pSV13 in the packaged ligation mixture.

Since pSV13 library DNA is propagated in E. coli as a plasmid by virtue of the pBR322 origin of replication in pSV13, standard plasmid preparations may be used to isolate library DNA. We usually recover approximately 1 mg of total library plasmid DNA per liter of bacterial cells grown in Superbroth containing 30 μg/ml chloramphenicol (see Fleischmann et al.[5]). This high copy number is due at least in part to an inactivation of the rom gene (RNAI inhibition modulator) of the ColE1 replicon,[13] a consequence of the construction of pSV2gpt[1,14]

Library plasmid DNA may be used to transfect animal cells in culture. In such experiments the goal is frequently to isolate an unidentified gene from the library by monitoring its phenotypic expression. There are several

[10] M. E. Gottesman and M. B. Yarmolinsky, J. Mol. Biol. **31**, 487 1968).
[11] C. Gorman and B. H. Howard, unpublished results (1983).
[12] Y. F. Lau and Y. W. Kan, Proc. Natl. Acad. Sci. U.S.A. **80**, 5225 (1983).
[13] J. Tomizawa and T. Som, Cell **38**, 871 1984).
[14] B. H. Howard and P. Berg, unpublished results (1980).

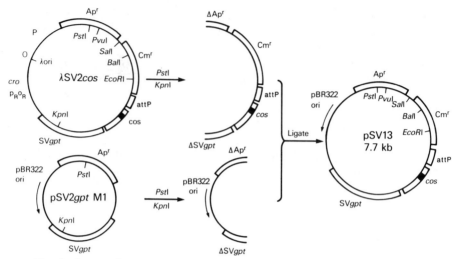

FIG. 3. Construction and map of pSV13. pSV13 was constructed from λSV2*cos* and pSV2*gpt*M1 as shown schematically in this figure and described in the text. pSV2*gpt*M1 contains a monkey repeated sequence between SV*gpt* and Ap[r11] but the portion of this plasmid that is used in the construction of pSV13 is identical to pSV2*gpt*.[1] pSV13 has unique *Pst*I and *Pvu*I sites in Ap[r], *Bal*I and *Eco*RI sites in Cm[r], and an intergenic *Sal*I site. The *Eco*RI site in λSV2*cos* is not unique in that vector. The *attP* site present in pSV13 is a result of its derivation from λSV2; *attP* is not relevant to the function of pSV13 as a cosmid vector. pSV14 (Ap[r], Cm[s]) and pSV15 (Ap[s], Cm[r]) are derivatives of pSV13 containing a *Sal*I to *Eco*RI, and a *Pvu*I to *Sal*I deletion, respectively.

alternative approaches which may include the option of using the SV*gpt* gene to select for transfected cells. First, sib selection may be employed using any assay or selection for the gene of interest (see McCormick,[15] this volume). Second, cosmids with inserts can be rescued from the transfected genome by *in vitro* packaging into λ phage particles (see Lau,[16] this volume). Third, the gene of interest can be rescued after serial transfections by virtue of its linkage to cosmid sequences. To give an example illustrating this last approach, isolation of the mammalian gene thymidine kinase (TK), for which there is a selection, would involve the following steps. A pSV13 cosmid library presumed to contain the TK gene cloned into the *Pst*I site (rendering the vector Ap[s] and Cm[r]) is transfected into thymidine kinase deficient (TK[-]) cells, and recipients selected by a standard HAT selection procedure. Since up to 3×10^3 kb of DNA can be taken up in a

---

[15] M. McCormick, this volume [33].
[16] Y.-F. Lau, this volume [31].

typical animal cell transfection, many library sequences unrelated to TK will also be present in these primary transfectants. In order to eliminate these extraneous library DNA sequences, genomic DNA from the primary transfectants can be used to transfer the TK gene, again into TK⁻ cells. (In this secondary gene transfer, approximately 0.1% of the transfected DNA will be library sequences, as opposed to 100% library DNA in the primary transfer.) A library can then be made from the genome of these secondary transfectants using a derivative of pSV13 with a deletion in Cm^r, pSV14 (Ap^r, Cm^s). Any TK gene clone, closely linked to Cm^r from the original library constructed in pSV13, could then be isolated by selecting for Cm^r in *E. coli*.

pSV14 is identical to pSV13 (Fig. 3) except for a deletion between the *Sal*I and *Eco*RI sites. We have also constructed an Ap^s, Cm^r version of the cosmid, pSV15. pSV15 contains a deletion between the *Pvu*I and *Sal*I sites of pSV13 (Fig. 3). This vector can be used in the type of gene isolation procedure described above for the TK gene, when the construction of the original library leaves Ap^r intact.

Acknowledgments

We thank Rudy Pozzatti for critically reading the manuscript.

[29] Preparation of a Genomic Cosmid Library

By ROBERT FLEISCHMANN, MARY MCCORMICK, and BRUCE H. HOWARD

The cloning of large segments of DNA in a single recombinant molecule is desirable when investigating certain aspects of gene organization. These include the investigation of large genes, the relationship of regulatory and structural sequences, and the organization of linked genes. A major problem in cloning large recombinant molecules is that the introduction of large plasmids (> 15 kb) into host bacteria is very inefficient. The development of *in vitro* packaging systems and bacteriophage λ vectors has helped to circumvent this problem.[1,2] These vectors allow the packaging of recombinant molecules into λ phage particles which can be used to infect a variety of host bacterial strains efficiently. The insert DNA is then cloned through plaque lysate formation. However due to the size of λ

---

[1] M. Sternberg, D. Tieneier, and L. Enquist, *Gene* **1,** 255 (1977).

[2] B. Hohn, this series, Vol. 68, p. 299.

vectors a size limit in the range of 20–25 kb is still imposed on the insert DNA.

Cosmid vectors are plasmids containing the λ *cos* sequences (cohesive end sites) required for *in vitro* packaging of recombinant molecules into the λ capsid.[3] These vectors are relatively small (4.0–7.5 kb) and replicate as plasmids in the host bacteria. Insert genomic DNA on the order of 35–50 kb can be ligated to form long concatameric molecules along with vector DNA, thus mimicking the natural substrate of the λ packaging system. These concatamers can be packaged into λ phage heads and used to infect bacteria with high efficiency where the cosmids are established as large (50 kb) replicating plasmids.

The development and use of a number of cosmid vector systems have been described. These include pJB8,[4,5] pHC79,[6] as well as more sophisticated cosmid vectors such as pCV103, 107, 108,[7] and pGC*cos*3*neo*[8] which carry mammalian selectable markers as well. The preparation of lambda *in vitro* packaging systems has also been described.[1,2,5] For our purposes we will assume the use of commercially available packaging systems developed from this pioneering work.

The cosmid vector pSV13 (Fig. 1) (see chapter by M. McCormick *et al.* [28], this volume) has been used routinely in our laboratory in order to construct a variety of cosmid libraries. Briefly this 7.7 kb vector contains the genes for chloramphenicol resistance (Cm[r]) and ampicillin resistance (Ap[r]), the pBR322 origin of replication, the λ *cos* sequence for *in vitro* packaging, and a number of unique restriction sites. In addition, the vector contains the *E. coli* gene for xanthine–guanine phosphoribosyltransferase under control of simian virus 40 early region transcription signals (SV *gpt*). The presence of this gene allows for the selection of the plasmid in mammalian cells in medium containing MXHAT following transfection of the library or a cosmid clone into a mammalian cell line.[9]

This chapter will focus on the methodology (outlined in Fig. 2) used in our laboratory for the construction of cosmid libraries using the vector pSV13. The preparation of a successful library depends upon the prepara-

[3] J. Collins and B. Hohn, *Proc. Natl. Acad. Sci. U.S.A.* **75,** 4242 (1978).

[4] D. Ish-Horowicz and J. F. Burke, *Nucleic Acids Res.* **9,** 2989 (1981).

[5] F. G. Grosveld, H. M. Dahl, E. deBoer, and R. A. Flavell, *Gene* **13,** 227 (1981)

[6] B. Hohn and J. Collins, *Gene* **11,** 291 (1980).

[7] Y. F. Lau and Y. W. Kan, *Proc. Natl. Acad. Sci. U.S.A.* **80,** 5225 (1983); see also Y.-F. Lau, this volume [31].

[8] G. Urlaub, A. M. Carothers, and L. A. Chasin, *Proc. Natl. Acad. Sci. U.S.A.* **82,** 1189 (1985).

[9] B. H. Howard and M. McCormick, *in* "Molecular Cell Genetics" (M. M. Gottesman, ed.), p. 211. Wiley, New York, 1985.

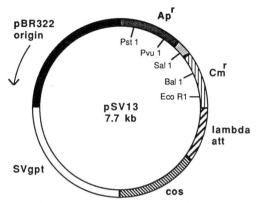

FIG. 1. Schematic diagram of the pSV 13 vector used to construct cosmid libraries. The diagram depicts the pBR322 origin of replication, the ampicillin (Ap^r) and chloramphenicol (Cm^r) resistance genes, the λ attachment site (lambda att), the λ cohesive end sequences *(cos)*, and the *E. coli* gene for xanthine–guanine phosphoribosyltransferase (SV*gpt*). In addition a number of unique restriction sites are indicated.

tion of high quality high-molecular-weight genomic DNA and appropriate 40 kb partial fragments from this DNA, appropriate preparation of the cosmid vector, and successful ligation, packaging, and infection of the recombinant molecules. The techniques presented in this chapter will describe the cloning into the unique *Pst*I restriction site of the cosmid vector; however, other appropriate unique restrictions sites may be used.

Preparation of High-Molecular-Weight Genomic DNA

When preparing high-molecular-weight genomic DNA (> 120 kb), care must be given to avoid shearing the DNA. A minimum amount of pipetting is done, extractions should be done with gentle rocking motions, and DNA bands are drawn with 16-gauge needles.

Prior to extraction of high-molecular-weight DNA, we routinely grow the cell lines of interest in four T-75 cm² tissue culture flasks. The total number of cells for a single DNA preparation should be on the order of (but not exceed) 3.0×10^7 cells. After the cells have been grown to the appropriate density the medium is removed and the cells are thoroughly washed 2× with Ca^{2+}/Mg^{2+}-free phosphate-buffered saline, pH 7.3. The cells are then washed briefly 1× with TE buffer [10.0 mM Tris–Cl, pH 7.9, 1.0 mM ethylenediaminetetraacetic acid (EDTA), pH 8.2]. Do not allow the TE buffer to sit on the cells for extended periods of time.

The cells are dissolved directly in the flasks by the addition of 5.0 ml of

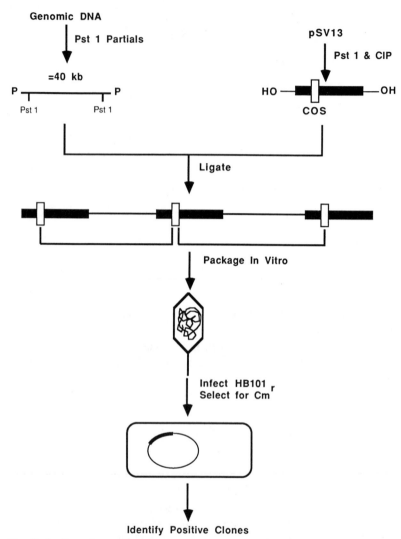

FIG. 2. Outline of procedures used to construct a cosmid library using high-molecular-weight genomic DNA and the vector pSV13.

proteinase K/EDTA/Sarkosyl solution [5.0 mg proteinase K (Boehringer-Mannheim), 23.8 ml of 0.5 M EDTA, pH 8.2, and 1.25 ml of 10.0% Sarkosyl] per flask. Incubate the cells for approximately 5 min at 37° in this solution. The digest forms a viscous solution which can then be scraped from the bottom of the flasks and poured into 50 ml conical tubes at 10.0 ml/tube. This cell extract is digested for 6 hr at 47°. At the end of 3 hr an additional 1/10 volume of the proteinase K/EDTA/Sarkosyl solution is added containing 5.0 mg/ml proteinase K.

At the completion of digestion the cell lysate is extracted 2× with 0.1 M Tris, pH 8.0 equilibrated phenol at a ratio of 1:1 (10 ml/tube). A gentle rocking motion should be used for the extractions in order to minimize shearing of the DNA. Due to the high EDTA concentration used the phenol separates as the top layer. Following the phenol extraction the samples are dialyzed. Ten milliliters of aqueous solution containing the DNA is removed from each tube and added to each dialysis bag, leaving room for a 3-fold increase in volume during dialysis. The aqueous sample is then dialyzed for 24 hr against 2 changes of 4.0 liters of TEN (5.0 mM Tris–Cl, pH 7.5, 2.5 mM EDTA, 20.0 mM NaCl) at 4°.

Following dialysis 32.48 g of DNA solution from each dialysis bag is transferred to a 50 ml tube, and 32.2 g of CsCl$_2$ is added. The final weight of the DNA–CsCl$_2$ solution should be 64.68 g. After the CsCl$_2$ has dissolved, ethidium bromide (0.84 ml/tube of 10 mg/ml ethidium bromide in TE buffer, pH 7.5) is added.

Once the ethidium bromide has been added to the DNA solution the remaining steps should be completed in a darkened room. The DNA solution is transferred to 40.0 ml Beckman Quick Seal Tubes or the equivalent, appropriately balanced, and centrifuged in a 60 Ti or equivalent fixed angle rotor at 34,000 rpm for 72 hr. Following this centrifugation a single fluffy DNA band can be visualized in each tube using long wave UV light. The band from each tube is carefully drawn with a 16-gauge needle and syringe and pooled from every two tubes. The combined bands are diluted to 40.0 ml with CsCl/EtBr solution (see Table I) in Beckman Quick Seal Tubes and again centrifuged at 34,000 rpm for 72 hr.

Following this final centrifugation step the DNA band is again drawn with a 16-gauge needle and syringe. The DNA solution is extracted at least 4× with an equal volume of isobutanol in order to extract the ethidium bromide from the DNA solution. Finally the DNA solution is dialyzed overnight at 4° against 4 liters of TEN. The final DNA solution should be

TABLE I
CsCl/EtBr SOLUTION FOR DILUTION OF DNA

Final volume (ml)		45.00	75.00	110.00	145.00
CsCl	(g)	33.30	55.50	81.40	107.30
TE buffer	(ml)	34.90	58.20	85.30	112.50
EtBr	(ml)	1.13	1.88	2.76	3.64

somewhat viscous. An approximate concentration can be determined by reading an OD_{260} (OD_{260} of $1 = 50 \mu g/ml$). Our yield is usually in the range of 200 $\mu g/ml$.

Preparation of 40 kb Partial Restriction Fragments

Having successfully prepared high-molecular-weight genomic DNA from the cell lines of interest, it is necessary to prepare 40 kb partial restriction fragments for ligation and packaging. The 40 kb genomic DNA fragments are prepared by partial cutting with the same restriction enzyme which will be used to cut the vector. After ligation the 40 kb genomic DNA fragments will be flanked by the cosmid vector, which will space the *cos* recognition sequences on the vector the appropriate 40–50 kb distance apart for efficient packaging into the λ phage heads.

Partial restriction digests can be conveniently controlled by varying either the restriction enzyme concentration or the length of time of the digestion. In this section we will describe the analytical procedures for determining an appropriate time of digestion at a fixed enzyme concentration that will yield preparative amounts of appropriately sized genomic DNA partials.

For the restriction enzyme *Pst*I approximately 0.5 U/μg of genomic DNA is used for each digestion. A typical analytical digest with *Pst*I is performed as follows. A single restriction digest reaction is assembled (Table II, "Analytical 40 kb Digest") containing 0.7 μg of genomic DNA in a final volume of 42 μl. The reaction is preincubated at 37° for 5 min. *Pst*I (0.35 U, diluted in 1× reaction mixture if necessary) is then added to the

TABLE II
REACTIONS FOR PREPARATION OF RECOMBINANT MOLECULES

Reaction	Enzyme	DNA	10× buffer[a]	H_2O[b]	0.1 M ATP	Final volume
Analyt. 40 kb digest	0.5 U/μg	0.7 μg	4.2 μl	—	None	42.0 μl
Prep. 40 kb digest	0.5 U/μg	20 μg	120.0 μl	—	None	1.2 ml
Linearize vector	1.0 U/μg	10 μg	5.0 μl	—	None	50.0 μl
Calf intestinal phosphatase	0.75 U/μg	10 μg	2.0 μl	—	None	20.0 μl
Ligation	0.1–1.0 U	1.5 μg[c] 3.0 μg[d]	2.0 μl	—	2.0 μl	20.0 μl

[a] See Maniatis *et al.*[11] for composition of these 10× buffers.
[b] Add appropriate volume of H_2O to bring to indicated final volume.
[c] Genomic DNA.
[d] Vector DNA.

reaction and the incubation is continued at 37°. At time points of 0, 2.5, 5.0, 10.0, 15.0, 30.0, 45.0, and 60.0 min 6-μl aliquots are removed and immediately brought to a final concentration of 20.0 mM EDTA in order to stop the reaction.

The samples are analyzed on a 0.2% agarose gel in Tris–borate buffer (5× stock = Tris base, 54 g/liters, boric acid, 27.5 g/liter, 0.05 M EDTA, pH 8.2, 20 ml/liter). Ethidium bromide is present in the agarose gel (5 μl/ 100 ml, 10 mg/ml stock solution) and in the Tris–borate buffer (50.0 μl/liter, 10 mg/ml stock solution). It is helpful to use FMC Gel Bond as a support material for the gel. Uncut λ and HindIII cut λ DNAs are used as markers. The agarose gel is run at 5.0 mA overnight and the ethidium bromide stained gel is photographed using UV light. The time of digestion is determined which gives the brightest staining band of DNA between the uncut λ and the 23.0 kb fragment of the HindIII cut λ DNA. Half this time is then used in the preparative digestion procedure (Figure 3).[10]

Approximately 20 μg of DNA can be digested per reaction in the preparative digest (Table II, "Prep. 40 kb Digest"). Care should be taken to duplicate all the conditions of the analytical digest, now scaled up for this preparative procedure using the single time point as previously determined. Following the digestion the reaction is immediately stopped with 20.0 mM EDTA, pH 8.2, the DNA is extracted 2× with an equal volume of phenol : chloroform (1 : 1) and 1× with an equal volume of chloroform. One-tenth volume of 3.0 M sodium acetate, pH 5.2, is added to the DNA solution along with 2.5 volumes of ice cold 100% ethanol. The DNA is precipitated in a dry ice–ethanol bath for approximately 15 min. The precipitated DNA is pelleted by centrifugation at 10,000 rpm for 30 min, washed with 80% ethanol, centrifuged again at 10,000 rpm for 30 min, and the ethanol is aspirated. The remaining DNA pellet is allowed to air dry. The DNA is then suspended in TE buffer, pH 7.5, at 10 μg/20 μl and allowed to completely dissolve overnight.

Linearization and Calf Intestinal Phosphatase Treatment of the Vector DNA

The cosmid vector pSV13 is prepared by linearization with the restriction enzyme PstI followed by cleavage of the 5′-phosphate groups with calf intestinal phosphatase (CIP, Boehringer-Mannheim). Treatment with CIP will minimize self ligation of the vector DNA during the ligation reaction.

Ten micrograms of the vector is incubated with 10 U of the enzyme PstI in a standard restriction digest reaction of 50 μl at 37° for 1 hr (Table

[10] B. Seed, R. Parker, and N. Davidson, *Gene* **19**, 201 (1982).

Time of Digestion

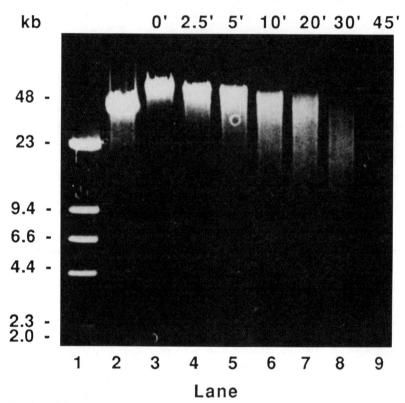

FIG. 3. Ethidium bromide-stained 0.2% agarose gel of a partial *Pst*I digestion of high-molecular-weight CHO genomic DNA. Approximately 0.5 U of enzyme per μg of DNA were used. Under these conditions between 20 and 30 min of digestion gives the brightest staining area between 46 and 23 kb. Therefore, a digestion time of approximately 12.5 min would be chosen for the preparative procedure. Lane 1: λ DNA digested with *Hin*dIII. Lane 2: uncut λ DNA. Lane 3: uncut high-molecular-weight genomic DNA. Lane 4: 2.5 min digestion. Lane 5: 5.0 min digestion. Lane 6: 10.0 min digestion. Lane 7: 20.0 min digestion. Lane 8: 30 min digestion. Lane 9: 45 min digestion.

II, "Linear Vector"). Following complete linearization of the vector the DNA is extracted 2× with an equal volume of phenol : chloroform (1 : 1) followed by 1× extraction with an equal volume of chloroform. One-tenth volume of 3.0 *M* sodium acetate, pH 5.2, is added to the DNA solution along with 2.5 volumes of ice cold 100% ethanol. The vector is precipitated in a dry ice–ethanol bath for approximately 15 min and pelleted in a microfuge at 4°. The ethanol is aspirated and the pelleted DNA is dried under vacuum.

The dried DNA is dissolved and treated with calf intestinal phosphatase (Table II, "Calf intestinal Phosphatase"). The 10× reaction buffer[11] contains 0.5 M Tris–Cl, pH 9.0, 10.0 mM MgCl$_2$, 1.0 mM ZnCl$_2$, and 10 mM spermidine. The time course of incubation is as follows: (1) 15 min at 37°, (2) 15 min at 56°. An additional 1.0 μl of CIP is added and incubations 1 and 2 are repeated. The reaction is terminated by the addition of 5.0 μl of 10× TNE (100 mM Tris–Cl, pH 8.0, 1.0 M NaCl, 10 mM EDTA, pH 8.2), 10.0 μl of 50.0 mM trinitriloacetic acid, and 15.0 μl of H$_2$O. The reaction is then heated at 68° for 15 min. The DNA is extracted 2× with phenol chloroform, 1× with chloroform, and ethanol pecipitated as previously described. The pelleted DNA is washed with 80% ethanol, pelleted again in a microfuge at 4°, and dried under vacuum. The vector DNA is resuspended in TE buffer, pH 7.5, at 10 μg/20 μl.

Ligation

The recombinant library is formed by ligating the pSV13 vector with the 40 kb genomic DNA restriction fragments so that every genomic fragment is flanked at both ends by a vector molecule. This results in the *cos* recognition sequences of the vector being spaced approximately 40–50 kb apart and allows for appropriate packaging in λ phage head particles. We use approximately a 10-fold molar excess of vector to genomic DNA in the ligation reaction to insure these results (Table II, "Ligation"). The total concentration of DNA in the ligation should be approximately 200–250 μg/ml. The 10× ligation buffer contains 0.66 M Tris–Cl, pH 7.5, 50.0 mM MgCl$_2$, 50.0 mM dithiothreitol, and 10.0 mM ATP. The ligation is carried out at approximately 12° for 12–14 hr.

Packaging, Infection, and Plating of the Genomic Cosmid Library

A number of λ *in vitro* packaging kits are available which produce satisfactory results. We routinely use the Amersham kits (product #N.334L) which provide two separate color-coded Eppendorf vials; (1) phage heads (PH) and (2) phage tails (PT). These kits must be stored at -70 to -140° as specified by the manufacturer and have a limited shelf life of at least 6 months. Before attempting to construct a complete library,

---

[11] T. Maniatis, E. F. Fritsch, and J. Sambrook, "Molecular Cloning: A Laboratory Manual." Cold Spring Harbor Lab., Cold Spring Harbor, New York, 1982.

an analytical library is prepared in order to determine the optimal packaging and infection conditions.

An analytical packaging assay is begun by thawing a PH vial and a PT vial on ice and briefly centrifuging in a microfuge. An aliquot of the ligation reaction (4.0 μl, approximately 1 μg of the DNA) is pipetted into the PH vial and 15 μl of the contents of the PT vial is immediately pipetted into the PH vial. The contents are very gently mixed and briefly centrifuged in a microfuge. The packaging reaction is incubated for 2 hr at room temperature. At the end of the incubation period 500 μl of a standard phage storage buffer (SM buffer = NaCl, 5.8 g/liter MgSO$_4$, 2.0 g/liter, 1 M Tris–Cl, pH 7.5, 50 ml/liter, 2.0% gelatin, 5 ml/liter, sterilize by autoclaving) and 10.0 μl of chloroform are added and gently mixed with the *in vitro* packaging reaction. Dilutions of the packaging reaction in SM buffer are prepared, 1 : 10 (20.0 μl packaged phage : 180.0 μl SM buffer) and 1 : 100 (10.0 μl of the 1 : 10 dilution : 180.0 μl SM buffer).

Prior to *in vitro* packaging of the recombinant molecules an overnight 5.0 ml culture of the HB101 host bacteria is prepared in LB broth + 0.2-0.4% maltose (to induce expression of phage receptors). Just prior to infection these cultures are pelleted at 5000 rpm for 10 min and resuspended in 2.5 ml sterile 10.0 mM MgSO$_4$. Analytical phage infections of the HB101 indicator bacteria using the undiluted and diluted packaged phage are now carried out in Falcon #2059 tubes (see Table III). The infection mixtures are incubated at 37° for 15 min to allow adsorption of the phage to the HB101 host bacteria. LB broth is added (0.9 ml/tube) and the incubation is continued for an additional 1 hr at 37° to allow expression of the antibiotic resistance gene. At the completion of incubation, 3.0 ml of top agar (7.0 g bacto agar/liter of LB broth) at 42° is added per tube and the contents of each tube are overlayed onto an LB chloramphenicol

TABLE III
ANALYTICAL COSMID LIBRARIES

HB 101 in 10 mM MgSO$_4$ (μl)	*In vitro* packaged phage in SM buffer
100.0	None
100.0	50.0 μl of undiluted phage
100.0	50.0 μl of a 1 : 10 dilution
100.0	50.0 μl of a 1 : 100 dilution
200.0	100.0 μl of undiluted phage
200.0	100.0 μl of a 1 : 10 dilution
200.0	100.0 μl of a 1 : 100 dilution

plate (30.0 μg/ml LB agar). The top agar is allowed to solidify at 4° and the plates are incubated overnight at 37°.

The following day the number of chloramphenicol-resistant bacteria (bacteria containing pSV13 recombinant molecules) is scored for each infection condition.

The number of colonies necessary for a complete library can be calculated from the following formula:

$$N = \ln(1 - P)/\ln(1 - f)$$

where N is the number of recombinants necessary for having a probability P of having any DNA sequence represented in the library. The term f represents the fractional proportion of the genome represented in a single recombinant. For a cosmid library f is usually $4 \times 10^4/3 \times 10^9$. From this equation, a library of approximately 2.25×10^5 colonies is necessary for a 95% probability of representing a single copy gene, while a library of approximately 5×10^5 colonies is necessary for a 99% probability of representing a single copy gene.

In order to construct a good preparative library three sets of criteria must be met. (1) The number of colonies/plate should be approximately $1.5-2.0 \times 10^4$. (2) The number of colonies/packaging should be approximately $7.5 \times 10^4 - 1.0 \times 10^5$. (3) The number of colonies/μg of vector DNA should be approximately $1-2 \times 10^5$. These conditions can usually be met by doing 5 packagings distributed over 25 100 mm LB + chloramphenicol plates. From each *in vitro* packaging mix a 1 : 10 and 1 : 100 dilution is prepared and plated for colony counts and calculation of library size. The 500 μl contents of each *in vitro* packaging mix are added to 1 ml of HB101 in 10 mM MgSO$_4$ (prepared as described above). Following phage attachment 4 ml of LB broth is added to each infection and incubated for 1 hr at 37°. At the end of the incubation 15 ml of top agarose (7 g of Sea Prep FMC agarose/1 of LB broth) at 42° is added per infection and 4 ml is overlayed onto each of five 100 mm LB + chloramphenicol plates. The dilutions from each *in vitro* packaging mix are plated as in the preparation of the analytical library. The plates are incubated at 4° for 1 hr in order to the top agarose to solidify and then incubated overnight at 37°. The following day the library can be scored from the appropriate dilution plates. The preparative library is collected by carefully washing and scraping with PBS the top agarose containing the transformed bacteria from the top of each plate. The pooled bacteria are dispersed by vortexing and diluted with 1 : 1 with 2× Hogness freezing solution (K$_2$HPO$_4$ 12.6 g/liter, KH$_2$PO$_4$ 3.6 g/liter, sodium citrate 0.90 g/liter, MgSO$_4$ 0.18 g/liter, (NH$_4$)$_2$SO$_4$ 1.8 g/liter, NaCl 3.0 g/liter, glycerol 35 ml/liter). The library is then pelleted at 3500 rpm for 10 min at 4°. The library is then resuspended in 50 ml of Hogness freezing solution and 1-ml aliquots of the library are frozen for storage at $-70°$ for later screening by transfection techniques (see M. McCormick [33], this volume) or by colony hybridization (see B. Troen [30], this volume).

Discussion

Cosmid libraries constructed from different genomic DNAs have been useful in studying the organization of a variety of genes including the human β-globin related genes,[5] the human α-globin gene cluster,[7] and wild-type and mutant CHO dihydrofolate reductase genes.[8] Successful cosmid cloning of large segments of genomic DNA is dependent upon careful preparation of high-molecular-weight genomic DNA and size fractionation of the genomic DNA, as well as complete dephosphorylation of the cosmid vector in order to prevent self-ligation of the vector molecules. In addition, frequent freeze-thaw cycles of the library should be avoided in order to prevent selection of bacterial transformants containing plasmids with smaller inserts.

We have used the cosmid vector pSV13 to prepare a number of genomic libraries from human, mouse, rat, and hamster DNAs. Using these libraries we have been able to study such diverse genes as those causing multiple drug resistance in cancer cells (Mdr genes), the gene encoding the major excreted protein in transformed mouse cells (MEP gene), and the genes encoding the RI and C subunits of the cAMP-dependent protein kinase in CHO cells.

[30] Colony Screening of Genomic Cosmid Libraries

By Bruce R. Troen

In the past decade, considerable progress has been made in the utilization of bacterial plasmid vectors and λ bacteriophage in the isolation, amplification, and subsequent analysis of eukaryotic DNA sequences. These cloning techniques have been used to generate genomic libraries which can then be screened at high density with specific probes for the DNA regions of interest. However these cloned regions are, of necessity, less than 20 kb, given the practical constraints limiting the size of the inserted DNA sequences. More recently, cosmid vectors have been established that allow DNA inserts of approximately 45 kb. As discussed elsewhere in this volume (see chapters by Fleischmann et al. [29]; McCormick

METHODS IN ENZYMOLOGY, VOL. 151

et al. [28]; and Lau [31]), cosmids are plasmid cloning vectors that contain the λ bacteriophage cohesive *(cos)* termini that permit *in vitro* packaging of recombinant DNA molecules between 37 and 52 kb. Therefore the advantages of screening cosmid DNA libraries include (1) isolation of larger genes than those that can be found in other vector cloning systems, (2) isolation of linked genes on one DNA segment, (3) recovery of flanking sequences of the gene of interest, and (4) reduction of the number of colonies that need to be screened.

This chapter describes a simple procedure used in our laboratory for the colony screening of cosmid libraries that have been constructed using the pSV13 vector (see chapter by Fleischmann *et al.* [29], this volume). It is a variation of those originally developed by Hanahan and Meselson[1,2] and Grunstein and Hogness.[3] It involves initial growth of the bacterial colonies directly on the agar plates, transfer of the colonies to filters, lysis of the colonies, and fixation of the DNA. The filters are then incubated with radiolabeled probe and analyzed by autoradiography. After further growth, colonies are then chosen by aligning the original plates with the autoradiographs, grown in medium for several hours before streaking onto agar plates, and after overnight growth, rescreened. At the point during serial screenings that a discrete individual colony can be localized to an appropriate hybridization spot on the autoradiograph, that colony is grown in medium overnight so that (1) an analytical extraction of the cosmid DNA (a "mini-prep") may be prepared for restriction enzyme digestion and Southern blot hybridization and (2) some of the suspension is frozen for future use. This procedure eliminates the need for growth of bacteria on nitrocellulose filters and the tedium of replica plating. The number of serial screenings can be reduced if the initial screening is performed at a moderate colony density.

Materials and Solutions

1. Either 85 or 150 mm LB agar plates with appropriate antibiotic
2. Millipore HATF (Triton-free) filters — 82 or 127 mm
3. One pair of flat-tipped forceps
4. 24 × 24 cm bioassay dishes (Nunc)
5. 10% SDS
6. 0.5 M NaOH – 1.5 M NaCl
7. 0.5 M Tris – HCl (pH 7.0) – 1.5 M NaCl

[1] D. Hanahan and M. Meselson, *Gene* **10,** 63 (1980).
[2] D. Hanahan and M. Meselson, this series, Vol. 100, p. 333.
[3] M. Grunstein and D. Hogness, *Proc. Natl. Acad. Sci. U.S.A.* **72,** 3961 (1975).

8. Proteinase K solution:
   1 mg/ml Proteinase K (Boehringer-Mannheim)
   0.5 M NaCl
   0.01 M Tris–HCl (pH 7.5)
   0.005 M EDTA
   0.1% SDS
9. 2× SSPE (diluted from 20× SSPE (pH 7.4) = 3.59 M NaCl, 0.2 M NaH$_2$PO$_4$, 0.022 M EDTA
10. Prewash solution:
    0.05 M Tris–HCl (pH 8.0)
    1 M NaCl
    0.001 M EDTA
    0.1% SDS
11. Hybridization (and prehybridization) solution:
    50% formamide (Fluka)
    5× Denhardt's solution [diluted from 50× = 1% Ficoll (Type 400, Sigma), 1% polyvinylpyrrolidone (Sigma), 1% bovine serum albumin]
    3× SSC [diluted from 20× SSC (pH 7.2) = 3 M NaCl, 0.3 M sodium citrate]
    0.1% SDS
    100 μg/ml salmon testes DNA (Sigma)
12. Posthybridization wash solutions:
    2× SSC–0.2% SDS
    0.2× SSC–0.2% SDS

Plating and Growth of Bacteria

Several approaches may be taken during the initial cosmid screening step, with the main goal being to limit the number of colony screenings necessary to obtain a single discrete clone. We have found that with a larger number of serial platings, the cosmid insert is more likely to be rearranged or deleted, thereby resulting in clones that contain incomplete sequences. The two major considerations include the size of the petri dish and the density at which the colonies are plated. We prefer using an 8.5 cm dish rather than the larger 15 cm culture dish, because colony transfer is easier and more reliable and the smaller filters are easier to handle during hybridization. The density of the colonies also affects the transfer to the nitrocellulose filters and, perhaps more importantly, will affect the number of screenings. If the colonies are too dense, then lack of physical separation may necessitate selection of several at one time, requiring an additional plating and hybridization step. Finally, colony diameters of 0.5 mm are preferable to insure a detectable autoradiograph signal above background. Given these considerations, we plate approximately 2500 to 5000 bacteria per 8.5 cm petri dish. There is a 75% chance of detecting a single copy gene of approximately 40 kb in a human genomic cosmid library by screening 20 plates (100,000 colonies). Two initial screenings of 20 plates each increase the overall success rate to about 94%. Employing more than 20 to 25 plates and corresponding filters makes the screening procedure unnecessarily difficult to manage. Therefore we recommend (1) plating between 20 to 25 LB agar 8.5 cm petri dishes containing the appropriate antibiotic with approximately 5000 bacteria per plate and (2) growing the bacteria overnight so that the final diameter of each colony is about 0.5 mm.

Preparation of Filters

We prefer to employ a colony lifting procedure rather than the more time-consuming technique of replica plating. First, autoclave 8.3-cm Millipore HATF filters and then cut a small triangular incision in the filters (see Fig. 1). Using two flat forcep tweezers, gently layer the filter over the culture plate, allowing the moisture from the agar to saturate the filter paper for about 30 to 60 sec. During this time, use a sterile 16-gauge needle to make two eccentrically placed holes in both the filter paper and the agar and also outline the area of the filter paper notch on the bottom of the culture plate with an indelible ink marker. Careully raise the filter paper to avoid smearing of the colonies on both the paper and the agar. Allow the HATF filters (colony side up) to dry at room temperature for about 20 min on Whatman 3MM paper. Label the filters with a ball point pen according

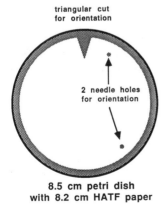

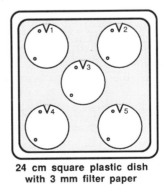

FIG. 1. Colony transfer to nitrocellulose filter and subsequent fixation in square Nunc dishes.

to their correspondingly numbered bacterial culture dishes. Return the culture dishes to the incubator for another 12 hr or until the colony size is once again 0.5 mm in diameter. At that time, wrap the dishes in Parafilm and store at 4° until ready to select colonies. Place a 23-cm-square piece of 3MM paper into each of four Nunc bioassay dishes (24-cm-square) and saturate the paper with the following solutions: (1) Dish 1: 10% SDS, (2) Dish 2: 0.5 M NaOH – 1.5 M NaCl, (3) Dish 3: 0.5 M Tris – HCl (pH 7.0) – 1.5 M NaCl, and (4) Dish 4: 2× SSPE. Leave a thin film of the solution over the Whatman 3MM paper in each dish.

Lay 4 or 5 dry filters onto the saturated paper in Dish 1 for 3 min so that the filters themselves become saturated with the SDS solution. Transfer the filters to Dish 2 for 5 min and then to Dish 3 for another 5 min. Repeat the above steps in an assembly-line fashion until all the filters have been processed. After neutralization in Dish 3, the filters are washed in the proteinase K solution at 37° for 40 min. All the filters can be washed in one dish using 100 ml of solution for each 10 filters. After the wash, transfer the filters to Dish 4 for 5 min and then allow them to dry at room temperature on Whatman 3MM paper for 20 to 30 min. Bake the filters in a vacuum oven for 1 to 2 hr at 80° and then store in a desiccator until they are hybridized with radiolabeled probe.

Prehybridization and Hybridization

Wash the filters in 300 ml of the "prewash" solution (1 filter/15 ml) for 1 hr at 42°. Place 5 filters on a one pint sealable plastic bag and prehybridize in 20 – 25 ml of "hybridization" solution (1 filter/4 – 5 ml) for at least 4 hr at 42° in a gently shaking water bath. After heating the radiolabeled probe at 80° for 10 min, place approximately 50 million cpm of the probe into 25 ml of hybridization solution (10 million cpm per filter). Then place the solution into a one pint sealable plastic bag containing a "blank" nitrocellulose filter that has been blotted with the plasmid vector into which the probe DNA sequence has been cloned (see below). Prehybridize the radiolabeled probe at 42° for 4 hr also. This accomplishes the dual purpose of reducing both the nonspecific binding of the radiolabeled probe to nitrocellulose and the binding between homologous sequences in the radiolabeled probe and cosmid vector. This is important when the cosmid vector and the plasmid vector in which the probe sequence is cloned contain common sequences. For example, our laboratory uses pSV13 as a cosmid cloning vector and also pBR322 as a cDNA cloning vector. Since pSV13 contains pBR322 sequences, we blot our "blank" nitrocellulose filters with pBR322 for prehybridization.

After prehybridization is complete, empty the solution from the bag containing the colony filters. Transfer the hybridization solution containing the radiolabeled probe to the plastic bag with the colony filters. Hybridize for 18 to 24 hr at 42° in a gently shaking water bath. If one begins with 20 or 25 nitrocellulose filters, then 4 or 5 hybridization bags will be needed. When hybridization is complete, remove the filters from the bags (after first aspirating the hybridization solution and cutting the bag open on three sides). The hybridization solution containing the radiolabeled probe may be used for a second batch of filters if hybridization is performed within several days. Wash the filters at 60° for 30 min in 2× SSC–0.2% SDS. Repeat once. Then wash at 60° for 30 min in 0.2× SSC–0.2% SDS. Repeat once for a total of 2 hr of posthybridization washing. Use at least 10 ml of solution for every filter. Stringency of washing can be decreased as necessary by increasing the concentration of SSC and/or decreasing the temperature.

Autoradiography and the "Comet Sign"

Blot the filters briefly on Whatman 3MM paper and arrange them on filter paper, holding the filters in position with a small piece of label tape. Wrap both the nitrocellulose filters and the filter paper backing in plastic wrap to prevent the filters from drying. Place in a film cassette along with a prefogged sheet of new X-ray film and a Dupont Cronex photosensitizing screen. Expose overnight at −70° and develop film. At this point, depending upon the extent of background development, it may be necessary to reexpose for either a shorter or longer period.

A positive signal should be an order of magnitude darker than the surrounding nonpositive background colonies. It is common to be able to distinguish almost all of the colonies on a filter as less intense grayish spots. The size and intensity of the positive radiosignals will vary according to the amount of hybridization between the probe and the colony nucleotides. This may reflect variation in either the colony size or the copy number of the cosmid per bacteria between clones. Frequently, there is an additional and striking clue as to whether an autoradiographic signal represents a truly positive clone. One can see an autoradiographic "tail" emanating from the circular signal. We call this the "comet sign." Sometimes this is manifest as just a small bulge on the periphery of the main signal. Other times it is a linear streak. This "comet sign" is probably a result of a small amount of smearing of the bacterial colony as the filter is both layered onto the agar and then lifted from the plate. In Fig. 2a, a single large positive autoradiographic signal is seen which is clearly darker than the still visible, but much less intense, background colonies. This is an excellent example of the "comet sign." Absence of the "comet sign" does not preclude a true positive signal. It is important to note that even under the best conditions, signals may appear to be "true" positives and yet fail to be positive upon subsequent colony screenings and/or Southern blot hybridizations. For example, Fig. 2b is an autoradiograph of a blot from a cDNA library. Many positive autoradiograph signals are present, some with typical "comet" appendages. However, of 35 positive autoradiograph signals from this library, only 14 were "true" positives upon subsequent analysis. The percentage of "true" positives may be as low as 20–50% during cosmid library screening with a homologous probe.

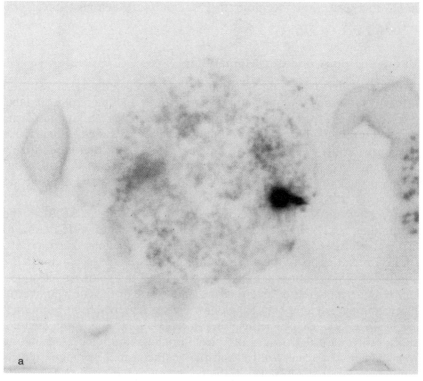

a

FIG. 2. (a) Single positive radiographic signal from an NIH genomic DNA cosmid library screened with radiolabeled MEP (major excreted protein) cDNA. This was the only "true" positive, containing MEP genomic DNA, of 5 positive autoradiographic signals from a screening of 30,000 colonies. (b) Single plate from the screening of a cDNA library. The "comet" signs in this autoradiograph are more typical than that seen in a. In both photographs, numerous nonpositive colonies can be visualized. A true positive signal should be much darker than the background colonies.

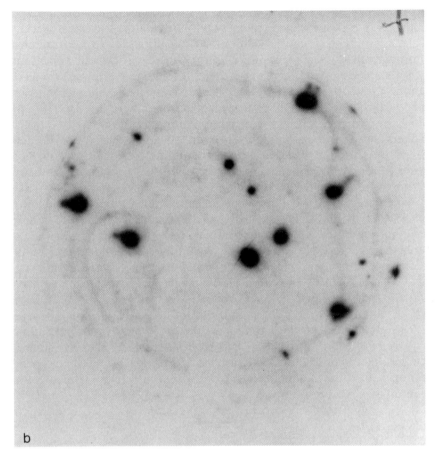

FIG. 2. *(Continued)*

Colony Selection and Subsequent Screening

In order to select appropriate colonies, it is necessary to accurately align the original culture dishes with the autoradiograph. First superimpose the autoradiograph upon the filters. Use a grease pencil to mark the two needle holes and the triangular notch for each filter (often the notch will be visible on the autoradiograph). Then superimpose the individual culture dishes upon the autoradiograph and align the needle holes in the agar and the notch outlined on the bottom of the plate with the appropriate pencil marks. Mark those areas of the plate that correspond to the "positive" autoradiographic signals. Depending upon the density of the bacterial colonies, between 1 and 10 colonies may underlie the signal. Use a sterile toothpick to pick individual colonies and inoculate the bacteria into 1 ml of the appropriate selective medium for 4 to 5 hr at 37° (place in microcentrifuge tubes—shaking is unnecessary). Streak the bacterial suspension with a sterile loop and then make three to four more cross-streaks after flaming the loop each time. Streak each colony over only one-third or one-fourth of a plate so that subsequent screenings require fewer filters. Grow the bacteria overnight in preparation for a second round of transfer to filters and subsequent hybridization with the radiolabeled probe. For the second (and third, if necessary) screening, discrete colonies of 1 mm or greater will make the subsequent correlation between autoradiograph and culture dish easier and more specific.

When only a single colony matches an apparent "positive signal," the following steps should be taken. Again use a sterile toothpick to pick a small portion of the colony for a 5 hr incubation in medium. Then use a sterile loop to inoculate two separate 15 ml tubes each containing 5 ml of medium with the appropriate selective antibiotic. Streak the bacteria from the 5 hr mini-incubation onto a plate for one final screening. The filter containing these colonies should exhibit 100% true positive signals upon hybridization. Grow the bacteria in the 15 ml tubes overnight at 37° in a shaking incubator. Use the suspension from one tube to prepare a "miniprep" (see below) of the cosmid DNA for restriction enzyme analysis and Southern blot hybridization. The suspension from the second tube should be frozen in 1-ml aliquots (0.7 ml of suspension plus 0.3 ml of 50% glycerol) at $-70°$ for future use.

Plasmid Blotting on Blank Nitrocellulose Filter

This procedure is designed to first linearize the plasmid and then blot it onto a nitrocellulose filter. The hybridization solution with the radiolabeled probe is then prehybridized with the filter in order to bind any remaining plasmid sequences and therefore reduce background and enhance specificity of the subsequent hybridization to the colony filters.

1. Begin with 5 μg of plasmid in 25 μl of solution (e.g., TE).
2. Add 25 μl of 0.25 M HCl and incubate at room temperature for 10 min.
3. Add equal volume (50 μl) of 0.5 M NaOH and incubate at room temperature for 10 min.
4. Add equal volume (100 μl) of 0.5 M NaH$_2$PO$_4$.
5. Dilute sample to 400 μl with 20× SSC (final concentration of 10× SSC).

6. Saturate a 4×4 cm nitrocellulose filter in H_2O and then in $10\times$ SSC. Air dry the filter.

7. Place the plasmid solution on plastic wrap and then place the dry filter on top of the solution.

8. Bake the filter for 1 to 2 hr at 80° in a vacuum oven.

9. Store filter in a desiccator.

Analytical Cosmid DNA Extraction ("Mini-prep")

This procedure permits the preparation of small amounts of plasmid/cosmid suitable for restriction enzyme analysis, Southern blot hybridization, and subcloning.

1. Grow bacteria overnight in 5 ml of LB broth with the appropriate antibiotic.

2. Centrifuge cells (full amount) for 10 min at 3000 rpm. Decant supernatant.

3. Resuspend cell pellet *gently* with 0.5 ml of the following solution: 50 mM Tris–HCl/50 mM EDTA/15% sucrose/1 mg/ml lysozyme (Cooper Biomedical). Make sure pH is ≥ 8.0 and add the lysozyme just prior to use.

4. Transfer to a microcentrifuge tube and incubate at room temperature for 10 min.

5. Add 25 μl of 10% SDS. Invert tube several times to mix (do *not* vortex).

6. Add 60 μl of 5 M potassium acetate (KAc). Invert to mix.

7. Incubate on ice for 30 min.

8. Centrifuge for 15 min in a microfuge in the cold.

9. Transfer supernatant into a fresh microfuge tube. Do *not* take any of the precipitate.

10. Add 2 μg of heat treated RNase (or μl of stock solution — 10 μg/ml) to each tube and incubate at room temperature for 15 min.

11. Add 0.5 ml of chloroform : isoamyl alcohol : phenol (24 : 1 : 25). Vortex well (15 sec).

12. Microcentrifuge for 1 min. Transfer the aqueous phase (top layer) to a new tube.

13. Repeat extraction with 0.5 ml chloroform : isoamyl alcohol.

14. Add 1 ml of cold ethanol to aqueous supernatant. Vortex and incubate on dry ice for 10 min. Microcentrifuge for 15 min in the cold.

15. Decant supernatant. Add 400 μl of H_2O and 20 μl of 5 M potassium acetate to the pellet and vortex well. Add 0.9 ml of cold ethanol and vortex again. Incubate on dry ice for 10 min and microcentrifuge for 15 min in the cold.

16. Decant supernatant and repeat step 15.
17. Wash pellet gently with 80% ethanol and then decant supernatant.
18. Dry pellet in a vacuum rotoevaporator (or lyophilize).
19. Resuspend pellet in 30–50 μl of TE buffer.
20. Use 2–5 μl for restriction enzyme analyses.

[31] Rescue of Genes Transferred from Cosmid Libraries

By YUN-FAI LAU

Cosmid vectors that can be used for the isolation, expression, and rescue of gene sequences are valuable tools in studying molecular genetics of mammalian cells. Recently, several such vectors have been developed that can serve as (1) highly efficient vehicles for complex genomic recombinant DNA library preparation, (2) expression vectors in mammalian cells, (3) shuttle vectors between bacteria and mammalian cells, and (4) cloning systems for the isolation of intact genes based on their expression in appropriate selections[1,2] (see also the chapter by McCormick et al. [28], this volume). Since the procedures for cosmid library construction and high density screening have been discussed elsewhere in the present volume (Fleischmann et al. [29] and Troen [30]) this chapter will focus on the shuttling aspects of these vectors.

Principle

Shuttle cosmids are very similar to conventional cosmids,[3] except that the mammalian selectable genes are covalently inserted in the vector. The inclusion of a small modular gene, such as SV2-gpt,[4] SV2-DHFR,[5] SV2-neo,[6] and HSV-TK,[7] in the cosmid prior to construction of the genomic library does not significantly alter either the cloning capacity or efficiency of these vectors. Recombinant cosmids with 30–45 kb inserts can be

[1] Y.-F. Lau and Y. W. Kan, Proc. Natl. Acad. Sci. U.S.A. 80, 5225 (1983).
[2] Y.-F. Lau and Y. W. Kan, Proc. Natl. Acad. Sci. U.S.A. 81, 414 (1984).
[3] J. Collins and B. Hohn, Proc. Natl. Acad. Sci. U.S.A. 75, 4242 (1978).
[4] R. C. Mulligan and P. Berg, Science 209, 1422 (1980).
[5] S. Subramani, R. Mulligan, and P. Berg, Mol. Cell. Biol. 1, 854 (1981).
[6] P. J. Southern and P. Berg, J. Mol. Appl. Genet. 1, 327 (1982).
[7] F. L. Graham, S. Bacchetti, R. McKinnon, C. Stanners, B. Cordell, and H. M. Goodman, in "Introduction of Macromolecules into Viable Mammalian Cells" (C. Croce and G. Rovera, eds.), p. 3. Liss, New York, 1980.

isolated using high density screening and specific nucleic acid probes. Once purified to homogeneity, these clones sequences can be transferred into mammalian cells using corresponding selection systems appropriate for the gene marker present within the vector. This procedure circumvents the necessity of cotransformation with an unlinked marker. Since in most cases the exogeneously acquired DNA is integrated tandemly in the host genomes, the cohesive end sites (COS) of the cosmid would be spaced approximately a λ length, about 50 kb, apart. Such concatenated cosmid sequences in the host genome, therefore, become suitable substrates in an *in vitro* bacteriophage λ packaging reaction. The packaged bacteriophages can be transferred into *E. coli* by infection and selected with appropriate antibiotics as in the usual cosmid cloning procedure.

The cosmid shuttling scheme can be readily modified to serve as a cloning system for the isolation of intact genes based on their expression in appropriate selections. Such modified protocols use the total DNA from a complex genomic cosmid library as donor sequences in a conventional DNA-mediated gene transfer to mutant cells. After selection with the appropriate system in which the target gene must be retained and be derived from the insert sequences of the cosmid library, the transformed sequences can be rescued from the host genome by cosmid shuttling. Alternatively, total cosmid library DNA can be transferred with either the built-in selectable gene, or another unlinked marker, to generate a large number of transformants that harbor a complete genome equivalent of cosmid sequences. Subpopulations of cells that express particular genes can then be selected by secondary means, such as fluorescence-activated cell sorting or UV irradiation and repair procedures. The transformed cosmids are then shuttled to bacteria by *in vitro* bacteriophage λ packaging.

Methods

The cosmid shuttling protocols described in this section represent our experience with the direct isolation of the human thymidine kinase gene using total cosmid library DNA as donor sequences to the mouse Ltk⁻ cells.[2] The rescue of cloned cosmids from mammalian cells is very similar and has been described elsewhere.[1]

Vectors, Bacterial Hosts, and Mammalian Mutant Cells.

The shuttle cosmids used in these studies were constructed by inserting several modular selectable markers in the cosmid pJB8 and have been described in detail previously.[1,8] Their general configuration is represented

[8] D. Ish-Horowicz and J. F. Burke, *Nucleic Acids Res.* **9**, 2989 (1981).

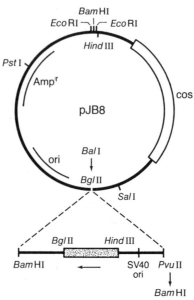

FIG. 1. Structure of shuttle cosmid vectors. The *Bal*I site in the cosmid pJB8 was converted to a *Bgl*II site. A *Bam*HI DNA fragment harboring either SV2-*gpt*, -DHFR, or -*neo* mammalian selectable gene marker is inserted in the *Bgl*II site, resulting in cosmids designated as pCV103, pCV107, and pCV108, respectively. pCV105 was constructed by inserting in pJB8 the 3.4 kb *Bam*HI DNA fragment containing the HSV-TK gene. pCV105 does not contain any SV40 sequences as indicated. All selectable gene markers are transcribed in the direction indicated by the arrow. The cloning site is *Bam*HI at the top position, and the inserts can be cleaved from the vector by *Eco*RI digestion.

in Fig. 1. The cosmid pCV108 containing the SV2-*neo* gene was used in most cases. This selectable marker confers resistance to the antibiotic G418 to mammalian recipient cells. The design of our TK cloning experiment is outlined in Fig. 2. The *E. coli* hosts used are

ED8767   *rec*A56, *sup*E, *sup*F, *hsd*S⁻R⁺M⁺, *met*⁻ [9]
DK1   *ara*D139, del(*ara,leu*)7697, del*lac*X74, *gal*U⁻, *gal*K⁻, *hrs*⁻, *hsm*⁻, del(*srl-rec*A)306, *str*A [10]

The mammalian host is the mouse L cell deficient in thymidine kinase activity (Ltk⁻).

[9] N. E. Murray, W. J. Brammer, and K. Murray, *Mol. Ben. Genet.* **150** 53 (1977).
[10] Gift from David Kurnit, Genetics Division, The Children's Hospital, Boston, Massachusetts.

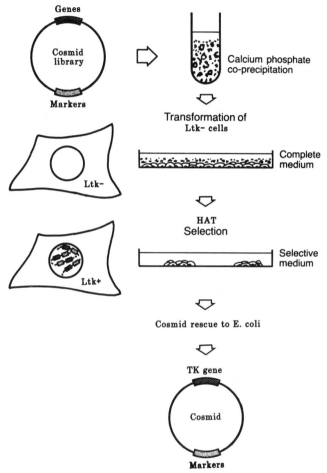

FIG. 2. Diagrammatic representation of the cloning strategy to isolate the human thymidine kinase gene using total DNA from a cosmid library as donor sequences.

Preparation of Total Cosmid Library DNA

About $2-5 \times 10^8$ bacteria harboring human cosmid libraries of 6–25 times genome-equivalent complexity are inoculated into 1 liter of L broth with 50–100 μg/ml ampicillin and are grown at 37° with good aeration. Cosmid sequences were amplified with 200 μg/ml of chloramphenicol at OD_{600} of 0.6–0.8 for 16–20 hr. Total cosmid DNA is isolated by a modified alkaline lysis method[11] and subjected to further purification with

---

[11] H. C. Birnboim and J. Doly, *Nucleic Acids Res.* **7**, 1513 (1979).

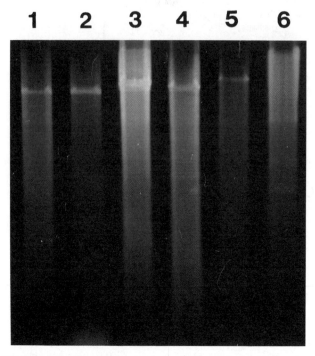

Fig. 3. Ethidium bromide staining patterns of six preparations of total cosmid library DNA after *Eco*RI digestion and agarose gel electrophoresis. Samples are isolated from human cosmid libraries constructed with different vectors. Lane 1 and 2, pCV107; lane 3 and 4, pCV108; lane 5, pCV001; lane 6, pCV007. pCV001 contains both SV2-*neo* and SV2-DHFR as selectable markers. pCV007 is a similar cosmid using the LTR from Rous sarcoma virus as promoters for both selectable genes [K. H. Choo, G. Filby, S. Greco, Y.-F. Lau, and Y. W. Kan, *Gene* **46,** 277 (1986)]. The brightly stained bands correspond to the vector DNA.

cesium chloride–ethidium bromide equilibrium ultracentrifugation. The final DNA preparations are dialyzed extensively against TE buffer (10 mM Tris–HCl 1 mM EDTA, pH 7.5) at 4°. The quality of DNA is examined by *Eco*RI digestion and electrophoretic analysis in an agarose gel. For a good representation of the genome, the ethidium bromide staining pattern of the total library DNA should resemble the smeared appearance of similarly digested genomic DNA plus the vector DNA band (Fig. 3).

Transformation of Mouse Ltk⁻ Cells

Five million Ltk⁻ cells are seeded in a T-150 culture flask for 8–12 hr in 20 ml of Dulbecco's modified Eagle (DME) medium with 10% fetal calf serum and penicillin/streptomycin. Fifty micrograms of total cosmid library DNA (in pCV108) is coprecipitated with calcium phosphate[12] in 2 ml volume and added onto the Ltk⁻ cells. Following an overnight incubation, the cells are treated with 15% glycerol in phosphate-buffered saline (PBS) for 2–3 min, washed with fresh PBS, and cultured in fresh medium for 24 hr. The cells are then selected with the HAT medium (complete DME medium containing 100 μM hypoxanthine, 0.5 μM aminopterin, and 20 μM thymidine). Visible colonies usually appear in about 8–14 days. The efficiency of transformation was 1 per 5 × 10⁶ cells for our TK cloning experiment. Individual cell colonies are singly picked and propagated stepwise to T-75 culture flasks in HAT with 200 μg/ml G418. Cells that are resistant to both HAT and G418 must have acquired the human TK and the SV2-neo genes in the inserts and the cosmid vector respectively from the donor sequences of the total cosmid library DNA.

Isolation of High-Molecular-Weight DNA from Transformed Cells

Ltk⁺ cells at 80–90% confluency in a T-75 culture flask are rinsed with PBS, and lysed *in situ* with 10 ml of proteinase K solution (50 mM Tris–HCl, pH 7.5, 1 mM EDTA, 100 mM NaCl, 0.5% SDS, and 100 μg/ml proteinase K), and incubated at 37° for 1–2 hr. The viscuous mixture is poured into a 30-ml Corex tube, and processed gently by the usual phenol/chloroform extraction and RNase digestion methods. After extensive dialysis against TE buffer, the DNA is precipitated with ethanol and dissolved completely in 50–100 μl of 0.1× TE buffer at 4°. About 50–100 μg of high-molecular-weight DNA (at least 800 kb in size) can be isolated from cells in a confluent T-75 culture flask.

Cosmid Rescue by in Vitro Bacteriophage λ Packaging

Highly efficient *in vitro* bacteriophage λ packaging extracts are prepared according to published protocols using the lysogenic *E. coli* strains BHB2688 and BHB2690.[13,14] One microgram of high-molecular-weight DNA isolated from the transformant cells is packaged in a final volume of 26 μl with lysogenic extracts at 23–25° for 1 hr. The resulting bacteriophages are diluted with 225 μl of SM buffer (50 mM Tris–HCl, pH 7.5, 100 mM NaCl, 8 mM MgSO₄, 0.01% gelatin). *E. coli* hosts (ED8767 or DK1) are grown overnight in LB with 0.2% maltose and concentrated 2 times in 10 mM MgCl₂. Bacteria (250 μl) are added to the bacteriophages,

[12] F. L. Graham and A. J. Van der Eb, *Virology* **52**, 456 (1973).

[13] F. G. Grosveld, H. H. Dahl, E. deBoer, and R. A. Flavell, *Gene* **13**, 227 (1981).

[14] B. Hohn, unpublished procedure.

and incubated at room temperature for 15 min. One milliliter of LB is added, and incubated at 37° for 30 min. The bacteria are then spread evenly onto 2–3 15 cm LB plates with 100 μg/ml of ampicillin, and incubated at 37° overnight. Depending on the number of cosmids per transformed cell, the efficiency of cosmid rescue varies from 1 to 1000 colonies per μg of total cellular DNA. Thirty-three cosmids per μg of Ltk⁺ transformant DNA had been rescued with this protocol. Restriction enzyme analysis of about 100 rescued cosmids revealed that there were only two basic patterns (Fig. 4). Secondary transformation with these two cosmids indicated that only one of them contained an intact human TK gene. The few species of cosmids rescued from the primary transformant cells is not surprising, since the initial HAT medium only selected for the cosmids containing the TK gene. Most of the nonessential cosmids would be lost before the integration events.

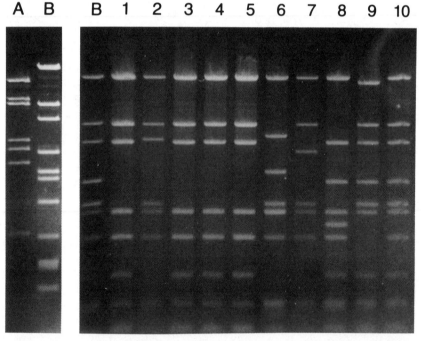

FIG. 4. Ethidium bromide staining of *Eco*RI-digested DNA from cosmids rescued from the primary (left) and secondary (right) transformants. Cosmid B contains an intact human thymidine kinase gene as indicated by secondary transformation and cosmid rescues (lanes 1–10).

Comments

The success of the present cloning strategy depends on several important factors. First, the size of the target gene is limited to 40 kb because larger genes are unlikely to be represented as intact units in the insert sequences. Second, the complexity and the recovery of the initial cosmid library determine the quality of the donor sequences. The most common problem encountered is preferential amplification of certain cosmids within the library during DNA preparation. This difficulty can be minimized by choosing a complex cosmid library with large number of independent recombinant cosmids and a good recovery rate after storage. Furthermore, this problem is easily identified by endonuclease digestion of the total library DNA and agarose gel electrophoreses. A good DNA preparation is represented by a smeared pattern plus the vector DNA band. Any additional DNA bands would suggest a preferential amplification of certain individual cosmids within the library. Third, as with the conventional DNA-mediated gene transfer, the specificity of the selection system is critical. Any leaky selection will produce a large number of revertants. Reversion background can be greatly reduced if secondary selection for the gene marker in the vector, such as SV2-*neo*, is also applied to the transformants after the primary selection. In such instances, the true transformant cells must retain the target gene as well as the cosmid vector. Any revertants without cosmid sequences are eliminated. Deletion mutants are most useful recipient cells in these cloning experiments. Fourth, in order to be rescued by the *in vitro* bacteriophage λ packaging method, the cosmids must be integrated in tandem and their COS sites spaced about 50 kb apart. Hence, isolation of high-molecular-weight DNA from the transformant cells is essential for sucessful cosmid rescue. Finally, under identical conditions, the efficiency of cosmid rescue is directly proportional to that of the *in vitro* bacteriophage λ packaging reaction. As with cosmid library construction, a minimum efficiency of 10^8 plaques per microgram of wild-type λ DNA is acceptable for such reaction. Recently, highly efficient packaging extracts have come on the market from several commercial vendors. Their availability has greatly alleviated some of the difficulties often encountered in cosmid cloning experiments.

Acknowledgments

The author thanks K. H. Choo for technical assistance and Y. W. Kan for support and encouragement. Y.-F. Lau is an Associate Investigator of the Howard Hughes Medical Institute.

[32] Phage-Mediated Transduction of cDNA Expression Libraries into Mammalian Cells

By HIROTO OKAYAMA

Genetic complementation has provided an extremely powerful tool to identify, isolate, and characterize genes of interest in prokaryotes and lower eukaryotes. The recent development of techniques for delivering DNA into cells and isolating appropriate mutant cells has made it possible to use the same approach in mammalian cells.[1,2] In fact, several mammalian genes have already been cloned solely on the basis of genetic complementation of mutant cells by means of genomic DNA transfection.[3-7] In mammalian cells, however, this approach has often been hampered by severe difficulties in recovering the transduced genes in *Escherichia coli*.[5,8] Species-specific repetitive sequences dispersed between genes in the mammalian genome are generally used as markers to clone the transduced genes in *E. coli*, but mammalian genes sometimes lack these sequences. In addition, mammalian genes are often large in size (due to the presence of lengthy and/or multiple introns) and they cannot be cloned intact in *E. coli*. As an alternative approach, the utilization of full length cDNAs for genetic complementation would avoid many of these problems. A system comprising techniques for efficient cloning of full length cDNAs and two vectors, a cDNA expression vector and a cDNA library transducing vector, which we have developed,[9-11] permits this approach.

In contrast to cellular genes, cloned cDNAs can be expressed in any desired host cells, bypassing the barrier of tissue and species-specificity, and in virtually any desired mode, including constitutive, inducible, reduced,

---

[1] L. Graham and A. J. van der Eb, *Virology* **52**, 456 (1973).

[2] L. H. Thompson, this series, Vol. 58, p. 308.

[3] M. Perucho, D. Hanahan, L. Lipsich, and M. Wigler, *Nature (London)* **285**, 207 (1980).

[4] L. Lowy, A. Pellicer, J. F. Jackson, G. K. Sim, S. Silverstein, and R. Axel, *Cell* **22**, 817 (1980).

[5] D. J. Jolly, A. C. Esty, H. U. Bernard, and T. Friedmann, *Proc. Natl. Acad. Sci. U.S.A.* **79**, 5038 (1982).

[6] L. C. Kuhn, A. MacClelland, and F. H. Ruddle, *Cell* **37**, 95 (1984).

[7] A. Westerveld, J. H. J. Hoeijmakers, M. van Duin, J. deWit, H. Odjik, A. Pastink, R. D. Wood, and D. Bootsma, *Nature (London)* **310**, 425 (1984).

[8] K. Takeishi, D. Ayusawa, S. Kaneda, K. Shimizu, and T. Seno, *J. Biochem.* **95**, 1477 (1984).

[9] H. Okayama and P. Berg, *Mol. Cell. Biol.* **2**, 161 (1982).

[10] H. Okayama and P. Berg, *Mol. Cell. Biol.* **3**, 280 (1983).

[11] H. Okayama and P. Berg, *Mol. Cell. Biol.* **5**, 1136 (1985).

or elevated expression, by choosing an appropriate promoter. This unique property of cDNA expression permits additional cloning applications using normal cell hosts. The possibility of elevated expression in heterologous hosts provides an opportunity to clone genes for various cell membrane receptors by screening for ligand-binding activity transduced in nonproducer host cells. It may also be possible to search for genes for a variety of cellular regulatory factors involved in gene expression, cell growth, cell differentiation, and cell transformation.

Detailed procedures for the construction of full length cDNA libraries for expression in mammalian cells are described elsewhere (this series, Vol. 154). In this chapter, I describe methods for the transfer of cDNA expression libraries into cultured mammalian cells.

Outline of Methods

A cDNA library is constructed with the SV40-based cDNA expression vector, pcD, according to the method of Okayama and Berg,[9] and recovered from *E. coli* host cells. These pcD recombinants (Fig. 1) are linearized by partial digestion with *Sal*I or *Sfi*I restriction endonucleases and inserted into λNMT (Figs. 2 and 3).[11] λNMT is a λ phage-based mammalian cell transducing vector containing as a dominant selectable marker gene the bacterial *neo* coding sequences inserted into the SV40 transcriptional unit (Fig. 2). After *in vitro* packaging and amplification, the phage recombinants are purified and then transfected into mammalian cells after coprecipitation with CaPO₄. Stably transfected cells are selected in medium containing the neomycin analog G418.[12] G418 specifically inhibits eukaryotic translation and kills cells, while cells expressing the *neo* gene are resistant to the drug. A second selection or screening procedure can then be used to isolate cells expressing the desired phenotype or protein, and the integrated cDNA is finally recovered and cloned in *E. coli* for chemical characterization.

Transfer of a cDNA Expression Library into λNMT

Described below is a protocol which uses the *Sal*I site for transfering a cDNA library into the phage vector. Two other restriction sites are available for this purpose: *Sfi*I and *Hind*III sites. The *Sfi*I site is located at the SV40 replication origin that resides in the SV40 promoter segment used in the both pcD and λNMT vectors. *Sfi*I sites consist of an 8 base-pair recognition sequence shared with *Bgl*I, and are found less frequently than *Sal*I sites in mammalian DNA. Digestion with *Sfi*I initially destroys the

---

[12] P. J. Southern and P. Berg, *J. Mol. Appl. Genet.* **1**, 327 (1982).

SV40 promoter in both vectors, but insertion restores both promoters. If these sites are used, the resulting recombinant phage can be amplified in C600.

Strains

C600   (*supE tonA lac*Y)[13]
JC8679   (*thr*-1 *leu*B6 *phi*-1 *lac*Y1 *gal*K2 *ara*-14 *xyl*-5 *mtl*-1 *pro*A2
           his-4 *arg*E3 *rps*L31 *tsx*-33 *supE44 recB21 recC22 sbcA23*
           his-328)[14]

Reagents

*Sal*I, Bethesda Research Laboratories; bacterial alkaline phosphatase, Sigma; T4 DNA ligase, Collaborative Research

λdil: 10 mM Tris–HCl (pH 7.5), 10 mM MgSO$_4$

3 M CsCl solution: 3 M CsCl, 10 mM Tris–HCl (pH 7.5), 10 mM MgSO$_4$, 0.1 mM EDTA

5 M CsCl solution: 5 M CsCl, 10 mM Tris–HCl (pH 7.5), 10 mM MgSO$_4$, 0.1 mM EDTA

7.2 M CsCl solution: 7.2 M CsCl, 10 mM Tris–HCl (pH 7.5), 10 mM MgSO$_4$, 0.1 mM EDTA

Procedure

cDNA libraries constructed in the pcD mammalian expression vector are recovered from *E. coli* cloning hosts by standard Triton X-100–lysozyme extraction followed by two cycles of CsCl equilibrium gradient centrifugation. After each centrifugation, all of the supercoiled plasmid DNA should be recovered to avoid potential loss of particular cDNA clones having slightly different buoyant densities due to high or low G/C content in the cDNA inserts or some difference in superhelicity.

Approximately 20 μg of the pcD-cDNA plasmid is digested with 40 U of *Sal*I at 37° in a 200 μl reaction mixture containing 150 mM NaCl, 6 mM Tris–HCl (pH 8.0), 6 mM MgCl$_2$, 6 mM 2-mercaptoethanol, and 100 μg of bovine serum albumin per ml. The duration of the reaction should be chosen so that it proceeds to about 50% completion. The digestion is terminated by adding 20 μl of 0.25 M EDTA (pH 8.0) and 10 μl of 10% SDS. After two extractions with 200 μl of phenol/CHCl$_3$, the DNA is precipitated by adding 40 μl of 2 M NaCl and 550 μl of ethanol and chilling on dry ice for 15 min followed by centrifugation for 10 min at 4° in

[13] R. K. Appleyard, *Genetics* **39**, 440 (1954).
[14] J. Gillen, D. K. Willis, and A. J. Clark, *J. Bacteriol.* **145**, 521 (1981).

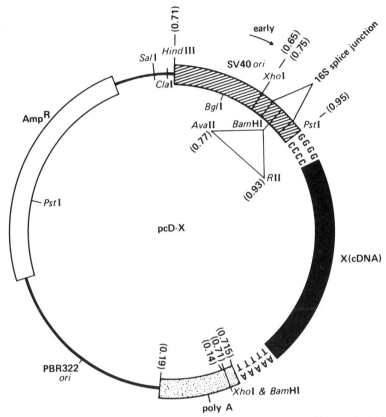

FIG. 1. Structure and components of the pcD-cDNA recombinant.[10] The principal elements of the pcD vector are a segment containing the SV40 replication origin *(ori)* and the early-region promoter oriented in the clockwise direction joined to a segment containing the 19 S and 16 S SV40 late-region splicing junctions (hatched area); the various cDNA segments (solid black area) plus the flanking dG/dC and dA/dT bridges to the vector produced in the cloning operation[9,10]; a segment containing the SV40 late-region polyadenylation signal [poly (A)] (stippled area); and the pBR322 segment containing the β-lactamase gene (Amp^R) and the pBR replication origin (pBR322 *ori*) (thin line and open area). Positions of the SV40 fragments are shown in map units.

an Eppendorf microfuge. The precipitated DNA is rinsed with 70% ethanol, dissolved in 50 μl of 50 mM Tris–HCl (pH 9.0), and then incubated with 0.05 U of bacterial alkaline phosphatase at 65° for 20 min. The reaction is terminated by adding 5 μl of 10% SDS, the solution is extracted twice with phenol/CHCl₃, and the DNA is recovered by ethanol precipitation as above.

The recovered linear pcD-cDNA is inserted into λNMT as follows. The

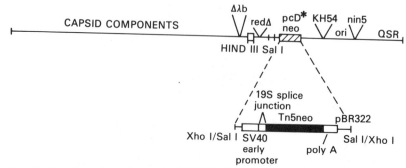

FIG. 2. Genetic and physical map of λNMT. The genotype is del*B*, del*red*, KH54, and *nin*5. At the *Xho*I site, λNMT contains a mammalian transcription unit that expresses the Tn*5 neo* coding sequences; this segment contains the *neo* gene (solid black area) joined at its 5′-end to a segment containing the SV40 early promoter and 19 S late splicing junctions and at its 3′-end to the SV40 late polyadenylation signal [poly (A)].

λNMT phage is amplified in *E. coli* strain C600 grown in LB soft agar on LB agar plates containing 0.2% glucose. The phage are eluted from plates by overnight incubation at 4° with λdil (5 ml/10 cm plate). After removal of cell debris by low-speed centrifugation, the phage particles are sedimented by high-speed centrifugation at 25,000 rpm for 90 min using an SW27 rotor. The phage pellet is resuspended in 1–2 ml of λdil. After

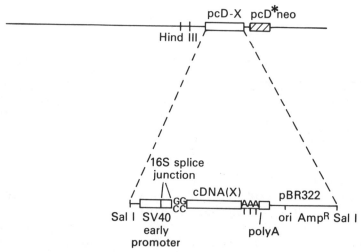

FIG. 3. General structure of λNMT-pcD-cDNA. The pcD-cDNA can be inserted into the λ vector in either orientation.

centrifugation for 2 sec in an Eppendorf microfuge to remove contaminating agar and cell debris, the supernatant is gently pipetted into a SW50 rotor tube containing 1 ml of 5 M CsCl solution overlaid with 3 ml of 3 M CsCl solution, and centrifuged at 36,000 rpm for 1 hr. The phage, which bands at the interface between the 3 and 5 M CsCl solutions, is collected by puncturing the side of the tube with a needle and withdrawing the band with a syringe. The collected phage (0.5–1 ml) are then mixed with an equal volume of 7.2 M CsCl solution, placed at the bottom of a SW50 tube, and carefully overlaid with 2–3 ml of the 5 M and 1 ml of the 3 M CsCl solutions. After centrifugation as above, phage, banding at the interface between 3 and 5 M CsCl, is collected with a syringe and phage DNA is extracted with formamide.[15] The purified phage suspension is mixed with 1/10 volume of 2 M Tris–HCl (pH 8.0) containing 200 mM EDTA, and then one volume of deionized formamide is added to the mixture. After incubation at room temperature for 2 hr, phage DNA released from the capsids is precipitated by adding one volume of H_2O and 6 volumes of ethanol, immediately centrifuged at room temperature for 2 sec in an Eppendorf microfuge, and finally rinsed with 70% ethanol. The phage DNA is digested to completion with SalI, and ligated to the linear pcD-cDNA as follows.

Four micrograms of the SalI–cut λNMT DNA is incubated at 43° for 1 hr in a reaction mixture (14 μl) containing 50 mM Tris–HCl (pH 7.5), 6 mM MgCl$_2$, and 8 μg of the bacterial alkaline phosphatase-treated, SalI-digested pcD-cDNA. The mixture is chilled on ice, and 2 μl each of 0.1 M dithiothreitol, 10 mM ATP, and T4 DNA ligase (3 U/μl) are added. Ligation is carried out at 12–16° overnight. The ligated DNA is packaged *in vitro* into phage capsids as described by Enquist and Sternberg[16] using a commercial packaging system (Gigapack, Vector Cloning Systems), with a yield of approximately 10^7 PFU of recombinant phage. Insertion of sequences at the SalI restriction site of the phage inactivates the phage's gamma function which protects the λ rolling circle replication intermediate from the host *recBC* exonuclease. To circumvent this problem, the packaged phage is amplified in a *recBC* mutant strain (JC8679) to prepare a stock of the library. JC8679 is grown in 14 ml of L broth containing 0.2% maltose at 37° overnight. Cells are collected by low-speed centrifugation, suspended in 7 ml of λdil and infected with the *in vitro* packaged phage. After incubation at room temperature for 20 min, 100 μl aliquots of the infected cells are plated on L broth agar plates (10 cm) together with 2.5 ml

---

[15] R. W. Davis, D. Botstein, and J. Roth, "Advanced Bacterial Genetics," p. 70. Cold Spring Harbor Lab., Cold Spring Harbor, New York, 1980.
[16] L. Enquist and N. Sternberg, this series, Vol. 68, p. 281.

of LB broth soft agar (total 80 plates). Plates are incubated at 37° for about 6 hr, and the phage is recovered from the plates as decribed above and stored at 4° following low-speed centrifugation and addition of 1 ml of CHCl₃.

Large-Scale Amplification of the Phage Library

Procedure

JC8679 is grown in L broth containing 0.2% maltose at 37° overnight, sedimented by low-speed centrifugation, and suspended in 1/2 volume of λdil. Twelve milliliters of the cells is infected with 5×10^7 PFU of the recombinant phage. After incubation at room temperature for 15–20 min, 100-μl aliquots of the infected cell suspension are plated on LB agar plates (10 cm) containing 0.2% glucose, together with 2.5 ml LB soft agar (total 120 plates). The plates are incubated at 37° for 5–6 hr until the plaques are just about to fuse. The amplified phage is recovered from the plates and purified by the two CsCl block gradient centrifugation as described above.

Transfection of Tissue Culture Cells with the Phage Library

Reagents
DMEM: Dulbecco's modified Eagle's medium
$0.25\ M$ CaCl₂
2× BBS buffer: 50 mM N,N-bis(2-hydroxyethyl)-2-aminoethanesulfonic acid (pH 6.87), 280 mM NaCl, 1.5 mM Na₂HPO₄

Procedure

Transfection of cells is carried out according to the method of Ishiura *et al.*[17] Cells are maintained at 35–37° in 5–7% CO₂ in DMEM containing 10% fetal calf serum. The purified recombinant phage is dialyzed against 1 liter of 10 mM Tris–HCl (pH 7.5) containing 10 mM MgCl₂, 100 mM NaCl, and 0.02% gelatin to remove CsCl. The dialyzed phage particles ($1–5 \times 10^{11}$ PFU) are mixed with 5 ml of $0.25\ M$ CaCl₂, and then added to 5 ml of 2× BBS buffer with gentle stirring. After incubation at room temperature for 10–20 min, 1-ml aliquots of the CaPO₄-precipitated phage suspension are added to culture dishes containing 10 ml of growth medium and cells that have been seeded at $0.5–1 \times 10^6$ cells/10 cm tissue culture

---

[17] M. Ishiura, S. Hirose, T. Uchida, Y. Hamada, Y. Suzuki, and Y. Okada, *Mol. Cell. Biol.* **2**, 607 (1982).

dish and incubated overnight (total 10 plates). After gentle swirling, the plates are incubated for 12–24 hr at 35°, washed, and refed with growth medium. Following incubation for 24 hr, cells are trypsinized, split 1:5 or 1:10, and replated on 50 or 100 plates. One day later, selection is started with medium containing G418(400 µg G418 of 100% potency/ml). Different cells require different amounts of the drug; this figure applies to LA9, Ltk⁻, and NIH3T3 cells. Medium is changed every 3 days for 2–3 weeks until neomycin-resistant cell colonies grow up. A second selection or screening method is then used to detect the cells which express the phenotype or protein encoded by the desired cDNA.

We have used this transducing system to complement hypoxanthine–guanine phosphoribosyltransferase (HPRT)-deficient mouse LA9 cells with a human fibroblast cDNA library, and obtained three neomycin-resistant colonies (per 10^7 transfected cells) that could continue to grow in HAT medium. Isoelectric focusing gel electrophoresis of cell extracts followed by *in situ* HPRT assay revealed that two of the three extracts contained a molecule with HPRT activity, the mobility of which is characteristic of the human enzyme (Fig. 4). The genome of the positive cells contained human HPRT cDNA sequences excisable with *Bam*HI endonuclease, characteristic of the cDNA inserted in the pcD vector. These results as well as those of the reconstitution experiments[11] indicate that the phage vector-mediated transduction of cDNA libraries is efficient enough to detect a particular functional cDNA clone as rare as one in 10^6 clones from the library.

Recovery of Transduced cDNA in *E. coli*

The transduced cDNA gene is recovered and cloned in *E. coli* by one of the following methods.

Fusion with COS Cells

This method relies entirely on the presence of the SV40 replication origin, the β-lactamase gene, and the pBR322 replication origin in the integrated λNMT-pcD recombinant. Breitman et al.[18] have used cell fusion between the transformed cells and COS cells to recover the pSV molecule from transformed cells. Large T-antigen and other permissive factors provided by COS cells drive the initiation of replication at the integrated SV40 origin and promote excision of the replicating molecule as an extrachro-

---

[18] M. L. Breitman, L. C. Tsui, M. Buchwald, and L. Siminovitch, *Mol. Cell. Biol.* **2**, 966 (1982).

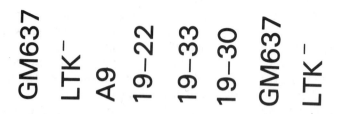

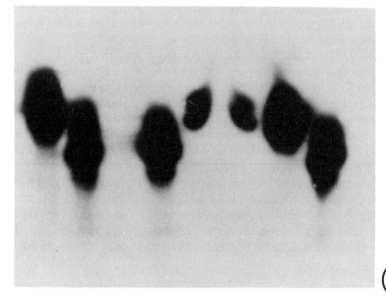

Fig. 4. Isoelectric focusing gel electrophoresis of extracts from HAT-resistant cells followed by *in situ* assay of HPRT. Approximately 10^7 HPRT-deficient A9 cells were transfected with 5×10^{11} PFU of the SV40-transformed human fibroblast (GM637) cDNA expression library in the λNMT vector. Three HAT-resistant colonies were obtained after sequential selection with G418 and HAT. Cell extracts from GM637, Ltk⁻, and A9 cells and three HAT-resistant colonies 19-22, 19-33, and 19-30 were electrophoresed in an isoelectric focusing polyacrylamide gel (pH 5.5 to 8.5) and assayed for HPRT activity *in situ* by the method of L. A. Chasin and G. Urlaub [*Somatic Cell Genet.* **2**, 453 (1976)].

mosomal circular DNA. Since such DNAs are likely to also contain the flanking pBR origin and β-lactamase gene as well as the cDNA, they can be cloned by direct transfection into *E. coli*.

Reagent

Polyethylene glycol (MW 1000), Sigma

Procedure. Approximately 10^6 transduced cells and 2×10^6 COS cells are mixed, and plated together on a 10 cm tissue culture plate in DMEM containing 10% fetal calf serum. After a 24-hr incubation, cells are washed twice with DMEM without serum, and treated with 2 ml of 50% polyethylene glycol/50% DMEM at room temperature for 1 min followed by washing twice with DMEM and once with DMEM containing 10% fetal calf serum. Cells are then refed with growth medium and incubated at 37° for 2–3 days. Extrachromosomally maintained small-molecular-weight DNA is recovered from the fused cells by the method of Hirt.[19] The fused cells are lysed with 0.5 ml of 10 mM Tris–HCl (pH 7.5) buffer containing 0.6% SDS and 10 mM EDTA, and after incubation at room temperature for 20 min, 0.2 ml of 5 M NaCl is added to the lysate, which is then transferred to a 15-ml Corex glass centrifuge tube and incubated at 4° overnight. Chromosomal DNA coprecipitates together with the SDS during this incubation and is removed by centrifugation at 15,000 g for 1 hr. The supernatant is treated with 20 μg of bovine pancreatic RNase at 37° for 20 min and extracted twice with phenol/CHCl$_3$. Small-molecular-weight DNA is recovered by ethanol precipitation. The precipitated DNA is dissolved in 100 μl of 10 mM Tris–HCl (pH 7.5) containing 1 mM EDTA and transfected into *E. coli* strain DH1 according to the method of Hanahan.[20] Transformants are selected on LB agar plates containing 50 μg ampicillin/ml.

We have used this procedure to recover the integrated pcD-human hprt cDNA from 19–33, one of our transduced cells (see above). The Hirt supernatant prepared from the fused cells yielded approximately 200 DH1 transformants, roughly 30% of which contained the pcD plasmid with a human hprt cDNA insert.

Construction of a λ Genomic DNA Library followed by Screening with Flanking pcD Vector Sequences

Although not rapid, this is a general, more established procedure and is suitable especially when the SV40 origin, the pBR322 origin, or the β-lactamase gene is inactivated in the integrated λNMT-pcD-cDNA recombinant, resulting in an inability to use the fusion method.

Reagents

Guanidine thiocyanate solution: 5.5 M guanidine thiocyanate, 25 mM sodium citrate, 1% sodium laurylsarcosine, 0.2 M 2-mercaptoethanol. After the pH is adjusted to 7.0, the solution is filtered with a Nalgene 0.45-μm filter.
3.8 M CsCl solution: 3.8 M CsCl, 0.1 M EDTA (pH 7.0)
5.7 M CsCl solution: 5.7 M CsCl, 0.1 M EDTA (pH 7.0)

[19] B. Hirt, *J. Mol. Biol.* **26,** 365 (1967).
[20] D. Hanahan, *J. Mol. Biol.* **166,** 557 (1983).

Procedure. Total genomic DNA is prepared from transduced cells by a modification of the guanidine thiocyanate method.[21] Approximately 10^8 cells are lysed wtih 20 ml of guanidine thiocyanate solution and loaded into a SW27 tube containing 20 ml of 5.7 M CsCl solution overlaid with 10 ml of 3.8 M CsCl solution. After centrifugation at 25,000 rpm for 24 hr at 20°, DNA, banding at the interface between the two CsCl solutions, is recovered by puncturing the tube with an 18-gauge needle and syringe. The DNA is dialyzed against 1 liter of 10 mM Tris–HCl (pH 7.5) containing 1 mM EDTA. The dialyzed DNA is concentrated by placing the bag in Ficoll powder and redialyzed against the same buffer. The DNA is partially digested with *Mbo*I and inserted into the EMBL3 or EMBL4 vector at the *Bam*HI site as described by Frischauf *et al.*[22] After packaging *in vitro,* approximately 10^6 independent recombinant phage clones are screened with nick-translated pcD vector sequences as a hybridization probe.

Comments

If difficulties in propagating the recombinant phage in JC8679 or some instability of the inserted pcD plasmid during this propagation are encountered, use the *Sfi*I site to transfer the library into the phage vector, and C600 to amplify the recombinant phage.

The conditions for tranfection of cells with phage particles have been optimized for mouse LA9, Ltk⁻, and NIH3T3 cells, and result in 0.1–0.5% of cells stably transformed to *neo* resistance, with approximately 10 pcD-cDNA recombinants integrated in the genome of each cell. Other cells may transform less efficiently, and optimization of each parameter (temperature, CO_2, phage dose, and G418 concentration) for each cell type may be required. Use of low potency serum of mycoplasma-infected cells may result in a marked decrease in transformation efficiency.

[21] J. M. Chirgwin, A. E. Przybyla, R. J. MacDonald, and W. J. Rutter, *Biochemistry* **18,** 5294 (1979).

[22] A.-M. Frischauf, H. Lehrach, A. Poustka, and N. Murray, *J. Mol. Biol.* **170,** 827 (1983).

[33] Sib Selection

By Mary McCormick

Many laboratories are interested in the problem of gene isolation from a library of DNA sequences. Sib selection is a method of sequential fractionation of a heterogeneous sample that can be applied to isolation of a sequence, gene, or gene family from a complete library. Sib selection is compatible with any assay system or selection for the sequence of interest; however, it is especially useful if there is no selectable phenotype or sequence information available. For example, a gene for which there is no selection or sequence information but which codes for an easily assayable enzyme may be isolated by sib selection. This could be done by transferring library DNA into a host cell deficient for that enzyme, and assaying for enzyme activity. Library fractions that are positive in that assay could then be further subfractionated until a single positive clone is obtained.

Cavalli-Sforza and Lederberg[1] originally described sib selection in the isolation of preadaptive mutants in bacteria, determining that drug-resistant mutants in *Escherichia coli* were spontaneous, not induced. Goubin *et al.*[2] used sib selection to isolate the B-*lym* oncogene from a B cell lymphoma induced by avian lymphoid leukosis virus. Colburn and colleagues[3] used sib selection to isolate sequences involved in tumor promotion.

Fractionation of a library in search of positive subfractions is a relatively straightforward concept. However, the probability of finding a positive sample with a given fractionation scheme must also be considered. One can perform such probability calculations using the equation

$$P = 1 - (1 - f)^N \tag{1}$$

or its alternative:

$$N = \ln(1 - P)/\ln(1 - f) \tag{2}$$

where P is the probability, N is the sample number, and f is the fraction of the total sample in each subfraction.[4] Initially, Eq. (2) can be used to calculate the number of colonies required to ensure a 99.9% probability that a unique sequence will be present in a mammalian cosmid library.

---

[1] L. L. Cavalli-Sforza and J. Lederberg, *Genetics* **41**, 367 (1955).

[2] G. Goubin, D. S. Goldman, J. Luce, P. E. Neiman, and G. M. Cooper, *Nature (London)* **302**, 114 (1983).

[3] M. I. Lerman, G. A. Hegamyer, and N. H. Colburn, *Int. J. Cancer,* **37**, 293 (1968).

[4] L. Clark and J. Carbon, *Cell* **9**, 91 (1976).

With a cosmid vector that accepts 40 kb inserts and a mammalian genome of 3×10^6 kb, $f = 40/3 \times 10^6$. Therefore

$N = \ln(1 - 0.999)/\ln(1 - 40/3 \times 10^6)$
$N = 5.3 \times 10^5$ colonies (assuming one cosmid vector plus insert per colony).

In another example, the probability of finding a particular unique sequence in a mammalian library composed of 1×10^5 bacterial colonies is

$$P = 1 - [1 - (40/30 \times 10^6)]^{1 \times 10^5}$$
$$P = 0.73$$

Simple Fractionation of a Cosmid Library

Probability Eqs. (1) and (2) can also be applied to library fractionation by redefining f as the fraction of the library represented in each sample and N as the number of samples. In the following examples we will consider a cosmid library where the colonies have been pooled and subfractions are taken as aliquots from that pool. In this case, the value of f equals the number of colonies in each sample divided by the number of colonies in the original library. If we assume that the sequence of interest is present only once in the original library (e.g., 5×10^5 colonies), and we take 10 random fractions, each consisting of one-tenth the size of the original library (5×10^4 colonies), the probability that any one of the 10 fractions would contain the clone of interest is

$$P = 1 - (1 - 0.1)^{10}$$
$$P = 0.65$$

The number of samples that must be analyzed to achieve a 90% probability that the clone of interest will be found using this fractionation scheme is

$$N = \ln(1 - 0.9)/\ln(1 - 0.1)$$
$$N = 23 \text{ samples}$$

The effect of finding the clone of interest in a subfraction of the library is to enrich for that clone. As described in Cavalli-Sforza and Lederberg,[1]

$$E = r_1/r_0 \tag{3}$$

where E is the enrichment factor, r_0 is the ratio of the number of colonies containing the clone of interest to the total number of colonies in the original library (using the cosmid library example), and r_1 is the ratio of the number of colonies containing the clone of interest to the total number of colonies in the positive subfraction.

Using Eq. (3) to calculate the enrichment factor in the example above,

$$E = (1/5 \times 10^4)/(1/5 \times 10^5)$$
$$E = 10$$

Therefore, a 10-fold fractionation of a library where the clone of interest is present once will yield a 10-fold enrichment of this clone in a positive subfraction. It is important to note that the enrichment factor is inversely related to the probability of finding the clone of interest in a given number of fractions of a particular size. For example, the probability of achieving a 100-fold enrichment by assaying 10 samples is low:

$$P = 1 - (1 - 0.01)^{10}$$
$$P = 0.1$$

Calculating the number of samples, N, to be analyzed for a 90% probability of finding a positive fraction with a one hundred-fold enrichment we find

$$N = \ln(1 - 0.9)/\ln(1 - 0.01)$$
$$N = 230 \text{ samples}$$

Therefore, it is more efficient to perform two rounds of sib selection with 10-fold enrichment each time than to attempt a 100-fold enrichment in one round of sib selection.

Serial Fractionation of a Cosmid Library

In many cases a sequence will be assayed for which the copy number in the genome is not known. In this event a serial fractionation, rather than a simple fractionation as described above, can be performed to determine at what number of colonies the standard sib selection should be initiated. For example, a series of samples consisting of unrelated 10^{-1}, 10^{-2}, 10^{-3}, and 10^{-4} fractions of a library of 5×10^5 colonies yields random samples of 5×10^4, 5×10^3, 5×10^2, and 50 colonies, respectively. If the copy number in the mammalian genome for the sequence of interest was 100, the P values for a cosmid library with 40 kb inserts for the above samples can be calculated. Since the sequence is present in 100 copies, one would use $f = 40 \text{ kb} \times 100/3 \times 10^6 \text{ kb}$, rather than $f = 40 \text{ kb}/3 \times 10^6 \text{ kb}$, in Eq. (1).

For the 5×10^5 colony sample (unfractionated)

$$P = 1 - [1 - (40 \times 100/3 \times 10^6)]^{5 \times 10^5}$$
$$P \approx 1$$

For the 5×10^4 colony sample (10^{-1} fraction)

$$P = 1 - [1 - (40 \times 100/3 \times 10^6)]^{5 \times 10^4}$$
$$P \approx 1$$

For the 5×10^3 colony sample (10^{-2} fraction)

$$P = 1 - [1 - (40 \times 100/3 \times 10^6)]^{5 \times 10^3}$$
$$P = 0.999$$

For the 500 colony sample (10^{-3} fraction)

$$P = 1 - [1 - (40 \times 100/3 \times 10^6)]^{500}$$
$$P = 0.49$$

For the 50 colony sample (10^{-4} fraction)

$$P = 1 - [1 - (40 \times 100/3 \times 10^6)]^{50}$$
$$P = 0.06$$

Therefore, the simple fractionation procedure described in the previous section would begin with a 10^{-2} or 10^{-3} fraction of the library. This would result in fewer sib selection cycles required for the isolation of a single clone.

Grid Analysis

Grid analysis is a two-dimensional fractionation procedure where samples are pooled so that more information is gained per assay than when each sample is tested individually. This procedure is best described by illustration.

	A	B	C	D	E
F	1	2	3	4	5
G	6	7	8	9	10
H	11	12	13	14	15
I	16	17	18	19	20
J	21	22	23	24	25

The 25 samples to be analyzed are pooled in groups of five according to the above grid. For example, pool A contains samples 1, 6, 11, 16, and 21; and pool F contains samples 1, 2, 3, 4, and 5. There are 10 such pools, A through J, altogether. Each of these 10 pools is then analyzed, and any positive samples from the original 25 can be identified from the grid. For example, if pools A and I are positive, and all others are negative, then sample 16 is positive, and all other 24 are negative. In this example, 10 samples were assayed, but information for 25 samples was obtained. In general, for a square grid

$$Y = (X/2)^2 \qquad (4)$$

where Y is the number of samples for which information is obtained, and

X is the number of pooled samples to be analyzed. By applying this type of grid analysis to sib selection, a larger number of samples can be analyzed per cycle, resulting in a fewer number of cycles required to obtain an isolated clone.

Summary

Isolation of a particular clone from a library by sib selection is based on fractionating the library and choosing a positive subfraction which is consequently enriched for the clone of interest. The goal is to repeat this cycle until the enrichment process results in purification. As discussed in this chapter, particular patterns of fractionating are more efficient than others depending on the relative abundance of the clone, and the size of the original library. Importantly, sib selection is an isolation process that can be used in conjunction with any assay system. In particular, it may be used to isolate a gene from a library where there is an assay for the gene product, but no selection or sequence information available.

Acknowledgments

I thank Bruce Howard for many informative discussions, and Rudy Pozzatti and Bruce Howard for critically reading the manuscript.

[34] Repetitive Cloning of Mutant Genes Using Locus-Specific Sticky Ends

By LAWRENCE A. CHASIN, ADELAIDE M. CAROTHERS, and GAIL URLAUB

Genetic analysis frequently requires the repeated cloning of the same gene. For example, one may want to clone multiple alleles from different individuals in a population, oncogenes from different tissues of the same organism, or mutant genes from different subclones of a cultured cell line. Once an initial isolation of a wild-type gene has been accomplished, this DNA sequence can be used as a probe to identify new alleles in subsequent cloning experiments. The availability of such a probe makes repeated isolations of the gene feasible using standard cloning procedures. However, in the case of the complex genomes of most higher organisms, this is a laborious process, since libraries of several hundred thousand recombinant molecules must be constructed and screened in order to ensure the inclusion of all single copy genes. Several strategies have been described that reduce the size of the libraries that need be screened for repetitive cloning once the wild-type gene is in hand. One method makes use of *in vivo* recombination in *Escherichia coli* to select phage or bacterial clones that contain sequences homologous to the original cloned gene.[1,2] A second method exploits the probability that a given sequence is devoid of sites for an empirically determined set of restriction enzymes; the desired fragment is enriched in the higher molecular weight fraction of a restriction digest.[3] In this chapter, we describe another method of preparing a recombinant DNA library that is enriched for one particular gene sequence. This method involves the use of vectors containing small regions derived from the flanks of the wild-type gene. The vectors are cut in such a way as to produce sticky ends that will preferentially ligate to the ends of the fragment to be cloned.

Cloning Strategy

The basic rationale is to use a restriction enzyme that makes a staggered cut at a nonspecific sequence either within or flanking its specific recognition site. The sequences at the sticky ends so generated are therefore characteristic of the locus rather than of the enzyme. For example, the enzyme *Bgl*I recognizes the sequence GCCNNNNNGGC, cutting this palindromic sequence to produce two ends with 3-base 3′ protrusions:

```
--G C C N N N N N G G C-- → --G C C N N N N +      N G G C--
--C G G n n n n n C C G--   --C G G n          n n n n C C G--
```

where N and n represent complementary pairs of any of the four bases. The fragments created by *Bgl*I thus have heterogeneous trinucleotide ends. Most of these molecules cannot be cloned into a given *Bgl*I site because their sticky ends will not hybridize to those of the vector. However, if a vector is constructed to include the particular *Bgl*I sites that flank the region to be cloned, the restriction fragment from this locus will be able to anneal to it.

The theoretical maximum enrichment in cloning a specific fragment using this method can be calculated for a restriction enzyme that produces a 3-base staggered cut, as follows. Assuming a random distribution of nucleotides, the chance that a particular trinucleotide will occur at the end of a particular restriction fragment is $(\frac{1}{4})^3$ or $\frac{1}{64}$. The probability that a molecule will fortuitously have the same two trinucleotides present at the

[1] B. Seed, *Nucleic Acids Res.* **11**, 2427 (1983).
[2] A. Poustka, H. Rackwitz, A. Frischauf, B. A. Hohn, and H. Lehrach, *Proc. Natl. Acad. Sci. U.S.A.* **81**, 4129 (1984).
[3] R. D. Nicholls, A. V. S. Hill, J. B. Clegg, and D. R. Higgs, *Nucleic Acids Res.* **13**, 7569 (1985).

two ends of the desired fragment is $\frac{1}{64} \times \frac{1}{64} \times 2 = \frac{1}{2048}$. Thus the use of a vector containing the very same restriction sites that bound the desired fragment should result in a 2000-fold enrichment compared to random cloning. In practical terms this difference should change the task of screening hundreds of thousands of recombinants into the more manageable task of screening hundreds.

This strategy has been used in conjunction with *Bgl*I and cosmid cloning vectors for the repetitive cloning of dihydrofolate reductase *(dhfr)* genes from mutant subclones of Chinese hamster ovary (CHO) cells. The procedure used in this application[4] is described below as an example of the method.

The Chinese Hamster *dhfr* Gene

The *dhfr* gene in Chinese hamster cells spans 26 kb and contains 6 exons.[5] The gene is flanked by two *Bgl*I sites, a 5' site located about 8 kb upstream of the transcriptional start site and a 3' site about 7 kb downstream of the last major polyadenylation site. There are no *Bgl*I sites within the gene, so digestion with this enzyme produces a 41 kb fragment. This size fragment is suitable for cloning into a cosmid vector.[6]

Construction of a *Bgl*I-Free Cosmid Vector

In order to specifically clone the *dhfr* fragment, it was necessary to construct two cosmid vectors, one containing the exact *Bgl*I site found in the 5' flank of the *dhfr* gene and one containing the exact 3' site. Two separate vectors are required so that the final recombinant molecules will exist as linear concatemers, the form of substrate required for packaging in λ. In addition, the vectors must contain no other *Bgl*I sites. These two vectors were constructed starting with the cosmid pGC*cos*3*neo,* constructed and kindly provided by Grey Crouse. An attractive feature of this vector is that it contains, in addition to the plasmid origin of replication and the β-lactamase gene, the *neo* gene under the control of the SV40 early promoter and followed by SV40 splicing and polyadenylation signals (Fig. 1). This gene confers resistance to neomycin and kanamycin in *E. coli;* at the same time it confers resistance to the drug G418 in most mammalian cells. Thus cloned genes inserted into this vector can subsequently be transferred into mammalian cells by selecting for G418 resistance. The

---

[4] G. Urlaub, A. M. Carothers, and L. A. Chasin, *Proc. Natl. Acad. Sci. U.S.A.* **82,** 1189 (1985).

[5] A. M. Carothers, G. Urlaub, N. Ellis, and L. A. Chasin, *Nucleic Acids Res.* **11,** 1997 (1983).

[6] J. Collins and B. Hohn, *Proc. Natl. Acad. Sci. U.S.A.* **75,** 4242 (1979).

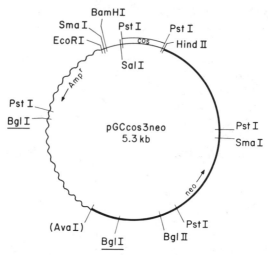

FIG. 1. Map of the cosmid vector pGCcos3neo. This vector was constructed by G. Crouse (Department of Biology, Emory University, Atlanta, Georgia, personal communication, 1983). It consists of the following (reading in a clockwise direction): (1) The *AvaI/EcoRI* fragment of pBR327 [X. Soberon, L. Covarrubias, and F. Bolivar, *Gene* **9,** 287 (1980)] containing the origin of replication and the β-lactamase gene (1.8 kb). The *Ava*I site was filled in and blunt end ligated to the *Pvu*II site of the origin of replication of pSV2*neo* [P. J. Southern and P. Berg, *J. Mol. Appl. Genet.* **1,** 327 (1982)], destroying both restriction sites. Note that pBR327 has had deleted all the sequences thought to "poison" replication in eukaryotes. (2) the *EcoRI/HindIII* polylinker region of pUC8, into which has been inserted the *BamHI/SalI* fragment of pBR327 and the 400 bp *cos* fragment of MUA-10 [E. M. Meyerwitz, G. M. Guild, L. S. Prestige, and D. S. Hogness, *Gene* **11,** 271 (1980)]. (3) The *BamHI/PvuII* fragment of pSV2*neo*. The *Bam*HI site was converted to a *Hind*III site by linker addition, the *Hind*III site of the SV40 *ori* was destroyed by filling in and reclosing, and the *Pvu*II site was destroyed by the ligation to the filled in *Ava*I site of pBR327. An analogous vector has been constructed containing the *E. coli gpt* gene instead of the *neo* gene. These cosmids have been designed so that the method of D. Ish-Horowicz and J. R. Burke [*Nucleic Acids Res.* **9,** 2989 (1981)] can be used to reduce the background of cosmid multimers. pGCB2 is a *Bgl*I-free derivative of pGCcos3*neo* that lacks the two *Bgl*I sites (shown underlined in the map). Its construction is described in the text.

resistance to kanamycin in *E. coli* is also an important feature of the vector. The β-lactamase gene conferring ampicillin resistance contains a *Bgl*I site that must be removed to make the vector useful here; the removal of this site leads to inactivation of this drug resistance marker. Selection in *E. coli* then depends on the presence of the *neo* gene. A second *Bgl*I site is present in the region of the SV40 replication origin; its removal does not interfere with the function of the SV40 early promoter. The *Bgl*I sites were removed in one step by cutting with *Bgl*I, trimming the 3' overhang with

T4 DNA polymerase,[7] and ligating the two blunt-ended fragments. Since one fragment contained the plasmid origin of replication and the other contained the *neo* gene, reassembled molecules could be readily selected. The ligated DNA was used to transform *E. coli* HB101 to kanamycin resistance, and transformants were screened for the absence of *Bgl*I sites by digesting minipreparations of plasmid DNA. The cosmid vector pGCB2 (see Fig. 1) was isolated in this way and served as the basis for all further constructions. This vector should be of general usefulness for the insertion of specific *Bgl*I sites in the application of this cloning method.

Construction of the Two Cosmid Cloning Vectors

The 5′ *Bgl*I site from the *dhfr* locus was isolated as a 2.0 kb restriction fragment from a previously isolated[8] cosmid clone carrying this region. It was inserted into a *Bam*HI site in the cosmid vector pGCB2. The 3′ *Bgl*I site was isolated as a 0.72 kb fragment and inserted into the *Bam*HI site in the second vector. The details of these constructions can be found in Ref. 4; the structure of the 5′ and 3′ vectors can be seen in Fig. 2. The desired recombinant vectors can be conveniently identified by screening minipreps for the presence of the cloned restriction site. There are two important general constraints in the construction of these cosmid vectors. First, the fragments containing the specific sites must be oriented such that when the two cosmid vectors are cut and ligated to the fragment to be cloned, the resulting ligation products will include molecules in which two cohesive end *(cos)* sites are in the same orientation and flank a region containing (1) the drug resistance marker, (2) the plasmid origin of replication, and (3) the desired fragment to be cloned. Second, the total distance between the *cos* sites should not exceed about 50–52 kb, the limit for packaging in λ phage heads.[6] It should be noted that the *Bgl*I-containing cloning sites need not be cloned as very small fragments to satisfy this size limit. Rather the critical factor is the size of the *Bgl*I to *cos* fragments contributed by each vector to the cosmid.

Ligation of Genomic and Vector DNA Fragments

Since the size of the *Bgl*I fragment to be cloned was known, and since it was larger than most *Bgl*I fragments in the digest, a size fractionation step was carried out prior to ligating the digested DNA. CHO DNA (350–

---

[7] T. Maniatis, E. F. Frisch, and J. Sambrook, "Molecular Cloning: A Laboratory Manual." Cold Spring Harbor Lab., Cold Spring Harbor, New York, 1982.

[8] J. D. Milbrandt, J. C. Azizkhan, and J. L. Hamlin, *Mol. Cell. Biol.* **3**, 1266 (1983).

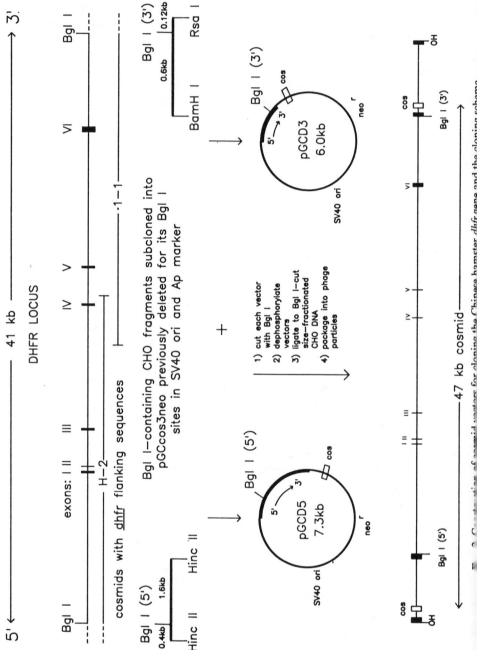

500 μg) was digested with *Bgl*I (0.5 units/μg, 2 hr at 37°) and divided among 3 or 4 tubes containing 13.5 ml of a 4–12% sucrose gradient in TEN buffer (TE containing 0.2 M NaCl; TE is 10 mM Tris–HCl, pH 8.0, 1 mM EDTA). The tubes were centrifuged at 22,000 rpm for 17 hr at 20° in a Beckman SW40 rotor. The pellet contains DNA of molecular size greater than approximately 35 kb. After removal of all but about 0.4 ml of the supernatant solution, an equal volume of TEN buffer was added and the DNA was solubilized and pooled. After precipitation with ethanol, the DNA was resolubilized in 0.5 ml of TE buffer, extracted three times with phenol and three times with ether, precipitated twice with ethanol, and resolubilized in 0.1× TE at a concentration of 1 mg/ml.

Vector DNA was cut with *Bgl*I and treated with calf intestinal phosphatase (for 1 hr at 55° because of the recessed 5' ends) to prevent self-ligation.[7] Each vector DNA was included at a weight ratio of about 1:25 compared to genomic DNA. For this experiment, this results in an approximate 4-fold molar excess of total genomic ends. However, the vectors are in great excess relative to the much smaller number of ligatable genomic molecules that contain matching sticky ends. We have found that this ratio of vector to genomic DNA is optimal for the production of recombinants; a greater excess of vector inhibits the packaging reaction. The ligation reaction was stopped by adding EDTA (pH 8.0) to 10 mM and heating at 68° for 15 min. The ligated DNA was then ethanol precipitiated and resolubilized to a concentration of 1 mg/ml in 0.1× TE buffer. The extent of the ligation can be monitored by Southern blot analysis at this stage, if desired. A substantial proportion, if not the majority, of the desired fragment should show evidence of ligation.

Packaging in λ Phage Heads

Packaging extracts were prepared and packaging was carried out as described in Ref. 7 (method II), except that putrescine was omitted. Putrescine allows the packaging of smaller molecules, so its inclusion or omission should be decided based on the known size of the recombinant molecule. Putrescine was not used here, since the molecule to be cloned was known to be large. Four micrograms of genomic DNA was included in each standard packaging reaction. Cosmid vector DNA (both vectors combined) totaled 350 ng per reaction. Amounts of vector greater than approximately 500 ng were found to inhibit the packaging reaction. This inhibition is likely to vary with different preparations of packaging extract and of vector DNA, so a titration (using a small amount of λ DNA and scoring for plaque-forming units) should be performed to determine the maximum amount of vector that can be added. Once this maximum is determined, the only way to obtain more packaged cosmids is to increase the number of packaging reactions. *E. coli* strain DH1[9] was used as the recipient strain for cosmid infection. Overnight cultures grown in Luria broth (LB) containing 0.2% maltose were centrifuged, resuspended in the original volume of 10 mM MgSO$_4$, and incubated at 37° for 30 min. The bacterial suspension (0.3 ml) and one-quarter of the packaged particles (0.12 ml) from a standard packaging reaction were mixed and incubated at 37° for 20 min. The packaged particles from each reaction were divided into four parts before adsorption in order to maximize the probability of isolating independent clones. After dilution with 1 ml of LB and shaking at 37° for 60 min to allow expression of the drug-resistant phenotype, each culture was selected on one plate of LB agar containing 20 μg/ml of kanamycin. Reconstruction experiments indicated that spreading this number of cells on a dish often led to lower colony yields. Maximum colony yield was obtained by inoculating the bacteria in 3 ml of top agar (0.7% containing 5 μg/ml kanamycin). The yield of kanamycin-resistant colonies per standard packaging reaction was usually 50 to 100.

Screening Drug-Resistant Colonies by Hybridization

Colonies were picked to duplicate grids on 150 mm agar dishes, one of which contained a gridded nitrocellulose sheet (10 × 10 cm, cut from a 137 mm circle of Schleicher and Schuell BA85/20) on its surface. After overnight growth, the filter (without replication) was processed for amplification and hybridization using the procedure of Hanahan and Meselson.[10] Colonies containing *dhfr* sequences were detected using radioactively labeled cloned sequences from the CHO *dhfr* gene as a probe. Positive colonies were recloned and screened for the presence of the 41 kb *Bgl*I fragment by digestion of minipreps with *Kpn*I and noting the characteristic fragment pattern upon gel electrophoresis.

Results of Cloning the *dhfr* Gene from a Variety of CHO Sublines Containing One Copy of the Gene

This method has been applied to cloning the *dhfr* gene from a CHO cell line that is hemizygous for this locus[11] and from six mutants derived from this line. Each cloning was carried out in duplicate. We usually try to generate many hundreds of kanamycin-resistant colonies in each experiment, and screen 350 by hybridization analysis. Occasionally, no positive

---

[9] D. Hanahan, J. Mol. Biol. **166,** 557 (1983).
[10] D. Hanahan and M. Meselson, this series, Vol. 100, p. 333.
[11] G. Urlaub, E. Kas, A. M. Carothers, and L. A. Chasin, *Cell* **33,** 405 (1983).

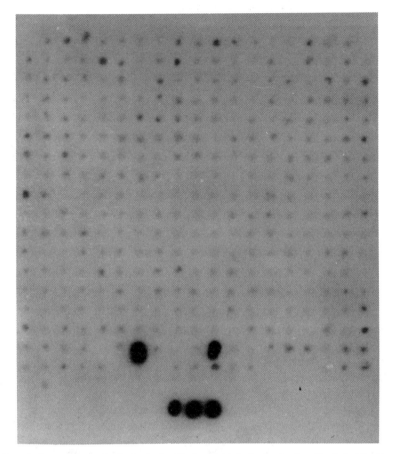

Fig. 3. Autoradiograph of a typical grid of cosmid-derived colonies after hybridization to a *dhfr*-specific probe. The three colonies at the bottom represent positive controls (transformants carrying a cosmid-cloned *dhfr* gene).

colony is present in this set, and another 350 are screened. Three hundred fifty colonies plus positive and negative control colonies conveniently fit on a single 10×10 cm filter sheet which itself fits into a 150 mm dish. A typical experimental filter is shown in Fig. 3.

The results of these 7 clonings are shown in Table I. The overall frequency of *dhfr*-containing cosmids was 17/6668 colonies screened, or 0.25%. The enrichment produced by this method can be estimated as follows. A pseudodiploid CHO cell contains about 5×10^6 kb of DNA. The initial size fractionation step resulted in a 5-fold purification, thus reducing this complexity to about 10^6 kb. If the average size of this DNA was 35 kb, then 29,000 different molecules would be present. Thus with no

TABLE I
SUMMARY OF *dhfr* CLONING EXPERIMENTS

Cloning experiment	Genomic DNA source	DNA[a] (μg)	Number of colonies screened	*dhfr*+ colonies found
1	UA21	46	453	3
2	DS11	24	363	2
3	DS21	110	2294	3
4	DS31	56	1330	2
5	RDS1	24	1054	3
6	RDS3	24	598	2
7	DU92	28	576	2
Total			6668	17 (0.25%)

[a] The μg of DNA refers to size-fractionated preparations.

enrichment, a frequency of 1/29,000, or 0.0035% would be expected. The observed frequency of 0.25% indicates that the use of the custom cloning vector has led to a 70-fold enrichment for the *dhfr* gene among recombinant colonies. Theoretically, one would predict a frequency of *dhfr*-positive colonies of 8%, since only 12 molecules with the appropriate ends (1/2048) should be present in the population of 29,000. That the method falls short of this maximum is probably due to illegitimate ligation of unmatched ends. Also, we have observed that cosmids carrying the *dhfr* gene grow more slowly and have a lower colony-forming efficiency than other cosmid clones. Thus the yield calculated for the *dhfr* gene may be an underestimate of the enrichment potential of the method in general. In any case, the important result is that a single copy gene can be reliably cloned from a complex genome by screening only several hundred colonies on a single nitrocellulose filter.

Discussion

In these experiments, locus-specific sticky ends were generated as a means of allowing the repetitive cloning of the *dhfr* gene. In this particular case, we were fortunate that sites for an enzyme producing such ends (*Bgl*I) were located in the flanks of this gene and that cutting produced a fragment of a size suitable for cloning in a cosmid vector. However, the method is not dependent on such a fortuitous placement in order to be of general use.

There are several restriction enzymes in addition to *Bgl*I that cut outside of their recognition sequence and so generate locus-specific sticky ends. A list of some of these enzymes compiled from the catalogs of

TABLE II
USEFUL RESTRICTION ENZYMES THAT PRODUCE LOCUS-
SPECIFIC STICKY ENDS[a]

Enzyme	Recognition and cutting(/) site	Vendors
*Bbv*I	GCAGCNNNNNNNN/ CGTCGNNNNNNNNNNNN/	NEB
*Fok*I	GGATGNNNNNNNNN/ CCTACNNNNNNNNNNNNN/	NEB, AM
*Hga*I	GACGCNNNNN/ CTGCGNNNNNNNNNN/	NEB
*Sfa*NI	GCATCNNNNN/ CGTAGNNNNNNNNNN/	NEB
*Bgl*I	GCCNNNN/NGGC	BM, BRL, NEB
*Bst*XI	CCANNNNN/NTGG	NEB
*Dra*III	CACNN/GTG	ANG, BM
*Sfi*I	GGCCNNNN/NGGCC	NEB

[a] Commercially available restriction enzymes that cut in a sequence other than their recognition sequence to produce locus-specific sticky ends with at least a 3-base overhang. The list was compiled from the catalogs of Amersham-Searle (AM), Anglian Biotechnology (ANG), Bethesda Research Laboratories (BRL), Boehringer-Mannheim (BM), and New England Biolabs (NEB).

commercial vendors is presented in Table II. The locus to be cloned could be screened with a number of enzymes to find the combination that yields the best placement of sites. One can anticipate that for many large genes, no enzyme will be found that does not interrupt the gene. In this case the region can simply be cloned in two pieces, since for many analyses, it is not necessary to isolate an intact gene.

The size of the fragment to be cloned will dictate the choice of a cloning vector. Cosmids were used here because of the large size of the fragment. Smaller fragments could also be cloned as cosmids if spacer DNA, devoid of the restriction sites used for cloning, is added to the cosmid vector to bring the final size into the range for packaging. Vectors other than cosmids should also be useful. Cloning in λ may present a problem in that the large arms are likely to include sites for the restriction enzyme chosen for cloning. If this is the case, it still may be possible to reassemble the molecule from several pieces, since all of the ends should preferentially anneal with their correct mates (and since the vector DNA is in excess). Plasmid and bacteriophage genomes have been successfully reconstructed in this way by the perfect reassortment of up to seven *Bgl*I or *Hga*I restriction fragments.[12,13] Unlike the case of cosmids, use of bacteriophage λ permits the two restriction sites conferring cloning specificity to be placed into a single cloning vector.

Plasmid vectors could also be used, with both cloning sites in one vector molecule. Although relatively low transfection efficiency has limited the usefulness of plasmids in the construction of single copy gene libraries, this problem may be overcome by the use of higher efficiency transformation protocols.[9] Moreover, the inclusion of a size fractionation step can provide enough of a purification to make the use of plasmids feasible even with modest transformation efficiencies. Size fractionation is a reasonable first step to include in any case, since it is relatively easy to carry out, and since a prototype gene will have been cloned and mapped to provide the information for a fractionation strategy. Treatment of the genomic DNA with a number of restriction enzymes that do not cut within the desired fragment is also a powerful way to enrich for the fragment to be cloned[3] and is a procedure that could be used in conjunction with the method described here.

Finally, the locus-specific sticky end method need not be used in its most selective form in order to be of value. The utilization of two distinctive cloning sites in the experiments reported here (one on each end of the fragment) led to an enrichment of about 100-fold. The use of a locus-specific cloning site on only one end of the molecule should therefore yield the square root of 100, about a 10-fold enrichment. This single order of magnitude can represent a significant savings in labor and expense when screening many large libraries. This one-selective-end approach could also be of use in cloning rearrangements where only one end of the locus remains constant, e.g., partial deletion mutations, immunoglobulins, T cell receptors, and translocated c-*myc* genes.

The method described here presents an alternative to methods based on homologous recombination in *E. coli*.[1,2] In comparison with those procedures, it provides some potential advantages. No special bacterial hosts are required. The use of *recA* hosts minimizes the chances that a cloned gene may be altered by the cloning regimen. The cloning sites precisely define the extent of the fragment that has been cloned. Finally, in principle it is not restricted to phage-based vectors.

Acknowledgment

This work was supported by USPHS Grant GM22629 to L.A.C.

[12] P. B. Moses and K. Horiuchi, *J. Mol. Biol.* **135,** 517 (1979).
[13] K. J. Burger and R. Schinzel, *Mol. Gen. Genet.,* **189,** 269 (1983).

[35] Strategies for Mapping and Cloning Macroregions of Mammalian Genomes

By Cassandra L. Smith, Simon K. Lawrance, Gerald A. Gillespie, Charles R. Cantor, Sherman M. Weissman, and Francis S. Collins

A haploid set of human chromosomes contains about 3×10^9 base pairs of DNA, corresponding to an average of about 1.3×10^8 base pairs per chromosome. A number of techniques have aided the understanding of the organization of DNA sequences within chromosomes. Recombinational studies have suggested that the human genome corresponds to about 3300 cM of genetic distance.[1] The correspondence between base pairs and centiMorgans, however, is not linear. It may be influenced by sex, the position of the DNA relative to the centromere, and fine structure of the genetic material. Cytogenetic studies of chromosomes, such as *in situ* hybridization, can, at best, resolve distances of the order of 10^6 base pairs.[2] Mapping of gene location and distance by conventional pedigree studies has been limited by the availability of polymorphic markers. The application of studies of restriction fragment length polymorphisms (RFLPs) has been a major advance for these studies, but in particular families the number of closely linked polymorphisms may be limiting. Even where polymorphic markers are available, the resolution that can be obtained is limited by the number of individuals in informative pedigrees. Under most circumstances, it is difficult to obtain linkage information in a single family of a resolution much finer than 1 cM, a figure on the average equal to about 10^6 base pairs (bp). Population studies are required to obtain finer resolution.

In contrast to the magnitude of the genome and the limitations of resolution of these approaches to structure, conventional cloning methods can only isolate DNA fragments of the order of 40 kilobases (kb) in a single step. In addition, the resolution of conventional gel electrophoretic methods is usually only adequate for fragments of about this size or smaller. Larger stretches of DNA have been cloned by chromosome walking procedures that involve iterated steps of cloning and probe isolation, but this process can be time consuming, and is complicated by the prevalence of repetitive sequences in the human genome. Walking requires isolating of unique sequence probes from the end regions of a cosmid and using these probes to identify new, overlapping, cosmid clones. On the average, the new clone will contain only 20 kb of additional DNA and the

---

[1] T. B. Shows, A. Y. Sakaguchi, and S. L. Naylor, *Adv. Hum. Genet.* **12,** 341 (1980).
[2] D. Robins, S. Ripley, A. Henderson, and R. Axel, *Cell* **23,** 29 (1981).

maximum distance covered per probe will be somewhat less than 40 kb. For higher eukaryotes, chromosome walking has been most successful when the region to be covered has contained multiple dispersed markers that can be probed for in parallel experiments.

Thus, as indicated in Fig. 1, there is a size range between 100 and 2000 kb which is too large to approach by standard molecular techniques and too small to resolve in cytogenetic and linkage analyses. There are, therefore, substantial applications for methods that can be applied in this size range to (1) fractionate DNA fragments, (2) order such fragments, and (3) enable the rapid cloning of DNA located at distances away from an available marker. Success with such methods would expedite progress in identifying and cloning end points of DNA deletions and chromosome crossover points, and would provide valuable assistance in proceeding from loosely linked polymorphic DNA restriction fragments, such as those which have been shown to be associated with a number of inheritable diseases (e.g., Huntington's chorea, cystic fibrosis), to DNA that is more closely linked to the trait of interest.

Progress in the development of longer range mapping and cloning methods has been made with the development of pulsed field gel (PFG) electrophoresis[3-6] and "chromosome hopping" techniques.[7,8] PFG electrophoresis is a method which separates DNA fragments in a linear relation with molecular weight over a size range of several hundred thousand to over two million base pairs. This method has made it possible to fractionate and display chromosome sized DNAs of organisms such as yeast, trypanosomes, and plasmodium,[4,9] as well as fragments, generated by infrequently cutting restriction enzymes, of E. coli, mouse and human chromosomes.[8] Application of DNA hybridization procedures, in conjunction with PFG electrophoresis, has made it possible to detect fragments complementary to short unique copy probes, throughout this size range. The data generated in these studies have been used to construct physical maps of the E. coli genome[10] and the human major histocompatibility complex (S. K. Lawrance, C. L. Smith, R. Srivastava, S. M. Weissman, and C. R. Cantor, Science, 235, 1387 (1987).

[3] D. C. Schwartz, W. Saffran, J. Welsh, R. Hass, M. Goldenberg, and C. R. Cantor, Cold Spring Harbor Symp. Quant. Biol. 47, 189 (1983).
[4] D. C. Schwartz and C. R. Cantor, Cell 37, 67 (1984).
[5] G. F. Carle and M. V. Olson, Nucleic Acids Res. 12, 5647 (1984).
[6] G. F. Carle, M. Frank, and M. V. Olson, Science 232, 65 (1986).
[7] F. S. Collins and S. M. Weissman, Proc. Natl. Acad. Sci. U.S.A. 81, 6912 (1984).
[8] C. L. Smith, P. W. Warburton, A. Gaal, and C. R. Cantor, Genet. Eng. 8, 145 (1986).
[9] L. H. T. Van der Ploeg, D. C. Schwartz, C. R. Cantor, and P. Borst, Cell 37, 77 (1984).
[10] C. L. Smith, J. Econome, A. Schutt, S. Kico, and C. R. Cantor, Science, 236, 1448 (1987).

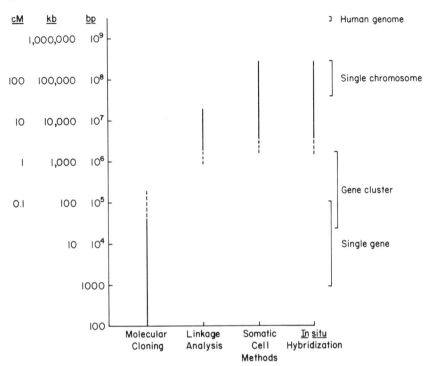

FIG. 1. Schematic representation of DNA regions analyzable by various standard techniques. Note that none of these methods is easily adaptable to the 100–2000 kb size range.

Chromosome hopping techniques include several strategies for the isolation and mapping of DNA fragments separated in the genome by distances ranging from a few thousand to greater than a million base pairs. Chromosome hopping, as originally described by Collins and Weissman,[7] is a technique that enables the isolation of DNA sequences within a defined range of distances in a specified direction from a given DNA probe. This is accomplished by circularizing large DNA fragments, generated either by random fragmentation or by complete digestion with infrequently cutting restriction enzymes, around a selectable marker and cloning the junction fragments so generated. These clones are referred to as jumping clones, or J-junctions. A similar approach has been described by Poustka and Lehrach.[11] Thus this technique may be employed to clone DNA throughout the resolution gap which can now be visualized in pulsed field gel electrophoresis and blotting experiments. For example, this procedure can

[11] A.-M. Poustka and H. Lehrach, *Trends Genet.* **2,** 174 (1986).

be used to isolate, in a single step, several unique sequence DNA probes interspersed in a distance of 50–200 kb to each side of an initial probe. These jumping fragments can then be used, in batch or in parallel, to probe a cosmid library to isolate extensive DNA sequences covering a region of 400 kb centered around the initial probe without resort to iterative cloning procedures.

A second type of library which is useful in the analysis of DNA fragments separated by distances within the resolution gap involves the selective isolation of genomic clones which contain internal sites for infrequently cutting restriction enzymes.[8] These fragments are denoted linking clones or L-junctions. L-junction libraries can be employed to determine the orientations of the fragments detected by PFG electrophoresis as well as those isolated from jumping libraries. As shown in Fig. 2, when jumping libraries are constructed from complete digests of genomic DNA with infrequently cutting restriction enzymes, the jumping and linking junction libraries are complementary and can be used in conjunction to walk rapidly along a chromosome.

The combination of PFG electrophoresis with chromosome hopping techniques, therefore, offers a considerable potential for generating restriction map data for significant portions of large genomes, including the human genome, and in addition, the cloning of gene family sized regions of DNA at preselected genomic locations.

Pulsed Field Gel Electrophoresis and Blotting

Several detailed discussions of the principles of PFG electrophoresis have recently been presented.[12] Here we will present a brief overview of the general methodology involved in this technique, and then we will focus more specifically on aspects of electrophoresis and sample handling particularly relevant to mammalian DNA samples.

Principles of PFG Electrophoresis

In conventional agarose electrophoresis of DNA molecules, the fractionation of different sized linear DNAs is based almost exclusively on the sieving properties of the gel matrix. DNA behaves like a free draining coil in electrophoresis. For such a coil, the friction is a linear function of size. The charge on DNA is also a linear function of size. Hence, the electrophoretic mobility, which depends on the ratio of charge to friction, should be independent of molecular weight. This is actually what is observed when

---

[12] C. L. Smith and C. R. Cantor, this series, Vol. 155, in press.

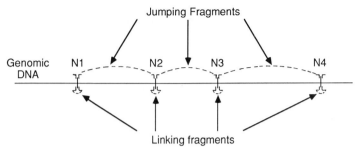

FIG. 2. Schematic representation of the relationship between *Not*I linking and jumping clones. The jumping fragments are formed by ligating the two ends of a fragment from a complete *Not*I digest of genomic DNA. The linking fragments are segments of chromosomal DNA containing an internal *Not*I site.

DNA in gels is extrapolated to zero gel concentration. At finite gel concentration while all molecules continue to move at the same local velocity, a small DNA can travel through the gel in a relatively straight path because most of the pores are accessible. A larger molecule, will have to take a longer path to find pores through which it can fit. Thus, the larger the molecule, the slower the net translation through the gel.

Sieving affords effective size separations of most types of molecules until the molecules become larger than the largest pores in the gel matrix. Then the molecules fail to migrate through the gel at all. However, DNA molecules behave like stiff coils. The shape of a DNA molecule is not constant. Under the influence of an electrical field, DNA molecules larger than the gel pores can distort their shape to enter the agarose gel. They can migrate through all the pores of the gel matrix by changing shape as the pores require. Once this process, called reptation, is initiated, all DNA molecules move through the gel with the same net velocity because the gel is no longer acting as a sieve with selective pore sizes.

In pulsed field gel electrophoresis, the ability of a gel to retard larger molecules selectively is restored by requiring that the molecules periodically change their direction of motion.[3,4] This is done by applying an electrical field in one direction for a fixed time period, the pulse time, then switching the field to a second direction, usually for the same pulse time, and then continuing to oscillate between the two directions. While no detailed mechanism for PFG electrophoresis has been proven, the following picture is consistent with the available facts.

When a field is applied, the DNA molecule must first distort its shape and orientation in the gel until it achieves a configuration that allows net translational motion. We will call this shape change DNA relaxation. The time required for relaxation is very sensitive to molecular weight; in fact over a wide size range it may scale roughly linearly in molecular weight. Thus, a larger DNA will require a larger fraction of the pulse time to relax, and have a smaller fraction of the pulse time for actual migration. This predicts that the resolution of PFG will actually improve as the molecular weight increases. Such increased resolution is observed in typical experiments until the DNA reaches a critical size where all larger molecules migrate with the same apparent velocity. What apparently happens is that these molecules have relaxation times longer than the pulse time. Thus, instead of continually relaxing in response to the changing fields, the molecules assume a time-independent conformation in the gel in response to the sum of the two applied fields. Then, since no relaxation occurs, the molecules respond to the applied field as in ordinary electrophoresis. Since all have the same mobility in the reptation limit, no separation can be achieved. Thus, choice of the pulse time sets an effective upper limit to the separation range. While the choice of a pulse time depends on other experimental conditions, the following rule of thumb applies for the typical PFG procedures described below. One to two second pulses are optimal for separations of DNA less than 50 kb; 10 second pulses are effective for DNAs in the size range of 50 to 200 kb; 60 second pulses work well for DNAs from 200 to 800 kb while 120 second pulses allow separations of DNAs as large as 1400 kb.

Practical Considerations in PFG Electrophoresis

While a variety of apparatus designs for pulsed field gel electrophoresis have been effective,[3-6] the device shown schematically in Fig. 3 seems most generally useful for a wide variety of DNA sizes. This is a submarine horizontal gel box containing independent electrodes on all four sides. The gel is placed at 45° relative to the square sides of the apparatus. Usually a 20-cm-square gel is employed while the actual box size can vary from 28 to 55 cm. The electrodes enter the gel vertically and are each connected to the power supply through a diode.[4] Thus, when one set of electrodes is energized, the other is essentially invisible. In practice only 3 to 6 electrodes on a side are sufficient to provide fairly smooth electrical fields. The electrode geometry is absolutely critical to high resolution PFG separations. DNA reptation times are very sensitive to the angle between the applied electrical fields. While no systematic survey has yet been reported, preliminary studies indicate that larger angles are better, at least for molecules up to 1000 kb. The particular geometry shown in Fig. 3 provides angles between alternate fields that range from about 100 to 150° and appears to have generally excellent performance for most DNA samples tested (A. Gaal, P. W. Warburton, C. L. Smith, and C. R. Cantor, unpublished results).

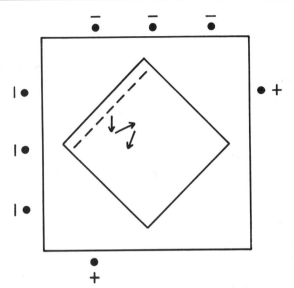

FIG. 3. Electrode arrangement used for pulsed field gel electrophoresis in the double inhomogeneous field configuration. The gel is placed at a 45° angle. Sample wells are indicated by the dashes.

In addition to the pulse time, and the angle, a number of other experimental variables are critical to high resolution PFG electrophoretic separations. Rather than go through these in detail, we present running parameters that appear optimal for mammalian DNA samples. We recommend an applied electrical field strength of 10 V/cm. If higher or lower fields are used, one will have to shorten or lengthen the pulse time proportionally. With fields of 10 V/cm the running time needed for excellent separations is roughly 1 hr/cm of box size. Temperature control is very critical for reasons that are completely unknown. The buffer in the submarine apparatus must be circulated to provide a uniform temperature. We have achieved excellent performance at 15° but it is not clear if this is optimal for all samples. A 1% agarose gel concentration works reasonably well for most samples. Raising the gel concentration can improve resolution but increases the run time needed considerably. Lowering the gel concentration usually leads to markedly decreased resolution. No extensive studies on buffer variations have been described. We generally use 1× TBE; however, others have reported that 0.5× TBE is quite effective.

Preparation of Mammalian DNA Samples

Intact mammalian chromosomal DNAs are generally 50 mb and larger except for organisms like chickens with microchromosomes. These molecules are much too large for current PFG technology. However, they can be broken into discrete fragments by treating unbroken chromosomal DNA with restriction nucleases that have very rare recognition sites in mammalian genomes. The usual objective in PFG electrophoresis is DNA fragments in the 100 kb to several thousand kb size range. Such large DNAs cannot be handled in aqueous solution because shear breakage caused by stirring or even thermal convection will be intolerable. To circumvent this problem we prepare DNA molecules directly in agarose as described below. All subsequent manipulations requiring the maintenance of high-molecular weight material are carried out inside the gel.

Preparation of DNA in agarose is actually much easier than solution procedures. Freshly grown mammalian cells are spun at 1000 rpm in a conical centrifuge tube and resuspended in 5–10 ml of room temperature phosphate-buffered saline (PBS). The cells are spun again and thoroughly resuspended in PBS at a final concentration of $1–2 \times 10^7$ cells/ml. One to two milliliters of the cell suspension is mixed with an equal volume of 1% Low Gelling Temperature Agarose (SeaPlaque, FMC) cooled to 45–50° and immediately distributed with a Pasteur pipet into a mold that makes 100 μl blocks. The agarose is allowed to solidify by placing the mold on ice for 5–10 min. The agarose blocks are removed from the mold and incubated in ESP [0.5 M EDTA (pH 9–9.5) 1% sodium lauroylsarcosine, and 1 mg/ml proteinase K] for 2 days at 50° with gentle shaking. The samples are stored at 4° in ESP and can be shipped at room temperature with no detectable damage.

In most cases, we use a gel mold that makes 50–100 blocks that are 2 by 5 by 10 mm. We usually aim for 10 μg of DNA per block. Thus, a million diploid mammalian cells are needed for each sample. A properly prepared DNA block sample will contain almost quantitative recovery of DNA. Analysis by ordinary gel electrophoresis or PFG should reveal little or no material entering the gel.

When first experimenting with new samples it is useful to make up inserts at several concentrations and to determine the best concentration empirically. We usually prepare agarose blocks at 2–4 times the DNA concentration that is used per run. Each block is cut into halves or quarters with a glass cover slip before loading onto a gel by simply pressing the piece of agarose into the well. The solid piece of agarose is sealed into place with liquid agarose. Thus, a single 10 μg insert can be used for several PFG experiments such as runs of different pulse times.

Restriction Endonuclease Digestions of Mammalian DNA

The DNA samples in agarose in ESP can be loaded directly onto a gel if one is looking for microchromosomes. Further treatment of the samples is necessary if the DNA is to be cut by restriction endonucleases. Each insert is treated twice with 1 ml of 1 mM phenylmethylsulfonyl fluoride (PMSF) in TE [10 mM Tris–Cl (pH 7.4), 0.1 mM EDTA] by slow rotation at room temperature for 2 hr. A fresh 0.1 M PMSF stock solution in 2-propanol is made up each day. This is followed up by three 1 ml washes with TE buffer alone for 2 hr each. Restriction enzyme digestions are carried out in 1.5 ml microcentrifuge tubes usually in buffers recommended by enzyme manufacturers supplemented with 100 μg/ml bovine serum albumin in a final volume of 250 μl. Inserts are usually added to the reaction buffer before enzyme is added. We routinely use 20 units of enzyme per μg of DNA. Reactions are usually allowed to proceed overnight at the appropriate temperature by gently shaking in a water bath. The next day the buffer is aspirated off carefully to avoid damage to the insert. One milliliter of ES (ES = ESP without proteinase K) is added and the samples are incubated for 2 hr at 50° with gentle shaking. The buffer is changed to 250 μl of ESP and the incubation continued for another 2 hr before samples are loaded onto a gel.

Several features of the above protocol deserve further comment. Some restriction nucleases are very sensitive targets for protease activity used in the original DNA preparation. Therefore, this must be eliminated prior to adding restriction nucleases. Some restriction nucleases appear to be easily inactivated by EDTA. Thus, it is important to remove the EDTA used in storage of DNA-agarose plugs before attempting restriction enzyme digestions. Small traces of restriction enzymes or other proteins left in the agarose plug may interfere seriously with PFG electrophoretic resolution. While the precise reason for this is unknown, a likely cause is that protein binding to the DNA will retard its mobility in PFG electrophoresis, just as is seen for ordinary electrophoresis. Thus, after any enzymatic treatment of the DNA, removal of DNA binding proteins by additional proteinase K treatment is recommended.[8,12]

In general we find that the effectiveness of agarose in allowing restriction nuclease digestion is extremely batch dependent and the best procedure is to assay different batches until a reliable one is found. It is likely that special grades of agarose optimized for restriction nuclease cutting *in situ* will soon be available from FMC, Inc. In our hands, with proper agarose, roughly the same amounts of enzyme that suffice to produce a complete digestion of DNA in solution will yield the same result in agarose. Once a good batch has been identified, it is effective for a wide range of different enzymes. However, it may be necessary to alter the buffer and temperature conditions found optimal in solution if optimal performance is desired in the gel.

The choice of restriction enzyme will undoubtedly depend on the particular DNA sample and experiments. However, overall we have had consistently good success in generating large fragments of human and mouse DNA with the enzymes *Not*I, *Sfi*I, *Mlu*I, and *Pvu*I[13,14] (see Table I). These enzymes consistently yield average DNA fragment sizes for mammalian DNA that range from 100 kb to greater than 1000 kb (Fig. 4). The relative size of fragments seen is quite consistent with relative frequencies observed for the appearance of these sites in the DNA sequence bank, GenBank. Thus, one can use this approach confidently to predict the behavior of enzymes. There are other restriction enzymes which have the potential to generate large fragments. In some cases we have tried these but have failed to detect bands upon hybridization with single copy mammalian probes (e.g., *Nru*I). In other cases the commercial enzyme preparations have not been active enough or have been contaminated with nucleases (e.g., *Nar*I, *Sca*I, *Nae*I, *Xma*III, *Sac*II). Still other potentially useful enzymes have not yet been tried, e.g., *Rsr*II.

The average size of restriction fragments generated by digestion of DNA with a particular restriction enzyme cannot simply be calculated on the basis of the size of the recognition sequence. The base composition, nearest-neighbor frequencies, and methylation pattern of the DNA sample as well as the methylation sensitivity of the restriction enzyme will influence the average fragment size obtained with a particular DNA sample. For instance, in human DNA the $G + C$ content is 40% while the CpG frequency is only 0.8%.[13] In principle, this allows one to use simple binomial statistics to attempt to calculate the expected restriction fragment sizes generated by enzymes with particular recognition sequences. Typical results are shown in Table II. However, these calculations do not agree very well with the observed results for mammalian DNA (Table I). There are two reasons for the discrepancy. First, although the sequence CpG is quite rare in mammalian DNA it occurs preferentially in HTF islands.[14] While the exact preference is not known, the result is that restriction sites containing multiple CpGs within six or eight base pairs of continuous $C + G$ are not nearly so rare as predicted by simple statistics. In contrast, CpGs embedded in $A + T$ rich recognition sequences can be expected to be extraordinarily rare (Table I).

The second problem with simple estimates of fragment sizes is DNA methylation. The major known methylation site in mammalian cells is the sequence CpG. It is estimated that over 50% of the CpGs are methylated.[14] However, the CpGs in HTF islands are generally not methylated.[15] Some restriction enzymes are inhibited by the presence of 5-methylcytosine.

[13] A. P. Bird and M..H. Taggart, *Nucleic Acids Res.* **8,** 1485 (1980).

[14] A. Bird, M. Taggart, M. Frommer, O. J. Miller, and D. Macleod, *Cell* **40,** 91 (1985).

[15] M. Ehrlich, A. Gama-Sosa, L.-H. Huang, R. M. Midgett, K. C. Kuo, R. A. McCune, and C. Gehrke, *Nucleic Acids Res.* **10,** 2709 (1982).

TABLE I

RESTRICTION ENZYMES THAT GENERATE MACRORESTRICTION FRAGMENTS

Enzyme	Recognition sequence[a]	Average fragment size (kb) based on sequence[b]	PFG[c]	Frequency in Genbank (%)[d]
NotI	5′ G C G G C C G C 3′	9750	1000	1.2
SfiI	G G C C N N N N N G G C C	390	250	1.4
MluI	A C G C G T	170	1000	1.0
SalI	G T C G A C	35	500	2.0
PvuI	C G A T C G	170	200	0.9
XhoI	C T C G A G	35	200	4.0
ApaI	G G G C C C	16	100	9.0
RsrII	C G G (A/T) C C G		nd	1.6

[a] † denotes inhibition of cutting by methylation; o denotes no effect of methylation on cutting; no mark indicates lack of knowledge; ↓ denotes site of cleavage. Methylation sensitivity is summarized from M. McClelland and M. Nelson [*Nucleic Acids Res.* 13 (Suppl.), r201 (1985)] and C. Kessler, T. S. Neumaier, and W. Wolf [*Gene* 33, 1 (1985)].

[b] Calculated by binomial statistics as shown in Table II.

[c] Weight average size seen by PFG electrophoretic resolutions of DNA from mouse and human cells.

[d] Mammalian sequences in Genbank were screened for the occurrence of each of the restriction enzyme recognition sequences. Shown is percent files, out of 1842 mammalian DNA sequence files, containing at least one occurrence of the particular sequence.

Digestion of mammalian DNA samples with these enzymes will result in incomplete cutting. However, the extent of the effect of methylation will depend whether the particular restriction sites are clustered in HTF islands. There is yet one further complication. There are both tissue-specific and cell-specific differences in levels of methylation.[14,16] Thus, it is difficult to predict a priori, the degree of incomplete digestion one may encounter with a particular sample. Table II summarizes what is known at present about the methylation sensitivity of particular restriction endonucleases. There is little available information about the occurrence of the sites of these enzymes in HTF islands. Thus, at present, not enough is known to

[16] A. Razin and A. D. Riggs, *Science* 210, 604 (1980).

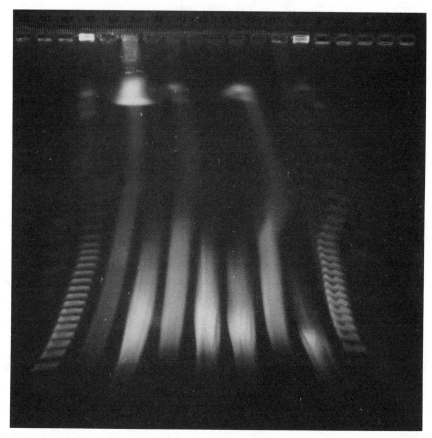

FIG. 4. Pulsed field gel electrophoresis of large DNA fragments. Samples were run in a 55-cm apparatus for 72 hr at 500 V using a 2-min pulse time. The lanes from left to right are yeast chromosomal DNAs (*S. cerevisiae,* strain DBY728), concatemers of λ *vir* (42.5 kb monomer), human DNA digested with *Not*I, *Sfi*I, *Sal*I, *Pvu*I, *Xho*I, *Mlu*I, *Apa*I, λ *vir* concatemer, and yeast.

predict the outcome of a particular restriction endonuclease digest on a particular tissue. One simple test one can perform is to determine whether a digest is complete by hybridizing such a digest with a single copy DNA probe. We have obtained single bands upon hybridization of single copy mammalian probes to PFG electrophoretic separations of digests with *Not*I, *Sfi*I, *Mlu*I, *Sal*I, *Pvu*I, and *Xho*I. This indicates either that the sites of these enzymes are not methylated, that a discrete subset of sites is stoichiometrically methylated, or that the enzymes are insensitive to methylation.

TABLE II
EXPECTED SIZES OF RESTRICTION FRAGMENTS FROM MAMMALIAN
GENOMES

Site size (bp)	Number of C + G	Number of CpG	Average fragment size[a] (kB)
6	6	0	16
6	6	1	78
6	6	2	390
6	6	3	1950
6	4	0	7
6	4	1	35
6	4	2	170
8	8	0	390
8	8	1	1950
8	8	2	9750
8	4	0	77
8	4	1	380
8	4	2	1920

[a] This was calculated by binomial statistics neglecting the clustering of CpG in HTF islands but including the observed occurrence of CpG at only 20% the frequency expected from the overall base composition.

While the three possibilities are indistinguishable at present the results indicate that the enzymes are clearly useful.

While the *in vivo* methylation of CpG described above leads to unfortunate complications, *in vitro* methylation can be used to improve the usefulness of certain restriction endonuclease to generate large fragments of DNA. For example, the inhibition of various restriction enzymes can be used to generate rare restriction enzyme cutting sites by increasing the apparent size of the site.[17,18] Alternatively, the unique requirement of *Dpn*I for methylated adenosines on both strands can be used to generate a large number of specific cutting sites by heterologous methylation of sequences that overlap the *Dpn*I recognition sequence.[19]

Length Standards for PFG Electrophoresis

Since mammalian DNA samples inevitably show a broad smear of DNA in most size ranges for PFG electrophoresis just as in ordinary electrophoresis, it is essential to have standards of defined DNA length

[17] C. Nelson, C. Christ, and I. Schildkraut, *Nucleic Acids Res.* **12**, 5165 (1984).
[18] M. McClelland, L. G. Kessler, and M. Bittner, *Proc. Natl. Acad. Sci. U.S.A.* **81**, 983 (1984).
[19] M. McClelland, M. Nelson, and C. R. Cantor, *Nucleic Acids Res.* **13**, 7171 (1985).

both to assess the performance of the apparatus as well as to provide length markers. As primary standards, the only current available samples are tandemly annealed λ DNA concatemers. These can be made by direct incubation of λ DNA inside agarose plugs as described elsewhere.[8] As secondary standards, yeast chromosomal DNAs like *S. cerevisiae* strains are recommended. These have discrete chromosomal DNAs that have been sized against concatemers. Use of λ DNA samples like those shown in Fig. 4 indicate that the resolution of PFG electrophoresis can be better than 5 kb for molecules as large as 1400 kb and much better than this for smaller DNAs.

Several types of evidence suggest that λ oligomers do provide accurate molecular weight markers applicable to other DNA species. When oligomer sets from two different sized λ DNA monomers are compared, the resulting patterns show the coincidences in mobility expected from simple arithmetic. When independent total *Not*I and *Sfi*I digests of *E. coli* are fractionated by PFG electrophoresis and sized, relative to λ standards, the resulting genome size calculated from the sum of all detected fragments is consistent to within a few percent.

Blotting and Detection of Fragments from PFG Electrophoresis

Large fragments of duplex DNA do not transfer efficiently from agarose gels to nitrocellulose or other media suitable for DNA immobilization and subsequent hybridization with cloned probes. We have experimented with a number of different protocols for DNA transfer. The object of these protocols is to break the large DNA into much smaller fragments which can be transferred more efficiently. In our experience, acid treatment has yielded unreliable results. It is possible that the size of the DNA fragments produced is extremely sensitive to the length of treatment. Introducing nicks into the DNA fragments by exposure of the DNA to short wavelength UV light in the presence of ethidium has been quite dependable. The DNA is stained by incubation in 1 μg/ml ethidium bromide for 10 min with gentle agitation on a platform shaker. We have then used 10 min exposures to a very weak 245 nm UV source during which time photographs of the gel are taken. It is, however, important to adjust the time of UV nicking to fit the intensity of the particular light source available. The gels are protected from light during subsequent manipulations prior to and during Southern blotting.[8] Denaturation is carried out for 1 hr in 0.5 N NaOH, 0.5 M NaCl and neutralization is carried out for 1 hr in 1.5 M Tris–Cl, pH 7.5, with gentle agitation. The gel is blotted to nitrocellulose (Schleicher and Schuell) or Zetapore (AMF-CUNO, Meriden, CT) membranes by ascending transfer overnight with 15× SSC. The filter is then baked for 2 hr *in vacuo* at 80°. Filters are stable for up to 6 months if stored in an air-tight containers.

Hybridization of DNA attached to filters is carried out for samples from PFG in much the same way as in ordinary agarose gel electrophoresis. An example is shown in Fig. 5. In this case, an HLA-DRα probe was hydridized with two filters: one contained an *Eco*RI digest of human DNA electrophoresed and blotted in a conventional manner, and the other contained *Not*I, *Sal*I, and *Mlu*I digests of human DNA electrophoresed in a pulsed field. The hybridization, washing, and autoradiographic conditions were identical. The signal is markedly weaker in the PFG blot, however. This may be a result of suboptimal transfer of DNA from the PFG gels. Thus, it is important to take steps to insure sensitivity in autoradiographic detection. We have accomplished this by using probes nick translated to high specific activity, minimally 10^7 cpm per hybridization at 10^8 cpm/μg for a single copy mammalian probe. Probes labeled by mixed oligo priming may also be used.[20] In general, we also find it advantageous to increase autoradiographic exposure times (e.g., 1 week in the case of the blot shown in Fig. 5) and to wash the filters at lower stringencies (e.g., 3× SSC at 50° in Fig. 5) than are commonly used in conventional Southern blotting experiments. To obtain accurate size measurements of hybridizing bands we have found it useful to probe the filters a second time with nick-translated λ DNA. As shown in Fig. 5, by referring to the illuminated λ ladder, the *Not*I, *Sal*I, and *Mlu*I bands detected with the HLA-DRα probe are readily assigned sizes of 920, 290, and 340 kb, respectively.

Chromosome Hopping: Generation of Jumping Libraries

Principles of the Circularization Method

As shown schematically in Fig. 6, if genomic DNA is broken into long linear fragments and each fragment is self-ligated to form a circle, the two ends of each linear fragment become covalently attached to one another. After digestion of the circularized fragments with a restriction endonuclease, a large number of restriction fragments are generated from sequences that were originally located internally in the large fragments. These restriction fragments are identical in sequence to restriction fragments that are obtained by digesting intact genomic DNA with the enzyme. However, digestion of the circular DNA produces an additional fragment, the jumping or J-fragment, that contains within it the sequence derived by ligation of the two ends of the original linear fragment. If the genomic DNA was

[20] A. P. Feinberg and B. Vogelstein, *Anal. Biochem.* **132**, 6 (1983).

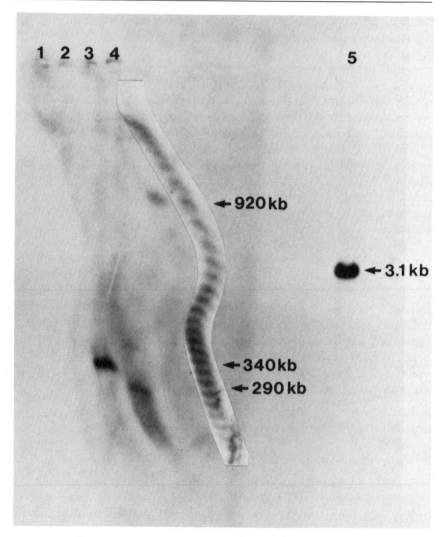

Fig. 5. Southern blot of DNA electrophoresed in a pulsed field. Autoradiogram showing the hybridization of the 3.1 kb *Eco*RI fragment of HLA-DRα [S. K. Lawrance, H. K. Das, J. Pan, and S. M. Weissman, *Nucleic Acids Res.* **13,** 7515 (1985)] with Southern blots of human DNA digested with (1) *Mlu*I, (2) *Sal*I, (3) *Not*I and electrophoresed in a pulsed field; and digested with (5) *Eco*RI and electrophoresed in a conventional manner. The sizes of the hybridizing fragments were determined by reference to the λ ladder (lane 4) illuminated by a subsequent hybridization of the filter with nick-translated λ DNA.

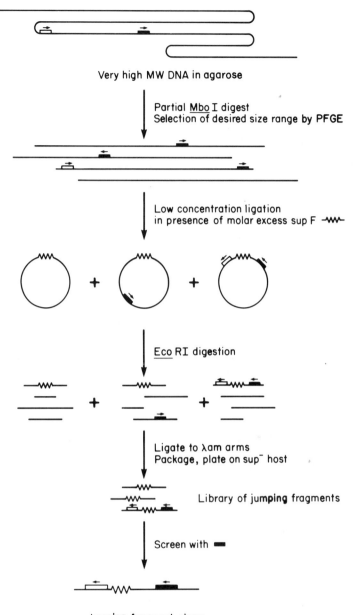

Very high MW DNA in agarose

Partial <u>Mbo</u> I digest
Selection of desired size range by PFGE

Low concentration ligation
in presence of molar excess sup F ⎯Ⱳ⎯

<u>Eco</u> RI digestion

Ligate to λam arms
Package, plate on sup⁻ host

Library of jumping fragments

Screen with ▬

Jumping fragment clone

FIG. 6. Scheme for chromosome hopping. Principle of the cloning procedure. The heavy bar represents the starting probe, which in the final jumping clone is present along with the marker gene and another segment of DNA (open box) that was initially many kilobases away in the genome. In this particular example, a suppressor tRNA gene *(supF)* is used as the marker. Other marker DNA segments allowing biological or physical separation of the jumping pieces may also be of use. Horizontal arrows show the orientation of the jumping fragment pieces relative to their original genomic arrangement.

initially broken into linear fragments of 100 kb, for example, then the J-fragments will each contain two segments of DNA that were originially separated in the genome by 100 kb of other sequences. Thus, a given J-fragment, isolated, for example, by hybridization with a unique sequence probe, also contains a segment of DNA which was originally located 100 kb to one side or the other of the probe's position in the genome. Unique sequence J-fragments, derived in this manner, can then be used to establish the linkage relationship of restriction fragments detected in PFG blotting experiments. They can also be used to screen cosmid libraries so as to isolate larger segments of DNA scattered over about 100 kb in either direction from the original probe. Used iteratively, they can allow "hopping" along a chromosome to move from a linked marker to a disease gene, or across a translocation or deletion breakpoint.

Use of Markers and Vectors

A technical difficulty in screening a library prepared simply by complete digestion of large circular DNA molecules is that most of the clones in the library are derived from internal rather than J-fragments. Thus, the majority of the clones isolated with a given probe will not serve the desired purpose. Additionally, the number of clones that must be screened for each J-fragment would be large and would increase with the size of the original fragments that were circularized. To overcome this problem, we have introduced a second step in which the linear DNAs are coligated with selective marker fragments to form circles in which the marker is incorporated between the two ends of the original linear fragment. The marker we have used to date is a synthetic tyrosine amber suppressor tRNA gene flanked by appropriate restriction cleavage sites.[21] The suppressor tRNA gene is sufficiently small (219 bp) that it is unlikely to undergo self circularization prior to ligating to other molecules. Following circularization, digestion, and cloning, one can select biologically for clones that have incorporated the sup tRNA gene. This can be done by cloning either into phage vectors that require the suppression of nonsense mutants to grow, or with plasmids containing selectable antibiotic resistance markers which can only be expressed if nonsense mutants are suppressed. In this way, libraries can be prepared which contain only J-fragments, reducing by one or more orders of magnitude the difficulty of screening for desired fragments.

Although we currently have less experience with their use, other selective markers could also be used. One possibility would be to enzymatically

[21] R. J. Dunn, R. Belagaje, E. L. Brown, and H. G. Khorana, *J. Biol. Chem.* **256,** 6109 (1981).

hemibiotinylate a short DNA fragment with a dNTP derivative linked to biotin through a reversible bond such as S–S, ligate it into the junctions of the genomic circles, cut with a second enzyme, and then use the physical selection provided by biotin–avidin interactions to purify the J-fragments. Biologic selection by *in vivo* recombination with a plasmid containing a suppressor tRNA gene[22] could then still be used to isolate the desired junction fragments. Another hypothetical possibility would be to use the *lac* operator sequences as the marker and *lac* repressor as the means of physical selection. These physical methods have the advantage that one only needs to clone and package the useful fragments and would consequently represent a considerable savings in materials. A disadvantage of this approach is that generally one wishes to obtain as extensive a library of junction fragments as possible and any additional preparative step prior to cloning would decrease the yield.

J-fragments can, in principle, be cloned in either plasmid, phage, or cosmid vectors. When the J-fragments are prepared from size selected partial digests of DNA the yield is sufficiently low, however, that the high efficiency of cloning with packaged DNA and the lack of bias for insert size over a fairly large range makes use of the phage system advantageous. The size of plaques produced with some of the established phage vectors that require suppression for growth but have incorporated a suppressor tRNA gene insert tends to be small. This can be a technical limitation but the use of λ Ch3A Δ *lac* (red⁺, gam⁺, provided by Dr. Fred Blattner) has provided a satisfactory solution. We find the most satisfactory sup⁻ host to be MC1061. The high efficiency transfection methods that have been developed in recent years may make plasmid cloning systems that depend on suppression competitive for some applications with phage vectors.

Preparation of Genomic DNA: Partial and Complete Digest Jumping Libraries

J-Fragments from Randomly Generated Genomic Fragments. The initial large linear fragments of genomic DNA for circularization can be prepared in several ways. A method which we have used successfully is to perform a partial digest of high-molecular-weight DNA using a restriction enzyme that has a recognition sequence which occurs frequently in mammalian DNA and that generates sticky ends which are suitable for the circularization step. For example, *Sau*3A1 or *Mbo*1, which cleave at the four base sequence 5′-GATC-3′, can be used in conjunction with the suppressor tRNA gene flanked by *Bam*HI linkers. To avoid shear damage

---

[22] B. Seed, *Nucleic Acids Res.* **11**, 2427 (1983).

of the partially digested high-molecular-weight DNA and to expedite the selection of the desired size ranges for hopping, PFG electrophoresis is the method of choice. DNA is prepared in agarose as described above. Standard inserts are tested with increasing amounts of enzyme or lengths of digestion to determine the optimal conditions for generating DNA of the desired size range. If any significant DNA enters the gel prior to restriction digestion, it is preferable to remove this by a short "pre-electrophoresis" of the blocks prior to enzyme digestion. This will remove sheared-end molecules which would be damaging to the protocol. These conditions are then scaled up for a preparative run. For preparative purposes, we have used 2 mm × 10 mm × 7 cm inserts containing approximately 5×10^7 cells (200 μg DNA). Using the λ ladders as a guide, the desired size range of DNA molecules is cut out of the gel, and electroeluted into 0.5× Tris–borate–EDTA (TBE) buffer. The electroelution is carried out in a dialysis bag in the PFG box. The gel fragment is then removed from the bag and the DNA is dialyzed twice at 4° against 10 mM Tris, pH 7.5, 1 mM EDTA. Great care must be taken in subsequent manipulations of the DNA to avoid exposing it to shearing forces. For example, centrifuging, vortexing, or pipetting must be avoided. The size of the DNA can be reexamined by running an analytical PFG gel. Its concentration is best assessed by evaporating a fixed volume of the solution to near dryness and comparing it with known standards on a conventional ethidium-stained agarose gel.

J-Fragments Generated by Infrequently Cutting Restriction Enzymes. An alternative approach to chromosome hopping is to circularize genomic DNA which has been digested to completion with a restriction enzyme, such as *Not*I, that cuts mammalian DNA very infrequently. A library of J-fragments produced in this fashion contains fragments corresponding to the two ends of the linear genomic fragments produced by complete digestion (see Fig. 2). In this instance, the DNA for circularization can be directly electroeluted from the inserts without size selection, resulting in a range of molecular sizes representing the range of genomic *Not*I fragments. For this approach, we have prepared a suppressor tRNA gene and an antibiotic resistance gene flanked by *Not*I linkers. J-fragments prepared in this way contain the two ends of a single *Not*I fragment of genomic DNA. Because of the limited number of *Not*I restriction sites in the human genome (perhaps 3000–4000), only a few thousand clones are required for a complete *Not*I hopping library. A limitation of such a library is that it is of use only if one begins that hop with a probe which contains DNA flanking a *Not*I site. An advantage of this approach is that it can be used in conjunction with a library of linking fragments. This procedure is discussed below.

Directionality

It is possible to employ chromosome hopping so as to clone in a predetermined direction from an initial probe. This can be readily seen if one considers the example of circles formed from partial *Sau*3A1 digests, which are then digested to completion with *Eco*RI to give *Eco*RI bounded jumping fragments. In the simplest case an *Eco*RI fragment from genomic DNA will have only one internal *Sau*3A1 site, and therefore, will be divided into a left half and right half by this enzyme. If the probe used for screening the jumping library was derived from the right half of the *Eco*RI genomic fragment, then all isolated J-fragments will contain this right half linked to a part of a different genomic *Eco*RI fragment. Prior to circularization, this other fragment would have been located at the distal end of a large linear fragment whose proximal end would have begun at the *Sau*3A1 cutting site within the original *Eco*RI fragment and would have extended from the *Sau*3A1 site to the rightward *Eco*RI site. The distal end could therefore only have been derived from DNA rightward in the genome. As shown in Fig. 7, provided that the probe is chosen so as to lie close to an *Eco*RI site this obligate direction of cloning will still be true regardless of the number of internal *Sau*3A1 sites.

Ligation Concentration Parameters

To avoid ligation of two long DNA fragments together while permitting circularization to occur, the simplest expedient is to perform the ligation at low DNA concentrations. The concentration of DNA at which circularization will occur is calculable from physicochemical estimates of the concentration of one end of the linear DNA molecule in the vicinity of the other end as a function of the length of the molecule. Theory predicts that this concentration should decrease as the square root of the length of the molecule.[23] To favor circularization, for example, at a ratio of one hundred to one over intermolecular ligation, the molar concentration of DNA molecules in the solution should be 100-fold less than the concentration of one end of any DNA molecule in the neighborhood of the other end. Theoretical calculations and experimental measurements appear to agree[7] and suggest that at a DNA concentration of $3.3/(kb)^{1/2}$ $\mu g/ml$, where kb is the length of the molecule in kilobases, 95% of the ligations will be circularizations.

The amount of DNA that must be used to obtain a complete hopping library increases with the length of the initial fragments to be circularized.

[23] H. Jacobson and W. H. Stockmayer, *J. Chem. Phys.* **18**, 1600 (1950).

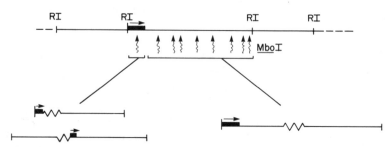

Fig. 7. Directionality of jumping. Since a jumping clone will only be represented in the library if an *Mbo*I site in that particular *Eco*RI fragment has been cut, choosing a probe which is immediately adjacent to an *Eco*RI site will make it very likely that the jump occurs in that direction. A jump to the other side can only occur (left side of figure) if the *Mbo*I cut occurs within the probe sequence.

This is true because a larger fraction of the total DNA is present in nonjunction fragments as the length of the circles increases. This consideration together with those discussed above indicate that the total volume of DNA solution in which the circularization is conducted must increase as the 3/2 power of the original DNA fragments, to produce libraries of various size hops, all of which contain similar numbers of J-fragments embedding each probe. For example, if one begins with 100 kb fragments and wishes to obtain 4×10^6 clones with greater than 95% true J-fragments, then it is necessary to ligate 5 μg of size selected DNA in a volume of 25 ml. These conditions are summarized in Table III. For these sorts of efficiencies to be present, packaging extracts must be of high quality, yielding 4×10^8 PFU/μg or better with wild type λ DNA.

To increase the size of the hop, for example, to 200 kb, while maintaining the number of junction clones produced, it is necessary to increase the amount of DNA to 10 μg, and the ligation volume to 70 ml. The DNA ligase concentration should be kept constant, at 1 unit/ml. To monitor the efficiency of ligation reactions and subsequent steps, we have found it useful to set aside small quantities of each reaction and to run these on 1.4% agarose gels. By probing Southern blots of such gels with the suppressor tRNA gene, the success of the reactions can be assessed. A successful ligation is indicated by ladder formation of the suppressor tRNA genes, and by the appearance of suppressor tRNA in the high-molecular-weight region of the gel.

Construction of hopping libraries, as outlined above, requires a 200 : 1 to 500 : 1 molar excess of marker DNA, such as the suppressor tRNA gene, in order to effect efficient recovery of J-fragments. To prepare large quantities of the marker, it is desirable to obtain a plasmid with multiple

TABLE III
PARAMETERS FOR 100 kb HOPPING LIBRARY

Parameter	Units
Starting amount of high MW genomic DNA	100 μg
Amount of size-selected DNA	5 μg
Range of sizes included	80–130 kb
Amount of *supF* gene (*Bam*HI ends)	2 μg
Molar excess of *supF* gene (220 bp)	200 : 1
Ligation volume	25 ml
Genomic DNA concentration	0.2 μg/ml
Amount of λ vector	150 μg
Total plaques on sup^+ host	4×10^8
Insert containing plaques on sup^+ host	1×10^8
Total plaques on sup^- host	4×10^6

tandem inserts of the tRNA gene, so that the molar yield per mass of plasmid DNA is increased. Plasmids containing 8 copies of this gene were obtained by ligating a large excess of suppressor tRNA gene monomer into plasmid. The plasmid has been passed several times without deletion of copies of the suppressor tRNA gene.

Avoiding Noncircular Ligations

A major potential hazard in the preparation of hopping libraries is that they may include a troublesome percentage of fragments derived from ligation of the ends of two separate long fragments. This can occur if the concentration of long fragments in the ligation is too high, if the ligation does not go to completion in the first stage and end fragments are ligated together during the process of ligating genomic restriction fragments with the vector DNA, or if the original long fragments were damaged at one end so that they could not be circularized and therefore the other ends could only ligate to different DNA molecules. The first possibility can be minimized by use of low concentrations. The second possibility can be minimized by use of ample amounts of ligase and reaction times. To some extent both possibilities two and three can be reduced if the circularized DNA is treated with alkaline phosphatase prior to digestion and cloning. Phosphatase treatment will inactivate unligated ends and prevent their aberrant joining during the circularization step. The most important precaution, however, is to begin with very high-molecular-weight DNA such as can be prepared as described above by cell lysis in gels, and to avoid steps that might break DNA molecules prior to ligation. As noted above, brief PFG electrophoresis of blocks containing undigested DNA in the PFG box can be employed to remove degraded material prior to partial digestion with *Sau*3A1 or complete digestion with enzymes such as *Not*I that cut sufficiently infrequently.

Even with these precautions, however, there is always a possibility that a given jumping clone will represent the connection of two genomic fragments which were not originally closely related in the genome. An independent means of assessing the relationships should be employed, especially if hopping is to be done iteratively. If an appropriate somatic cell hybrid or chromosome sorted DNA is available, one can determine whether the new fragment is on the same chromosome as the starting fragment. An alternative method is to prepare the library itself from a somatic cell hybrid which contains a single human chromosome on a background of another species. Hopping can be performed along the human chromosome, and any noncircular ligations will be recognizable by the presence of nonhuman sequences. A further advantage of this method is that it permits the general validity of the library to be assessed by using human-specific repeats before an extensive screening is carried out. For example, in a 100 kb hopsize hamster–human chromosome 4 hopping library, by screening for clones containing human repeat sequences and then rescreening with hamster repeats, we were able to rapidly establish that greater than 85% of the clones arose from circularizations (F. S. Collins, unpublished results).

Cloning Fragments with Internal Rare Restriction Sites: Linking Libraries

Principles and Construction of Linking Libraries

Linking libraries[8] consist of all those clones from the genome containing a particular rare enzymatic cutting site of interest (see Fig. 2). We refer to these as linking fragments, or L-fragments. A variety of techniques can be employed to construct linking libraries. One such strategy is outlined schematically in Fig. 8. As indicated, the suppressor tRNA with *Not*I linkers described above may be used to isolate plasmid, λ, or cosmid clones containing genomic fragments with internal *Not*I sites. Linking libraries specific for a chromosome of interest can be isolated by ligating selective markers into DNA prepared from a library of DNA prepared by chromosome sorting.[24]

---

[24] Available from National Laboratory Gene Library Project, Lawrence Livermore and Los Alamos National Laboratories.

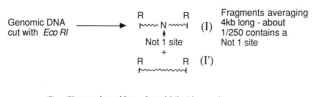

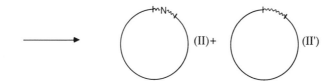

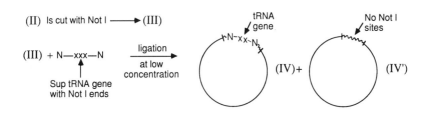

(IV) + (IV') Are cloned into bacteria containing P3 plasmid with suppressible mutations in *tet* and *amp* genes on *tet* + *amp* selection (IV) grows, (IV') does not ———▶ library of linking fragments

Fig. 8. Scheme for construction of a linking library. Linking fragments are isolated by preparing a total genomic library or a library from sorted individual chromosomes in a plasmid vector with suppressible mutations in the tetracycline or ampicillin genes. DNA is isolated from an amplification of this library. The preparation is digested with *Not*I (which does not cleave the vector) and the suppressor tRNA with *Not*I linkers is inserted. Upon retransformation, only plasmids which contain the suppressor tRNA and, therefore, also a linking fragment, convert the recipient bacteria to tetracycline or ampicillin resistance.

An alternative scheme is to circularize *Eco*RI or partial *Sau*3A1 digests of chromosome-specific DNA around a suppressor gene. Subsequent digestion with *Not*I will only linearize those circles with a *Not*I site, which can then be selected by ligation into an amber-mutated phage with a *Not*I cloning site, and plating on a sup⁻ host.

Determining the Organization and Relationship of Fragments Detected with Linking Clones

When linking clones are used to probe a PFG electrophoretic separation of DNA fragments generated with the enzyme, each clone should reveal two DNA fragments, and these must be adjacent in the genome. In principle, with just a single library and digest, one should be able to order the rare cutting sites, although it will not generally be possible to distinguish between two fragments of the same size. Some ambiguities will be resolvable by performing a second digest with another rare cutting enzyme. Other ambiguities should be resolvable by examining a partial digest with the original rare cutter. However, in general it will be more efficient to use several different libraries, each for a particular enzyme and overlap the resulting patterns just as in ordinary restriction fragment analysis.

The jumping and linking libraries can be used in a complementary fashion (Fig. 2). Jumping libraries, if constructed by digestion to completion with a rare cutter (see above), consist of clones containing the ends of each large fragment generated by this particular enzyme. Each jumping clone will detect only a single fragment on a PFG electrophoretic separation of fragments generated with the same enzyme. However, each linking clone, since it spans a restriction site, should cross-hybridize with two different jumping clones, and each jumping clone should likewise cross-hybridize with two different linking clones. Thus, by cross-screening the two libraries, it should be possible in principle to walk from clone to clone and generate a complete ordered map without use of DNA blotting techniques. No information will be provided directly about the physical distances between the clones, but that information is readily available from PFG electrophoresis. Thus, it is clear that the two libraries outlined above are complementary and in combination with DNA blotting provide the necessary redundancy to allow error detection.

A major potential difficulty with the schemes outlined above is repeated DNA. This difficulty can be minimized in several ways but ultimately some regions of mammalian genomes will still contain too much highly repeated DNA to be mapped easily.

Once ordered large DNA fragments are available, it should be possible to generate finer restriction maps for each by an analog of the Smith–Birnstiel procedure.[25,26] One performs a total digest of genomic DNA with the same enzyme used in generating jumping or linking libraries. Then one performs a partial digest with an enzyme that cuts somewhat more frequently so that each large fragment will be nicked on the average of once or less to produce smaller fragments that average roughly half the original size. The pattern of fragments generated is viewed selectively by indirect end labeling through hybridization with one half of a junction clone. This will detect only these fragments extending from the rare cutting site in that

[25] H. O. Smith and M. L. Birnstiel, *Nucleic Acids Res.* **3**, 2387 (1976).
[26] R. B. Saint and J. B. Egan, *Mol. Gen. Genet.* **171**, 103 (1979).

clone to a site of partial digestion with the more frequently cutting enzyme as shown schematically in Fig. 9. Thus, the pattern of fragment sizes reveals the order of the more frequent cutting sites. This technique has already been used successfully to examine the structure of large fragments of the *E. coli* genome. There is no fundamental reason why it should not be effective for mammalian DNA although it does require an order of magnitude more sensitive detection than ordinary Southern blots.

It may even be possible to map a larger region by using a "double barrelled" Smith–Birnstiel approach, as diagrammed in Fig. 10. In this protocol, a partial digestion is done with a rare cutter such as *Not*I, and parallel lanes are separately probed with the two halves of as linking clone. The pattern observed should allow mapping of several *Not*I sites to either side of the linking fragment probe.

None of the methods described above actually provides pure DNA internal to the large fragments. That DNA can be obtained in an ordered manner by preparing a jumping library where the jump starts from a rare enzyme cutting site and proceeds to a more frequent site. The same half-junction or half-jumping fragments needed for PFG analysis as described above can also be used to screen jumping libraries for the desired sets of clones adjacent to particular sites of interest.

The linking library approach would only require 50–300 clones to cover each human chromosome. The identification of such clones and

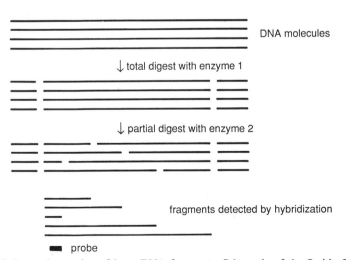

FIG. 9. Internal mapping of large DNA fragments. Schematic of the Smith–Birnstiel method[25] for rapid restriction mapping as modified for large DNA fragments by detection through indirect labeling.

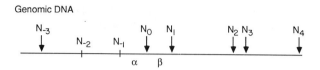

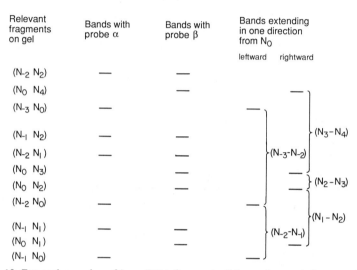

FIG. 10. External mapping of large DNA fragments. Scheme for restriction mapping by partial digestion without end-labeled DNA. N_0 represents a NotI site embedded in a linking fragment. To map NotI sites proximal to N_0, a partial NotI digest of genomic DNA is prepared, run on PFGE, and blotted. Probes β and α are, respectively, the left and right halves of the NotI linking fragment. The bands detected with these probes are designed by NotI sites at their two ends, e.g., $(N_{-1} \ N_{-2})$ is the band derived by cleavage at position N_{-1} and N_{-2} during partial digestion.

their use in creating a physical map of chromosomes could certainly set the stage for subsequent total genomic sequencing.

Conclusion

Mapping of complex mammalian genomes, especially those of mouse and man, has progressed rapidly over the past few years. New DNA probes, many of them polymorphic, are appearing in large numbers, and can now be readily mapped to a particular part of a chromosome using *in situ* hybridization or somatic cell techniques. An enlarging panel of such probes will soon allow the mapping of genes for most Mendelian disorders, through the method of linkage analysis. While such techniques are extremely powerful, they are not well suited for molecular physical mapping in the size range from 100 to 2000 kb, as shown in Fig. 1. Standard molecular cloning methods are also limited in their application to this size range. The ability to work in this macroregion is essential if one wishes to use map location to clone genes whose normal protein product is unknown; it is also necessary for mapping relatively large gene complexes or for identifying the specific breakpoints of chromosomal deletions or translocations, many of which have major biologic consequences.

By combining pulsed field gel electrophoresis with the generation of jumping and linking libraries, it is now possible to outline strategies that will allow the complete physical mapping of large segments of mammalian genomes. Placing preexisting polymorphic DNA probes on this map can then be readily accomplished by PFG blots, which will disclose the relationship between recombinational and physical distances. These mapped probes, together with the linking libraries, can be used in combination with random-ended jumping libraries (Fig. 6) to generate other clones in interesting regions of the genome in a truly directed manner. Thus, although the technical aspects of many of these methods are not trivial, the necessary tools are in place for a real conquest to be made of the detailed physical structure of mammalian genomes.

Acknowledgements

This work was supported in part by Grants GM34960 to F.S.C., GM14825 and CA39782 to C.R.C., and from the National Cancer Institute to S.M.W. Support from the Hereditary Disease Foundation is also gratefully acknowledged. Also, the authors wish to express their appreciation to Ann M. Mulvey for preparing this manuscript.

[36] Insertional Mutagenesis to Isolate Genes

By STEPHEN P. GOFF

Insertional mutagenesis has long been recognized as a powerful method for the inactivation and identification of genes in prokaryotes.[1,2] The technique of transposon tagging to mutate and subsequently clone selected genes has recently been extended to eukaryotes; transposable elements

---

[1] A. L. Bukhari, J. A. Shapiro, and S. L. Adhya, "DNA Insertion Elements, Plasmids, and Episomes." Cold Spring Harbor Lab., Cold Spring Harbor, New York, 1977.
[2] N. Kleckner, *Cell* **11,** 11 (1977).

have been demonstrated to generate mutations in yeast,[3,4] and the transposable P elements of *Drosophila* have been used to clone several new genes.[5,6] The analogous elements of choice for the mutagenesis of mammalian cells are the retroviruses: agents which upon infection direct the synthesis of a provirus, a DNA copy of the viral genome, and insert that DNA at randomly selected sites in the host genome.

Numerous studies have shown that the insertion of retroviral DNA can result in mutation of cellular genes. Early studies showed that oncogenesis by the leukemia viruses occurs by insertional activation of cellular oncogenes; examples include the activation of genes known as c-*myc*,[7,8] c-*myb*,[9] and c-*erbB*.[10] Insertion events in tumorigenesis have also allowed the identification of novel putative oncogenes as common target genes for retroviral insertions.[11-16] Infection of early embryos, or microinjection of DNAs into eggs, can result in the formation of germline insertions; surprisingly often these insertions cause mutations with observable phenotypes.[17,18] In tissue cultures, infection has been used to induce mutations in several known loci, including a resident v-*src* gene,[19] and a gene for a nucleotide salvage enzyme.[20] It is reasonable to expect that new genes will soon be identified and cloned by retroviral tagging. This review will describe procedures for the generation of mutations by retroviral infection and give some examples of applications.

[3] G. S. Roeder and G. R. Fink, *Cell* **21**, 239 (1980).

[4] S. J. Silverman and G. R. Fink, *Mol. Cell. Biol.* **4**, 1246 (1984).

[5] L. L. Searles, R. S. Jokerst, P. M. Bingham, R. A. Voelker, and A. L. Greenleaf, *Cell* **31**, 585 (1982).

[6] W. Engels, *Annu. Rev. Genet.* **17**, 315 (1983).

[7] W. S. Hayward, B. G. Neel, and S. M. Astrin, *Nature (London)* **290**, 475 (1981).

[8] D. Steffen, *Proc. Natl. Acad. Sci. U.S.A.* **81**, 2097 (1984).

[9] G. L. C. Sheng-Ong, M. Potter, J. F. Mushinski, S. Lavu, and E. P. Reddy, *Science* **226**, 1077 (1984).

[10] Y.-K. T. Fung, W. G. Lewis, L. B. Crittenden, and H.-J. Kung, *Cell* **33**, 357 (1983).

[11] R. Nusse and H. E. Varmus, *Cell* **31**, 99 (1982).

[12] G. Peters, S. Brookes, R. Smith, and C. Dickson, *Cell* **33**, 369 (1983).

[13] C. Dickson, R. Smith, S. Brooke, and G. Peters, *Cell* **37**, 529 (1984).

[14] P. N. Tsichlis, P. G. Strauss, and L. F. Hu, *Nature (London)* **302**, 445 (1983).

[15] G. Lemay and P. Jolicoeur, *Proc. Natl. Acad. Sci. U.S.A.* **81**, 38 (1984).

[16] H. T. Cuypers, G. Selten, W. Quint, M. Zijlstra, E. R. Maandag, W. Boelens, P. van Wezenbeek, C. Melief, and A. Berns, *Cell* **37**, 141 (1984).

[17] A. Schnieke, K. Harbers, and R. Jaenisch, *Nature (London)* **304**, 315 (1983).

[18] E. F. Wagner, L. Covarrubias, T. A. Stewart, and B. Mintz, *Cell* **35**, 647 (1983).

[19] H. E. Varmus, N. Quintell, and S. Oritz, *Cell* **25**, 23 (1981).

[20] W. King, M. D. Patel, L. I. Lobel, S. P. Goff, and M. C. Nguyen-Huu, *Science* **228**, 554 (1985).

Retroviral Infection

Choice of Virus as Mutagen

The mammalian retrovirus now in widest use as a genetic tool in molecular biology laboratories is the Moloney murine leukemia virus[21] (M-MuLV). The virus grows to high titers in a variety of murine cells, infectious DNA clones are available,[22,23] the entire genome has been sequenced,[24] and numerous variants have been constructed that allow sophisticated manipulation of the life cycle. The standard host cell for the preparation of viral stocks is the NIH/3T3 cell line, a hardy and commonly used line.

Wild-type M-MuLV can be used to induce insertional mutations in cells, and indeed the earliest procedures did use the unaltered virus.[19] Stocks of wild-type virus are obtained by collecting the medium from a producer line; the best are infected NIH/3T3 cells, growing in Dulbecco's modified Eagle's medium (DMEM) containing 10% calf serum. A particularly high-titer producer is the Clone 1 isolate.[25] For maximum titer the virus should be collected from exponentially growing cells, about 24 hr after a medium change. The harvested medium is filtered through 0.22- or 0.45-μm filters (Nalgene) and can be stored at −70° indefinitely. A serious disadvantage of the use of the wild-type virus in murine cells is that the newly inserted proviruses cannot be readily detected by hybridization, e.g., by Southern blots. The reason is that the mouse genome contains approximately 50–100 copies of endogenous sequences exhibiting strong homology to the exogenous MuLVs, and the presence of these sequences interferes with the detection of the new proviruses (for a review, see Coffin[26]). Probes with some specificity for M-MuLV have been prepared,[27] and we have had some success with these probes in hybridization experiments, but the signals are quite weak and there are problems with background hybridization.

A more useful virus is a viable variant of M-MuLV, *in*31SuIII, that

[21] J. B. Moloney, *J. Natl. Cancer Inst.* **24,** 933 (1960).

[22] C. Shoemaker, S. P. Goff, E. Gilboa, M. Paskind, S. W. Mitra, and D. Baltimore, *Proc. Natl. Acad. Sci. U.S.A.* **77,** 3932 (1980).

[23] I. Chumakov, H. Stuhlmann, K. Harbers, and R. Jaenisch, *J. Virol.* **42,** 1088 (1982).

[24] T. M. Shinnick, R. A. Lerner, and J. G. Sutcliffe, *Nature (London)* **293,** 543 (1981).

[25] H. Fan and M. Paskind, *J. Virol.* **14,** 421 (1974).

[26] J. Coffin, *in* "RNA Tumor Viruses" (R. Weiss, N. Teich, H. Varmus, and J. Coffin, eds.), p. 1109, Cold Spring Harbor Lab., Cold Spring Harbor, New York, 1982.

[27] I. M. Verma, personal communication.

carries a bacterial suppressor tRNA gene.[28] This virus was constructed by the insertion of a 220-bp fragment containing the *E. coli SuIII*[+] tyrosine suppressor tRNA gene into the retroviral long terminal repeat (LTR) sequence. The insertion, near the left (transcriptionally 5′-most) edge of the LTR sequence, does not disrupt any viral function and in particular does not prevent the integration of the proviral DNA. The inserted DNA at this position is stably retained through many infectious cycles by the virus. This virus has no disadvantages in comparison with the wild-type M-MuLV: it is replication competent, grows to equally high titers, and can infect the same spectrum of cells. Virtually identical viruses have been constructed in Jaenisch's laboratory.[29] The advantage of the use of these viruses is that the integrated proviral DNAs can be readily recovered in the form of recombinant phage (see below). The *SuIII*[+] gene can also serve as a hybridization probe, as there are no homologous sequences in the mammalian genome.

Other virus constructs are also useful for particular purposes. Replication-defective constructs, carrying foreign DNA, are potentially valuable as insertional mutagens. Constructs carrying and expressing a number of marker genes have been described, including genes encoding resistance to the aminoglycoside G418,[30] the thymidine kinase gene conferring resistance to HAT medium,[31] the gene for hypoxanthine–guanine phosphoribosyltransferase (HGPRT) conferring resistance to HAT medium,[32] and many others. These viruses are as a rule replication-defective, and the transmission of the viral constructs through virion particles requires the presence of a helper virus genome. Wild-type virus can be used as helpers, but in many ways the ideal helpers are the Psi2[33] or Psi-am viruses.[34] These helper genomes encode all the viral proteins needed for formation of virions but contain mutations that prevent the incorporation of the helper RNAs into those virions. Thus, NIH/3T3 cells expressing these genomes assemble and release empty virions. If a viral construct containing a selectable marker gene is introduced into such cells by transfection, the transcribed RNA can be packaged into the virions; application of these virions to target cells results in the synthesis and insertion of a DNA copy of the

[28] L. I. Lobel, M. Patel, W. King, M. C. Nguyen-Huu, and S. P. Goff, *Science* **228**, 329 (1985).

[29] W. Reik, H. Weicher, and R. Jaenisch, *Proc. Natl. Acad. Sci. U.S.A.* **82**, 1141 (1985).

[30] C. L. Cepko, B. E. Roberts, and R. C. Mulligan, *Cell* **37**, 1055 (1984).

[31] C. J. Tabin, J. W. Hoffmann, S. P. Goff, and R. A. Weinberg, *Mol. Cell. Biol.* **2**, 426 (1982).

[32] A. D. Miller, D. J. Jolly, T. Friedmann, and I. M. Verma, *Proc. Natl. Acad. Sci. U.S.A.* **80**, 4709 (1983).

[33] R. Mann, R. Mulligan, D. Baltimore, *Cell* **33**, 153 (1983).

[34] R. D. Cone and R. C. Mulligan, *Proc. Natl. Acad. Sci. U.S.A.* **81**, 6349 (1984).

construct. The disadvantage is that the titer of these constructs is often very low; it is frequently difficult to infect a population of cells so as to produce even one provirus per cell. These constructs, however, have the compensating advantage that successfully infected cells can be isolated by selection for the marker gene on the virus. Another advantage is that there is no replicating virus in the target cells; only the defective construct genome is transmitted to the target cell and no helper virus genome is transferred.

Choice of Target Cell Line

Numerous cell lines can be infected by retroviruses and thus can serve as targets for insertional mutagenesis. The majority of the murine cells in culture can be infected with M-MuLV; receptor is present on a wide spectrum of cell types, and there are no genetic backgrounds known that can block provirus insertion of the NB-tropic M-MuLV intracellularly. There is very little toxicity and normally no detectable increase in the doubling time for most murine cells. Any cell line that has previously been infected with an ecotropic MuLV, or that expresses an ecotropic MuLV envelope protein, will be resistant to infection. This is by far the most common limitation. The popular L cell line, for example, is resistant because it is chronically infected by and is a producer of the L cell virus (LCV), an ecotropic MuLV related to M-MuLV. Rat cells are generally somewhat less sensitive to infection with MuLVs than mouse cells, but are perfectly usable as targets. These cells have the enormous advantage that the endogenous retroviral sequences are only very weakly homologous to MuLVs, and the presence of new MuLV proviruses can be readily detected by hybridization with MuLV probes.

One murine cell type deserves particular mention. Embryonal carcinoma (EC) or teratocarcinoma cell lines (see chapter by Jakob and Nicolas [6], this volume) can be infected with ecotropic viruses like M-MuLV, and proviral DNAs are inserted normally into the DNA of these hosts. The inserted viral DNA, however, is not actively transcribed in these cells to form viral mRNAs or genomic RNAs.[35,36] The result is that there are no viral gene products and no progeny virus formation. Insertional mutagenesis of these cells is particularly attractive because there can be no complications in selection procedures that virus progeny might cause.

Other species are generally resistant to infection by ecotropic MuLVs. Cells from an enormous spectrum of species, however, can be infected by

[35] C. L. Stewart, H. Stuhlmann, D. Jahner, and R. Jaenisch, *Proc. Natl. Acad. Sci. U.S.A.* **79**, 4098 (1982).
[36] O. Niwa, Y. Yokota, H. Ishida, and T. Sugahara, *Cell* **32**, 1105 (1983).

so-called amphotropic viruses[37,38]; these are mouse viruses which utilize a different receptor than the ecotropic viruses, one that is apparently common on cells of most mammalian species. These include human cells, numerous primate cells, canine and feline cells, and most rodent cells. Significant exceptions are the popular Chinese hamster ovary (CHO) cell line, and the L cell line. The viruses of choice for the permissive cells are a virus called 292A,[39] and one called 4070A.[37] Biologically cloned isolates of both viruses are available, and the latter virus has been molecularly cloned.[40] The same spectrum of cells that can be infected with these amphotropic viruses can be infected with virus constructs based on the M-MuLV genome, if the constructs are transferred with the Psi-am helper virus.[34] This helper was constructed simply by inserting the 4070A *env* gene into a M-MuLV backbone, conferring on that virus the broad host range of the amphotropic parent.

Infection Protocols

The simplest procedure for the transfer of virus from a producer cell to an adherent target cell is by infection with a cell-free virus harvest. The virus can be collected from the producer cell line immediately before infection, if the producer and target cells are both ready at the same time. Under normal conditions the virus is collected whenever it is convenient, stored at $-70°$, and thawed at the time of infection at room temperature or $37°$. The virus titer drops rapidly if the thawed stock is left warm, and it should be put on ice as soon as possible. Polybrene (Sigma; stock solution of 8 mg/ml in phosphate-buffered saline) is added to a final concentration of 8 μg/ml to enhance the infectivity. The polybrene is not harmful to the virus stock and can be added at any time. Some cells are rather sensitive to polybrene and lower concentrations (2 μg/ml) or none at all may be used.

The recipient cells should be subconfluent and rapidly growing at the time of infection; typically they would be at a density of 10^6 cells per 10-cm dish (10 ml of medium). With adherent cells, the infection is simply carried out by removing the medium and replacing it with 1/10th volume of undiluted virus. For suspension cultures, the cells are spun down and resuspended in the virus. The virus is allowed to adsorb to the cells for 1–2 hr with occasional mixing (either rocking plates of adherent cells or stirring tubes of suspension cells), then 1 volume of medium is added to the

[37] J. W. Hartley and W. P. Rowe, *J. Virol.* **19,** 19 (1976).
[38] S. Rasheed, M. B. Gardner, and E. Chan, *J. Virol.* **19,** 13 (1976).
[39] M. B. Gardner, *Curr. Top. Microbiol. Immunol.* **70,** 215 (1978).
[40] S. K. Chattopadhyay, A. I. Oliff, D. L. Linemeyer, M. R. Lander, and D. R. Lowy, *J. Virol.* **39,** 777 (1981).

cells. Normally we leave the virus in the medium; if the cells are sensitive to polybrene, the virus preparation and the polybrene it contains are removed before addition of the medium.

The cells are allowed to grow for at least 48 hr before any selection is imposed. Retroviral integration begins in perhaps 12 hr, but circular preintegrative forms persist for several days, and there may be integration continuing during this long period. Integration is thought to require cell division, and it is probably wise to keep the cells rapidly growing for at least 2 days. If low titers of replication-competent virus are used, or if the cells are rather resistant to infection, it may be desirable to wait several days until all the cells are infected.

When replication-competent virus is used, the mulitplicity of virus used is not very critical. If an moi of 10 is achieved, the target cells will probably receive about 10 proviral inserts per cell. If an moi of 1 is used, the cells initially receive one provirus, but such infected cells begin to produce progeny virus that can superinfect the producer cell themselves for a period of about a day or two. At that time envelope protein, the product of the *env* gene, appears on the cell surface in substantial quantity, and the cell becomes resistant to further superinfection. The result is that usually 5–10 proviruses are formed before superinfection resistance appears. If a very low moi is used, virus will spread through the culture until essentially all the cells are infected; the result is that about 10 insertions per cell still occur.

An exception to this rule occurs when EC cells are the targets. These cells do not express virus or envelope protein; they never exhibit superinfection resistance.[35,36] Thus, the moi directly determines the number of insertions per cell. One can increase the effective moi by applying virus repeatedly to the same cultures, adding a few proviruses each time.

Cocultivation

When replication-defective constructs in Psi helper virions are used, the titers are often very low. The infected cells can often be selected with appropriate drugs, selecting for the gene on the virus; but it may be difficult to infect enough cells with enough virus to generate the desired number of independent insertion events. The solution is to use cocultivation of the target cells with the producer cells. In general, the target cells are simply replated onto a lawn of producer cells; about 10^6 target cells could be plated on 10^6 producer cells in a 10-cm dish. Direct contact seems to promote highly efficient transfer of virus, and extended contact allows continuous infection to generate a large number of inserts in the target cells. Plating EC cells as targets, onto a lawn of producer cells, results in an essentially unlimited transfer; the EC cells do not become resistant to superinfection, and as many as 100 proviral inserts per cell can be achieved.[35] Our experience with the F9 EC cell showed that cocultivation for 1 day yielded about 5–10 copies per cell, and for three days yielded 20–50 copies.[20] These numbers are certain to vary with choice of producer and target cell lines. Polybrene seems to be unnecessary.

When cocultivation is used, it may be problematical to separate the donor from the target cells. Normally a selection for the desired insertional mutation is imposed soon after cocultivation; if the donor line shows a low background frequency of spontaneous mutation, then there is no problem. If the cells show obvious morphological differences, one could discard any survivors of the donor morphology. A more general solution is to kill the donor cells with an agent that does not immediately affect virus release from the cells. NIH/3T3 cells, for example, can be killed with a brief exposure to UV light. The medium is removed, the cells are washed once with PBS, and the uncovered dishes are exposed to the germicidal UV lights in a tissue culture hood (distance of about 50 cm) for 10–20 sec. Such cells continue to release virus for several days but cannot divide and eventually die. Alternatively, growing producer cells can be killed by treatment with mitomycin C (10 μg/ml for 2–4 hr), washed, and overlaid with target cells.[35]

Selection of Insertional Mutants: Requirements and Frequencies

Applicable Selection Methods

The key step in the successful recovery of insertional mutants is the selection procedure. The problem is to devise protocols which allow the growth only of those cells *lacking* the expression of the target gene; two classes of such selections are described below. The difficulties are enormously enhanced by the fact that most cultured cells are diploid or aneuploid. This fact means that there are usually two codominantly expressed copies of each autosomal gene, and both must be mutated before the cell is free of the gene products. Our experience has been that insertions into even a single copy gene are not trivial to detect (see below). Two options exist. One can attack genes known to be located on the X chromosome and present in only one active copy; or one must select for insertion into only one copy of the two present. An important step for the identification of inserts into autosomal genes, therefore, is to devise schemes that will allow selection for an insertion into a single allelic copy of the two present in each cell.

A mammalian cell is usually considered to contain about $2-3 \times 10^6$ kb of DNA per haploid genome. An ordinary gene might span 2–30 kb of DNA, and if we assume that a proviral insertion anywhere in this region will inactivate the gene, then a single-copy gene is a target that represents 10^{-5} or 10^{-6} of the total DNA. If retroviral integrations were truly randomly distributed throughout the mammalian genome, then the introduction of one provirus per cell should cause a mutation in 10^{-5} or 10^{-6} of the cells. In the limited number of cases tested to date, this expectation has not been realized; the insertion of about 25 proviruses per cell generated mutations in the HGPRT locus in slightly less than 10^{-6} of the cells.[20] The most likely reason is that retroviral insertion is not truly random, and that the HGPRT gene is a "cold spot" for integration; it seems to be affected about 50× less often than would be predicted for random insertions. It is not clear that this is a general phenomenon for all target genes, but it seems appropriate to be prepared for the possibility.

The use of transposon tagging to identify new genes requires that one or at most a very few proviruses be inserted in each cell. If there are too many proviruses, it becomes problematical to determine which one is causing the mutation by being in the desired locus and which ones are unimportant. The number of proviruses present after infection by replication-competent viruses (perhaps 5 or so) is probably in the right range. If the target gene behaves like the HGPRT locus, then insertional mutation frequency will be about 10^{-7}. This is a low frequency, placing two important constraints on the selection schemes if the majority of the selected cells are to contain insertional mutations. First, the gene must exhibit a very low spontaneous mutation rate, below the insertional mutation rate; the rate should probably be in the range of 10^{-7} to 10^{-8}. Second, the selection must be very powerful, allowing the survival of a very small fraction of the parent cells; again the fraction of survivors should probably be in the range of 10^{-7} to 10^{-8}. If less favorable conditions are used, the problem is that most of the mutant cells will not bear insertional mutations. In early studies of insertions into a v-*src* gene this was precisely the case: only 2 out of 60 mutants arising after M-MuLV infection were due to insertion at the v-*src* locus.[19] If there were no preexisting *src* probe it would not have been easy to determine that these 2 mutants were the relevant ones.

These constraints are compounded if autosomal condominant alleles are to be targetted. The frequency of two independent insertions into the two alleles would presumably be the square of the frequency of insertion into one, i.e., in the totally unworkable range of 10^{-14}. In fact insertion into one allele might result in gene conversion of the second allele relatively often, giving a combined frequency of, say 10^{-9}, but even this number is impractical. This pessimistic expectation is supported by our observation that insertional mutants of the APRT locus arose at a frequency of less than 10^{-8}. We have not made further attempts at targetting such loci.

Selective Methods

It is rare that a procedure can be devised that selects for the *absence* of expression of a particular gene. We will describe two such methods. More detailed discussions of this topic are available elsewhere.

Toxic Metabolites (Suicide Selections). In particular cases it may be possible to kill selectively the parent cells with toxic compounds that are metabolized by these cells but not by mutants (Pouysségur and Franchi [11], Patterson and Waldren [10]). Sometimes radioactive, isotopically labeled substrates can be used. The classic drugs in this category are the nucleoside or base analogs, which selectively kill those cells able to convert them to the nucleotide form. Examples include 5-bromodeoxyuridine (BrdUrd); 8-azaguanine (8-AG) and 6-thioguanine (6-TG), and azaadenine (AA). Mutant cells lacking the salvage pathways for the utilization of these compounds are unaffected by exposure. Unfortunately, most of the enzymes of these pathways are encoded by autosomal genes and so are not useful targets. The exception is the HGPRT locus on the X chromosome, which we have studied.[20] Exposure of infected cells to a mixture of two drugs (3 μg/ml 8-AG and 6 μg/ml 6-TG in ordinary medium) kills the majority in a few days and leaves the HGPRT$^-$ survivors unaffected. The spontaneous mutation frequency depends on the cell line but in EC cells it is below 10^{-7}. Introduction of 20–50 proviruses per cell gives a marked stimulation in the mutation frequency above the spontaneous background.

Not many situations will be found to be as favorable as for the mutagenesis of the HGPRT locus. In some cases insertion into one allele may be selectable; the reduction in the level of the gene product by a factor of two can cause significant changes in sensitivity to particular drugs.[41,42]

Immunoselection. The most powerful solution to the problem of detecting inserts into one of two alleles of a locus is to mutagenize target cells that are heterozygous at the target locus, and to use allele-specific antibodies to select against cells expressing one of the two alleles. The target cells are killed by exposure to the antibody in the presence of rabbit complement; the selection is strictly limited to genes whose gene products are cell-surface proteins. The net effect is to counterselect for expression of a single-copy gene, eliminating the ploidy problem inherent in the mutagenesis of most autosomal genes. The utility of the method has been demonstrated by the isolation of insertions into the β_2-microglobulin locus.[43]

[41] G. Urlaub and L. A. Chasin, *Proc. Natl. Acad. Sci. U.S.A.* **77**, 4216 (1980).
[42] G. Urlaub, E. Kas, A. M. Carothers, and L. A. Chasin, *Cell* **33**, 405 (1983).
[43] W. Frankel, T. A. Potter, N. Rosenberg, J. Lenz, and T. V. Rajan, *Proc. Natl. Acad. Sci. U.S.A.* **82**, 6600 (1985).

Immunoselection depends heavily on the quality and specificity of the antibody preparation. The antibody source will vary with target gene, but monoclonal antibodies will be the most commonly used. There must be available strains of animals, normally mice, which show strong allelic differences in their response to the antibody. Conditions of exposure to antibody must be devised so as to kill nearly all of the sensitive cells of a cell population without effect on cells of the resistant strain. Different dilutions of the antibody are simply tested in parallel on the two cell types to determine the appropriate level; typical exposure conditions are for 3 hr at 37°, after which the survivors are allowed to grow out and are counted. The ratio of survival rates between the two strains should be in the range of 10^6.

If antibodies of sufficient selectivity are available, F_1 heterozygous animals are prepared by mating the two distinctive strains. Permanent cell lines must then be established from these animals, and the expression of the cell surface marker on these cells must be demonstrated. There are many ways to establish immortal cell lines from mice, including immortalization by exposure to chemical carcinogens; one of the most straightforward is transformation by any of a number of retroviruses, including the Abelson MuLV, as originally described by Rosenberg and Baltimore.[44] The method of choice will be most probably determined by the tissue distribution of the surface marker. The sensitivity of the established, heterozygous cell line to antibody and complement, and the spontaneous mutation frequency at the sensitive allele, should be determined. These cells are then infected with the mutagenic virus construct, exposed to selection, and the survivors cloned and grown into large cultures.

Analysis and Recovery of Mutant Alleles

The value of insertional mutagenesis in the identification of new target genes is that the inserted DNA provides a physical and genetic tag for the isolation of the target region. It is here that the use of marked retroviral genomes provides a most important advantage over wild-type genomes.

Counting Inserts

If the spontaneous mutation rate at the target locus were zero, if the selection procedure were perfect, and if there were only one retroviral insertion per mutant cell, it would be possible to clone one provirus region from one mutant and be confident that the target gene was in hand.

---

[44] N. Rosenberg and D. Baltimore, *J. Exp. Med.* **143**, 1453 (1976).

Unfortunately these conditions are never met in reality. Usually, there is more than one provirus per cell, and many of the cell lines surviving the selection procedure are spontaneous mutants, bearing proviruses at irrelevant loci. One must therefore isolate many independent potential insertional mutants, and determine whether there are any insertion sites in common among them. Such common sites must then be the site of the target gene.

Under normal circumstances one should isolate 20–50 independent surviving cell lines after infection. Each clone is grown to about 10^7 cells, and genomic DNA is prepared by SDS-proteinase K lysis and repeated phenol extractions (e.g., see Goff et al.[45]). The DNAs are cleaved with restriction enzymes that do not cleave within the mutagenic retroviral genome, and analyzed by blot hybridization[46] after electrophoresis on very low percentage agarose gels. For wild-type M-MuLV in mouse cells, the only useful probes are small plasmids such as pMS2 and pMS4 derived by Verma[27]; these show fair specificity for M-MuLV over the endogenous viruslike DNAs. For variant in31SuIII[28] and related viruses,[29] the SuIII gene of plasmid piVX[47] is the appropriate probe. For constructs like pZIP-NeoSV(X)1, the bacterial neomycin resistance gene can be used.[30] The number of bands seen by blot hybridization should reflect the number of insertions per cell. As noted above, it is helpful if this number is small: one, or at most a few.

If DNAs of several cell lines, cut with enzymes that cut outside the provirus, show the same pattern of hybridizing bands, then this is indicative of repeated hits of the same region. The provirus is cloned from these cells. The results of such surveys, however, are usually ambiguous, due to the fact that the relevant fragments are large (all are bigger than the virus) and that the gels have limited resolution. If the target gene is large, two hits of that gene may not even yield similar patterns. In these cases all the proviruses must be cloned from each of many cell lines.

Cloning the Proviruses and Flanking DNA

Isolation of Clones. The next step is the isolation of the proviruses from several genomic DNAs. The procedure must be rapid and easy, since clones must be obtained from as many as 20–50 genomic samples. Two methods have been used to facilitate the cloning of these elements.

Selection for SuIII⁺. If the mutagenic provirus carries the bacterial SuIII⁺ gene, all the proviruses can be recovered from a genomic library in

[45] S. P. Goff, E. Gilboa, O. N. Witte, and D. Baltimore, *Cell* 22, 777 (1980).
[46] E. M. Southern, *J. Mol. Biol.* 98, 503 (1975).
[47] B. Seed, *Nucleic Acids Res.* 11, 2427 (1983).

phage vectors. The procedure is similar to the recovery of transforming genes artificially linked to the *SuIII* gene by cotransformation.[48] A phage library is prepared by conventional means in one of the Charon phage vectors[49] carrying amber mutations in essential phage genes. We have used Charon 30A, but 4A, 16A, and 21A work as well. Partial cleavage of the genomic DNA with *Sau*3A, and insertion into vectors cut with *Bam*HI, is a good procedure in that this ensures good odds of cloning the provirus no matter what the structure of the flanking DNA. The DNA is ligated to the vector, packaged into phage coats, and amplified on a suppressing bacterial host such as LE392. At least 10^6 independent recombinant phage must be formed. About 10^8 of the amplified phage are plated on a nonsuppressing, amber *lac*⁻ host, including the indicator dye XGal[50] (40 μg/ml final concentration) in the plate. Any phage carrying the *SuIII*⁺ gene can grow on the nonsuppressing host, and will also suppress the amber *lac*⁻ mutation of the host, forming blue plaques. These plaques are simply picked and grown up for further analysis. There will also be white plaques; these are phage that have lost the amber markers by recombination with DNA in the packaging extracts, but do not carry the desired *SuIII* gene.

Selection for plasmid replicon and drug marker. If the retrovirus contains a complete functioning plasmid replicon, the provirus can be cloned directly. The genomic DNA is cleaved to completion with a restriction enzyme that does not cleave in the provirus, and 5–10 μg are ligated at low concentration (5–10 per ml; total volume 1 ml) to promote circularization. The ligated DNA is then used to transform *E. coli* to drug resistance, selecting for the marker on the provirus. Standard procedures can be used, so long as the efficiency transformation is in the range of 10^7 transformants per μg of DNA. No more than about 50 ng can be used per 0.1 ml of competent cells, plated on each 10-cm plate. By way of example, the pZIP-NeoSV(X)1 vectors[30] contain a ColE1 replicon and confer resistance to kanamycin (50 μg/ml in L plates). The resulting colonies are picked and plasmid DNA is isolated from the cultures.

Isolation of Target Gene: Comparison of Flanking DNAs.. In the final step, the clones obtained from the genomic DNAs must be compared to determine if any represent insertions in common, shared target loci. The appearance of such sites pinpoints that common locus as a significant one, for it is extremely unlikely that any two independent insertion events would gratuitously occur in the same region; "hot spots" for insertions are probably rare. The recognition of two insertions in the same area, however, is not simple even with clones in hand. All the clones will contain portions

[48] M. Goldfarb, K. Shimizu, M. Perucho, and M. Wigler, *Nature (London)* **296,** 404 (1982).
[49] D. L. Rimm, D. Horness, J. Kucera, and F. R. Blattner, *Gene* **12,** 301 (1980).
[50] XGal represents 5-bromo-4-chloro-3-indolyl-β-D-galactoside.

of the same retrovirus DNA, and most flanking DNA will contain repetitive elements that will hybridize to other flanking DNAs. There are few easy shortcuts here. One screen to apply to the clones is simply to digest them all with enzymes which cleave often, producing numerous small fragments; examination on gels might in favorable cases reveal common size fragments. The fragments from the retrovirus and vector itself, which will be common to all clones, have to be ignored.

The ultimate test for homology between flanking clones is to use each clone as a probe for Southern blot hybridizations against all the other clones. The DNAs are digested with enzymes so as to produce at least a few fragments free of vector and provirus, and multiple copies of a blot are prepared. Hybridization with individual clones as probes are performed. If any probe cross-hybridizes with a flanking DNA fragment of another, more detailed studies are made. The possibility of repetitive elements being responsible for the cross-reaction must be eliminated. To accomplish this, unique sequence (nonrepetitive) probes must be prepared; subclones of short subregions of the flanking DNA must be screened by hybridization with total, labeled, genomic DNA for the absence of such repetitive sequences. If a unique flanking probe still shows cross-reaction, the gene is identified. This probe can then be used to clone the wild-type locus, to prepare cDNAs, and to examine other mutants for alterations in the locus.

Corroborative evidence

If any other handles are available for the target gene, screening of the 20–50 proviral clones for the relevant ones can be considerably facilitated. For example, if the chromosomal assignment of the gene is known, candidate clones can be tested on Southern blots of hybrid cells containing only the relevant chromosome to determine if they are derived from that chromosome. Another example: if the gene can be transferred by DNA-mediated transformation, DNA from the recipient cells can be tested for the presence of the flanking DNA.

In general it is this step—determining which clones of many candidates are the relevant ones—that is most daunting in the use of insertional mutagenesis of mammalian cells to identify new genes. In fact only a few genes, all potential oncogenes, have been identified by virtue of commonality of their use as targets by retroviruses; these are the *int* series of genes activated in oncogenesis by mouse mammary tumor virus (MMTV),[11–13] and analogous genes activated in leukemogenesis by MuLVs.[14–16] The directed use of infection *in vitro* to identify new genes is only just now being attempted. The success of the method to attack known target genes, however, as described above, suggests that these efforts will soon succeed.

[37] Microcloning of Mammalian Metaphase Chromosomes

By JAN-ERIK EDSTRÖM, ROLF KAISER, and DAN RÖHME

DNA cloning techniques typically require genomic DNA quantities in the range down to 100 ng in reaction volumes of at least 10 μl for a single preparation. With λ phage as a cloning vector 10^7 recombinant phages may be obtained from such an amount of DNA. This is because of the efficiency of the *in vitro* packaging technique[1] which may be brought to a level of at least 10^8 infecting phage per μg λ phage DNA, i.e., an efficiency of about 0.5×10^{-2} when compared to the DNA content of λ (2×10^{10} phage per μg DNA). Restricted genomic DNA inserted into such vector is, however, propagated with an efficiency that is lower by an order of magnitude due to the consequences of restricting and religating the vector molecules, i.e., of the order of 0.5×10^{-3}. This value is interesting because it shows that a reasonable number of hand collected mitotic chromosomes can give a useful library. From 100 identical chromosomes in G_2 it is theoretically possible to obtain a library representing about 10% of this chromosomal DNA. Such a collection of a small mammalian chromosome or chromosome arm representing approximately 1% of the G_2 genome would give about 12 pg which should, optimally, result in 1000–2000 clones.

This is not to say that a cloning experiment can be performed only by decreasing the amount of DNA from 100 ng to the 1–10 pg level, i.e., by a factor of 10^4–10^5. This is because the rate of ligation of vector arms to genomic fragments decreases with decreasing concentration of reactants. On the other hand, a compensatory increase in vector concentration is not possible because of the resulting increased background of false positives.

The solution to this problem lies in decreasing reaction volumes in parallel with the amounts of reactants. Such an approach, based on work in an oil chamber with a micromanipulator[2] and instruments prepared in a microforge,[2] is available for handling nanoliter volumes.[3] This technology was further developed for the cloning of DNA from microdissected parts of *Drosophila* polytene chromosomes[4] and later for mammalian chromo-

[1] B. Hohn and K. Murray, *Proc. Natl. Acad. Sci. U.S.A.* **74**, 3259 (1977).
[2] P. de Fonbrune "Technique de Micromanipulation," Monographies de l'Institut Pasteur, Masson, Paris, 1949.
[3] J.-E. Edström, *Methods Cell Physiol.* **1**, 417 (1964).
[4] F. Scalenghe, E. Turco, J.-E. Edström, V. Pirrotta, and M. Melli, *Chromosoma* **82**, 205 (1981).

somes.[5] For polytene chromosomes a few chromosome pieces are sufficient for comparatively complete libraries. Due to more dissection work required for mammalian libraries it is hardly realistic to plan to have more than a fraction of a selected part of a genome represented in a library.

Preparation of Chromosome Spreads

Standard protocols for preparation of metaphase chromosomes are not useful for microcloning due to acid-induced damage to DNA leading to depurination and very short inserts. The following procedure minimizes exposure to acetic acid and gives large inserts (up to 7 kb).[6] After hypotonic treatment the material is fixed in 70% ethanol. Aliquots of the suspension, 200 μl in size, are centrifuged for a few seconds in an Eppendorf centrifuge and resuspended in the same volume of methanol : acetic acid (3 : 1), immediately dropped onto cold clean coverslips, and frozen by immersion in liquid nitrogen. Coverslips are stored in methanol in the refrigerator until used.

Equipment and Microinstruments

The Oil Chamber

The oil chamber (Fig. 1) is formed by a 25 mm wide, 3 mm deep rectangular groove in a $70 \times 35 \times 6$ mm glass slide over which $12 \times 30 \times 0.17$ mm coverslips are placed. All work is done on the lower surface of the coverslips. The space between the coverslips and the bottom of the chamber is filled with liquid paraffin (Merck no. 7161), kept in place by the capillary forces.

Chambers are made from one piece of glass, as solvents in glued chambers may interfere with enzyme reactions. Glass chambers and coverslips are washed in hydrochloric acid and distilled water before use. The liquid paraffin used in the oil chamber is stored over RI buffer, i.e., the buffer used for EcoRI digestions, in order to reduce the amount of water-soluble contaminants.

The Microscope

A phase-contrast microscope has to be used as focusing is done with the stage. It should have a long working distance condensor in view of the

[5] D. Röhme, H. Fox, B. Herrmann, A.-M Frischauf, J.-E. Edström, P. Mains, L. M. Silver, and H. Lehrach, Cell **36**, 783 (1984).
[6] R. Kaiser, J. Weber, K.-H. Grzeschik, J.-E. Edström, A Driesel, S. Zengerling, M. Buchwald, L. C. Tsui, and K. Olek, Mol. Biol. Rep. in press.

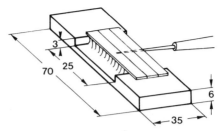

FIG. 1. Oil chamber used for microdissection–microcloning, here covered with three narrow coverslips.

6-mm-thick oil chamber. Dissection is best performed with a high-power dry objective (at least 40X). Low-power objectives (10X and 20X) are used for micropipetting work.

The Micromanipulator

The de Fonbrune micromanipulator (Fig. 2) is a flexible instrument for microwork.[7] It can be used for dissections, with one single needle or with two needles hooked onto the same manipulator, and for micropipetting work. This manipulator was originally produced by Etablissements Beaudouin, Paris, but is no longer made by this firm. Manipulators from this source are, however, excellent in the smoothness with which hand-directed movements are transferred to the working instrument. de Fonbrune manipulators from other sources are not always suitable for chromosome dissections. The manipulator consists of the hand-operated part which is a joy-stick arrangement with the aid of which there is pneumatic transfer of movements to a receptor part. Here aneroid membranes oriented in the three dimensions are influenced by air pressure changes to move an axis onto which the working instrument can be fastened (Fig. 2).

The Microforge and the Microinstruments

The instruments used for microcloning are glass needles and volumetric nanoliter pipettes.[4,8] They can be made in the de Fonbrune microforge (Fig. 3) which, like the de Fonbrune micromanipulator, is produced by Bachofer Laboratorium Geraete, Postfach 7089, D-7410 Reutlingen, F. R. Germany. Glass needles are preferably made from soda glass rods, 3 mm in diameter. An 8-cm glass rod is heated in the middle and pulled about

[7] B. Timm, C. Kondor-Koch, H. Lehrach, H. Riedel, J.-E. Edström, and H. Garoff, this series, Vol. 96, p. 496.
[8] V. Pirrotta, H. Jäckle, and J.-E. Edström, *Genetic Eng.* 5, 1 (1983).

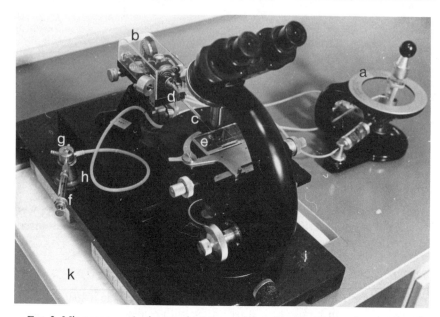

Fig. 2. Microscope and micromanipulator used for microdissection–microcloning. The joy-stick arrangement (a) is connected with three rubber tubes to the receptor (b). Here a pipet (c) fastened with a holder (d) has been introduced into the oil chamber (e). The pressure in the pipet is controlled by means of a syringe (f) connected with a three-way valve (g) and rubber tube to the pipet. The valve is fastened by means of a wing to a holder (h) fastened to a wooden stand (i) on which microscope and receptor are placed. The wooden stand in turn rests on a heavy concrete table (k).

10 cm, to about 1 mm diameter. It is then cut off in the middle to give material for two needles. The cut-off end of the thin part of the glass rod is brought in contact with the electrically heated platinum wire of the micro-forge to melt the glass. A cone of glass can then be pulled out at a right angle from the glass rod. A tip is formed when the molten glass is pulled free from the platinum wire. To obtain a sharp tip one should—shortly before the tip is formed—turn on the air source for cooling the platinum wire and increase the heating just enough to keep the glass molten. This makes the range of heat more restricted and helps to shape a sharp tip. A cone that is too long and slender gives too little resistance for removing chromosome fragments from the coverslip. On the other hand a short cone and high-angled tip, even if sharp, can be difficult to use if it obscures the light path too much. A compromise has to be tried out.

The starting material for nanoliter volumetric pipets is 10-mm thin-

Fig. 3. Microforge in which a micropipet is being made. A glass tube bent at a right angle (a) holds a capillary (b) which is heated with a platinum wire while it is being pulled with a weight (c). From Timm *et al.*[7]

walled Duran or Pyrex glass tubing. It is pulled in an oxygen gas flame to give 1–2 mm diameter capillaries. It is important to avoid increasing the relative wall thickness. Capillaries cut in 10 cm pieces are further pulled close to one of the ends over an ethanol flame to produce a thinner part about 1 cm long and 0.2–0.4 mm in diameter. The capillary distal to this thinning is bent over an ethanol flame to give a hook. The prepared capillary is inserted into a glass holder made of 4-mm soda glass tubing, bent at 90° angle to give two arms, 3 and 4 cm long. It is sealed with picein or histological paraffin into the long arm of the holder without blocking the open end of the capillary. The hook is turned in the same direction as the short arm of the holder. The assembly is then placed vertically in the holder of the microforge, hook and short arm pointing to the right. A suitable weight, about 1 g, is hung in the hook. The assembly is then tilted counter-clockwise about 30° and a bend is made at the upper part of the thinning by bringing the heated platinum wire close to the capillary, but not touching it (Fig. 4). The heated wire is then brought close to the upper part of the thinning to allow the weight to pull it to form a future pipet neck. Finally the heated wire is placed close to the capillary, 1–2 mm further down and the weight is now allowed to pull the glass until it breaks. Through this procedure a bulb for a volumetric pipet has been shaped. The size of the pipet opening increases with increased weights. A 1 g weight gives a suitable diameter of 5–7 μm.

A whole series of pipets of different volumes in the range between 0.5 and 5 nl is useful for microcloning work. These volumes cannot easily be predesigned. Suitable pipets have to be selected from a collection of various sizes. All pipets are first given a light silicone treatment. One percent dichlorodimethylsilane in CCl_4 is drawn into the apical part of a pipet and the liquid kept there for about 10 sec before being removed by aspiration. A second wash is made with 1 mM EDTA. Too strong a treatment makes smooth aspiration and ejection impossible, too light a treatment causes aqueous solution to remain on the pipet walls.

Pipets are calibrated with a concentrated solution of tritiated uridine. The bulb is filled (Fig. 5) and the contents delivered onto a coverslip in the oil chamber. The coverslip is lifted off with adhering oil and put in a

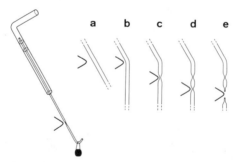

FIG. 4. Steps to make a micropipet. Starting position to the left with heated platinum wire approaching the thinning of the capillary from left. In steps (a) to (e) only the thinning is shown. It is further reduced in diameter in (c) to give a pipet neck and in (d) until it breaks in (e) to give the opening of the pipet. In step (b) the pipet is given an angle to permit it to reach the coverslip in the oil chamber from below. From Pirrotta et al.[8]

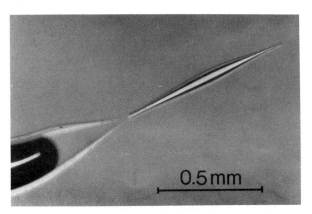

FIG. 5. A 4-nl volumetric pipet as seen from the side in the oil chamber (turned 90° for demonstration). The bulb (a) is here filled with aqueous solution, then follows liquid paraffin (b) in the shaft and the air (c) which transmits pressure changes from the air-filled syringe. From Scalenghe *et al.*[4]

suitable scintillation cocktail and measured together with reference dilutions of the uridine. If several pipets calibrated this way are available, they can be used for calibrating others.

For use in the oil chamber the shaft of a micropipet is connected with rubber tubing to one of the outlet ports of a syringe fitted with a valve that can be switched between this port and another port which is left open for passage of air to or from the syringe. The valve has a wing that is mounted onto a holder fastened to the working table. By tilting the syringe sideways when operating it it can be connected to one or the other of the outlet ports (Fig. 2). This makes it easy to aspirate and eject liquid into and out of the pipet.

Dissection Procedure

A coverslip on which mitotic chromosomes have been spread is placed over the oil chamber, the material facing downward. Since the chromosomes are often easier to distinguish before oil has been added to the chamber one can first identify suitable metaphase plates in a chamber without oil and note the coordinates before adding liquid paraffin. Identification in an oil-filled chamber can be facilitated by prior light staining of the chromosomes with methylene blue. A second, narrow (12 × 30 × 0.17 mm) coverslip is also placed over the chamber. On the lower surface of this glass drops of buffered glycerol (4 parts 87% glycerol, 1 part of

0.05 *M* sodium potassium phosphate buffer, pH 6.8) have been arranged. This is most simply done by first putting a small drop of buffered glycerol on the glass. Drops in the picoliter range are then formed by pulling out liquid onto the surrounding glass surface with the tip of a needle before the coverslip is placed over the chamber. Drops of suitable size are then localized in the microscope and their coordinates determined. The dissecting needle is then inserted, the cone of the dissecting needle tilted sideways about 45° from the vertical plane. Chromosomes or chromosome fragments are scratched away from the coverslip (Fig. 6). An isolated fragment is then transferred to a suitable drop of buffered glycerol. Chromosome fragments tend to stick better to the tip of the glass needle for transport through the oil once the tip has become wetted through contact with a buffered glycerol drop. A large number of chromosome pieces are collected in the same drop (e.g., a hundred pieces) (Fig. 6). Before the pieces are used for microcloning as much excess glycerol as possible is removed with the aid of a micropipet with a very fine tip. It is then possible to push the pieces together into a ball of material which can be picked up with the tip of a glass needle (Fig. 6) in order to have it introduced into a drop of proteinase K – SDS for DNA extraction.

DNA Extraction

After the biological material has been assembled on a narrow coverslip the whole coverslip is lifted off the chamber and placed over a new chamber where it is placed closest to the manipulator. A second narrow glass is treated with 1% dichlorodimethylsilane for 60 sec and then briefly with 1 m*M* EDTA and placed side by side with the coverslip carrying the biological material. Finally a third coverslip is placed, adjoining the second one, with a supply drop of 0.5 mg proteinase K per ml 0.1% sodium dodecyl sulphate (SDS), 0.01 *M* NaCl, 0.1 *M* Tris–Cl, pH 7.4. Volumes of about 0.5 nl are taken from the supply drop of extraction buffer and placed at even distances on the siliconized coverslip in its midline. The drops should be spherical but should still stick well to the glass surface. There is a risk when chromosome material is transferred with a microneedle into such a drop that the extraction volume leaves the siliconized glass surface and disappears along the shaft of the needle. To eliminate this risk a drop may first be tested by introduction of an empty tip into it. If there seems to be a risk of loosing drops the siliconized coverslip should be discarded and a new one tried after shorter silicone treatment.

Chromosome material, brought together onto the tip of a glass needle, is introduced into an extraction drop in which it disperses and dissolves within a couple of minutes. The chamber is then placed in a petri dish with a filter paper moistened with the Tris–NaCl buffer, which is in turn placed

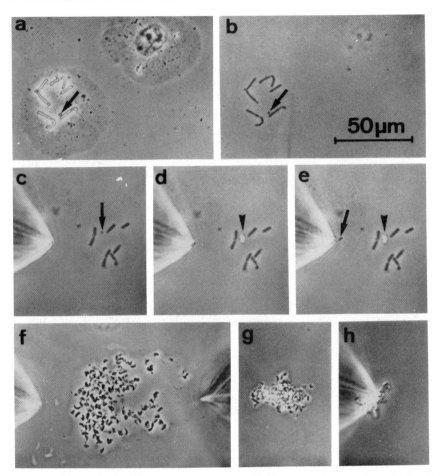

FIG. 6. Spread mitotic chromosomes from the Indian muntjac. Cells are shown in the oil chamber before (a) and after (b) addition of liquid paraffin. The Y chromosomes are indicated by arrows, in (c) before removal with the microneedle which is approaching from the left. In (d) the arrowhead shows the empty area after removal of the chromosome (with surrounding cytoplasm). In (e) the material has been released from the tip of the needle. In (f) 100 collected chromosomes are shown in a drop of buffered glycerol. In (g) most of the glycerol has been sucked away and the chromosomes brought together to a clump of material that has been taken up on the tip of a needle in (h). A second needle which can be used as a precaution against loss of chromosomes on the shaft of the dissecting needle is seen in (f).

in a larger petri dish with moistened filter paper. The whole assembly is then placed at 37° for 90 min.

A new supply coverslip with a drop containing phenol saturated with the Tris–NaCl buffer (without SDS) is then substituted for the old supply coverslip. It is necessary to bring some of the aqueous phase into the phenol supply to guarantee its saturation during the pipetting work. The phenol phase is almost invisible under oil but can usually be discerned due to small drops of aqueous phase that separate out of the phenol at the border of the supply drop. The phenol is taken up in a micropipet with a volume 3–4 times larger than the extract. It is much more visible in the pipet where it has a slightly bluish tinge. When ejected onto the extract the phenol interposes itself between the liquid paraffin and the extract which is often released as a sphere swimming in the phenol (Fig. 7). The phenol phase is pipetted away from the extract after 3–4 min and a new volume added. Altogether three extractions are performed. The phenol can be removed very efficiently provided the pipet has been given a light silicone treatment. After phenol treatment of the DNA extract(s) all discarded phenol is removed with the supply. The microcloning glass is then transferred to a new chamber. This is best done by sliding it over to a chamber with some liquid paraffin already added to ensure that paraffin always covers the microcloning glass surface. Change of chambers is made twice to remove phenol that has become dissolved in the oil during the extractions. Although the residual phenol in the extract can be removed with chloroform[4] we now prefer to leave the chamber in a double humid petri dish in the refrigerator overnight to have residual phenol extracted by the liquid paraffin.

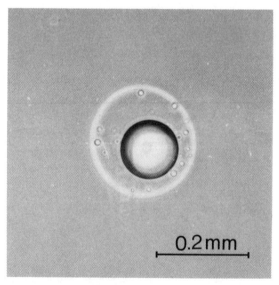

FIG. 7. A 1 nl DNA extract (inner sphere) during treatment with 4 nl buffer-saturated phenol (outer sphere) as seen in the oil chamber.

Restriction Enzyme Digestion

For restriction of the extracted DNA a supply drop of *Eco*RI is placed on a fresh coverslip in the oil chamber. The concentrations used for microcloning are of the order of 50–100 U/μl. Because of the high concentrations it is advisable to eliminate traces of *Eco*RI* activity (i.e., recognition of only the central four base pairs of the *Eco*RI site). This is best done by performing restrictions at high ionic strength, in an "RI buffer" which is 100 mM NaCl, 10 mM MgCl$_2$, 100 mM Tris–Cl buffer, pH 7.4, and 5 mM dithiothreitol (DTT). Furthermore, the restriction enzyme stock solution is diluted sufficiently to reduce the glycerol concentration to 5% or less. We are presently using the same volume (but another micropipet) for restriction enzyme solution as for the proteinase K extraction solution, i.e., 0.6 nl. The supply solution is therefore 2$\times$ the final concentrations.

After the restriction enzyme has been added, the bottom of the oil chamber is covered with a piece of filter paper wetted with RI buffer and the chamber placed in a double humid petri dish where humidification occurs with RI buffer. Digestion takes place for 2 hr at 37°. Like Fisher *et al.*,[9] we now prefer to use phenol instead of heat for enzyme inactivation. We carry out two extractions and change oil chambers twice afterward. Traces of residual phenol are eliminated by keeping the oil chamber in the refrigerator in a double humid petri dish overnight.

In work on genomes with high contents of repetitive sequences it might be of interest to obtain *Eco*RI* restricted clones. In this case the λ vector should permit convenient isolation of the insert, as in λgt10 where *Hin*dIII sites surround the *Eco*RI insertion site.

Ligation to Vector

Ligation is performed at 10-fold molar excess of vector to reduce the frequency of multiple inserts. Ten picograms of restricted genomic DNA would consequently require 1 ng of vector DNA (assuming a molecular weight difference, restricted fragment: λ DNA of 1 : 10), e.g., 5 nl at 200 μg/ml. T4 ligase is then added to a final concentration of 0.2–0.5 U/μl, ATP to 0.5 mM, DTT to 10 mM, MgCl$_2$ to 7 mM, and Tris–Cl, pH 7.4, to 30 mM. The oil chamber is then placed in a double humid chamber wetted with water overnight at 15°.

Recovery of Ligated Extracts and *in Vitro* Packaging

After completion of ligation the chamber is submerged into liquid paraffin in a petri dish, the two flanking coverslips removed, and the

[9] E. M. C. Fisher, J. S. Cavanna, and S. D. M. Brown, *Proc. Natl. Acad. Sci. U.S.A.* **82**, 5846 (1985).

central coverslip placed on the bottom of the oil chamber after being turned around (under oil).

A 1 μl drop of 0.01 M Tris–Cl, 0.001 M EDTA, pH 8.0, is added under oil to each ligation mixture which is then taken up with a Gilson pipet and placed in an Eppendorf tube. After this 2 μl of sonic extract and 10 μl of freeze-thaw lysate are added, prepared as described by Scherer et al.,[10] the mixture is shaken and kept at room temperature for 60 min. Phage dilution buffer (10 mM MgCl$_2$, 20 mM NaCl, 10 mM Tris–Cl, pH 7.4) in a volume of 100 μl is then added and aliquots of the extract mixed with 100 μl plating bacteria for phage absorption at 37° for 15 min in the presence of 10 mM Mg^{2+}. After addition of 3 ml of top layer agar the phage–bacterial suspension is applied on agar plates.

Vector and Host

For cloning EcoRI restricted fragments λNM641[11] has mostly been used. It is an insertion vector with a single EcoRI site in the cI gene. The inactivation of this gene by an insert changes plaque morphology from turbid to clear in a strain such as Q358,[12] $r_k^-m_k^+$. Alternatively, recombinants can be selected on bacterial strains with mutations such as hfl or lyc7 which do not allow the lytic growth of phage with an intact cI gene. We have been using POP13b, $r_k^-m_k^+$ which was derived by N. Murray from a lyc7 strain.[13] For both selection techniques the background of false positives should be low, preferably well below 50 per ng of vector DNA. Also HindIII can be used for microcloning with the λNM1149[11] strain as vector. It contains a single HindIII and a single EcoRI site in the cI gene. A useful vector–host combination for microcloning is λgt10 for which a C600 hflA$^-$ host was used.[9] Since the EcoRI site in the cI gene is surrounded by HindIII sites it is also useful for EcoRI* digestions.

Applications

Some studies of microcloned mammalian chromosomes have now been published. Röhme et al.[5] isolated the proximal part of chromosome 17 which could be identified by being fused to chromosome 8 in a Robertsonian translocation. From 270 t-region-containing fragments 212 cI$^-$

[10] G. Scherer, J. Telford, C. Baldari, and V. Pirrotta, Dev. Biol. **86**, 438 (1981).
[11] N. E. Murray, W. J. Brammar, and K. Murray, Mol. Gen. Genet. **150**, 53 (1977).
[12] J. Karn, S. Brenner, L. Barnett, and G. Cesarini, Proc. Natl. Acad. Sci. U.S.A. **77**, 5172 (1980).
[13] R. Lathe and J. P. Lecocq, Virology **85**, 204 (1977).

phages were recovered. About 20% of the clones contained sequences repeated at least 100-fold in the mouse genome. Of the remainder 71% of those having inserts larger than 200 bp hybridized to mouse DNA. In a sample of 15 clones the origin from chromosome 17 could be ascertained for 14 clones. The average size of the inserts hybridizing to mouse DNA was 1.15 kb, clearly below the statistical average which is about 4 kb. The DNA content of the 270 fragments is unknown but might be of the order of 10 pg indicating a cloning efficiency of about 20 per pg which is in agreement with results on polytene chromosomes.

Fisher et al.[9] microdissected 100 pieces of a proximal region of the mouse X chromosome, estimated to contain about 2 pg of DNA. They found 85% of 650 cI^- clones to be true recombinants, a cloning efficiency almost an order of magnitude larger than for the chromosome 17 work. In this case the average insert size was very small: thus 40% of the inserts were less than 100 bp long and the average size of the remaining clones only 400 bp. The acid treatment used for preparing chromosome spreads may cause depurination. Statistically large DNA pieces run a larger risk of being depurinated than small ones, which could result in a preferential recovery of small inserts. Weith et al.[14] isolated homogeneously staining regions (HSR) from the mouse chromosome no. 1. From several hundred pieces estimated to represent 4 pg DNA about 1000 recombinants were obtained, a sample of which contained inserts in the 0.9–3.7 kb range.

Kaiser et al.[6] isolated the middle part of human chromosome no. 7 in a human–mouse hybrid cell line (Fig. 8). Although the efficiency of cloning was low in this case (20 recombinants from 100 pieces) the inserts were large (in the range of 1–7 kb for nine of the clones).

Summary and Prospects

Microcloning can be used to obtain libraries of clones from individually dissected mammalian chromosomes as well as chromosome regions. The numbers of chromosomes needed for linkage studies and for starts of chromosome walks are moderate, on the order of 100–500 genome equivalents. Microcloning is also the method of choice when the amount of material is limited or where sufficient amounts of bulk isolated specific chromosomes cannot easily be obtained.

With the presently used equipment for microdissection relatively coarse dissections can be made (half of the mouse chromosome 17, one fourth of the X chromosome). For work with higher resolution it will be necessary to develop the manipulator, e.g., with piezoelectric control, and

[14] A. Weith, H. Winking, B. Brachmann, B. Boldyreff, and W. Traut, *EMBO J.* **6**, 1295 (1987).

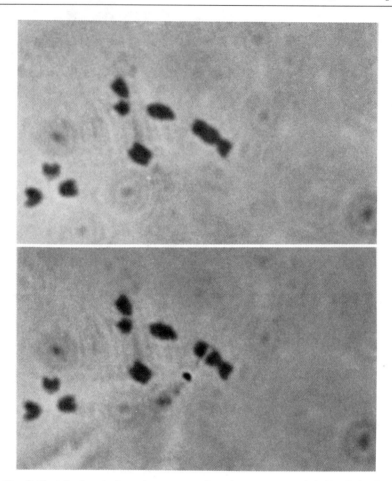

FIG. 8. Unstained metaphase chromosomes from human–mouse hybrid cell line including human chromosome 7. Upper picture shows chromosomes before dissection and lower picture after dissection with dissected fragment lying close to the chromosome. From Kaiser et al.[6]

television monitor. Another line to pursue is to investigate to what extent banding techniques can be combined with microcloning work. Such developments are highly desirable, among others in view of increasing needs for polymorphic DNA markers in human linkage studies.

Acknowledgments

The authors were supported by the Swedish Cancer Society and D. R. also by the Swedish Science Foundation (NFR).

Section V

Gene Regulation in Tissue Culture

[38] Antisense RNA Inhibition of Endogenous Genes

By J. TIMOTHY STOUT and C. THOMAS CASKEY

The development of recombinant DNA techniques has made it relatively easy to isolate genes that are expressed in a tissue-specific or developmentally regulated fashion; however, it is often difficult to assign a functional role to these gene products. To do so, geneticists have classically relied on naturally occurring mutants that fail to express these genes in a normal fashion. The *in vitro* generation of cellular mutants has been primarily limited by a lack of adequate mutant selection methods and the diploid nature of most genes. Even more difficult has been the production of mice with recessive mutations in structural or regulatory genes.

A recent series of experiments suggests that nucleic acid sequences complimentary to endogenous transcripts can be used to establish dominant "null" phenotypes specific for these genes. This approach is based on the assumption that these "anti-sense" sequences can hybridize to primary RNA transcripts and disrupt normal RNA processing, transport, stability, or translation, resulting in the prevention of normal gene expression. This inhibition of RNA processing or translation can theoretically be achieved for any gene by the dominant action of antisense RNA.

We report the successful modeling of this strategy for an X-linked recessive trait, HPRT deficiency. The high degree of inhibition of endogenous HPRT expression suggests the method will be useful for other expressed genes and may be applicable to the development of transgenic mice with dominant antisense inhibition of selected genes. If successful, these techniques may augment the study of developmental genes as well as the pathophysiology of diseases such as the Lesch–Nyhan syndrome (HPRT deficiency).

Background

Regulation of gene expression by antisense inhibition was first described as a natural prokaryotic control mechanism.[1-7] Several of these

[1] R. M. Lacatena and G. Cesareni, *J. Mol. Biol.* **170**, 635 (1983).
[2] J. Tomizawa, T. Itoh, G. Selzer, and T. Som, *Proc. Natl. Acad. Sci. U.S.A.* **78**, 1421.
[3] J. I. Tomizawa and T. Itoh, *Cell* **31**, 575 (1982).
[4] J. I. Tomizawa, *Cell* **38**, 861 (1984).
[5] D. R. Moser, D. Ma, C. D. Moser, and J. L. Campbell, *Proc. Natl. Acad. Sci. U.S.A.* **81**, 4465 (1984).
[6] T. Mizuno, M. Y. Chou, and M. Inouye, *Proc. Natl. Acad. Sci. U.S.A.* **81**, 1966 (1984).
[7] R. W. Simons and N. Kleckner, *Cell* **34**, 683 (1983).

examples are listed in Table I. Control of ColE1 plasmid replication by DNA polymerase I in *E. coli* depends on the hybridization of a primer transcript (RNA II) to the template DNA at the replication origin.[1-5] The concentration of RNA II molecules able to form priming hybrids is titrated by the production of antisense RNA II. These antisense transcripts hybridize to RNA II molecules and prevent the formation of the primed complex. In this way the cell uses overlapping (bidirectional) transcription units to prevent lethal, run-away plasmid replication.

A similar control mechanism is used in the production of osmoregulatory outer membrane proteins *(OmpF* and *OmpC* in *Escherichia coli).*[6] The *OmpC* locus is transcribed bidirectionally under conditions of high osmolarity. Transcribed sequences, upstream of the *OmpC* gene, are complementary to the 5' region of the *OmpF* gene and result in the down-regulation of the *OmpF* gene product. Simons and Kleckner have reported genetic evidence which suggests that the transposable element of *E. coli*, IS10, can negatively control expression of its own transposase protein at the translational level via antisense RNA inhibition.[7] An RNA transcript, complementary to a 36 nucleotide (nt) region that includes the start codon for the transposase gene, is synthesized and presumably prevents ribosomal binding of the transposase message through the formation of an RNA–RNA hybrid.

The first experimental manipulation of antisense RNA in mammalian cells involved the coinjection of plasmids containing the herpes simplex virus (HSV)–thymidine kinase (TK) gene in sense and antisense orientation into TK⁻ cells.[8] Izant and Weintraub demonstrated that when antisense constructions were coinjected with wild-type TK genes at a 100:1 (antisense:sense) ratio, a 4- to 5-fold reduction in TK activity was observed when compared with neighboring cells coinjected with the TK gene and an excess of control plasmids. Results with calcium phosphate or DEAE dextran mediated transformation of cells with antisense TK plasmids provided similar evidence that inhibition of TK activity was sequence specific and would occur with a 10- to 20-fold excess of antisense molecules.[9,10]

This inhibitory strategy was first used by Rosenberg to disrupt normal development of an organism when sequences complementary to the *Drosophila* Kruppel mRNA were microinjected into developing *Drosophila* embryos.[11] Embryos injected with antisense Kruppel RNA developed

[8] J. G. Izant and H. Weintraub, *Cell* **36,** 1007 (1984).
[9] J. G. Izant and H. Weintraub, *Science* **229,** 345 (1985).
[10] S. K. Kim and B. J. Wold, *Cell* **42,** 129 (1985).
[11] U. B. Rosenberg, A. Preiss, E. Seifert, H. Jackle, and D. C. Knipple, *Nature (London)* **313,** 703 (1985).

TABLE I

REGULATION OF GENE EXPRESSION BY ANTISENSE INHIBITION

Gene	Method of introduction	Length of complementarity[a]	Outcome	Reference
Prokaryotic examples				
ColE1 primer (RNA II)	Natural regulatory mechanism	108 nt	α-sense RNA inhibits priming reaction Necessary for DNA replication	1–5
Tn10 transposase	Natural regulatory mechanism	36 nt	α-sense RNA inhibits translation of transposase message	7
OmpC/OmpF	Natural regulatory mechanism	174 nt	Bidirectional transcription at OmpC locus produces α-sense RNA that inhibits translation of OmpF mRNA.	6
β-Galactosidase	In vitro transfection	831 nt	α-sense RNA inhibits β-galactosidase synthesis	25
Eukaryotic examples				
Thymidine kinase (herpes, chicken)	Microinjection, CaPO₄-mediated gene transfer	As small as 52 nt	α-sense RNA inhibits expression of endogenous or exogenous TK genes	8–10
β-Actin	Microinjection, CaPO₄-mediated gene transfer	500 nt	α-sense RNA inhibits cell viability and leads to microfilament disarray	9
β-Globin	Microinjection	As small as 45 nt	SP6 generated α-sense RNA inhibits translation of Xenopus β-globin mRNA	13
Gap junction protein	Microinjection	14-mer and 17-mer oligonucleotides	α-sense oligonucleotides alters normal development of Xenopus blastocyst cells.	N. B. Gilula, personal communication
Kruppel	Microinjection	2300 nt	α-sense RNA produces phenocopies of the Kruppel mutation when microinjected into wild-type prosophila embryos	11
Discoidin-1	Dictyostelium transformation	320 nt	α-sense RNA inhibits synthesis of discoidin-1 and produces phenocopies of disc-1⁻ mutants	12
HPRT	CaPO₄-mediated gene transfer	469 nt	α-sense RNA inhibits HPRT activity and produces resistance to 6-TG	This chapter

[a] nt, Nucleotide.

lethal segmentation Kr phenocopies at high frequencies. Kumar, Warner, and Gilula have shown that when oligonucleotides (14 and 17 nt in length), complementary to coding sequences of the *Xenopus* gap junction protein gene, are microinjected into developing *Xenopus* embryos normal development of these cells is altered (N. B. Gilula, personal communication). Similarly, Crowley *et al.* have reported the establishment of discoidin 1-minus mutant phenocopies in *Dictyostelium* transformed with vectors that express antisense discoidin RNA.[12]

While it is clear that antisense molecules can inhibit normal gene expression, the exact mechanisms of inhibition remain uncertain. Melton demonstrated that the formation of sense:antisense hybrids could arrest translation of *Xenopus* β-globin mRNA. When both sense and antisense molecules were coinjected into *Xenopus* oocytes, nuclease resistant RNA:RNA hybrids could be recovered from these cells. In addition, deletion studies showed that antisense RNAs must be complementary to the 5′ end of the message in order to block translation. These studies demonstrate that antisense RNA complementary to the 5′ end of globin mRNA can prevent globin synthesis at the translation level. Kim and Wold have produced cell lines that express high levels of antisense TK RNA, in a stable manner, leading to an 80–90% reduction in endogenous TK activity.[10] RNA:RNA duplexes were detected in the nuclear fraction of these cells suggesting that antisense molecules hybridize to nascent transcripts in the nucleus and that these duplexes fail to enter the cytoplasm. Furthermore, when Crowley *et al.* produced *Dictyostelium* mutants, nuclear run-on assays showed that both the endogenous gene and the transfected antisense gene were transcribed.[12] Since few endogenous (sense) transcripts were detected in Northern blots, these authors suggested that antisense:sense hybrids form in the nucleus and are rapidly degraded. It is clear that there are many potential steps in the biogenesis of RNA and protein at which antisense nucleic acids can be inhibitory. In eukaryotic cells that express both sense and antisense RNAs, it appears likely that the most profound inhibitory effects occur in the nucleus—the site of RNA synthesis and requisite processing.

We have employed a vector developed by Kaufman and Wong to promote abundant transcription of mouse HPRT genomic fragments or a mouse HPRT cDNA in COS cells.[14,15] This vector facilitates the study of

[12] T. E. Crowley, W. Nellen, R. H. Gomer, and R. A. Firtel, *Cell* **43**, 633 (1985).
[13] D. A. Melton, *Proc. Natl. Acad. Sci. U.S.A.* **82**, 144 (1985).
[14] R. J. Kaufman, *Proc. Natl. Acad. Sci. U.S.A.* **82**, 689 (1985).
[15] G. G. Wong *et al.*, *Science* **228**, 810 (1985).

antisense inhibition of HPRT in both transient and stable expression systems and is a convenient method to determine which regions of the HPRT gene, when transcribed into antisense RNA, are most effective in depressing endogenous levels of HPRT enzyme activity.

Constructions

The vector p91023B (kindly provided by R. Kaufman) has many features that make it suitable for the evaluation of transient and stable expression of antisense HPRT RNA.[14,15] This vector contains the pBR322 origin of replication and the tetracycline resistance gene which allows maintenance and selection in *E. coli;* in addition, this vector lacks plasmid sequence that have been shown to be inhibitory to replication in COS cells. The SV40 origin of replication/enhancer segment is included to allow the plasmid to achieve a high copy number in COS cells. Fragments of the HPRT gene were transcribed under the control of a promoter comprised of the adenovirus major late promoter (AdMLP) spliced to a cDNA copy of the adenovirus tripartite leader. A hybrid intron, consisting of the 5' splice site from the tripartite leader and a 3' splice site from a mouse immunoglobulin gene, immediately precedes the unique *Eco*RI cloning site. This vector includes the coding region from the dihydrofolate reductase (DHFR) gene (reported to enhance mRNA stability in COS cells) and the SV40 polyadenylation signal 3' to the cloning site. Thus, transcription begins within the AdMLP, proceeds through the tripartite leader, the hybrid intron, the HPRT gene inserts, the DHFR gene, and terminates with the addition of a poly(A) tail near the SV40 poly(A) site. The combination of a strong promoter and a high extrachromosomal copy number in COS cells makes this expression system particularly useful in the study of antisense inhibition.

The constructions pmHPT-1 and pmHPT-2 (Fig. 1) include genomic DNA fragments from the 5' end of the mouse HPRT gene in sense and antisense orientations.[16] The insert in pmHPT-2 is a 1.4 kb *Eco*RI fragment that contains the first exon, 400 bp from the first intron and 858 bp of 5' flanking sequence. pmHPT-1 includes the same intron and exon sequences as pmHPT-2, but lacks the region 5' to the cap site. A full-length mouse HPRT cDNA was used to construct pmHPT-cDNA.[17]

[16] D. W. Melton, D. S. Konecki, J. Brennand, and C. T. Caskey, *Proc. Natl. Acad. Sci. U.S.A.* **81,** 2147 (1984).

[17] J. Brennand, A. C. Chinault, D. S. Konecki, D. W. Melton, and C. T. Caskey, *Proc. Natl. Acad. Sci. U.S.A.* **79,** 1950 (1982).

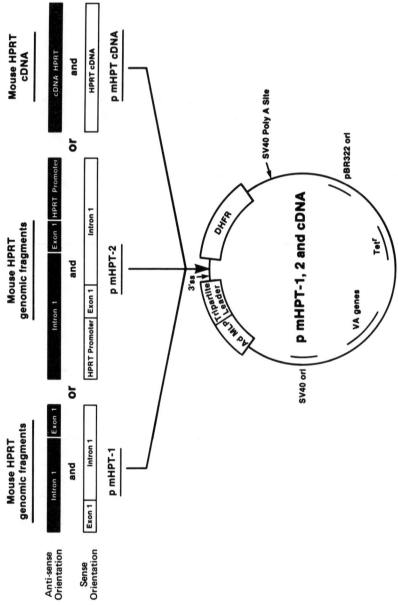

Fig. 1. Sense and antisense constructions. Solid and open boxes represent antisense and sense orientations, respectively.

Transfections

COS cells (M6 clone, R. Kaufman) were transfected by the calcium phosphate precipitation method in the following manner. Approximately 12 hr before transfection COS cells were plated in DMEM containing 10% heat-inactivated fetal calf serum (Gibco Inc.) 0.2 mg/ml gentamycin, 1 μg/ml Fungizone, and 0.35% NaHCO$_3$ (5×10^5 cells/100 cm^2 dish). Sense and antisense pmHPRT-1, 2 and cDNA plasmids were prepared by two rounds of purification through CsCl gradients. Supercoiled plasmids were isolated from low melting point agarose gels (Sea Plaque agarose, FMC Inc.). Ten micrograms of plasmid DNA were added to 500 μl of HBS (150 mM NaCl, 5 mM KCl, 0.5 mM Na$_2$HPO$_4$, 5 mM dextrose, 20 mM HEPES buffer, pH 7.05). A 2 M CaCl$_2$ solution was added to yield a final concentration of 125 mM. This mix was allowed to incubate at room temperature until a visible precipitate formed ($\approx$ 45 min). After removal of media from cells the DNA coprecipitate was added to each plate (10 μg DNA/plate). After 20 min at room temperature 10 ml of media was added and each plate was transfered to 37°. After 4 hr, the media was removed and replaced with 3 ml of HBS + 15% glycerol. After 3 min the HBS + glycerol was removed, cells were washed twice with Hanks' buffered saline (Gibco Inc.), and fresh media added.

Evaluation of Expression

Transient Expression

At 24, 48, 72, and 96 hr after transfection, duplicate plates were harvested, and cell extracts or total RNA were obtained. To produce cell extracts, cells were recovered by brief trypsinization, washed once with 5 ml of Hanks' buffered saline, centrifuged at 1500 g for 5 min, and resuspended in 100 μl of 20 mM Tris–HCl (pH 7.5). Crude cell suspensions were subjected to five freeze-thaw cycles to lyse cells and the cellular debris was cleared by centrifugation at 15,000 g for 10 min. Supernatants were assayed for HPRT and APRT activities as described by Olsen and Milman[19] and Holden *et al.*[20]

RNA was isolated by a modification of the method described by Chirgwin *et al.*[21] Media was aspirated from duplicate plates, the cells washed

[18] F. L. Graham and A. J. van der Eb, *Virology* **52**, 156 (1973).
[19] A. Olsen and G. J. Milman, *J. Mol. Biol. Chem.* **249**, 4030 (1974).
[20] J. A. Holden, G. S. Meredith, and W. N. Kelley, *J. Mol. Biol. Chem.* **254**, 6951 (1979).
[21] J. M. Chirgwin *et al., Biochemistry* **18**, 5294 (1979).

once with Hanks' saline, and lysed *in situ* by the addition of 1 ml of a solution containing 4 M guanidine isothiocyanate, 25 mM sodium citrate (pH 7.0), 0.5% sarkosyl, 0.1 M 2-mercaptoethanol. Genomic DNA was sheared by passing this lysate through a 22-gauge needle 7–10 times. RNA was purified from these crude extracts by centrifugation through a solution of 6.2 M cesium chloride, 0.1 M EDTA (pH 7.0). RNA pellets were resuspended in 100 μl of 1× TE (10 mM Tris, 1 mM EDTA) and ethanol precipitated. Approximately 15 μg of total RNA from each sample was electrophoresed through a 1.5% formaldehyde–agarose gel, and transferred to nitrocellulose. Blots were prehybridized for 12 hr at 42° in a solution containing 50% formamide, 5× SSC (1× SSC = 0.15 M NaCl, 15 mM sodium citrate, pH 7.4), 0.5% SDS, 5× Denhardt's solution (100× = 2% BSA, 2% Ficoll, 2% polyvinylpyrrolidone), and 100 μg/ml heat-denatured shearing herring sperm DNA.[22] Hybridizations included 2 × 10^6 cpm/ml of ^{32}P-labeled sense HPRT RNA, generated by the SP 6/T7 transcription system (pGEM-Riboprobe, Promega Biotech Inc.).[23] Blots were hybridized for 14 hr at 42°, washed twice for 1–1.5 hr in 5× SSC, 0.5% SDS at 65°, and autoradiographed for 36 hr at −70° with an intensifying screen (Quanta III, DuPont Inc.).

Stable Expression

After 48 posttransfection hr in nonselective media, cells were split (1 : 4) and placed in media containing 5 × 10^{-5} M 6-thioguanine (6-TG) to select for HPRT$^-$ cells. Media was changed every 48 hr and the selection was allowed to proceed for 14 days. 6-TG-resistant cells were pooled and assayed for HPRT and APRT activities (as above).

Results and Comments

Transient Expression

HPRT and APRT specific activities in cells that received sense or antisense plasmids were determined and compared with the values obtained from mock-transfected cells. Figure 2 demonstrates a specific reduction of HPRT activity in cells that received antisense plasmids. This diminution was not observed in APRT activities in these cells or for either

[22] T. P. Yang, T. I. Patel, A. C. Chinault, J. T. Stout, L. G. Jackson, B. M. Hildebrand, and C. T. Caskey, *Nature (London)* **310,** 412 (1984).
[23] D. Melton, P. Krieg, M. Rebagliati, T. Maniatis, K. Zinn, and M. Green, *Nucleic Acids Res.* **12,** 7035 (1984).

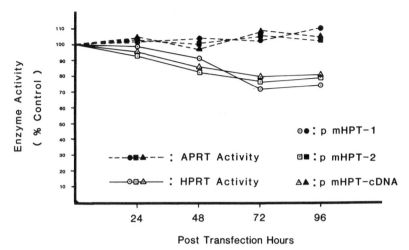

FIG. 2. HPRT and APRT activities of cells transfected with HPRT antisense RNA expressing vectors. Assays were performed twice with total cellular extracts produced from pooled, duplicate plates.

enzyme in cells that received sense-orientation plasmids. The fact that this reduction is observed in extracts from mixed population cells suggests that the sub-population of cells that actually received plasmid DNA was profoundly inhibited. No significant difference in the level of HPRT inhibition was noted between cells transfected with vectors containing antisense genomic fragments (with or without the mouse promoter) and those that received antisense cDNA constructs.

Northern analysis of RNAs isolated from these cells is shown in Fig. 3. Production of RNA that is able to hybridize to labeled sense HPRT RNA is specific for those cells that received antisense plasmids. There is a quantitative increase during the first 72 hr of antisense transcripts synthesized in cells that received pmHPT-1 or pmHPT-2. These levels decline slightly by 96 hr. This roughly correlates with the time course of enzyme inhibition as shown in the second figure.

These results suggest that the production of antisense HPRT transcripts can transiently inhibit HPRT synthesis in COS cells. While this inhibition occurs across species boundaries (mouse antisense RNA production in monkey cells), Konecki et al. have shown that mouse and human HPRT sequences are highly conserved.[24] Homology between human and mouse HPRT is 95% in coding sequences and 85% in non translated regions.

[24] D. S. Konecki, J. Brennand, J. C. Fuscoe, C. T. Caskey, and A. C. Chinault, *Nucleic Acids Res.* **10**, 6763 (1982).

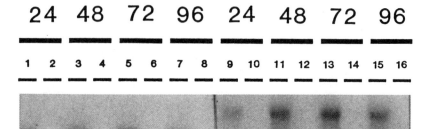

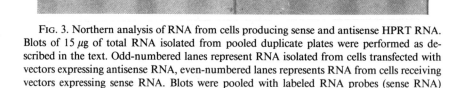

FIG. 3. Northern analysis of RNA from cells producing sense and antisense HPRT RNA. Blots of 15 μg of total RNA isolated from pooled duplicate plates were performed as described in the text. Odd-numbered lanes represent RNA isolated from cells transfected with vectors expressing antisense RNA, even-numbered lanes represents RNA from cells receiving vectors expressing sense RNA. Blots were pooled with labeled RNA probes (sense RNA) produced by the SP6/T7 transcription system.

Stable Expression

To determine if the transient inhibition of HPRT described above could be maintained in cells that express these antisense plasmids in a stable fashion, COS cells were transfected with sense and antisense plasmids and allowed to grow in the presence of 6-thioguanine. As is evident in Fig. 4, numerous 6-TG-resistant colonies were produced from cells transfected with antisense plasmids; cells that received sense plasmids (or no DNA) did not survive selection. Table II shows that 6-TG resistance in cells transfected with antisense pHPT-2 is due to a reduction in HPRT activity. Thus, the stable expression of these antisense constructions leads to continued depression of endogenous HPRT activity.

Many questions remain about the optimal conditions for antisense inhibition of endogenous genes. Our studies suggest that with this vector and host combination, there are no effective differences between the ability of antisense genomic fragments (from the 5′ end of the gene) or cDNA constructs to inhibit HPRT synthesis. The ratio of antisense RNA to endogenous sense RNA needed for inhibition is unknown and may vary

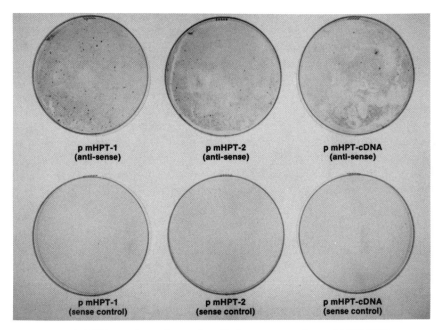

FIG. 4. 6-Thioguanine resistance in cells that exrpess antisense HPRT RNA.

from system to system; certainly the specific mechanism of inhibition affects this ratio. Tissue-specific variations in the rates of transcription, hnRNA processing, and message degradation, as well as the amount of inhibition required for phenotypic alteration, will similarly influence the optimal conditions for useful antisense inhibition.

TABLE II
HPRT ACTIVITY IN STABLE ANTISENSE HPRT CELLS

	Activity[a]		
Cell Line	HPRT	APRT	6-TG[r]
Normal lymphoblast control	100	100	No
Cos + antisense RNA	1.7	99.2	Yes
Cos + sense RNA	97.6	101.2	No
GM 6804 (Lesch–Nyhan patient)	0.0	102.4	Yes
COS (untreated)	94.6	104.6	No

[a] All enzyme activities normalized to HPRT[+] control cells.

These experiments demonstrate that the availability of cloned sequences, complementary to cellular genes, enables investigators to selectively inhibit proper expression of these genes. This will undoubtedly be useful in a wide range of studies. Antisense techniques may yield information about the functional role of many expressed sequences which are currently uncharacterized. In conjunction with transgenic mouse techniques, antisense inhibition may offer a unique opportunity to study the pathophysiologic significance of metabolic imbalances, and their consequences on development and morphogenesis. A better understanding of the specific mechanisms of inhibition and improvements in our abilities to deliver antisense DNA and RNA will be important in making this technique applicable to a wide range of organisms.

Acknowledgments

We thank R. Kaufman for the gift of expression vector p91023b. J. T. S. was supported by the Michael Philip Berolzheimer Medical Scientist Fellowship. C. T. C. is an investigator of the Howard Hughes Medical Institute and is a recipient of NIH Grant AM31428.

[25] S. Pestka, B. L. Daugherty, V. Jung, K. Hotta, and R. K. Pestka, *Proc. Natl. Acad. Sci. U.S.A.* **81,** 7525 (1984).

[39] *In Situ* Detection of Transcription in Transfected Cells Using Biotin-Labeled Molecular Probes

By GILBERT H. SMITH

Our understanding of molecular genetics has expanded dramatically over the past decade with the development of recombinant DNA techniques and sensitive methods for analyzing the function, organization, regulation, and intracellular location of specific genes and their transcripts. The method of *in situ* hybridization has been a powerful tool for the localization of specific DNA sequences on chromosomes,[1] to detect RNA transcripts within individual cells during development and differentiation,[2-4] or to detect the activity of specific transcriptional promoters or

[1] J. G. Gall and M. L. Pardue, this series, Vol. 38, p. 470.
[2] D. G. Capso and W. R. Jeffrey, *Dev. Biol.* **67,** 137 (1978).
[3] L. M. Angerer and R. C. Angerer, *Nucleic Acids Res.* **9,** 2819 (1981).
[4] G. H. Smith, P. J. Doherty, R. B. Stead, C. M. Gorman, D. E. Graham, and B. H. Howard, *Anal. Biochem.*, **156,** 17 (1986).

enhancer sequences in cells transfected with recombinant DNA plasmids designed for expression.[5,6] Most of these studies have employed single- or double-stranded polynucleotide probes labeled to high specificity with radioactive nucleotide analogs. These autoradiographic methods are highly sensitive and reproducible. However, they do have the limitations associated with the use of radioisotopes many of which possess short functional half-lives. In addition, radioisotopes are costly in terms of procurement, personnel safety, disposal problems, and reduced probe stability. With regard to tritium-labeled probes, which are relatively stable and relatively safe, the inherent lack of resolution due to the track of decay particles and the thickness of the autoradiographic emulsion reduces the desirability of this approach for precise morphological work.

Langer *et al.*[7] have developed a biotinylated dUTP that can be incorporated into DNA by nick-translation. The biotin molecule, which is attached by an alkylamine linker arm to the pyrimidine ring, can then be detected by cytochemical methods using avidin complexed with signal molecules such as horseradish peroxidase, rhodamine, alkaline phosphatase, or colloidal gold. More recently, a method for coupling biotin directly to the nucleic acid probe via photoactivation has become commercially available from Spector Laboratories, Burlington, California. Biotinylated molecular probes provide stable, sensitive, and utilitarian tools for the detection of gene transcription *in situ*. We have used them to detect *in situ* the transient expression of recombinant DNA vectors following their transfection into mammalian tissue culture cells.

Sources of Reagents

Biotinylated dUTP and a nick-translation kit for preparing biotin-labeled DNA probes were purchased from Enzo Biochem, Incorporated, New York, New York. Biotinylated avidin DH-biotinylated horseradish peroxidase complex (Vecta-Stain ABC kit) was purchased from Vector Laboratories, Incorporated, Burlington, California. Pronase was obtained from Calbiochem-Behring Corporation, La Jolla, California. Diaminobenzidine tetrahydrochloride and paraformaldehyde flakes were purchased from Polysciences, Incorporated, Warrington, Pennsylvania. Dextran sulfate (~500,000 MW) was purchased from Pharmacia, Piscataway, New

[5] D. J. Brigati, D. Myerson, J. J. Leary, B. Spalholz, S. Z. Travis, C. K. Y. Fong, G. D. Hsiung, and D. C. Ward, *Virology* **126,** 32 (1983).

[6] R. H. Singer and D. C. Ward, *Proc. Natl. Acad. Sci. U.S.A.* **79,** 7331 (1982).

[7] P. R. Langer, A. A. Waldrop, and D. C. Ward, *Proc. Natl. Acad. Sci. U.S.A.* **78,** 6633 (1981).

Jersey. Formamide was obtained from Fluka, A. G., Switzerland. Standard nucleotide triphosphates were from P-L Biochemicals, Milwaukee, Wisconsin. Plasmid DNAs, pRSVcat, and pRSV-csnα were supplied after cesium chloride–ethidium chloride equilibrium gradient purification by Drs. Cornelia M. Gorman and Paul J. Doherty, Laboratory of Molecular Biology, NCI, Bethesda, Maryland.

Transfection of Cells

Preparation of plasmid DNAs, DNA fragments, assay of CAT activity, and transfection procedures, etc., are described in detail elsewhere in this volume.[8] Briefly, transfections were carried out on CV-1 (monkey kidney) cells by using calcium phosphate (10 μg DNA per 100 mm petri dish with confluent monolayer) followed by glycerol shock. Following transfection and shock, the cells were removed from the tissue culture dishes by trypsinization and plated (1 × 10⁵) onto 35-mm Falcon plastic petri dishes where they were allowed to recover for 48 hr at 37°. They were then prepared for hybridization while still attached to the tissue culture dishes.

Optimization of Cell Transfection Conditions

DNA-mediated introduction of specific genes into mammalian cells is a methodology which has substantially increased our knowledge of and understanding of DNA sequence involved in cell growth, terminal differentiation, and regulatory functions. One limitation to this technique, however, is the absence of a generally available method for determining one of the most important parameters of the transfection protocol, i.e., the fraction of cells that transiently express recombinant vector sequences following exposure to the DNA, a parameter that is complementary to overall transfection efficiency. Transfection efficiency is often monitored by a reporter sequence such as the *Escherichia coli* chloramphenicol acetyltransferase (CAT) gene, which provides a product which is conveniently and rapidly assayed biochemically in extracts of transfected eukaryotic cells, a characteristic which has proved useful in optimizing DNA transfection procedures and in making preparatory comparisons of promoter efficiency in various eukaryotic cell types.[9] Our approach was to transfect CV-1 African green monkey cells with the CAT gene under the control of Rous sarcoma virus long terminal repeat (RSV-LTR). Transfection effi-

---

[8] C. M. Fordis and B. H. Howard, this volume [27].

[9] C. M. Gorman, G. T. Merlino, M. C. Willingham, I. H. Pastan, and B. H. Howard, *Proc. Natl. Acad. Sci. U.S.A.* **79**, 6777 (1982).

ciency was monitored by the standard biochemical assay for the expression of CAT sequences. Cells for *in situ* hybridization to biotin-nucleotide labeled RSVcat plasmid DNA were harvested at periods optimized by CAT biochemical expression. For comparison, CV-1 cells transfected with plasmid DNAs where mouse caseinα coding sequences were inserted (RSV-csnα) in place of CAT were monitored for csnα expression by *in situ* hybridization. We found that comparable hybridization signals were obtained in CV-1 cells transfected with either RSVcat or RSVcsnα expression vectors when transfection efficiency, and RSV-LTR promoter activity were optimized in separate cultures using the CAT biochemical assay protocol. Thus, examination of transiently expressing transfected cells by *in situ* hybridization with biotinylated plasmid DNA provides a rapid, relatively simple procedure for evaluating the efficiency of transfection and expression of recombinant plasmids where biochemical markers are unavailable (pRSVcsnα).

Preparation of Biotinylated Probes

DNA probes were nick-translated by the method of Rigby *et al.*,[10] using biotinyl-dUTP derivative that contains an 11-atom spacer arm between the five position of the pyrimidine ring and the carboxyl group of the biotin molecule. The Bio-11-dUTP (EPB-806) was obtained from Enzo Biochem, Incorporated, as was a nick-translation kit (EBP-803). If desired the entire Enzo Bio-probe System (EBP) can be obtained from Enzo Biochem, Incorporated, 325 Hudson Street, New York, New York 10013. The contents of the kit include all the ingredients essential for nick translation and a relatively simple description of the steps necessary for the preparation of a biotinylated DNA probe. The complete Bioprobe system also includes material for detection of the biotinylated molecular probe after hybridization and stains for cells. In our experience we found that the Enzo nick-translation kit provided reproducible incorporation of the biotinyl-dUTP into double-stranded DNA. For *in situ* hybridization biotin attached with 11 or 16 atom long linker molecules to dUTP should be used. In our hands biotin attached to dUTP with 4 and 6 atom linker arms did not provide sufficient stereochemical flexibility to produce a detectable signal after hybridization to tissue sections or cellular monolayers.

Briefly, incorporation of biotinyl-11-dUTP or biotinyl-16-dUTP into DNA was accomplished by replacing TTP with an equimolar concentration of the biotinylated dUTP analog. Total DNA concentration in each nick-translation was kept constant at 20 μg/ml. For example, a typical

---

[10] P. W. J. Rigby, M. Dieckmann, C. Rhodes, and P. Berg, *J. Mol. Biol.* **113**, 237 (1979).

reaction consisted of 30 μM biotinyl-dUTP, 20 μg/ml plasmid DNA, 5 mM MgCl$_2$, 30 mM each of dCTP, dGTP, and dATP, 6 μCi [^{3}H]ATP, 2.5 μg/ml DNase I, and 80 units DNA polymerase I in 50 mM Tris–HCl, pH 7.50, final volume 50 μl. Approximately 40% of the dATP is replaced by [^{3}H]dATP after 2 hr at 14°.

Recently a new method for covalently labeling nucleic acids with biotin was reported by Forster et al.[11] In this method a photoactivatable biotin is mixed directly with the nucleic acid of choice and irradiated with visible light. This methodology can be used for double- and single-stranded DNA or for RNA. This photoactivatable biotin is available presently from Vector Laboratories, Burlington, California, as Photoprobe biotin.

Undenatured nick-translated biotinylated DNA probes were stable at 0–4° for upward of 6 months. Recovery of the biotinylated nucleic acid probe regardless of the method of labeling should avoid the use of phenol because biotinylated nucleic acids are often lost at the phenol/aqueous interphase. To recover the labeled biotinylated probe, the labeling reaction is stopped by heating the reaction mixture for 10 min at 65°, then cooled and loaded onto a Sephadex G-50 column preequilabrated in 10 mM Tris–HCl, pH 7.5, 1 mM EDTA in a 1 ml tuberculin disposal syringe plugged with glass wool (~ 1.2 ml resin fills syringe). The syringe is fitted into the puncture cap of a 15 ml plastic conical tube so that it can be held over a 1.5 ml Eppendorf tube placed in the bottom of the empty contrifuge tube, so as to articulate with the opening to the Eppendorf tube. The tube–syringe assembly is spun at low speed in a bench top centrifuge for 3–4 min to pack the resin whose final volume will be between 0.85 and 0.9 ml. The fluid in the Eppendorf tube is removed; the tube syringe is reassembled and 50 μl of the nick-translated reaction is pipetted onto the top of the G-50. After centrifugation is repeated, the fluid found in the Eppendorf tube will contain ~ 50 μl of biotinylated DNA probe free of unincorporated nucleotides. Other methods are equally useful for recovering the biotinylated probe; this protocol has provided sufficiently reproducible results in our hands.

Preparation of Cells for in Situ hybridization

When examining transfected cells with immunocytochemical methods for the transient expression of gene products by recombinant DNA expression vectors, we found that the density of the cellular monolayer was of extreme importance in obtaining optimal results. This is also true for in

---

[11] A. C. Forster, J. L. McInnes, D. C. Skingle, and R. H. Symons, *Nucleic Acids Res.* **13**, 745 (1985).

situ molecular hybridization. Empirically, the best results are obtained when cells cover 65–85% of the culture surface. In the case of CV-1 monkey kidney cells, plating 1×10^5 cells in a 35 mm Falcon petri dish following transfection and glycerol shock gives optimal cell densities after 48 hr of incubation. In most cases, the mock-transfected control cells produce higher densities under these conditions. Since the rate of growth and the morphological characteristic of cell lines vary, it is necessary to determine the optimal seeding density for each recipient cell line individually.

There are several reasons why cell density in the monolayer is critical to *in situ* hybridization analysis. First, the optimum geometry for morphological analysis of the RNA content in the cellular cytoplasm is attained when the cells are flat and tightly adherent to their substrate. Second, the absence of cell overlapping improves the signal-to-noise ratio and enables accurate determination of the numbers of positive to negative cells. Finally, preparation of single-cell thicknesses provide samples which can be directly analyzed by a variety of microscopic techniques simultaneously (e.g., brightfield, fluorescence, phase, or interference microscopy).

Fixation of the cell monolayer is accomplished by rinsing the cells twice with culture medium without serum, then adding 2 ml (35-mm plate) of 3.3% formaldehyde in phosphate-buffered saline. This solution is freshly prepared from paraformaldehyde powder, e.g., 3.3 g of paraformaldehyde polymer (Polysciences, Incorporated, Warrington, Pennsylvania) is added to 33 ml or distilled H_2O warmed on a hot plate, and a small amount of 0.1 N NaOH is added, to clear the solution. Ten milliliters of 10× phosphate-buffered saline (0.3 M Na phosphate, 1.5 M NaCl, pH 7.4) (PBS) is added and the final volume adjusted to 100 ml with distilled H_2O. Fixation proceeds for 5–10 min (hybridization signals are sometimes reduced by longer fixation times) whereupon the monolayers are rinsed briefly in PBS and permeabilized with 0.1% Triton X-100 (scintillation grade from Polysciences, Incorporated, Warrington, Pennsylvania) in PBS for 5 min. After rinsing the monolayers with PBS several times, the cells are dehydrated by the addition of absolute ethanol for 2 min, and then the plates are inverted and allowed to air-dry. Cell monolayers prepared in this way are suitable for *in situ* hybridization directly or may be kept for periods up to 6 months at 4° if they are kept completely dry.

Hybridization Conditions

In preparation for hybridization the cells are rehydrated by incubation at room temperature for 1 hr with several changes of 1× SSPE (360 mM NaCl, 30 mM sodium phosphate, and 1 mM Na EDTA, pH 7.0). All procedures unless otherwise noted are performed directly at room temperature on the cells which remain attached to the 35 mm petri dish. The cells are then treated for up to 45 min with Pronase (100 μg/ml) dissolved in 1× SSPE. The proteolytic digestion of the cells is an extremely critical step; the digestion must facilitate efficient intracellular penetration of the hybridization probe without excessive disruption of cell morphology, degradation of the RNA, or reduction of the adhesive forces anchoring the cells to their substrate. Incubation time and protease concentration will vary depending on the cell line employed and upon the particular lot of enzyme used. These parameters should be optimized by evaluating parallel samples for a fixed time with serial 2-fold dilutions. If cells round up or fall off the substrate at a set concentration of Pronase the next lower concentration should be used and the time left constant. To reduce the possibility of RNA degradation by contaminating nuclease(s), we routinely autodigested (preincubated) our pronase solutions for 45 min at 37°. We have also used proteinase K for digestion with equally good results. Following digestion the cells are rinsed in SSPE and treated with 0.1 N HCl for 10 min after which they are rinsed again and fixed a second time in 3.3% formaldehyde (prepared from paraformaldehyde). This second fixation serves to stabilize cellular morphology and block any residual proteolytic activity.

To determine that cytoplasmic hybrids are the result of interaction between cellular RNA and the biotinylated probe, two methods are available, both of which degrade cellular RNA. The simplest and most reproducible method is to treat the monolayers with 0.1 N NaOH for 15–30 min at 60° following protease digestion just preceding the treatment in 0.1 N HCl. In our hands, incubation in 0.1 N NaOH at room temperature did not affect hybridization, in fact, the signal was slightly enhanced. Alternatively, transfected monolayers can be treated with pancreatic RNase I (100 μg/ml Sigma) for 60 min at 37° following digestion of the sample with pronase or proteinase K as described above, just preceding exposure to 0.1 N HCl.

Prehybridization of the cell monolayers (1.0 ml per 35 mm petri dish) is carried out for 3 hr at 42° in 50% formamide (Fluka), 10% dextran sulfate (Pharmacia), 2× SSPE, pH 7.0, 12 mM vanadyl sulfate (Fischer), and 250 μg/ml denatured salmon sperm DNA. We find that preincubation with hybridization mixture and carrier is important in reducing nonspecific labeling particularly in the nucleus. At the end of the preincubation time, the excess hybridization mixture is removed by pipette and 50 μl hybridization mixture with 2–4 μg/ml of biotin-labeled probe is added to the center of the plate and covered with a 22-mm² glass coverslip. In preparation for hybridization, an aliquot of the probe is denatured by mixing with an equal volume of 0.1 N NaOH and immediately neutralizing the mixture with a one-third volume of 0.5 M NaH$_2$PO$_4$. This operation is carried out concomitant with the preparation of the final hybridization mixture and immediately preceding its addition to the cell monolayers. The hybridization solution and coverslip are then covered with molten dental wax at 70° (Polysciences, Incorporated, Warrington, Pennsylvania). Hybridization is then allowed to proceed for 48 hr at 42°.

Following hybridization, the dental wax (now solid) and coverslip are removed from the petri dish by forceps, and the monolayers are washed in 50% formamide in 2× SSPE at 37°, then in 2× SSPE and finally in 1× SSPE at room temperature. The cultures are then immediately prepared for detection of biotinylated probe as described below.

Detection and Visualization of Hybrids

Following removal of the nonspecific biotinylated hybrids as above, the monolayers are covered with normal rabbit serum (166 μl in 2 ml SSPE–3% serum) or 4.0% bovine serum albumin in 1× SSPE for 10 min at room temperature. This step is necessary to prevent nonspecific binding of the avidin–biotin complex to the cellular monolayers. After rinsing with 1× SSPE, biotinylated hybrids are detected by adding avidin–biotin–horseradish peroxidase (ABC-Vector Laboratories, Burlington, California) diluted 1:80 in PBS directly to the dish (2 ml/dish) for 90 min. After rinsing three times in PBS, 2 ml of diaminobenzidine tetrahydrochloride (600 μg/ml in PBS containing 0.03% H_2O_2) is placed in the petri dish and the peroxidase reaction is allowed to proced for 20 min at room temperature. The monolayers are then washed three times in PBS, and then counterstained with Mayers hematoxylin [hematoxylin crystals, 4.0 g; $NaIO_3$, 0.8 g; $KAl(SO_4)_2 \cdot 12H_2O$ (K alum), 200 g; citric acid, 4.0 g; chloral hydrate, 200 g; and 2000 ml distilled H_2O; dissolve alum in H_2O, without heating, add hematoxylin, sodium iodate, then citric acid and chloral hydrate with constant agitation]. Two milliliters of hemotoxylin stain is placed in the petri dish to cover the cells. After 2 min the stain is decanted and replaced with warm tap water; the dark red color will become blue as washing continues for several minutes. Thereafter, the monolayers are dehydrated in alcohol 50, 70, 90, 100% for 2 min each, and allowed to air dry while inverted. A 22-mm² coverslip is placed over the dry monolayer, and the petri dish can be observed directly under a standard light microscope fitted with a 20× apochromatic objective lens and a 15× ocular. Optimal magnification for scanning is approximately 250 diameters. The cells will appear light blue with darker blue nuclei. Biotinylated molecular DNA–RNA hybrids are seen as puncate cytoplasmic droplets of reddish-brown diaminobenzidine (DAB) precipitates (Figs. 1–4). These precipitates will not dissolve or recrystallize; therefore ABC-peroxidase preparations are permanent and can be evaluated several times under the

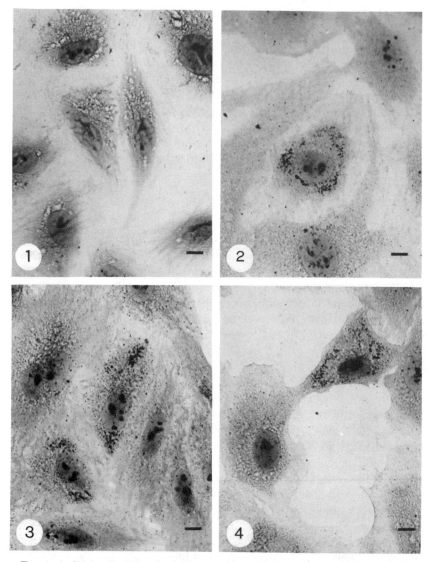

Figs. 1–4. CV-1 cells 48 hr after either mock transfection (Fig. 1) or transfection with pRSVcat-DNA (Figs. 2–4). *In situ* hybridization was then performed using a biotinylated pRSVcat-DNA probe. RNA–DNA hybrids were detected using avidin–biotin conjugated HRP followed by incubation with DAB. Bars = 10 μm.

microscope without loss of detail such as is experienced with fluorescent preparations.

Sensitivity of *in Situ* Hybridization for Evaluation of Transfection Efficiency

Quantitation of *in situ* hybridization signals for the detection of actin gene expression during the differentiation of chicken embryonic muscle cultures has been carefully analyzed recently by Lawrence and Singer.[12] The reader is encouraged to consult this work for clues to the preparation of samples suitable for quantitation. In our hands, *in situ* hybridization with biotinylated plasmid DNA probes produces signals whose sensitivity compared favorably with the detection of CAT-antigen-positive cells in transfected cultures transiently expressing pRSVcat. In general, for any given CAT expression vector, e.g., pSV_2cat or pRSVcat, the number of antigen-positive CAT-expressing cells reflects the efficiency of transfection as measured by CAT biochemical activity, i.e., lower CAT activity was reflected in a smaller number of CAT-antigen-positive cells and higher activity by greater numbers. We have been able to detect as low as 0.05% (1:2000) positive cells in CV-1 cell cultures transiently expressing pRSVcat by either immunological or *in situ* hybridization techniques. In addition, following optimization of transfection efficiency in CV-1 cells using CAT expression vectors, we were able to detect transcription of a pRSV vector containing a cDNA cloned fragment of mouse α-casein by *in situ* hybridization. In these experiments, α-casein antigen content could not be detected by immunoperoxidase with polyclonal antibodies directed against total mouse casein.[4]

[12] J. B. Lawrence and R. H. Singer, *Nucleic Acids Res.* **13,** 1777 (1985).

[40] RNA Detection and Localization in Cells and Tissue Sections by *in Situ* Hybridization of 35S-Labeled RNA Probes

By MARY E. HARPER AND LISA M. MARSELLE

In situ hybridization has proven to be an important tool for cellular and molecular biological studies. With this technique, it is possible to detect as well as localize specific nucleic acid sequences within particular cells and tissues. In regard to gene expression, *in situ* hybridization allows the study of specific mRNA level and distribution between various cell types, or of comparative levels of RNA throughout development.[1-3] Furthermore, this method can be used to detect exogenous nucleic acid sequences, such as viral RNA or DNA,[4,5] helping to identify the types of infected cells as well as their distribution within particular tissues. *In situ* hybridization is direct, can give quantitative information, and requires small tissue samples. Because of these advantages, we decided to utilize this approach for specific studies involving human retroviruses, which required very high sensitivity. Using several methodologic advances developed by others, most notably the use of single-stranded RNA probes[1] and more efficient fixation and hybridization steps,[6] we developed a highly sensitive and rapid *in situ* hybridization method which is surprisingly simple.[5] While originally used as an assay for viral expression (see Section on Data Analysis), this method can be easily applied to many studies of cellular gene expression.

Preparation of Cell Specimens onto Microscope Slides

A variety of cell preparations can be analyzed for expression of specific gene sequences by *in situ* hybridization. Cells in suspension, either primary samples or cultures, are most evenly deposited on slides by cytocentrifugation. However, tissue touch-preps or cell smears can also be used. Adherent cells can be prepared in several different ways. If previously grown on coverslips, gluing the coverslips to slides with strong glue such as Super Glue will facilitate handling throughout the hybridization and autoradiography procedures. An easier alternative is to seed the cells into 4- or 8-well chamber slides, allow the cells to adhere, and then remove the plastic chambers either prior to or after fixation. In the case of whole tissue blocks, frozen sections result in a relatively high hybridization efficiency, as opposed to formalin-fixed, paraffin-embedded sections. While the optimal procedure is to embed small pieces of tissue and then snap freeze at −70°, if embedding materials are not available, tissue may be snap-frozen in small pieces by laying on dry ice or immersion into a −70° alcohol bath. In either case, tissue should be wrapped tightly after freezing to minimize dehydration.

[1] K. H. Cox, D. V. DeLeon, L. M. Angerer, and R. C. Angerer, *Dev. Biol.* **101**, 485 (1984).
[2] R. C. Angerer and E. H. Davidson, *Science* **226**, 1153 (1984).
[3] D. J. Anderson and R. Axel, *Cell* **42**, 649 (1985).
[4] M. Brahic and A. T. Haase, *Proc. Natl. Acad. Sci. U.S.A.* **75**, 6125 (1978).
[5] M. E. Harper, L. M. Marselle, R. C. Gallo, and F. Wong-Staal, *Proc. Natl. Acad. Sci. U.S.A.* **83**, 772 (1986).
[6] J. B. Lawrence and R. H. Singer, *Nucleic Acids Res.* **13**, 1777 (1985).

After the cells or tissue sections are placed onto slides, the preparations should be air-dried in order to minimize loss of cellular material during the hybridization steps. This drying time should be kept to a minimum in order to reduce RNA degradation before fixation. It is also advantageous, particularly in cases where tissue samples are very small, to mount several sections from one block together on each slide.

During the basic fixation procedure, cell or tissue preparations are fixed very briefly in 4% paraformaldehyde in PBS, and then stored in 70% ethanol at 4°. This fixation method results in both superior maintenance of cell morphology and relatively high hybridization efficiency, as noted previously.[6] Comparison of paraformaldehyde fixation with ethanol–acetic acid fixation[4] indicated 2- to 3-fold higher grain counts on cells fixed with paraformaldehyde.[7] Furthermore, we have found that storage of slides in 70% ethanol after fixation provides excellent preservation of RNA for at least 6 months.

As in all procedures involving RNA, precautions to prevent RNase contamination should be strictly followed. Gloves must be worn, all water should be autoclaved (or treated with diethyl pyrocarbonate and autoclaved[8]), glassware baked at 250° for at least 4 hr, and disposable plasticware used whenever possible.

Reagents
Cytospin preparations:
  Tissue culture media
  Fetal calf serum
Frozen tissue sections:
  2-Methylbutane (95% ethanol can be substituted)
  Dry ice
  Cryomolds of various sizes (e.g., Lab-Tek #4557, 4565)
  O.C.T. embedding compound (Miles Scientific #4583)
  Staining boxes, either glass or plastic, with accompanying slide holders
    (e.g., Lab-Tek #4457, 4465)
  1 N HCl
  Poly(L-lysine), 100 μg/ml

Fixation:
  Paraformaldehyde, reagent grade such as Baker #S898
  10× phosphate-buffered saline, Ca^{2+}- and Mg^{2+}-free

[7] M. E. Harper, L. M. Marselle, K. J. Chayt, S. F. Josephs, P. Biberfeld, L. G. Epstein, J. M. Oleske, C. J. O'Hara, J. E. Groopman, R. C. Gallo, and F. Wong-Staal, *in* "Biochemical and Molecular Epidemiology of Cancer" (C. C. Harris, ed.), p. 449. Alan R. Liss, New York, 1986.
[8] T. Maniatis, E. F. Fritsch, and J. Sambrook, "Molecular Cloning: A Laboratory Manual," p. 190. Cold Spring Harbor Lab., Cold Spring Harbor, New York, 1982.

70% ethanol
Coplin jars or other slide holders such as VWR #48429

Protocol

Cytospin preparations:

1. To clean microscope slides, immerse in 70% ethanol. Dry each slide individually with lint-free tissues or gauze to remove oil (or use cleaning procedure described below for tissue sections).
2. Suspend cells at 10^6/ml in media containing 10% serum.
3. Prepare cytospin machine and load 200 μl cell suspension per slide. Spin at 500 rpm for 5 min.
4. Remove slides from machine holders; dry at room temperature for 5 min, and fix according to procedure described below.

Frozen tissue sections:

1. Prepare bath of 2-methylbutane with dry ice to $-70°$.
2. Place specimen (cut to small pieces if necessary) in mold of proper size so that margin of space surrounds tissue.
3. Place O.C.T. compound around tissue, covering tissue and filling mold almost to top.
4. Lower filled mold with tongs into $-70°$ 2-methylbutane bath for 15–30 sec.
5. Wrap filled, frozen mold tightly in Parafilm and then in foil, set in dry ice to keep frozen, and store at $-70°$.
6. To prepare microscope slides for sections, clean as follows. Place slides in staining rack and immerse in 1 N HCl for 30 min. Rinse briefly in 2 changes distilled water. Immerse in 95% ethanol for 30 min and then dry in air or with jet. Immerse in poly(L-lysine) for 30–60 min. Rinse very briefly in two changes distilled water and dry overnight at 37°. Store treated slides at 4° until used.
7. Section frozen tissue according to standard procedures, place sections onto slides, and let dry for 3–5 min. Fix immediately according to procedure described below.

Fixation:

1. Prepare fixative by dissolving paraformaldehyde at a concentration of 4% in 1× PBS. Heat to 60° for approximately 30 min until solution clears. Filter after cooling using a disposable filtering unit and vacuum. Store at room temperature.
2. To fix cell or tissue preparations, immerse in paraformaldehyde fixative for 1 min only. Transfer to Coplin jars or slide holders containing 70% ethanol and store immersed at 4°.

Transcription of ^{35}S-Labeled RNA Probes

Specific RNA probes with high specific activity can be easily made using transcription systems such as SP6 and T7. We have found that if the concentration of the radioactive nucleotide triphosphate, e.g., UTP, is increased to 25 μM and adequate levels of RNA polymerase are used as suggested,[9] 80% incorporation of the labeled UTP can be routinely attained, resulting in synthesis of 0.5 μg RNA per reaction. For *in situ* hybridization, ^{35}S is superior to other radioisotopes such as ^{3}H, ^{125}I, or ^{32}P, due to the high specific activity attainable, high autoradiographic efficiency, minimal problems with nonspecific sticking, reasonable half-life, and safety. We routinely utilize only one ^{35}S-labeled nucleotide triphosphate, which results in a specific activity of approximately 10^9 dpm/μg and adequate stability. Because the polymerization reaction proceeds very rapidly, the cold chase with unlabeled UTP which is included in the protocol likely results in minimal strand extension and is optional. Sizing of the RNA transcripts by formamide–formaldehyde denaturing gel electrophoresis[10] indicated lengths of 1–2 kb.[5] We and others[6] have noted high hybridization efficiencies when native transcripts are used without hydrolysis into smaller sized fragments.

In order to prevent synthesis of vector-specific RNA, plasmid templates must be linearized with an appropriate restriction endonuclease, then phenol–chloroform extracted and ethanol precipitated. Most of the reaction reagents can be prepared in advance and stored at −20°. Following transcription, purification of ^{35}S-labeled RNA is carried out according to standard procedures. Two protocols utilized successfully in our laboratory, using either column chromatography or multiple extraction–precipitation steps,[9] are described. As discussed in the previous section on slide preparation, all appropriate precautions against RNase contamination must be followed.

Reagents

[^{35}S]UTP, 1000 Ci/mmol in tricine (New England Nuclear #NEG-039H)

Dithiothreitol (DTT)

RNasin RNase inhibitor (Promega Biotec #P2111)

Cold nucleotide triphosphates:

Mixture of 2.5 mM each of ATP, CTP, and GTP

5 mM UTP

5× transcription buffer:

200 mM Tris–HCl, pH 7.5

30 mM MgCl$_2$

---

[9] M. T. Johnson and B. A. Johnson, *BioTechniques* **2**, 156 (1984).

[10] H. Lehrach, D. Diamond, J. M. Wozney, and H. Boedtker, *Biochemistry* **16**, 4743 (1977).

10 mM spermidine

50 mM NaCl

RNA polymerase (SP6, NEN #NEE-151; T7, Promega Biotec #P2051)

DNase I, ribonuclease-free (Worthington #06333), reconstituted in normal saline at 1 mg/ml and frozen in aliquots

Carrier DNA, such as salmon sperm DNA, sheared and purified (see protocol in Step 10 below)

10% TCA

BSA, 10 mg/ml

Protosol tissue solubilizer (NEN #NEF-935)

Liquid scintillation solution (e.g., Econofluor, NEN #NEF-969)

Protocol

1. Place 250 pmol [^{35}S]UTP (approximately 25 μl) in microcentrifuge tube. Dry completely in Speed-Vac centrifuge or with N_2 stream.

2. Assemble transcription reaction in new tube at room temperature in order shown:

2 μl 5× transcription buffer

1.5 μl water

1 μl 100 mM DTT (prepared fresh)

0.5 μl RNasin (40 units/ul)

2 μl ATP, CTP, and GTP, 2.5 mM each

2 μl linearized DNA template (0.5 μg/ul)

Mix together and transfer to tube containing dried [^{35}S]UTP. Add 1 μl RNA polymerase (15 units).

3. Incubate at 40° for 30 min.

4. Add 1 μl unlabeled UTP, 5 mM, and 1 μl (15 additional units) RNA polymerase. Incubate at 40° for 30 additional min.

5. Add 1 μl (40 additional units) RNasin, 6 μl DNase I (diluted to 100 μg/ml), and 11 μl water. Digest DNA template at 37° for 10 min.

6. Purify ^{35}S-labeled RNA by column chromatography or precipitation according to standard methods. The following two procedures have been used successfully in our laboratory.

7. Purification protocol using Worthington Mini-Spin Column (#LS04404):

a. Bring reaction volume to 75 μl with TE buffer. Remove 1 μl accurately, e.g., using capillary pipet, and dilute with 49 μl water. TCA precipitate 5 μl of the diluted reaction mixture, count in liquid scintillation counter, and compare cpms with those obtained from counting unprecipitated sample to calculate incorporated ^{35}S, as described below (Step 9).

b. Extract remaining 74 μl reaction mixture with an equal volume of

1:1 phenol:chloroform–isoamyl alcohol (24:1). Extract one time with chloroform–isoamyl alcohol (24:1).

c. Add 2 μg carrier salmon sperm DNA (1 μl) to extracted RNA sample, load onto prepared mini-spin column, and process according to manufacturer's instructions. Increase eluted volume to 100 μl with water. Remove 1 μl from final sample, dilute with 9 μl water, and count 1 μl of diluted sample. Final probe concentration should be $1-4 \times 10^6$ cpm/μl. Aliquot 5 μl per tube, freeze, and store at $-70°$.

8. Purification protocol using extraction–precipitation:

a. Bring reaction volume to 75 μl with water; add 50 μl 2% SDS and 100 μl 0.1 M sodium acetate, pH 5.0.

b. Remove 1 μl accurately, e.g., using capillary pipet, and dilute in 49 μl water. TCA precipitate 5 μl of the diluted reaction mixture, count in liquid scintillation counter, and compare cpms with those obtained from counting unprecipitated sample to calculate incorporated ^{35}S, as described below (Step 9).

c. Extract remaining 224 μl reaction mixture with 200 μl 1:1 phenol:chloroform–isoamyl alcohol (24:1), followed by two extractions with chloroform–isoamyl alcohol (24:1).

d. To aqueous layer, add 10 μg carrier salmon sperm DNA and 750 μl cold 100% ethanol. Precipitate in dry ice–ethanol bath for 10 min.

e. Centrifuge for 15 min at 4°, remove supernatant, and resuspend in 200 μl 0.2% SDS, 2 mM EDTA, 0.3 M ammonium acetate, pH 5.2. Add 400 μl cold absolute ethanol; precipitate and resuspend as described previously, then precipitate a third time.

f. Dry pellet and resuspend in 100 μl water. Remove 1 μl, dilute with 9 μl water, and count 1 μl of diluted sample. Final probe concentration should be $1-4 \times 10^6$ cpm/μl. Aliquot in 5-μl aliquots, freeze, and store at $-70°$.

9. TCA precipitation:

a. Mix together in microcentrifuge tube: 5 μl sample, 30 μl BSA (10 mg/ml), 400 μl cold 10% TCA.

b. Chill on ice for 30 min.

c. Centrifuge for 5 min. Decant supernatant.

d. Rinse with 400 μl 10% TCA, centrifuge, decant supernatant, and tap out excess TCA.

e. Dissolve pellet completely in 300 μl Protosol; heat at 37–45° for approximately 30 min if necessary.

f. Add dissolved pellet in Protosol to 5 ml Econofluor LCS solution, mix well, and count.

g. Add 5 μl sample (without precipitation) to 5 ml Econofluor containing 300 μl Protosol, mix, and count. Compare cpms from precipitated sample with those from unprecipitated sample to calculate percentage [^{35}S]UTP incorporated.

10. Protocol for shearing carrier DNA by depurination[11]:

a. Dissolve 100 mg carrier DNA in 20 ml 20 mM Tris–HCl, pH 7.5. Stir overnight at 4°.

b. Add 1 ml 2 M sodium acetate, pH 4.2; place at 70° for 40 min.

c. Add 1 ml 4 N NaOH; place in 100° water bath for 20 min.

d. Neutralize with HCl to pH 7.6. Add 1 M Tris–HCl, pH 7.5, to final concentration of 80 mM.

e. Ethanol precipitate and redissolve in 20 ml TE buffer. Phenol-chloroform extract until interface is clean. Ethanol precipitate and redissolve in TE to desired concentration. Size fragments by gel electrophoresis; peak size should be approximately 300–600 bp.

In Situ Hybridization

The *in situ* hybridization method described is straightforward, relatively rapid, and highly sensitive. We have found that pretreatment steps which increase tissue permeability and probe diffusion are not necessary, as noted previously by others.[6] However, such treatments may be helpful when hybridizing to paraffin-embedded tissue sections. The two brief pretreatment steps included help to minimize nonspecific binding of the probe. The two ethanol dehydration series prior to hybridization and to RNase digestion were included to facilitate rapid handling of the slides and also serve to "rebind" tissue sections to the slides, thereby helping to minimize tissue section loss.

Hybridization at standard stringency is carried out at 50°, as suggested previously.[1] If lower stringency is desired, hybridization and rinsing at reduced temperatures, e.g., 45°, are possible without a substantial increase in nonspecific binding.[12] At standard stringency, sensitivity of the method was determined by comparison of *in situ* hybridization results with Northern blot analysis using HTLV-III-infected cells. Following hybridization with the HTLV-III probe, it was determined that each grain observed after a 2-day exposure represented 1–3 copies of RNA.

Prior to hybridization, all precautions to avoid RNase contamination described previously in the section on slide preparation must be followed. Therefore, all staining boxes, slide holders, and Coplin jars, other plastic-

---

[11] M. Wilson, personal communication.

[12] H. Koprowski, E. C. DeFreitas, M. E. Harper, M. Sandberg-Wollheim, W. A. Sheremata, M. Robert-Guroff, C. W. Saxinger, M. B. Feinberg, F. Wong-Staal, and R. C. Gallo, *Nature (London)* **318,** 154 (1985).

ware, glassware, and solutions should be kept separate between the pre- and posthybridization steps.

Reagents

Staining boxes, either glass or plastic, with accompanying slide holders (e.g., Lab-Tek #4457, 4465)

20× SSC

0.1 M triethanolamine, pH to 8.0 with HCl

Acetic anhydride

PBS (Ca^{2+}, Mg^{2+} free)

1 M Tris–HCl, pH 7.0

1 M glycine

Formamide, reagent grade

500 mM DTT, stored at $-20°$

Yeast tRNA, purified (e.g., BRL #5401SB)

Salmon sperm DNA, sheared and purified (see previous section, step 10)

BSA, nucleic acid grade (e.g., BRL #5561UA)

Slide warmer

Coverslips, 18 × 18 mm², 22 × 22 mm²

Rubber cement

Ribonuclease A (Sigma #R4875)

Ribonuclease T₁ (Boehringer-Mannheim #109–193)

Moist chamber (tray with strips of thick rubber tubing to hold slides above thin layer of water)

Protocol

1. All pretreatment solutions are poured into staining boxes appropriate for the slide holder. With the holder immersed in 2× SSC, transfer slides to be hybridized from 70% ethanol to the 2× SSC. Agitate slides gently for 1 min. Rinse slides in additional change of 2× SSC for 1 min.

2. Acetylate by immersing slides in 0.1 M triethanolamine, pH 8.0, contained in box with stir bar. Start to stir vigorously and add acetic anhydride to 0.25%. Stir for 10 min.[13]

3. Rinse slides in 1 change 2× SSC, then 1 change PBS, 1 min each with agitation.

4. Immerse in 0.1 M Tris–HCl, pH 7.0, 0.1 M glycine for 30 min.

5. Rinse slides in 2 changes 2× SSC, 1 min each. Dehydrate through 70, 80, and 95% ethanol, 1 min each, and dry with air jet.

---

[13] S. Hayashi, I. C. Gillam, A. D. Delaney, and G. M. Tener, *J. Histochem. Cytochem.* **26,** 677 (1978).

6. Mix probes, allowing 10 μl per slide with use of 18 × 18 mm coverslips. Per 10 μl:

  5 μl formamide

  1 μl 20× SSC containing 100 mM DTT (made by mixing 25× SSC and 500 mM DDT in ratio of 4:1)

  1 μl yeast tRNA, 10 mg/ml

  1 μl sheared salmon sperm DNA, 10 mg/ml

  1 μl BSA, 20 mg/ml

  1 μl ^{35}S-labeled RNA (usually 10^6 cpm/μl)

7. Heat probes at 90° for 5 min, mix again, and transfer to 55°.

8. Lay slides on preheated slide warmer. Apply 9 μl probe to each slide, mount coverslips, and seal edges with rubber cement using pasteur pipet cut to larger bore. During hybridization, count 1 μl of remaining hybridization mix in liquid scintillation counter to determine actual probe concentration used.

9. Hybridize for 3 hr. Remove rubber cement with forceps. Transfer slides to 50% formamide–2× SSC in Coplin jar at 52° and agitate gently up and down until coverslips slide off.

10. Transfer slides to slide holder immersed in 50% formamide–2× SSC at 52° and rinse with frequent agitation for 5 min. Transfer to second box of 50% formamide–2× SSC at 52° and rinse with frequent agitation for 20 min.

11. Rinse slides well in 3 changes of 2× SSC at room temperature, 1 min each. Dehydrate through 70, 80, and 95% ethanol, 1 min each, and dry with air jet.

12. Mix RNase solution, consisting of 100 μg/ml RNase A, 500 units/ml RNase T$_1$ in 2× SSC. Apply 30 μl RNase to each slide and mount 22 × 22 mm coverslips (no rubber cement). Place slides in moist chamber and incubate at 37° for 30 min.

13. Rinse slides in 2× SSC in Coplin jar, again allowing coverslips to slide off. Transfer slides to slide holder in box containing 2× SSC and rinse briefly. Transfer slides to second box of 50% formamide–2× SSC at 52° (used previously for 20 min rinse) and agitate frequently for 5 min. Rinse slides briefly in 2× SSC, then transfer to box with stir bar containing 2× SSC. Stir gently for 10 min. Dehydrate in 70, 80, and 95% ethanol, 1 min each, and dry with air jet.

Autoradiography and Development

Hybridized slides are autoradiographed with a nuclear track emulsion, exposed at 4° for an appropriate time (generally 2–10 days), and developed. Autoradiography with Eastman NTB2 emulsion and development with Kodak Dektol are recommended. Since a detailed procedure for autoradiography and slide development has been recently described,[14] the reader is referred to the previous protocol. However, for convenience, a reagent list is included here.

Reagents

Kodak type NTB2 nuclear track emulsion (Eastman Kodak DC Special Products #165-4433)

Pipet filler, 10 ml (Markson #D-13666)

Slide racks (e.g., Fisher #12-587-20)

Drying box

Light-proof containers, such as cardboard or metal specimen holders

Black electrical tape

Drierite desiccant

Black Bakelite slide boxes

Large dessicators, such as air-tight boxes

Dektol developer (Kodak #146-4726)

Kodafix solution (Kodak #146-4080)

Staining boxes, either glass or plastic, with accompanying slide holders (e.g., Lab-Tek #4457, 4465)

Staining and Data Analysis

Developed preparations may be stained with a variety of counterstains, the choice depending on the type of cells or tissue. In addition to utilizing Wright–Giemsa stain for a variety of hematopoietic cell populations, we have also found this stain to be adequate for certain frozen tissue sections. One advantage of Wright–Giemsa stain in ease of use. Typically, 2 ml of Wright–Giemsa stain (Harleco #742) is placed on each slide for 5 min. Buffer (Volu-Sol #VWB-032) is then added to the stain (2 ml per slide) for an additional 5 min. Slides are rinsed well with water and dried in air or with a jet. Destaining briefly in 70% ethanol followed by thorough rinsing with water and drying lightens the stain if too dark and also results in better differentiation. The staining procedure is also facilitated by use of a staining rack (e.g., Fisher #12-597-10).

Slides are most easily analyzed by microscopy using high dry objectives without coverslips, particularly if staining intensity requires alteration by destaining and/or restaining. High quality objectives such as Zeiss Epiplan 40×, Plan 63×, or Epiplan 80× STM should be considered.

Critical and necessary components of data analysis are positive and negative hybridization controls, in regard to both cells and probes. Experiments should include positive and negative cell controls for each probe utilized. We also routinely hybridize each cell sample with a negative probe

---

[14] M. E. Harper and L. Chan, this series, Vol. 128, p. 863.

control in order to show specificity of positive label. Analysis should be carried out in a blinded fashion whenever possible. Only with use of such controls can the proper interpretation be made.

Examples of *in situ* hybridization results utilizing the method described are shown in Fig. 1. While the HTLV-III infected cell line H9/HTLV-IIIB is highly positive for HTLV-III RNA (panel A) after only 2 days exposure, both the uninfected H9 cell line hybridized with the same HTLV-III probe

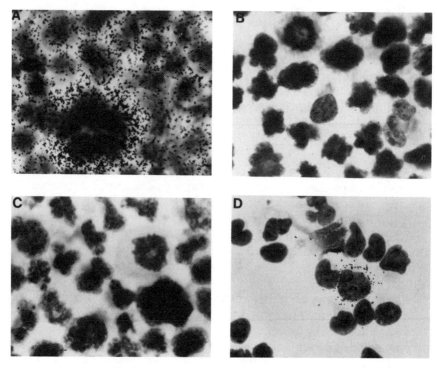

FIG. 1. Sensitivity and specificity of *in situ* hybridization for HTLV-III RNA in cytospin preparations of infected and uninfected cells. (A) *In vitro* infected H9/HTLV-IIIB cell line, expressing an average of 100–300 copies of HTLV-III RNA per cell. (B) Uninfected H9 cell line hybridized and autoradiographed under the same conditions utilized in A, and negative for HTLV-III RNA expression. (C) Infected H9/HTLV-IIIB cell line as in A and hybridized under the same conditions as in A, but with a ^{35}S-labeled RNA probe specific for bacteriophage λ sequences, demonstrating lack of nonspecific binding of probe. (D) Mononuclear cell preparation from primary peripheral blood of patient with AIDS-related complex, demonstrating rare cell expressing HTLV-III RNA. In all panels, hybridization was carried out at a probe concentration of 10^8 dpm/ml of ^{35}S-labeled RNA; autoradiographic exposure was for 2 days. From Harper *et al.*[5]

(cell control; panel B) and the infected cell line hybridized with a λ probe (probe control; panel C) are negative for label, even when treated under the same conditions. Panel D shows a rare positive cell expressing HTLV-III RNA in a primary peripheral blood mononuclear cell sample from a patient with AIDS-related complex. Such positive cells were consistently seen in repeated experiments, but at very low frequency (approximately 0.001%).[5] However, the same sample hybridized with the λ probe was consistently negative, as were peripheral blood mononuclear cell samples from uninfected individuals hybridized with the HTLV-III probe.[5]

With continuing use of *in situ* hybridization, further improvements in methodology will continue to be made. Nonisotopic methods of labeling, while currently not as sensitive as isotopic tags, continue to show great promise. Combined use of *in situ* hybridization and immunocytochemistry will also reveal much information, but again, highly sensitive methods must be developed. As these developments are made, *in situ* hybridization will offer even more potential as an important means to determine how cells function and interrelate to each other.

Acknowledgments

The authors thank Kelli Carey and Minerva Krohn for critical reading of the manuscript.

[41] Electrophoretic Assay for DNA-Binding Proteins

By ALEXANDER VARSHAVSKY

The development over the last two decades of increasingly powerful methods for the detection and analysis of protein–DNA interactions has been essential for achieving our current understanding of molecular genetic mechanisms. This chapter describes a generally applicable assay for specific DNA-binding proteins in crude extracts. The assay is based on gel electrophoretic separation of protein–DNA complexes from each other and from free DNA. Electrophoretic analysis of *in vitro*-formed complexes between purified DNA-binding proteins and their cognate DNA sequences in low ionic strength polyacrylamide gels was introduced in 1981 by Fried and Crothers[1] and Garner and Revzin.[2] Similar electrophoretic systems had previously been used for the analysis of *in vivo*-formed nucleoprotein

[1] M. G. Fried and D. M. Crothers, *Nucleic Acids Res.* **9,** 6505 (1981).

[2] M. M. Garner and A. Revzin, *Nucleic Acids Res.* **9,** 3047 (1981).

complexes such as nucleosomes.[3,4] In 1984, Strauss and Varshavsky[5] showed that a version of the above essay can be used both for the detection of specific DNA-binding proteins in crude extracts and for monitoring their subsequent purification. This assay has since been employed in several laboratories to detect a variety of DNA-binding proteins. In spite of our relatively brief experience with the electrophoretic assay as a detection tool, its exceptionally high sensitivity has already made this assay the method of choice in many experimental settings. The electrophoretic assay has been successfully applied to the detection of both highly sequence-specific, nonabundant DNA-binding proteins and abundant DNA-binding proteins of relatively low nucleotide sequence specificity.

After briefly describing some of the alternative techniques suitable for the detection of DNA-binding proteins in crude extracts, I shall discuss the electrophoretic assay, its recent modifications, and consider other applications of the assay.

An Overview of Alternative Assays for DNA-Binding Proteins

Zonal Sedimentation and Gel Exclusion Chromatography of DNA–Protein Complexes.[6-13] In these assays, DNA–protein complexes are separated from free DNA and proteins by either zonal sedimentation or gel chromatography, respectively. Because of their relatively low sensitivity and resolution, these assays are less suitable for detection of specific DNA-binding proteins in crude extracts than are some of the more recent assays described below. The sedimentation assay was used in the first successful attempts to detect and purify bacterial repressors.[11,12]

Protein/DNA Blotting Assay.[14-16]. In this assay, electrophoretically fractionated proteins are transferred from a gel onto a solid support such as

[3] A. Varshavsky, V. V. Bakayev, and G. P. Georgiev, *Nucleic Acids Res.* **3**, 477 (1976).

[4] V. V. Bakayev, T. G. Bakayeva, and A. Varshavsky, *Cell* **11**, 619 (1977).

[5] F. Strauss and A. Varshavsky, *Cell* **37**, 889 (1984).

[6] K. R. Yamamoto and B. M. Alberts, *J. Biol. Chem.* **249**, 7076 (1974).

[7] D. E. Jensen and P. H. von Hippel, *Anal. Biochem.* **80**, 267 (1977).

[8] A. Revzin and P. H. von Hippel, *Biochemistry* **16**, 4769 (1977).

[9] D. E. Draper and P. H. von Hippel, *Biochemistry* **18**, 753 (1979).

[10] T. M. Lohman, C. G. Wensley, J. Cina, R. R. Burgess, and M. T. Record, Jr., *Biochemistry* **19**, 3516 (1980).

[11] W. Gilbert and B. Muller-Hill, *Proc. Natl. Acad. Sci. U.S.A.* **56**, 1891 (1966).

[12] M. Ptashne, *Nature (London)* **214**, 232 (1967).

[13] A. D. Frankel, G. K. Ackers, and H. O. Smith, *Biochemistry* **24**, 10100 (1985).

[14] B. Bowen, J. Steinberg, U. K. Laemmli, and H. Weintraub, *Nucleic Acids Res.* **8**, 1 (1980).

[15] L. Levinger and A. Varshavsky, *Proc. Natl. Acad. Sci. U.S.A.* **79**, 7152 (1982).

[16] S. B. Patel and P. R. Cook, *EMBO J.* **2**, 137 (1983).

a nitrocellulose filter, followed by incubation with a radioactively labeled DNA probe, washes of varying stringency, and autoradiography. Although the initial electrophoretic fractionation involves either a near-complete denaturation of proteins (SDS–gel electrophoresis)[14,16] or at least a partial denaturation (low-pH gel electrophoresis),[15] some DNA-binding proteins renature sufficiently during or after transfer to bind added DNA specifically. A promising modification of the blotting assay which has not yet been tried is to use a nondenaturing electrophoretic system for the protein fractionation.

Immunoassay for DNA-Binding Proteins.[17–19a] In this assay, a complex between a specific DNA-binding protein and a DNA fragment is selectively recovered by immunoprecipitation with an antibody against the protein. The assay allows detection of specific DNA binding by a protein in a crude extract, and, if antibodies of sufficient specificity and avidity are available, a more detailed equilibrium study of the protein–DNA interaction.

DNA Affinity Chromatography.[20–22] This method is used primarily for preparative-scale work (see, however, Refs. 23 and 24).

Filter Binding Assay.[25–30a] This sensitive assay is based on the selective retention of protein–DNA complexes but not of free DNA by nitrocellulose filters. Several variations of the assay have been and continue to be extensively used both in studies of purified DNA-binding proteins and for the detection of new ones in crude extracts. Some of the shortcomings of the assay are discussed in Refs. 28–30.

Footprinting Assays. Footprinting assays are based on the ability of a DNA-bound protein to modulate, at or near the protein binding site,

[17] R. D. G. McKay, *J. Mol. Biol.* **145**, 471 (1981).
[18] A. Johnson and I. Herskowitz, *Cell* **42**, 237 (1985).
[19] C. Desplan, J. Thies, and P. H. O'Farrell, *Nature (London)* **318**, 630 (1985).
[19a] C. K. Lee and D. M. Knipe, *J. Virol.* **54**, 731 (1986).
[20] B. M. Alberts, F. J. Amodio, M. Jenkins, E. O. Gutmann, and F. L. Ferris, *Cold Spring Harbor Symp. Quant. Biol.* **33**, 289 (1969).
[21] P. deHaseth, T. M. Lohman, and M. T. Record, Jr., *Biochemistry* **16**, 4783 (1977).
[22] P. J. Rosenfeld and T. J. Kelly, *J. Biol. Chem.* **261**, 1398 (1986).
[22a] J. T. Kadonaga and R. Tjian, *Proc. Natl. Acad. Sci. U.S.A.* **83**, 5889 (1986).
[23] P. Gaudray, C. Tyndall, R. Kamen, and F. Cuzin, *Nucleic Acids Res.* **9**, 5697 (1981).
[24] B. M. Alberts, *Cold Spring Harbor Symp. Quant. Biol.* **48**, 1 (1984).
[25] M. Yarus and P. Berg, *Anal. Biochem.* **28**, 479 (1967).
[26] A. D. Riggs, H. Suzuki, and S. Bourgeois, *J. Mol. Biol.* **48**, 67 (1970).
[27] D. C. Hinkle and M. J. Chamberlin, *J. Mol. Biol.* **70**, 157 (1972).
[28] G. G. Johnson and E. P. Geiduschek, *Biochemistry* **16**, 1473 (1977).
[29] H. S. Strauss, R. R. Burgess, and M. T. Record, Jr., *Biochemistry* **19**, 3496 (1980).
[30] C. P. Woodbury, Jr., and P. H. von Hippel, *Biochemistry* **22**, 4730 (1983).
[30a] J. F. X. Diffley and B. Stillman, *Mol. Cell. Biol.* **6**, 1363 (1986).

chemical modifications of DNA that are produced by the footprinting agent. Different versions of the footprinting assay employ nucleases such as DNase I[31-36] and exo III,[37-41] or low-molecular-weight reagents such as dimethyl sulfate[37,42] and methidiumpropyl-EDTA · Fe(II).[43] In recently developed *in vivo* footprinting techniques, dimethyl sulfate[44,45] and UV light[46] are used to map protein–DNA contacts within intact cells.

Prior to the development of the electrophoretic assay, footprinting and filter binding assays were the clear first choices for the *in vitro* detection of DNA-binding proteins. Although the electrophoretic assay is more often than not a superior choice for initial detection, *in vitro* footprinting continues to be used for the same purpose, and remains essential for the precise localization and analysis of protein binding sites on the DNA.

Electrophoretic Assay for DNA-Binding Proteins

The assay is based on gel electrophoretic separation of protein–DNA complexes from each other and from free DNA. The exceptionally high sensitivity of the electrophoretic assay[1,2,5,47] is due at least in part to the apparent stabilization of DNA–protein complexes once they enter a gel. The apparent decrease in the rate of complex dissociation is not well understood but appears to be due at least in part to a "caging" effect of the

[31] D. Galas and A. Schmitz, *Nucleic Acids Res.* **5,** 3157 (1978).

[32] W. Ross and A. Landy, *Proc. Natl. Acad. Sci. U.S.A.* **79,** 7724 (1982).

[33] W. S. Dynan and R. Tjian, *Cell* **35,** 79 (1983).

[34] P. C. Fitzgerald and R. T. Simpson, *J. Biol. Chem.* **260,** 15318 (1985).

[35] M. J. Solomon, F. Strauss, and A. Varshavsky, *Proc. Natl. Acad. Sci. U.S.A.* **83,** 1276 (1986).

[36] W. S. Dynan, S. Sazer, R. Tjian, and R. Schimke, *Nature (London)* **319,** 246 (1986).

[37] U. Siebenlist, R. T. Simpson, and W. Gilbert, *Cell* **20,** 269 (1980).

[38] D. T. Shalloway, T. Kleinberger, and D. M. Livingston, *Cell* **20,** 411 (1980).

[39] P. T. Chan and J. Lebowitz, *Nucleic Acids Res.* **11,** 1099 (1983).

[40] X. Y. Zhang, F. Fittler, and W. Horz, *Nucleic Acids Res.* **11,** 4287 (1983).

[41] C. Wu, *Nature (London)* **317,** 84 (1985).

[42] D. Gidoni, W. S. Dynan, and R. Tijan, *Nature (London)* **312,** 409 (1984).

[43] M. W. Van Dyke, R. P. Hertzberg, and P. B. Dervan, *Proc. Natl. Acad. Sci. U.S.A.* **79,** 5470 (1982).

[44] G. M. Church and W. Gilbert, *Proc. Natl. Acad. Sci. U.S.A.* **81,** 1991 (1984).

[45] A. Ephrussi, G. M. Church, S. Tonegawa, and W. Gilbert, *Science* **227,** 134 (1985).

[46] M. M. Becker and J. C. Wang, *Nature (London)* **309,** 682 (1984).

[47] R. Schneider, T. Dorper, I. Gander, R. Mertz, and E. L. Winnacker, *Nucleic Acids Res.* **14,** 1303 (1986).

gel matrix[1,2,48,49] (see, however, Ref. 49a). The following is a step-by-step description of the essential components of the assay together with comments about possible modifications.

Crude Nuclear Extract. Example[5]: confluent monolayers of green monkey CV-1 cells are rinsed with cold 0.1 M NaCl, 50 mM KCl, 10 mM Na-phosphate (pH 7.2), scraped with a rubber policeman into a small volume of the same buffer and centrifuged at 500 g for 3 min. The cell pellet is resuspended, washed once, and resuspended in 3 volumes of 0.23 M sucrose in buffer A [60 mM KCl, 15 mM NaCl, 0.25 mM MgCl$_2$, 0.5 mM Na-EGTA, 0.5 mM spermine, 0.15 mM spermidine, 1 mM dithiothreitol (DDT), 15 mM Na-HEPES (pH 7.5)], containing freshly added proteinase inhibitors phenylmethylsulfonyl fluoride (PMSF, 0.2 mM) and 10 μg/ml each of antipain, leupeptin, chymostatin, and pepstatin A (Sigma Chemical Co., St. Louis, MO). Cells are disrupted in a motor-driven Potter-Elvehjem homogenizer (3000 rpm, $\sim$12 strokes); phase-contrast microscopy is used to determine the optimal number of strokes. The lysate (60 ml, $\sim$5 $\times$ 10^9 nuclei for a large-scale preparation[5]) is diluted 3-fold with 2.0 M sucrose in buffer A, layered onto a 9 ml cushion of 1.7 M sucrose in buffer A, and centrifuged at 25,000 g for 45 min at 4°. The nuclei are resuspended, washed in 0.23 M sucrose in buffer A, and resuspended again by vortexing to a final DNA concentration of $\sim$0.5 mg/ml in 0.35 M NaCl, 5 mM Na-EDTA, 1 mM DTT, 10 mM Na-HEPES (pH 7.5) containing the above five proteinase inhibitors. After 30 min at 0° with periodic stirring, the suspension is centrifuged at 10,000 g for 15 min. The supernatant may be used either immediately or after storage at -70° in the presence of 10–20% glycerol. If the desired DNA-binding activity is assayable in a crude extract it is advisable to check that freezing and thawing do not decrease the activity. The procedure described above was used as the initial step for isolation of an (A + T) − DNA binding protein (α-protein; Figs. 1 and 2 and Refs. 5 and 35). Extraction with higher than 0.35 M NaCl concentrations was shown not to increase the yield of α-protein.[5] An additional reason to avoid using much higher ionic strength buffers for extraction of specific DNA-binding proteins (unless the activity of interest is not extracted efficiently with 0.35 M NaCl) is that higher salt concentrations remove most of the histone H1 and at least some of the core histones. (Most histones and most but not all of the nonhistone proteins are extracted from nuclei in 2 M NaCl.) The

[48] M. G. Fried and D. M. Crothers, *J. Mol. Biol.* **172**, 241 (1984).
[49] M. G. Fried and D. M. Crothers, *J. Mol. Biol.* **172**, 263 (1984).
[49a] A. Revsin, J. A. Ceglarek, and M. M. Garner, *Anal. Biochem.* **153**, 172 (1986).

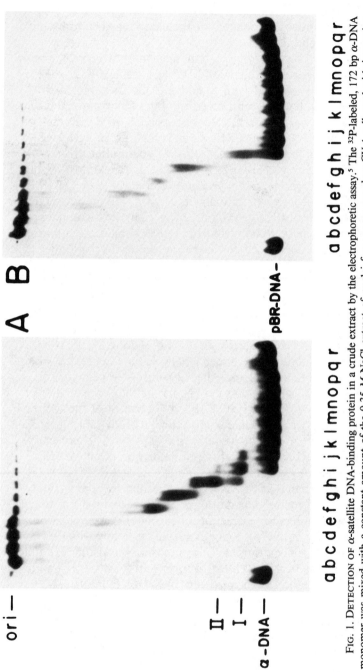

Fig. 1. Detection of α-satellite DNA-binding protein in a crude extract by the electrophoretic assay.[5] The ^{32}P-labeled, 172 bp α-DNA monomer was mixed with a constant amount of the 0.35 M NaCl extract of nuclei from green monkey CV-1 cells and with increasing amounts of unlabeled sonicated E. coli competitor DNA before electrophoresis in a 4% polyacrylamide gel (see main text). Lane a: α-DNA fragment (172 bp, 25,000 cpm, ∼2.5 ng of DNA) in the absence of added extract. Lanes b–r: same but in the presence of the extract (5 μl, containing ∼1 μg of total protein) together with 0, 5, 11, 22, 43, 87, 173, 340, 680, 1.4 × 10³, 2.8 × 10³, 5.5 × 10³, 1.1 × 10⁴, 2.2 × 10⁴, 4.4 × 10⁴, 8.8 × 10⁴, and 17.6 × 10⁴ ng of E. coli DNA per assay, respectively. I and II indicate discrete α-DNA–protein complexes observed in the presence of an optimal excess of competitor DNA (see also Fig. 2). Complexes I and II contain one and two α-protein molecules per particle, respectively [J. McCartney, M. J. Solomon, F. Strauss, and A. Varshavsky, unpublished results (1986)]. For a detailed study of DNA-binding properties of the purified α-protein, see Ref. 35. Reproduced with permission from Strauss and Varshavsky.[5]

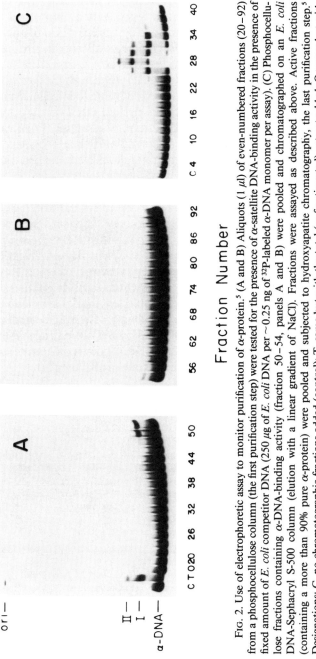

Fraction Number

FIG. 2. Use of electrophoretic assay to monitor purification of α-protein.[5] (A and B) Aliquots (1 μl) of even-numbered fractions (20–92) from a phosphocellulose column (the first purification step) were tested for the presence of α-satellite DNA-binding activity in the presence of fixed amount of E. coli competitor DNA (250 μg of E. coli DNA per ~0.25 ng of ^{32}P-labeled α-DNA monomer per assay). (C) Phosphocellulose fractions containing α-DNA-binding activity (fraction 50–54, panels A and B) were pooled and chromatographed on an E. coli DNA-Sephacryl S-500 column (elution with a linear gradient of NaCl). Fractions were assayed as described above. Active fractions (containing a more than 90% pure α-protein) were pooled and subjected to hydroxyapatite chromatography, the last purification step.[5] Designations: C, no chromatographic fractions added (control); T, same but with the total (unfractionated) extract added; O, same but with the flow-through fraction from the phosphocellulose column added. Other designations are as in Fig. 1. Reproduced with permission from Strauss and Varshavsky.[5] Overexposure of the autoradiograms (A) and (B) reveals the existence of other distinct, minor α-satellite DNA–protein complexes, in addition to the major ones seen in fractions 50–54.[72]

resulting unfolding of higher order chromatin structures would then require either centrifugation at much higher g forces or an altogether different means to remove DNA from the extract. For preparation of analogous extracts from either prokaryotic or lower eukaryotic organisms such as yeast, see Refs. 50–53. It is obviously preferable to produce a crude extract in which the desired DNA-binding activity can be assayed directly, which comprises as small as possible a fraction of the total cellular protein, and at the same time contains most of the desired DNA-binding activity. While preferable, neither of these conditions is essential. For instance, when the DNA-binding assay is fast and simple, it should be feasible to detect the desired DNA-binding activity in fractions from the first purification step without necessarily being able to detect such an activity in the crude extract.

DNA Probes. DNA fragments from ~30 bp to several kbp have been used as probes in the electrophoretic binding assay[5,47-64] (see Figs. 1–3). In fact, the use of synethetic double-stranded oligodeoxyribonucleotide probes less than ~50 bp long enhances the sensitivity of detection of specific DNA–protein complexes,[47] apparently by decreasing the contribution of nonspecific DNA–protein interactions outside of the site of specific binding (Fig. 3). The minimal useable length of a synthetic oligonucleotide probe is determined both by the length of a protein-binding site on the DNA (typically ~10 to ~50 bp), and by the decreasing stability of duplex DNA with decreasing DNA chain length. These constraints suggest an optimal size of a duplex oligonucleotide probe between ~25 and

[50] W. Hendrickson, *BioTechniques* **3**, 198 (1985).

[50a] W. Hendrickson and R. Schleif, *Proc. Natl. Acad. Sci. U.S.A.* **82**, 3129 (1985).

[51] B. Aracangioli and B. Lescure, *EMBO J.* **4**, 2627 (1985).

[52] J. Huet, P. Cottrelle, M. Cool, M. L. Vignais, D. Thiele, C. Marck, J. M. Buhler, A. Sentenac, and P. Fromageot, *EMBO J.* **4**, 3539 (1985).

[53] J. Berman, C. Y. Tachibana, and B. K. Tye, *Proc. Natl. Acad. Sci. U.S.A.* **83**, 3713 (1986).

[54] H. Singh, R. Sen, D. Baltimore, and P. A. Sharp, *Nature (London)* **319**, 154 (1986).

[55] R. W. Carthew, L. A. Chodosh, and P. A. Sharp, *Cell* **43**, 439 (1985).

[56] J. Topol, D. M. Ruben, and C. S. Parker, *Cell* **42**, 527 (1985).

[57] L. F. Levinger, *J. Biol. Chem.* **260**, 14311 (1985).

[58] N. Marzouki, S. Camier, A. Ruet, A. Moenne, and A. Sentenac, *Nature (London)* **323**, 176 (1986).

[59] J. Weinberger, D. Baltimore, and P. A. Sharp, *Nature (London)* **322**, 846 (1986).

[60] J. R. Greene, L. M. Morrissey, L. M. Foster, and E. P. Geiduschek, *J. Biol. Chem.* **261**, 12820 (1986).

[61] J. R. Greene, L. M. Morrisey, and E. P. Geiduschek, *J. Biol. Chem.* **261**, 12828 (1986).

[62] R. Treisman, *Cell* **46**, 567 (1986).

[63] R. Sen and D. Baltimore, *Cell* **46**, 705 (1986).

[64] I. Kovesdi, R. Reichel, and J. R. Nevins, *Cell* **45**, 219 (1986).

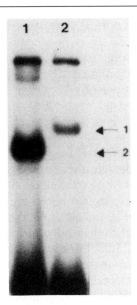

FIG. 3. Use of chemically synthesized duplex oligodeoxynucleotides as probes in the electrophoretic assay.[47] A 0.2 M NaCl extract of nuclei from porcine liver was incubated in 1 mM Na-EDTA, 5 mM DTT, 0.15 M NaCl, 0.5 mM PMSF, 10% glycerol, 25 mM Na-HEPES (pH 7.5) for 15 min at room temperature with ^{32}P-labeled DNA fragments A (lane 1) or B (lane 2), followed by electrophoresis in an 11% polyacrylamide gel (acrylamide/bisacrylamide weight ratio of 55) in 0.375 M Tris-glycine (pH 8.8). Buffer chambers contained 40 mM Tris-glycine (pH 8.5). DNA fragments A and B are 34 bp long, chemically synthesized duplex oligonucleotides, with two 5'-overhangs 4 bases long. Fragment A contains a binding site for nuclear factor I (arrow 2), fragment B lacks such a site but contains a binding site for another, apparently less abundant nuclear protein (arrow 1). Note that specific signals in this version of the electrophoretic assay are obtained in the absence of added competitor DNA. Reproduced with permission from Schneider et al.[47]

~ 50 bp. Both restriction endonuclease-produced DNA fragments and synthetic double-stranded oligonucleotides can be end-labeled with ^{32}P either by the phosphatase/kinase method[5] (kinase treatment alone is sufficient for synthetic oligonucleotides) or by the Klenow fragment of E. coli DNA polymerase I.[65] End labeling with ^{35}S using ^{35}S-labeled dideoxynucleoside triphosphates and either terminal transferase or Klenow polymerase fragment is also possible.[66]

Competitor DNA. A heterologous unlabeled DNA is used to decrease

[65] T. Maniatis, E. F. Fritsch, and J. E. Sambrook, "Molecular Cloning," pp. 112–113. Cold Spring Harbor Lab., Cold Spring Harbor, New York, 1982.
[66] DuPont/New England Nuclear kit (Cat. No. NEG-057H).

the relative contribution of nonspecific protein binding to the labeled DNA probe in the electrophoretic assay, analogous to the use of competitor DNA in other DNA-binding assays. Purified total *E. coli* DNA sheared (e.g., sonicated) to an average length of ~ 1 kbp is often a good choice (Figs. 1 and 2). However, it has recently been shown by Singh *et al.*[54] and by Carthew *et al.*[55] that the use of a simple alternating copolymer duplex such as poly(dI-dC) · poly(dI-dC) as a competitor produces a significant increase in sensitivity, presumably by minimizing the binding of the relevant protein to the competitor DNA (Fig. 4). For duplex oligonucleotides as probes, the sensitivity of the electrophoretic assay may be high enough so that no competitor DNA is required at all (Ref. 47 and Fig. 3). If, however, an oligonucleotide-based electrophoretic assay does require addition of a nonspecific competitor DNA, an ideal (but relatively expensive) choice would be to use unlabeled duplex oligonucleotides containing sequence alterations relative to the initial, "wild-type" oligonucleotide sequence. In this approach, one can compare the electrophoretic patterns observed using "mutated" oligonucleotides as competitors with the patterns obtained by employing the "wild-type" oligonucleotide as *both* the probe and

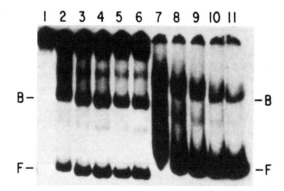

FIG. 4. Use of a simple-sequence duplex DNA as competitor in the electrophoretic assay.[54] A ^{32}P-labeled, ~ 300 bp DNA fragment, derived from the upstream region of the mouse κ light-chain immunoglobulin gene was incubated with a nuclear extract of a human lymphoma cell line in the absence (lane 1) or presence of two different competitor DNA (lanes 2–11). Binding reactions (25 μl) contained 10 mM Tris–HCl (pH 7.5), 50 mM NaCl, 1 mM DTT, 1 mM EDTA, 5% glycerol, and the nuclear extract (8 μg of total protein). Reactions 2–6 also contained 800, 1600, 2600, 3200, and 4000 ng, respectively, of poly(dI-dC) · poly(dI-dC). Reactions 7–11 contained 300, 600, 900, 1200, and 1500 ng, respectively, of *Hin*fI-digested *E. coli* DNA. After 30-min incubation at room temperature, the resulting complexes were electrophoresed in a 4% polyacrylamide gel (see main text). F and B indicate the positions of free and protein-bound DNA fragments, respectively. Reproduced with permission from Singh *et al.*[54]

the competitor. The use of "mutated" competitor oligonucleotides may also provide information about the detailed sequence requirements of the DNA-binding protein.

Preparation of Samples for Electrophoresis. Example[5]: The ^{32}P-labeled 172 bp α-satellite DNA fragment, unlabeled *E. coli* DNA competitor, and nuclear extract are mixed in 25 μl of 70 mM NaCl, 0.1% Triton X-100, 4% glycerol, 1 mM Na-EDTA, 1 mM DTT, 10 mM Na-HEPES (pH 7.5) (final concentrations). Nuclear extract (1 to 5 μl, centrifuged before use at 12,000 g for 1 min) is added last. The mixture is incubated for 10–20 min at room temperature and thereafter loaded onto a low ionic strength 4% polyacrylamide gel described below. While the presence of Triton X-100 (or Nonidet P-40) in the binding buffer is unnecessary for crude extracts (Fig. 1), it has been found necessary to prevent losses of purified α-protein (Fig. 2 and data not shown). Properties of DNA-binding proteins *in vitro* (e.g., their tendency to aggregate or to stick to vessel's walls, stability of DNA–protein complexes as a function of the ionic stength and multivalent ion content of solution) vary sufficiently widely between different DNA-binding proteins so that no given buffer formulation should be assumed to be optimal for a particular DNA-binding protein. One illustration of the diversity in specific solvent requirements by different DNA-binding proteins *in vitro* is provided by a recent study[67] of TFIIIA, a zinc-containing transcription factor specific for the *Xenopus* 5 S RNA genes.[67-70] Detection of specific complexes of purified TFIIIA with its cognate DNA sequences by the electrophoretic assay required the presence of Zn^{2+} in the sample buffer and of spermine and DTT in the gel.[67] (A Tris-glycine electrophoretic system[47] described below allows the detection of specific TFIIIA-DNA binding in the absence of spermine and DTT[71].) At the same time, for a majority of the known DNA-binding proteins from both prokaryotes and eukaryotes the binding buffer described above is an acceptable compromise.

Electrophoretic Conditions. For probe DNA fragments of ~ 100 to ~ 500 bp in size, 4 or 5% polyacrylamide gels (acrylamide : bisacrylamide weight ratio of 30 : 1 or 30 : 1.12, respectively) containing 1 mM Na-EDTA, 3.3 mM sodium-acetate, 6.7 mM Tris (pH 7.5) have been used successfully.[5,35,54,55] Both the gel and the electrophoretic chambers contain the same buffer. An alternative, equally suitable gel buffer is 1 mM Na-EDTA,

[67] D. R. Smith, I. J. Jackson, and D. D. Brown, *Cell* **37**, 645 (1984).
[68] A. B. Lassar, P. L. Martin, and R. G. Roeder, *Science* **222**, 7610 (1983).
[69] S. Sakonju and D. D. Brown, *Cell* **31**, 395 (1982).
[70] J. Miller, A. D. McLachan, and A. Klug, *EMBO J.* **4**, 1609 (1985).
[71] L. Peck, P. Kolodziej, and A. Varshavsky, unpublished results (1986).

10 mM Na-HEPES (pH 7.5).[57,72] Replacement of EDTA with at least
0.5 mM MgCl$_2$ does not present any problems in the detection of partially
purified α-protein by the electrophoretic assay.[72] The gels (0.15×16 cm)
are preelectrophoresed for 2 hr at ~ 12 V/cm. Electrophoresis is carried out
at the same voltage gradient at either 4° or room temperature with the
buffer recirculated between the compartments. A fairly rapid recirculation
is essential in view of the relatively low buffering capacity of the solutions
used. Autoradiographic detection of labeled DNA after electrophoresis is
carried out using standard procedures.[73] Recently, higher ionic strength,
higher pH buffers have been successfully employed in the electrophoretic
assay (375 mM Tris-glycine, pH 8.8 in the gel and 40 mM Tris-glycine, pH
8.5 in the buffer compartments; no preelectrophoresis).[47] This formulation
works at least as well as the low ionic strength electrophoretic system for
several DNA-binding proteins.[47,71,74] However, until the influence of varying electrophoretic conditions (pH, ionic strength, presence or absence of a
reducing agent, specific divalent cations, etc.) on the apparent stability of
DNA–protein complexes in the gel is better understood, it is advisable to
begin with both the low and the high tonic strength systems to decrease the
possibility of missing a relevant DNA-binding activity that may be undetectable with one of the systems. In another recent modification of the
electrophoretic assay, both agarose[53] and mixed agarose-polyacrylamide
gels[56] have been successfully used for detection and analysis of specific
DNA-binding proteins.

Other Applications of the Electrophoretic Assay

*Equilibrium and Kinetic Studies of DNA–Protein Interactions Using
Purified Proteins.* This set of important applications of the electrophoretic
assay (described in Refs. 1, 2, 48, 49a, 60, 61, and 75–78) preceded its use
for detection of new DNA-binding activities in crude extracts[5] (only the
latter application is the subject of the present review). In addition, since the
mobility of a protein–DNA complex changes upon binding of another

[72] F. Strauss, R. Pan, and A. Varshavsky, unpublished results (1984).
[73] T. Maniatis, E. F. Fritsch, and J. E. Sambrook, "Molecular Cloning," pp. 150–185. Cold
Spring Harboar Lab., Cold Spring Harbor, New York, 1982.
[74] H. Singh, R. W. Carthew, and P. A. Sharp, personal communication (1986).
[75] S. H. Shanblatt, and A. Revzin, *Nucleic Acids Res.* 12, 5287 (1984).
[76] D. C. Straney and D. M. Crothers, *Cell* 43, 449 (1985).
[76a] H. N. Liu-Johnson, M. R. Gartenberg, and D. M. Crothers, *Cell* 47, 995 (1986).
[77] M. G. Fried and D. M. Crothers, *Nucleic Acids Res.* 11, 141 (1983).
[77a] M. M. Garner and A. Revzin, *Trends Biochem. Sci.* 11, 395 (1986).
[78] A. Spassky, S. Rimsky, H. Garrean, and H. Buc, *Nucleic Acids Res.* 12, 5321 (1984).

protein to the same complex, the electrophoretic assay is also a powerful tool for detecting and dissecting multiple protein–protein and protein–DNA interactions within a specific stretch of DNA.[50,56,77,77a] A stable nucleoprotein complex such as an isolated nucleosome can itself be used as a ligand in the electrophoretic assay to detect and study nucleosome-binding proteins.[15,77a,79–81.] Furthermore, discrete stages in the initiation of transcription, e.g., formation of open and closed complexes of RNA polymerase with its cognate promoter DNA can also be analyzed using the electrophoretic assay.[49a,75,76] Another recent development is the use of a combination of the dideoxynucleotide-terminated primer extension reaction with the electrophoretic assay.[76a] This generally applicable approach allows one to define in thermodynamic terms the DNA region responsible for the equilibrium stability of specific DNA–protein complexes.[76a] Note that at least some of these applications can be informative even when used with partially purified protein fractions.

Electrophoresis of Nucleoprotein Complexes as an Initial Step in Footprinting and Binding-Interference Assays. In footprinting assays, nucleases or chemical probes are used to detect DNA sequences that are occupied by the bound protein. In binding-interference assays, on the other hand, the DNA is first chemically modified, and the modification sites that interfere with subsequent binding of a protein are determined.[35,37,50,50a,69] In these assays, a protein-bound modified DNA must be separated from free DNA before the analysis. It is at this step that the nondenaturing gel electrophoresis is superior to earlier methods by allowing both binding interference mapping in crude extracts and the resolution of multiple protein–DNA complexes.[35,50,50a] Although a footprinting assay does not in principle require prior separation of protein-bound from free DNA, the ability to do so by gel electrophoresis often increases the power of the assay. For example, if several specific DNA–protein complexes can be formed by a DNA fragment in a given sample, the intermediate step of separation of DNase I-nicked, end-labeled complexes from each other and from free DNA by gel electrophoresis allows independent footprinting of these complexes.[55,56]

The Use of Electrophoretic Assay to Detect Specific DNA Binding by in Vitro-Synthesized Proteins. In this generally applicable approach developed by Hope and Struhl,[82,83] a DNA sequence encoding the protein of interest is cloned into a transcription vector such as SP6 and transcribed *in*

[79] J. K. W. Mardian, A. F. Paton, G. J. Bunick, and D. E. Olins, *Science* **209,** 1534 (1980).
[80] G. Sandeen, W. I. Wood, and G. Felsenfeld, *Nucleic Acids Res.* **8,** 3757 (1980).
[81] P. Swerdlow and A. Varshavsky, *Nucleic Acids Res.* **11,** 387 (1983).
[82] I. A. Hope and K. Struhl, *Cell* **43,** 177 (1985).
[83] I. A. Hope and K. Struhl, *Cell* **46,** 885 (1986).

vitro to produce messenger RNA, which is then translated *in vitro* to produce the desired (radioactively labeled) protein. To see if the *in vitro*-synthesized protein is capable of specific DNA binding, it is incubated with a mixture of unlabeled DNA fragments one or more of which are thought to interact specifically with the protein. Protein–DNA complexes are then separated from the free labeled protein by gel electrophoresis. Among the useful features of this approach[82,83] is the ability to test for specific DNA binding by a protein of interest without first isolating the protein from cells. This ability becomes particularly important in studies of mutant DNA-binding proteins whose primary structures have been manipulated by *in vitro* mutagenesis of the corresponding DNA clones, because it bypasses the necessity of purifying mutant proteins produced *in vivo*. The method also solves the recurring problems of addressing the DNA-binding properties of extremely scarce proteins for which the genes have been cloned.

Electrophoretic Assay for RNA-Binding Proteins. Detection and analysis of specific RNA–protein complexes using the gel electrophoretic assay is yet another useful application of the assay that has been developed recently by Konarska and Sharp.[84]

Electrophoretic Assay for Protein–Protein Interactions? An interesting application of the electrophoretic assay which to my knowledge has not yet been developed is to use the assay for detection and analysis of specific protein–protein interactions. In one version of such an assay, a purified, radioactively labeled protein of interest is incubated with either a crude or partially fractionated extract followed by electrophoretic detection of protein–protein (as opposed to nucleic acid–protein) complexes containing a labeled protein probe.

Terminology. Repeated attempts to evolve a name for the gel electrophoretic assay that would be short, pronounceable, and at the same time would invoke an important feature of the assay have not yielded a clear winner. With minor reservations, I suggest a "band-shift assay," to denote any version of the gel electrophoretic assay, and a "competition band-shift assay," to denote a version of the assay that is used in particular for detection of specific DNA-binding proteins in the presence of other DNA-binding proteins.

Although nondenaturing electrophoretic systems have existed for almost as long as the method of gel electrophoresis itself, the arrival of powerful analytical applications of this technology is quite recent, and other useful variations of the theme are still to come.

---

[84] M. M. Konarska and P. A. Sharp, *Cell* **46,** 845 (1986).

Acknowledgments

I thank my colleagues at MIT and elsewhere for permission to reproduce their published results in Figs. 3 and 4, and for providing preprints describing their unpublished work. I also thank Daniel Finley, Peter Kolodziej, Edward Winter, and especially Mark Solomon for their comments on the manuscript. Work in the author's laboratory is supported by grants from the National Institutes of Health.

Author Index

Numbers in parentheses are footnote reference numbers and indicate that an author's work is referred to although the name is not cited in the text.

Subject Index

A

B

Hybridization (*cont'd.*)
  of chromosomes, 279–292
    problems encountered, 289–292
    data analysis, 288
    gene mapping, 313
    procedures, 284, 285
      formamide, 285
      sodium hydroxide/ethanol, 285
    sample preparation, 280–282
    scheme for probe and chromosomes, 279, 280
    sensitivity of, for transfection, 539
    of ^{35}S-labeled RNA probes, 539–551
    somatic cells, 119–121
    spot-blot. *See* Spot-blot hybridization
Hybridoma
  growth, 51–53
  immunoplate assay, 55–61
  somatic cell genetic analysis, 50–65
  toxin–antitoxin production, 250
  variants
    cell sorting, 65
    sib selection assay, 61–65
Hydroxyurea, as inhibitor of ribonucleotide reductase, 93, 95, 99
Hypoxanthine phosphoribosyltransferase, 15, 21
  activity, antisense RNA inhibition, 525–530
  assay, isoelectric focusing gel electrophoresis, 441, 442
  deficiency, antisense inhibition, 519–530
  mRNA detection
    by methylmercuric hydroxide gel, 378, 379, 380
    by microinjection and complementation of mutant cells, 377, 378
Hypoxanthine phosphoribosyltransferase gene, 315
  antisense RNA, construction, 523

I

Immunofluorescence
  HLA antigens, 161–165
  L-cell transfectants, 159, 160
  of markers in cell differentiation, 78, 79

Immunoglobin
  production, from myelomas, 51
  screening, immunoplate assay, 60, 61
Immunoglobulin A, myeloma mutant, 61
Immunoglobulin G, subclasses, 60
  detection of switching, ELISA, 61–65
Immunoglobulin M
  detection of somatic cell variants, 65
  myeloma mutant, 61
Immunoplate assay, detection of mutants, 54–61
  antibody overlay, 55
  antigen overlay, 57–59
Immunoprecipitation
  assay for DNA-binding protein, 553
  polysome purification, 329
Immunoselection, 498, 499
Immunotoxin
  construction using *Pseudomonas* exotoxin A, 139–145
    buffer preparation, 14, 143
    conjugation procedure, 139–141
    materials, 140
  cytotoxic activity, 143
  evalution, 143, 144
  inhibition of protein synthesis, 143, 144
Inosine triphosphatase, as marker for chromosome 20, 186
Insulin gene, mapping, 292
Iodoacetamide, cell selection, 238
Isocitrate dehydrogenase, as marker for chromosome 2, 186
Isoelectric focusing gel electrophoresis, 441, 442
Isozyme
  as chromosome markers, 169–194
    microcell hybrids, 323, 324
    loci, regions of human genome marked by, 193, 194
  separation, 171–176
    electrophoretic run, 173–176
    gel preparation, 172, 173
    sample preparation, 171, 172
    by starch gel electrophoresis, 171–176
      buffer systems, 172
  staining, histochemical, 176–192
    colorimetric reactions, agar overlay, autoradiography, 176, 177